VERY HIGH ENERGY COSMIC RAY INTERACTIONS

PROCEEDINGS SUPPLEMENTS
NUCLEAR PHYSICS B

Nuclear Physics B (Proc. Suppl.) 97 (2001) APRIL 2001

VERY HIGH ENERGY COSMIC RAY INTERACTIONS

Proceedings of the 11th International Symposium
on Very High Energy Cosmic Ray Interactions

'Gleb Wataghin' Institute of Physics
State University of Campinas
Campinas, Brazil
17–21 July 2000

Edited by

Carola DOBRIGKEIT and Reinaldo RIGITANO
Instituto de Física Gleb Wataghin
Universidade Estadual de Campinas (UNICAMP)
Campinas, Brazil

NORTH-HOLLAND

∞The paper used in this publication meets the requirements of ANSI/NISO Z39.48-1992 (Permanence of Paper).

Printed in the Netherlands

FOREWORD

The 11th International Symposium on Very High Energy Cosmic Ray Interactions (ISVHECRI) was held at the 'Gleb Wataghin' Institute of Physics of the State University of Campinas, in Campinas, Brazil, on July 17–21, 2000, under the auspices of the Cosmic Ray Commission of the International Union of Pure and Applied Physics and its Emulsion Chamber Committee. Previous meetings of this biennial series had been held in Nakhodka (1980), La Paz and Rio de Janeiro (1982), Tokyo (1984), Beijing (1986), Lodz (1988), Tarbes (1990), Ann Arbor (1992), Tokyo (1994), Karlsruhe (1996), and Assergi (1998).

The objectives of the 11th ISVHECRI included a focus on specific topics concerning high-energy cosmic ray interactions as studied by emulsion chamber techniques and through the observation of extensive air showers with surface arrays, as well as on the interconnection between cosmic ray observations and experiments with particle accelerators. Due to the current interest in extremely high energy cosmic rays, one of the invited talks addressed specifically this subject. Special attention was also given to the results of underground experiments, and to studies of the primary cosmic ray spectrum and composition, especially in the energy region of the so-called "knee" of the spectrum (1–10 PeV). Besides these main topics, results of cosmic ray phenomenology and astroparticle physics were also reported.

The Organizing Committee is grateful to the Director of the IFGW, Professor C.H. de Brito Cruz, for the continuous and unremitting support. We would like to thank the International Advisory Committee for their advice in the preparation of the scientific program; in particular, special thanks are due to Professor G. Schatz, Professor H. Rebel, and especially to Professors Lawrence W. Jones and Antonella Castellina, who shared with us their experience in the previous symposia.

This Symposium was made possible through the financial support of *Fundação de Amparo à Pesquisa do Estado de São Paulo (FAPESP), Conselho Nacional de Desenvolvimento Científico e Tecnológico (CNPq), Centro Latino-americano de Física (CLAF), Fundo de Apoio ao Ensino e Pesquisa (FAEP)* from the State University of Campinas, and *Coordenação de Aperfeiçoamento de Pessoal de Nível Superior (CAPES)*.

At its July 20 meeting, the Emulsion Chambers Committee gratefully accepted the proposal made by Professors Peter Grieder and Lawrence Jones that the next ISVHECRI Symposium be held at the CERN Laboratory in Geneva, Switzerland, in 2002.

January 22nd, 2001
Carola Dobrigkeit and Reinaldo Rigitano
Editors

COMMITTEES

International Advisory Committee

C. Aguirre	ABC, La Paz, Bolivia	C.M.G. Lattes	CBPF, Brazil
J.D. Bjorken	SLAC, USA	S. Miyake	Kanagawa, Japan
J.-N. Capdevielle	LPC Paris, France	G. Navarra	Torino, Italy
A. Castellina	Torino, Italy	A. Ohsawa	ICRR, Tokyo, Japan
J.W. Cronin	Chicago, USA	O. Saavedra	Torino, Italy
J.D. de Deus	IST, Portugal	G. Schatz	Karlsruhe, Germany
A. Etchegoyen	Tandar, Argentina	S.A. Slavatinsky	Lebedev, Moscow, Russia
Y. Fujimoto	Waseda, Japan	A. Turtelli	Unicamp, Brazil
T.K. Gaisser	Delaware, USA	A. Watson	Leeds, United Kingdom
L.W. Jones	Michigan, USA	T. Yuda	ICRR, Tokyo, Japan
J. Kempa	Lodz, Poland	G.T. Zatsepin	INR, Moscow, Russia

National Advisory Committee

J. Anjos	CBPF, Rio de Janeiro	A. Natale	IFT-UNESP, São Paulo
C.O. Escobar	UNICAMP, Campinas	H. Portella	CBPF, Rio de Janeiro
E. Ferreira	UFRJ, Rio de Janeiro	A. Santoro	CBPF, Rio de Janeiro
Y. Hama	IFUSP, São Paulo	R.C. Shellard	CBPF, Rio de Janeiro
T. Kodama	UFRJ, Rio de Janeiro	E.H. Shibuya	UNICAMP, Campinas

Local Organizing Committee

J. Bellandi	UNICAMP, Campinas	M.J. Menon	UNICAMP, Campinas
J.A. Chinellato	UNICAMP, Campinas	C. Navia	UFF, Rio de Janeiro
C. Dobrigkeit (Chair)	UNICAMP, Campinas	R.C. Rigitano	UNICAMP, Campinas
P. Gouffon	IFUSP, São Paulo	R. Rosenfeld	IFT-UNESP, São Paulo
M.M. Guzzo	UNICAMP, Campinas	E.H. Shibuya	UNICAMP, Campinas

INSTITUTO DE FISICA

SCIENTIFIC PROGRAM

Monday, July 17, 2000

Opening session – Gleb Wataghin Memorial Lecture
C.H. de Brito Cruz, I.E. Chambouleyron, C. Dobrigkeit, R.A. Salmeron

Detection of ultra high energy cosmic rays
J. Cronin

Models of ultra high energy cosmic ray sources
A. Olinto

Arrival directions and chemical composition of ultra-high energy cosmic rays
A.A. Mikhailov

Self-organized criticality in atmospheric cascades
G. Wilk et al.

Muons from strangelets
Z. Włodarczyk et al.

Cosmic ray issues for accelerator experiments
L.W. Jones

Recent results of KASCADE: phenomenology of extensive air showers
G. Schatz et al.

Towards the energy spectrum and composition of primary cosmic rays in the knee region: methods and results at KASCADE
A. Haungs et al.

Tuesday, July 18, 2000

Interconnection between cosmic ray and accelerator results
L.W. Jones

NESTOR neutrino telescope status report
P.K.F. Grieder et al.

The project of comprehensive cosmic ray investigations with mountain elevation experimental setups in the energy range 10^{15}–10^{18} eV
S.A. Slavatinsky et al.

A new detector for the measurement of the energy spectrum of cosmic ray nuclei in the TeV region
J.R. Hörandel et al.

Characteristics of hadron-induced showers observed by Pamir thick lead chambers
M. Tamada et al.

Study of hadronic component in air showers at Mt. Chacaltaya
N. Kawasumi et al.

LAAS network observation of air showers
N. Ochi et al.

Search for coincident air showers in the network observation
N. Ochi et al.

Anisotropy of successive showers
N. Ochi et al.

A study on penetrating nature of gamma- and hadron-induced cascade showers in two-storey emulsion chambers
M. Tamada

The L3+Cosmics experiment
H. Wilkens et al.

New approach to separation of electromagnetic and hadron cascades and to energy measurement in detection of primary particles
R.A. Mukhamedshin et al.

On the nature of gamma-hadron family alignment
A.S. Borisov et al.

Phenomenological approach to the problem of alignment
R.A. Mukhamedshin

Multiple production and coplanar emission
J.-N. Capdevielle

High energy gamma-families with halo and mass composition of primary cosmic rays in energy region above 10 PeV
A.S. Borisov et al.

Wednesday, July 19, 2000

Nuclear interactions by means of emulsion chamber technique – nuclear interactions at 10^{14}–10^{20} eV
A. Ohsawa

The distributions of $E{\cdot}R$ in e–γ families
J. Kempa et al.

Inelasticity fluctuations and the efficiency of emulsion chambers for hadrons
G. Schatz et al.

Absorption of gamma-families in the atmosphere at the level of the Pamirs
S. Slavatinsky et al.

Angular distributions of hadrons and gammas in Pamir emulsion calorimeter
J. Malinowski et al.

Mass composition of primary cosmic ray for energies of 10–1000 TeV on the basis of the Pamir experimental results
J. Malinowski

Features of observed non-linearity in the arrival times of air showers at Mt. Chacaltaya
N. Ohmori

Presentation ceremony of the Pierre Auger Research Project
C.H. de Brito Cruz, J. Cronin, C. Escobar, J.F. Perez

Thursday, July 20, 2000

Cosmic ray primary composition and energy spectrum: direct and EAS measurements
A. Castellina

Test of high-energy interaction models with data of the emulsion chamber experiment PAMIR
A. Haungs et al.

Using TOP-C for commodity parallel computing in cosmic ray physics simulations
J. Swain et al.

Air shower simulation using GEANT4 and commodity parallel computing
J. Swain et al.

Test and analysis of hadronic interaction models with KASCADE event rates
M. Risse et al.

Results from CACTI experiment
G. Yodh

Centauro phenomena and QCD
A. Pemmaraju

Results of the CLUE experiment
M. Ciocci et al.

Future of the Ultra Violet Experiment
A. Menzione et al.

Identification of showers observed by emulsion chambers (I)
A. Ohsawa et al.

Identification of showers observed by emulsion chambers (II)
A. Ohsawa et al.

Hadronic structures of extensive air shower cores
J. Kempa et al.

Anomalous extensive air showers of ultrahigh energy cosmic rays and their relation to pulsars
A.A. Mikhailov et al.

The TeV gamma-ray emission from point sources: galactic and extragalactic
V.G. Sinitsyna et al.

Evidence of TeV gamma-ray radiation in supernova remnants CYGNUS X-3
V.G. Sinitsyna et al.

At the same time and at the same meridian: simultaneous observations of galactic and extragalactic sources of very high energy gamma-quanta by imaging mirror telescopic systems located on one meridian
V.G. Sinitsyna

Lateral distributions, localization methods, $\rho(600)$, size and energy determination in giant EAS
J.-N. Capdevielle et al.

Friday, July 21, 2000

Underground experiments: Results and new projects
S. Cecchini

Cosmic rays tells us on supernova explosion in the nearby interstellar space
Y. Stozhkov et al.

Fractal electromagnetic showers
J.D. Swain et al.

Simulation of water Čerenkov detectors using GEANT4
J.D. Swain et al.

A pot of gold at the end of the cosmic "raynbow"?
J.D. Swain et al.

Concluding remarks
C. Dobrigkeit and L.W. Jones

Posters

Diffuse reflectivity of Tyvek in air and water, and anisotropical effects
J.C. Arteaga Velásquez et al.

Leading nucleon and total pp cross sections in VHE
J. Bellandi et al.

The influence of using plastic scintillators in the determination of extensive air showers parameters
J.A. Chinellato et al.

Muon bremsstrahlung and muonic pair production in air showers
A.N. Cillis et al.

The acceptance of fluorescence detectors for quasi horizontal showers induced by weak interacting particles
J.C. Díaz et al.

Geometrical reconstruction of the events seen by the fluorescent detectors of the Pierre Auger Observatory
J.C. Díaz et al

Analysis of the Čerenkov time profile as a technique to distinguish the primary particle
C. Dobrigkeit et al.

Status of the solution to the solar neutrino problem based on non-standard neutrino interactions
P.C. de Holanda et al.

Application of Markov model of random medium for the interpretation of experimental results on muon intensity measurements at underground installations
A.A. Lagutin et al.

The "knee" in the primary cosmic ray spectrum as consequence of the anomalous diffusion of the particles in the fractal interstellar medium
A.A. Lagutin et al.

Lateral distribution of the electrons in EAS at superhigh energies: predictions and experimental data
A.A. Lagutin et al.

Mean square scattering angle of muons at large depths in rock and water
A.A. Lagutin et al.

Multiplicity spectra in muon bundles deep underground
A.S. Lidvansky et al.

The response of the EAS muon component in the GAMMA installation
R.M. Martirosov et al.

Possibility to determine the mass composition from the GAMMA experimental data
R.M. Martirosov et al.

Behaviour of the EAS electron–photon and muon component characteristics in the knee region at mountain altitudes
J. Procureur et al.

Artificial Neural Networks for the shower reconstruction of gamma-showers in the energy range 20–300 GeV
J. Procureur

Study of the influence of redshift on the pathlengths of UHE gammas
R.C. Rigitano

UHECR and supermassive particles in the primordial universe: a possible connection
R. Rosenfeld et al.

CASTOR: a detector for CENTAURO and strange object research
Z. Włodarczyk et al.

CONTENTS

(Abstracted/Indexed in: Current Contents: Physical, Chemical & Earth Sciences/INSPEC)

Review Talks

ELSEVIER

Nuclear Physics B (Proc. Suppl.) 97 (2001) 3–9

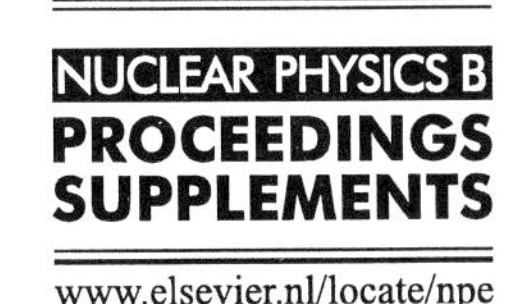

www.elsevier.nl/locate/npe

Ultra high energy cosmic rays

James W. Cronin[a]

[a]Enrico Fermi Institute, University of Chicago,
5640 S. Ellis Ave., Chicago, IL, 60637, USA

The evidence for the existence of cosmic rays with energies in excess of 10^{20} eV is now overwhelming. There is so far no indication of the GZK cutoff in the energy spectrum at 5×10^{19} eV. This conclusion is not firm for lack of statistics. A cutoff would be expected if the sources of the cosmic rays were distributed uniformly throughout the cosmos. The sources of cosmic rays with energy above the GZK cutoff must be at a distance $\leq$ 100 Mpc, and if they are protons they are very likely to point to these sources. There are no easy explanations how known astrophysical objects can accelerate protons (or atomic nuclei) to these energies. The fluxes of these cosmic rays is very low and large instruments are required to observe them even with modest statistics. One such instrument, the Pierre Auger Observatory, is described. It is designed for all-sky coverage and the construction of its southern site in Argentina has begun.

1. The cosmic ray energy spectrum above 10^{18} eV

In recent years the interest in extremely high energy cosmic rays (EHECR), those with energy $\geq 10^{18}$ eV (EeV), has revived because of a number of discoveries. Therefore there are many excellent reviews, books, and conference proceedings to which the reader is referred [1]. The energy spectrum of cosmic rays is quite well measured up to 10^{19} eV. Above the knee (3×10^{15} eV) it falls as a power law in energy, $dN/dE \sim E^{-\alpha}$, with an index α=3.

Above 10^{17} eV the cosmic ray spectrum has significant structure, which is displayed in Fig 1, where the differential spectrum has been multiplied by E^3 to better display the observed structures. These data are the contribution of four experiments which have operated over the past 20 years. These experiments observe the cosmic rays indirectly by means of the air showers they produce. They are from the Haverah Park surface array in England [2], the Yakutsk surface array in Siberia [3], the Fly's Eye fluorescence detector in Utah [4], and the AGASA surface array in Japan [5]. Before plotting, the energy scale of each experiment has been adjusted by amounts $\leq$ 20% to show most clearly the common features. The method of energy determination in each of these experiments is quite different, and the fact that they agree within 20% is remarkable.

The spectrum continues with an index of 3.0 until about 5×10^{17} eV where it steepens with an index of about 3.3. Above an energy of 10^{18} eV it is difficult for the galaxy to contain even iron nuclei and galactic accelerators that can produce such energies cannot be imagined. If cosmic rays at these energies continue to be produced in the galaxy, they should show a strong anisotropy which correlates with the distribution of matter in our galaxy. Above this energy such a correlation is not observed probably due to lack of statistics. Above 5×10^{18} eV the spectrum hardens to a spectral index of 2.7. This hardening of the spectrum may be due to a new component that is extragalactic.

The composition of the cosmic rays is notoriously difficult to measure with the indirect air shower methods. Such evidence as does exist suggests that the composition is moving towards a lower mean atomic number as the energy increases from 10^{17} eV to 10^{19} eV [6].

2. The Difficulty of Acceleration

Above 10^{19} eV the precision of the spectrum measurement suffers from lack of statistics. There have been about 60 events recorded with energy

PII S0920-5632(01)01179-3

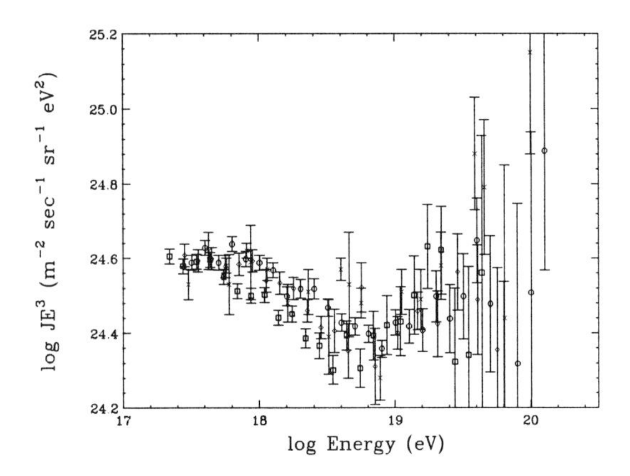

Figure 1. Upper end of the cosmic ray spectrum. Haverah Park [2] points (crosses) serve as a reference. Yakutsk [3] points [diamonds] have been reduced by 20%. Fly's Eye [4] points (squares) have been increased by 10%. AGASA [5] points (circles) have been reduced by 10%.

greater than 5×10^{19} eV. Yet it is above this energy that the scientific mystery is the greatest. There is little understanding how known astrophysical objects can produce particles of such energy. At the most primitive level, a necessary condition for the acceleration of a proton to an energy of 10^{20} eV requires that the product of the magnetic field B and the size of the region R be much larger than 3×10^{17} gauss-cm. This value is appropriate for a perfect accelerator such as might be scaled up from the Tevatron at Fermilab. The Tevatron has a BR=3×10^9 gauss-cm and accelerates protons to 10^{12} eV. The possibility of acceleration of cosmic rays to energies above 10^{19} eV seems difficult and the literature is filled with speculations. Two reviews which discuss the basic requirements are given by Greisen [7] and Hillas [8]. While these were written some time ago, they are excellent in outlining the basic problem of cosmic ray acceleration. Biermann [9] has recently reviewed all the ideas offered to achieve these high energies. Hillas in his outstanding review of 1984 presented a plot which graphically shows the difficulty of cosmic ray acceleration to 10^{20} eV. Figure 2 is a reproduction of his figure. Plotted are the size and strength of possible acceleration sites. The upper limit on the energy is given by:

$$E_{18} \leq 0.5\beta Z B_{\mu g} L_{kpc}.$$

Here the E_{18} is the maximum energy measured in units of 10^{18} eV. L_{kpc} is the size of the accelerating region in units of kilo-parsec, and $B_{\mu g}$ is the magnetic field in μgauss. The factor β was introduced by Greisen to account for the fact that the effective magnetic field in the accelerator analogy is much less than the ambient field. The factor β in the Hillas discussion is the velocity of the shock wave (relative to c) which provides the acceleration. Lines corresponding to a 10^{20} eV proton with β=1 and 1/300 are plotted. A line is also plotted for iron nuclei (β=1). With Z=26, iron is in principle easier to accelerate, but in a realistic situation it is difficult to avoid the disintegration of the nucleus during the acceleration process. Real proton accelerators should lie *well* above the solid line. The figure is also relevant for "one shot" acceleration as it represents the emf induced in a conducter of length L moving with a velocity β through a uniform magnetic field B. Synchrotron energy loss is also important. For protons the synchrotron loss rate at 10^{20} eV requires that the magnetic field be less than 0.1 gauss for slow acceleration (the accelerator analogy)[7]. The conclusion from this figure is that the acceleration of cosmic rays to 10^{20} eV is not a simple matter [10].

A review of the ideas as to how these cosmic rays are produced is contained in a companion paper submitted to the conference by Professor Angela Olinto [11]

3. Diagnostic Tools

There are some natural diagnostic tools which make the analysis of the cosmic rays above 5×10^{19} eV easier than at lower energies. The first of these is the 2.7K Cosmic Background Radiation (CBR). Greisen [12] and Zatsepin and Kuz'min [13] pointed out that protons, photons, and nuclei all interact strongly with this radiation (GZK effect).

The collision of a proton of 10^{20} eV colliding

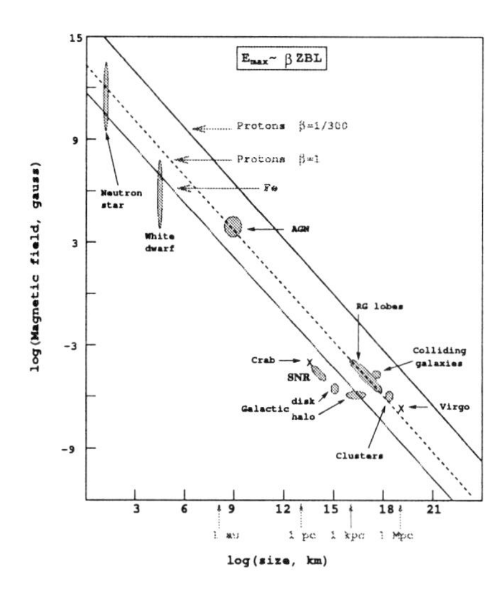

Figure 2. Plot adapted from Hillas [8]. Size and magnetic field of possible sites of acceleration. Objects below the solid line cannot accelerate protons to 10^{20} eV.

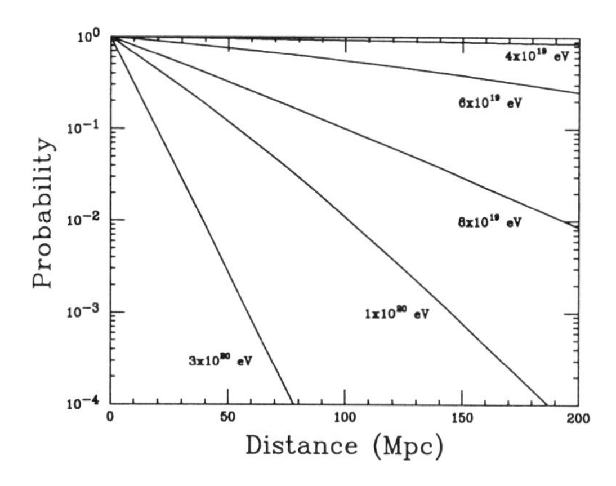

Figure 3. Probability that cosmic ray proton has traveled further than the indicated distance. Sources are assumed to be distributed uniformly throughout space and emit cosmic rays with a differential spectral index α=2.5. This figure is based on calculations by Paul Sommers, University of Utah.

with a CBR photon of 10^{-3} eV produces several hundred MeV in the center of mass system. The cross section for pion production is quite large so that collisions are quite likely, resulting in a loss of energy for the primary proton. In Fig. 3 we plot the probability for selected proton energies that the source is further away than the indicated distance. This probability assumes that the sources are distributed uniformly and the spectrum has a powerlaw dependence with α=2.5.

Almost independent of the initial energy of a proton, it will be found with less than 10^{20} eV after propagating through a distance of 100 Mpc (3×10^8 light years). Thus the observation of a cosmic ray proton with energy greater than 10^{20} eV implies that its distance of travel is less than 100 Mpc and that its initial energy at its source had to have been much greater. This distance corresponds to a red shift of $\sim$0.025 and is small compared to the size of the universe. Similar arguments can be made for nuclei or photons in the energy range considered. There are a limited number of possible sources which fit the Hillas criteria (Fig. 2) within a volume of radius 100 Mpc about the earth. The fact that the cosmic rays, if protons, will be little deflected by galactic and extragalactic magnetic fields serves as a second diagnostic tool [14]. The deflection of protons of energy 5×10^{19} eV by the galactic magnetic field (~ 2 μgauss) and the intergalactic magnetic fields ($\leq 10^{-9}$ gauss) is only a few degrees, so that above 5×10^{19} eV it is possible that the cosmic rays will point to their sources. We thus approach an astronomy where the "light" consists of cosmic rays and the sources visible are dominated by those less than 100 Mpc away.

4. Cosmic Ray Astronomy

The energy 5×10^{19} eV represents a lower limit for which the notion of an astronomy of charged particles from "local" sources can be applied. For a universal distribution of sources with energy $\geq 5 \times 10^{19}$ eV, about half of the events will come from a distance of less than 100 Mpc due to attenuation of the more distant sources by the GZK effect. Above this energy the deflection of protons by galactic and extragalactic magnetic fields

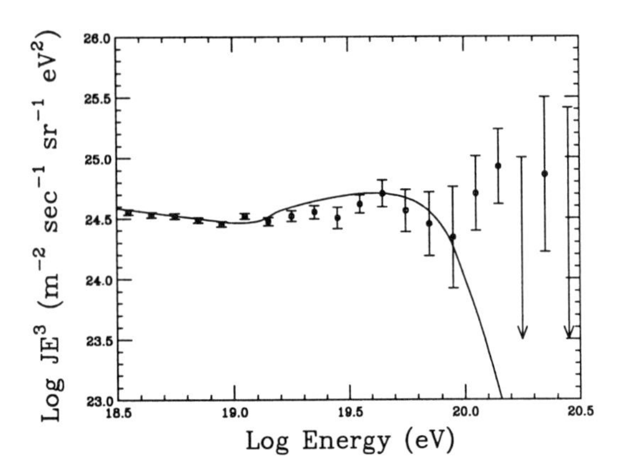

Figure 4. Cosmic ray spectrum observed by the AGASA experiment during seven years of operation [15] [19]. The solid line is the spectrum expected for a universal distribution of sources showing the GZK cutoff, including the effects of instrument resolution.

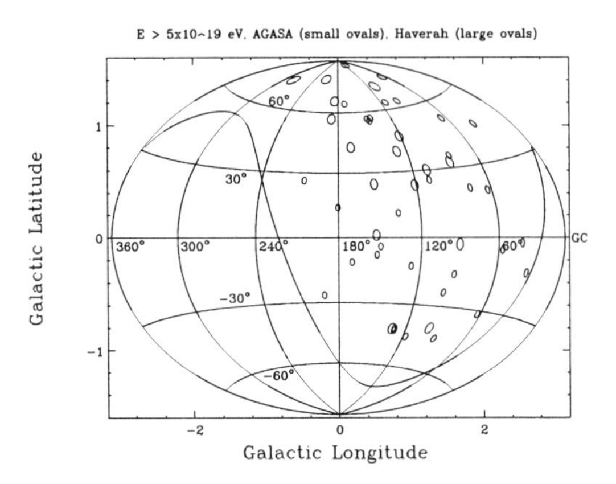

Figure 5. Plot in galactic coordinates of arrival directions of cosmic rays with energy $\geq 5 \times 10^{19}$ eV. Large symbols, Haverah Park [18]; small symbols, AGASA [19]. The size of the symbols indicate the resolution of the experiments (63% of the events within the symbol). The empty region bounded by the solid line is the part of the sky not seen by the experiments which are located in the Northern Hemisphere.

is expected to be no more than a few degrees. Recently the AGASA group [15] has published the spectrum of EHECR based on seven years of operation of their 100 km^2 surface array. Their spectrum is plotted in Fig 4. There is no evidence of a GZK cutoff in this spectrum. To date about 13 events worldwide have been observed with energies in excess of 10^{20} eV. Ninty percent of these must have come from a distance less than 50 Mpc. Of these events, two particularly stand out with energies reported to be 2×10^{20} eV by the AGASA experiment [16] and 3×10^{20} eV by the Fly's Eye experiment [17].

The events above 5×10^{19} eV are too few to derive a spectral index. It is not clear that a single spectrum is even the proper way to characterize these events. Since they must come from "nearby" the actual number of sources may not form an effective continuum in space, so the spectrum observed may vary with direction. The distribution of matter within 100 Mpc is not uniformly distributed over the sky. It is probably more fruitful to take an astronomical approach and plot the arrival directions of these events on the sky in galactic coordinates.

Arrival direction data are available for the Haverah Park experiment [18] and the AGASA experiment [19]. In Fig. 5 we plot the arrival directions of 27 AGASA events and 16 Haverah Park events. The size of the symbols corresponds to the angular resolution. What is remarkable in this figure is the number of coincidences of cosmic rays coming from the same direction in the sky. Of the 43 events reported by AGASA and Haverah Park, there are two triplets. The probability of a chance coincidence for this occurance is about 0.1%. From astrophysical experience this small probabiliy is not sufficient to rule out chance, but the possibility that these clusters may be real cannot not be ignored. The cluster in the southern galactic hemisphere contains the largest AGASA event of 2×10^{20} eV, a Haverah Park event of about 1×10^{20} eV, and an AGASA event of 5×10^{19} eV. This cluster is particularly interesting as it contains cosmic rays separated by a factor of four in energy. The intervening magnetic fields have not separated the cosmic rays in space by more than a few degrees. This is an encouraging prospect for future experiments where, with many more events, one may observe point sources, clusters, and larger scale anisotropies in the sky. The crucial questions will be: Does the distribution of cosmic rays in the sky follow the distribution of matter within our galaxy or the distribution of "nearby" extragalactic matter, or is there no relation to the distribution of matter? Are there point sources or very tight clusters? What is the energy distribution of events from these clusters? Are these clusters associated with specific astrophysical objects? If there is no spatial modulation or no correlation with observed matter, what is the spectrum? This situation would imply an entirely different class of sources which are visible only in the "light" of cosmic rays with energy $\geq 5 \times 10^{19}$ eV. Of course there may be a combination of these possibilities. If even crude data on primary composition is available, it can be divided into catagories of light and heavy components which may have different distributions. Crucial to these considerations is uniform exposure over the whole sky. And a final and perhaps most fundamental question is: Is there an end to the cosmic ray spectrum?

5. The Pierre Auger Observatory

The discussion so far makes clear that the EHECR's are a mystery that will require even larger detectors than the present AGASA detector which has an aperture of $\sim$ 125 km^2-sr.

The rate of cosmic rays with energy $\geq 5 \times 10^{19}$ eV is about 4 km^{-2}-sr^{-1}-century^{-1} and 1 km^{-2}-sr^{-1}-century^{-1} for energies $\geq 10^{20}$ eV. Thus detectors with very large acceptances are required to gather even modest statistics.

Table 1
Events per year

Energy	HiRes	Auger
≤ 10 EeV	$\sim$ 500	10000
≥ 10 EeV	387	5150
≥ 20 EeV	170	1590
≥ 50 EeV	63	490
≥ 100 EeV	15	103
≥ 200 EeV	4	32
≥ 500 EeV	1.4	10

The next improvement in the sensitivity of cosmic ray detectors is the High Resolution Fly's Eye (HiRes) [20]. It anticipates a time-averaged aperture of 300 km^2-sr at 10^{19} eV and 1000 km^2-sr at 10^{20} eV. The completed HiRes experiment has begun operation. At the most recent International Cosmic Ray Conference [21] the first preliminary data from a portion of the HiRes experiment was presented. They find a spectrum in agreement with the AGASA experiment with approximately 7 events above 10^{20} eV.

We will briefly describe a more ambitious approach to the problem, the Pièrre Auger Observatories named after the French physicist who discovered extensive air showers [22]. In 1938 Auger demonstrated that particles were arriving from outside the earth with energies $\geq 10^{15}$ eV[23].

The Auger project is a comprehensive experiment designed to study cosmic rays with energy $\geq 10^{19}$ eV with the least possible bias concerning theories of their origin. Since the cosmic rays are likely to point to the sources, a comprehensive study requires that the entire celestial sphere be observed.

Two instruments will be built at mid-latitude sites in the southern and northern hemispheres. Each instrument will observe cosmic rays at zenith angles up to 60°, so that as the earth turns the whole sky is nearly uniformly observed. Each instrument is a hybrid consisting of a surface array to measure the lateral distribution of the shower particles on the ground, and a fluorescence detector to measure the longitudinal development of the shower. The surface array consists of 1600 water tanks (10 m^2 x 1.2 m deep) spread over 3000 km^2. The configuration of the fluorescence detectors is such that when conditions permit their operation (dark, moonless nights) they will register $\geq 90\%$ of the showers which trigger the surface array. Approximately 10% of the showers will be observed by both detectors. This subset of events will permit a cross check of the energy and provide the maximum possible information on the composition of the primary. In Table 1 we present the expected number of events for a single site for one year based on the AGASA spectrum [15]. For comparison the expected number for HiRes is also presented.

An international collaboration consisting of 19 countries is pooling resources to build two cosmic ray observatories each with an aperture of 7000 km^2-sr for energies $\geq 10^{19}$ eV. The observatories are to be built in Mendoza Province, Argentina (35.2° S, 69.2° W, altitude 1400m) and the state of Utah in the United States (39.1°, 112.6° W, altitude 1400m). Construction has begun at the southern site and is expected to be complete by 2003. Construction is expected to begin in the northern site in 2002.

6. Acknowledgements

This work was supported by NSF grant PHY0070761.

REFERENCES

1. Yoshida, S. and Dai H., 1998, J. Phys. **G 24**, 905; Sokolsky, P., P. Sommers, and B. R. Dawson, 1992 Physics Rep. **217**, 225; *Proceedings of the Paris Workshop on the Highest Energy Cosmic Rays*, 1992, Nucl. Phys. lo(Proc. Supp.) **B 28**, 213; *Proceedings of the International Symposium on Extremely High Energy Cosmic Rays: Astrophysics and Future Observations*, 1996, ed M. Nagano, (Institute for Cosmic Ray Research, University of Tokyo); Swordy, S., rapporteur talk, 1994, *Proceedings of the 23rd International Cosmic Ray Conference, (Calgary)* 243; Watson, A. A., 1991, Nucl. Phys. (Proc. Supp.) **B 22**, 116; V. S. Berezinskiĭ, S. V. Bulanov, V. A. Dogiel, V. L. Ginzburg (editor) and V. S. Ptuskin, 1990, *Astrophysics of Cosmic Rays*, North Holland, Elsevier Science Publishers, The Netherlands;Bhattachargee, P. and G. Sigl, astro-ph/9811011, submitted to Physics Reports; Cronin, J. W., 1990, Rev. Mod. Phys. **71**, S165.
2. Lawrence, M. A., et al., 1991, J. Phys. **G 17**, 773.
3. Afanasiev, B. N., et al., 1995, *Proceedings of the 24th International Cosmic Ray Conference (Rome)* **2**, 756.
4. Bird, D. J., et al., 1994, Ap. J. **424**, 491.
5. Yoshida S., et al., 1995, Astroparticle Physics **3**, 105.
6. Yoshida, S. and H. Dai , 1998, J. Phys. **G 24**, 905.
7. Greisen, K., 1965, *Proceedings of the 9th International Cosmic Ray Conference (London)* **2**, 609.
8. Hillas, A. M., 1984, Ann. Rev. Astron. Astrophys. **22**, 425.
9. Biermann, P., 1997, J. Phys. **G 23**, 1.
10. Blandford, R., *Acceleration of Ultra High Energy Cosmic Rays* 1998, paper presented at Nobel Symposium, August 1998, Stockholm, Sweden.
11. Olinto, A., 2000, talk presented at this conference.
12. Greisen, K., 1966, Phys. Rev. Letters **16**, 748.
13. Zatsepin, G. T. and V. A. Kuz'min, 1966, JETP Letters **4**, 78.
14. Kronberg, P. P., 1994, Rep. Prog. Phys. **57**, 325; Kronberg, P. P., 1994, Nature **370**, 179; Cole, P., Comments Astrophys. **16**, 1992, 45.
15. Takeda, M., et al., 1998, Phys. Rev. Letters, **81**, 1163.
16. Hayashida, N., et al., 1994, Phys. Rev. Letters, **73**, 3491.
17. Bird, D. J., et al., 1995, Ap. J., **441**, 144; Elbert, J. W. and P. Sommers, 1995, Ap. J., **441**, 151.
18. Watson, A. A., 1997, University of Leeds, private communication.
19. Takeda, M. et al.,1999, Ap. J. **522**, 225; update of catalog in astro-ph/0008102.
20. Abu-Zayyad, T., et al., 1997, in *Proceedings of the 25th International Cosmic Ray Conference*, Durban, edited by M. S. Potgeiter, B. C. Raubenheimer, and D. J. van der Walt (World Scientific, Singapore), Vol 5, p321; this paper and the eleven that immediately follow describe various aspects of the HiRes detector.
21. Abu-Zayyad, T. et al., 1999, in *Proceedings of the 26th International Cosmic Ray Conference*, Salt Lake City, edited by D. Kieda,

M. Salamon, and B. Dingus, paper OG 1.3.06, Vol 3, p264.
22. Pièrre Auger Design Report, 2nd ed. March 1997, Fermilab; Design Report and more than 200 technical notes available at Auger web site, www.auger.org.
23. Auger, P., et al., 1938, Comptes Rendus **206**, 1721; Auger, P., 1939, Rev. Mod. Phys. **11**, 288.

ELSEVIER

Nuclear Physics B (Proc. Suppl.) 97 (2001) 10–15

NUCLEAR PHYSICS B
PROCEEDINGS
SUPPLEMENTS
www.elsevier.nl/locate/npe

The Interconnection between Cosmic Ray Results and Accelerator Results

Lawrence W. Jones[a]

[a]University of Michigan, Department of Physics, Ann Arbor, MI 48109-1120

The employment of the large detectors at the CERN electron-positron collider, LEP, for cosmic ray muon studies is reviewed, and results from observations with the ALEPH detector are discussed. The muon multiplicity distributions do not agree with simulations for large multiplicities. Possible future extensions of this program are noted.

1. Introduction

The four detectors at the electron-positron collider (LEP) at the European Laboratory for Particle Physics (CERN) in Geneva, Switzerland and neighboring France each employ elegant muon detection systems and each represent an investment of hundreds of millions of Swiss francs; more than could be justified for a facility devoted exclusively to the study of cosmic ray muons. And yet, with an investment of the order of a percent of the original detector costs, these detectors can be utilized for cosmic ray muon physics research with no sacrifice of their continuing utilization for e^+e^- physics. In recent years, two of the four detectors have been utilized in this way, the ALEPH and the L3 detectors [1]. An earlier summary of these programs is included in the October, 1999 issue of the CERN Courier [2].

2. CosmoLEP

The use of the large underground detectors at CERN for cosmic ray studies had been suggested some time ago in the context of the UA1 detector [3]. More recently, H. Wachsmuth [6] suggested that it would be interesting to explore possible time-coincident cosmic ray muons in two or more of the LEP detectors. These four detectors are equally spaced around the 27 km LEP ring (and hence are at the corners of a square approximately 6 km on a side) and are located under between 30 mand 140 m of rock overburden [4]. He was stimulated, in part, by a paper of Carrol and Martin, reporting marginally-significant coincidences between cosmic ray events in 4 surface scintillation detectors separated by over 100 km [5].

As a member of the ALEPH collaboration, Wachsmuth initiated cosmic ray muon studies with that detector, and communicated with interested members of the other LEP collaborations. The ALEPH detector is located under about 140 meters of rock overburden, corresponding to a vertical muon detection threshold of 70 GeV. Wachsmuth and his collaborators have pesented results of their studies of the correlation between the observed muon flux and barometric pressure as well as with time of day (reflecting the dependence of the pion decay probability with atmospheric density in the upper atmosphere). They also have reported decoherence studies; the muon coincidence rate as a function of separation between the central ALEPH detector and scintillation counters located at varying distances up to over 500 m (in the detector assembly hall and in the LEP tunnel) and at (approximately) the same rock depth [6].

3. L3 + Cosmics

Members of the L3 collaboration also became interested in using that detector for cosmic ray muon studies. Located at a shallower depth than the other LEP detectors, it has a vertical muon detection threshold of 15 GeV. The muon drift chamber spectrometer system has a projected area of over 100 m^2, and is in a magnetic field of 0.5 tesla, providing a momemtum resolution of a few percent at 50 GeV/c. The emphasis of this research program to date has been on collecting a high-statistics inclusive muon spectrum

PII S0920-5632(01)01180-X

with small systematic uncertainties. These data are highly desired by those calculating the atmospheric muon-neutrino flux in the context of the recent experimental evidence, from Super-Kamiokande and elsewhere, for neutrino oscillations. Other topics which are being studied include the muon charge ratio as a function of energy, multimuons, searches for possible point sources, and searches for possible evidence of WIMP decays or interactions in this detector. A separate, more comprehensive report on the L3 cosmic ray muon research program is being presented by H. Wilkins at this Symposium [7].

4. Multimuon studies with ALEPH

K. Eggert had been intrigued by reports of muon bundles in the cosmic ray literature [8] and has assembled a group at CERN to explore cosmic ray multimuon events with the ALEPH detector. The ALEPH detector has a time projection chamber (TPC) of about 16 m^2 area as its central tracking detector; it is located in a 1.5 tesla superconducting solenoid and has excellent multitrack resolution. The outer hadron calorimeter, about 11 m long and of comparable diameter, also has imaging capabilities, although with no magnetic field and with more limited multitrack resolution. The ALEPH detector is "live" for about 2.5 μs (encompassing the LEP beam-crossing time window) each 22 μs (the interval between successive beam crossings), or about 10% of real time. Over the recent years - between 1997 and 2000 - it has been possible to collect data for an integrated time of over 10^6 seconds [9].

Cosmic ray muons have been studied extensively near the earth's surface with detectors having a muon threshold energy of about 2 GeV, and also in large underground detectors (such as MACRO) with a threshold energy of a TeV or more. However, there has been more limited quantitative study of muons of energies around 100 GeV (order of magnitude). It is in this energy range that the parent pions, produced in the first interaction of primary cosmic rays in the upper atmosphere, have comparable probabilities of decay and of interaction. And high muon multiplicity events produced by primary cosmic rays in the "knee" region of the spectrum, 1 PeV to 10 PeV, include muons of these energies.

The ongoing issue of primary composition in the knee energy region is receiving very considerable attention at this conference as it has at cosmic ray meetings over the past two decades; there are still strong disagreements over the correct values of $< lnA >$ vs. energy at these energies. As discussed in another presentation at this Symposium, the observable consequences of the physics of the first interaction and of the primary composition are interrelated, so that it is necessary to study many different observables at our observation depth in the atmosphere and to compare with a variety of Monte Carlo predictions in order to reach the correct conclusions about both the composition and the interaction physics [10].

Ideally, it would be desirable to collect data on EM showers (density vs. radius from the shower core), muon energy and radial distributions, and hadron energy and radial distributions simultaneously for a statistically large sample of cosmic ray interactions, through this energy range where direct observation of the primaries is not possible. Lacking such a comprehensive observation capability (the Karlsruhe "Kascade" comes as close as any detector to realizing this), the next best thing is to fit each individual data set to different Monte Carlo interaction models with different composition assumptions.

As one example of the application of muon data to these problems, the radial density distributions of muons of over 70 GeV are significantly narrower for proton primaries as compared with iron primaries over the energy range between 3 PeV and 100 PeV, with the density ratio (relative to the density at the core axis) almost independent of primary energy, based on CORSIKA simulations. Muon data of the sort collected with the LEP detectors may add a new and valuable dimension to the measurements against which the Monte Carlos and composition models may be checked. Of course there is the real possibility that none of the first interaction simulations currently in use is correct; they are all based on accelerator data at energies orders of magnitude lower than a PeV. The discovery of 'new physics' is an exciting possibility.

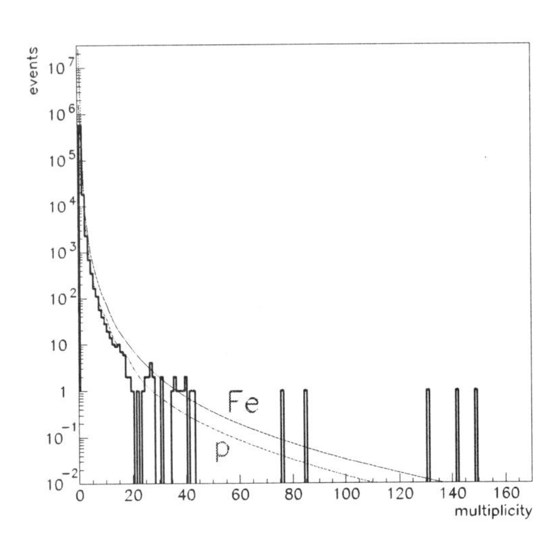

Figure 1. The multiplicity distribution of muons in the TPC compared to CORSIKA simulations for proton and iron primaries.

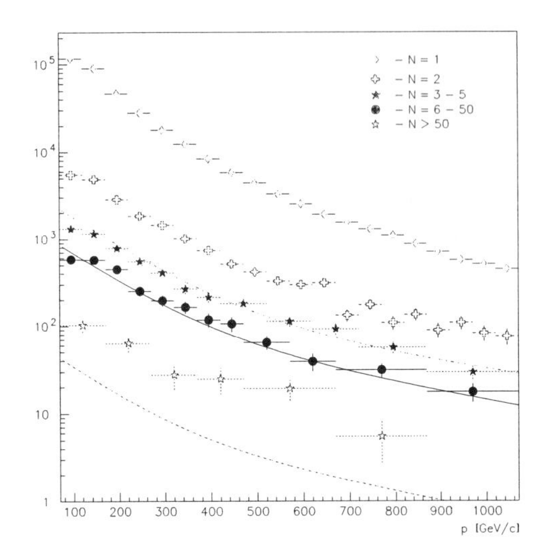

Figure 2. Momentum distributions of muons in the ALEPH TPC for several multiplicity bins. The energy loss in the overburden has been compensated for by adding 70 GeV/cosθ to the measured momenta. The line for the multiplicity bin n$\geq$3 represents the simulation for proton primaries, absolutely normalized.

A great deal has also been said about possible point sources of neutral primary cosmic rays, and there have been numerous reports of such sources in the literature, even as observed by underground muons. As the real time is automatically recorded with each event, it is straightforward to search the accumulated data set for such point sources.

5. ALEPH data

The results of the muon studies with ALEPH data were reported at the Sodankyla Workshop last year [11]. The most interesting consequence of these data is the apparent excess of high multiplicity muons relative to that predicted by the Monte Carlo programs used. The data were compared with the CORSIKA simulation of the air shower cascade development and the QGS-JET interaction model. The observed muon multiplicity distribution is presented on Figure 1, with the model predictions for pure proton and pure iron composition included for comparison. These multiplicities are those observed in only the central ($16\,m^2$) TPC detector. The data, taken during the 1997 through 1999 LEP operation, are from about 580,000 events, corresponding to an effective data taking time of 1.7×10^6 seconds. Characteristics of the five highest multiplicity events are summarized in Table 1. Note that the "primary energy", entered in the last column, is the energy assuming a proton-initiated event with its core centered on the detector and assuming the CORSIKA simulation.

The highest multiplicity event in this study is shown in Figure 3, where the two orthogonal views of the event are shown as well as a reconstructed projection normal to the event axis. This event has about 150 tracks visible in half of the Time Projection Chamber.

Of interest also are the momentum distributions within events sorted according to muon multiplicity; these are plotted in Figure 2. It is perhaps not surprizing that the momentum distributions are somewhat flatter for higher muon multiplicities.

Table 1
Characteristics of the Five Highest Multiplicity Events. The primary energy was estimated by assuming the shower center to be close to the TPC and taking into account the zenith angle.

event	muon density (m^{-2})	zenith angle (°)	primary energy (PeV)
97-a	4.75	40.8	30
97-b	5.3	37.7	30
97-c	8.9	40	60
98-a	8.2	48.6	70
98-b	18.6	27	100

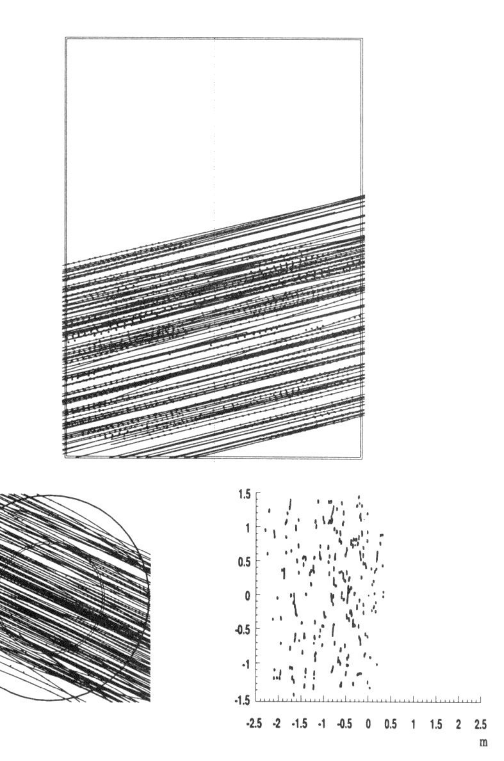

Figure 3. The ALEPH event with the highest muon multiplicity (about 150 tracks in half of the TPC) in three different views: longitudinal, transverse, and along the shower axis. Reconstructed tracks are shown as lines.

6. Future plans

It is planned to expand the muon detection area in the ALEPH detector area by installing 8 modules of drift tube tracking chambers in the assembly hall near the ALEPH detector. Each module has an area of $20\,m^2$, giving a total active detector area of $160\,m^2$. The modules, used in earlier CERN experiments, consist of two (in some cases, three) vertically-spaced layers of drift chambers and associated trigger counters, making possible the measurement of muon numbers and angles over a significantly larger detector area in high-multiplicity events. The floor-plan, including the existing ALEPH detector and these added chambers, is illustrated in Figure 4 [12]. This figure is, unfortunately, a little out-of-date; as the LEP running ends in early November, 2000, the ALEPH will be decommisioned and will very probably not be operational for these cosmic ray muon studies.

There has also been discussion of the addition of a surface air shower array above the ALEPH pit, to provide complementary data on the electromagnetic component of the cascades accompanying observed muon events. Again, existing counters from completed experiments would be salvaged and utilized.

Other, additional tracking chambers could be added in the LEP tunnel at varying distances from the ALEPH detector to sample muon densities further from the event core (which, for events of greatest interest, will be located in or near the ALEPH detector). As the LEP electron-positron experimental program is to be completed this year (2000), the installation and operation of these additional counters would not interfere with the CERN accelerator physics program, nor would there be background from the accelerated beam particles to interfere with the cosmic ray measurements.

7. Conclusions

Cosmic ray muon observations with the LEP detectors at CERN show promise of making significant contributions to the understanding of particle interactions and of primary composition at energies above a PeV. Early data are cer-

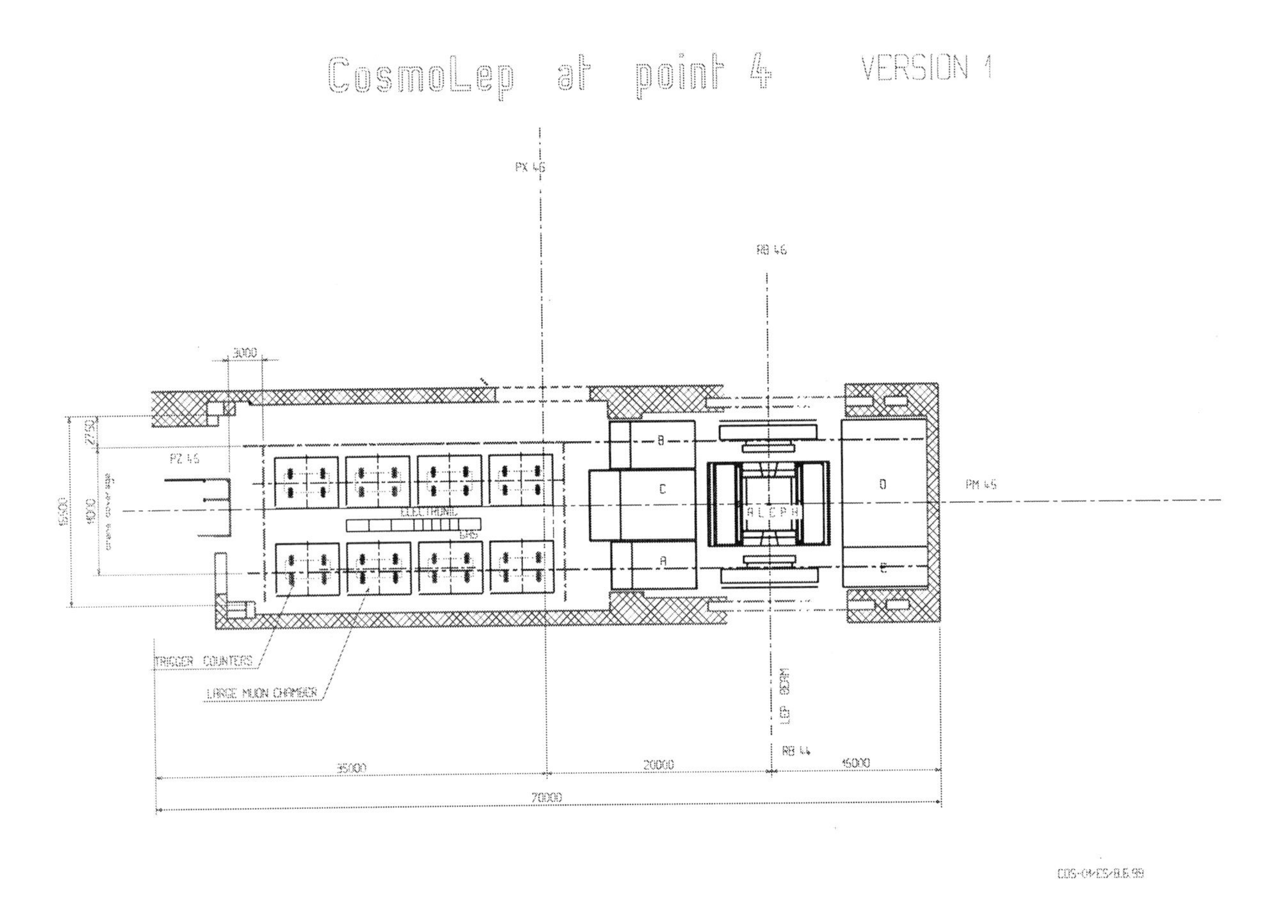

Figure 4. Proposed array of added muon chambers in the ALEPH underground cavern. The ALEPH detector is between the areas denoted by ABC and DE.

tainly interesting, and have stimulated members of the ALEPH and the L3 collaborations to expand these studies and to exploit these unique detector capabilities.

REFERENCES

1. Note: Karsten Eggert from CERN was originally scheduled to speak at this Symposium on this topic, but scheduling conflicts prevented his attendence. He has graciously supplied much of the material presented below.
2. CERN Courier **39**, 8 (1999) 29-33.
3. O.C. Allkofer, K. Eggert, P. Erhard *et al.*, *The UA1 detector as a possible cosmic ray device*, Proc. XVII International Cosmic Ray Conference, Paris **10** (1981) 401.
4. H. Wachsmuth, *Proposal to search for Cosmic Ray coincidences in the LEP detectors*, (draft) 16 Dec. 1993 (CosmoLEP-note 94.000), and A. Ball *et al.*, CERN/LEPC Note 94-10 (1994).
5. O. Carrol and M. Martin, Phys. Lett. **B325** (1994) 526.
6. A.-S. Muller, Nucl. Phys. B (Proc. Suppl.) **52B** (1997) 261-263 and H. Wachsmuth *et al.*, Proc. XXV International Cosmic Ray Conference, Durban **7** (1997) 249-252.
7. H. Wilkins (for the L3 Collaboration), *First results about atmospheric muons from the L3 + C experiment at CERN*, presentation at this Symposium.

8. S. Miyake *et al.*, J. Phys. Soc. Japan **18** (1963).
9. Note: The term "CosmoLEP" was originally used by Wachsmuth to refer to the possible search for coincident events in the 4 LEP detectors, and "CosmoALEPH" has been used to refer to cosmic ray observations with the ALEPH detector. The L3 group, meanwhile, refers to their cosmic ray program with the term "L3 + Cosmics". Recently, CosmoLEP has also been used to refer to any cosmic ray work with the ALEPH detector and to future cosmic ray observations utilizing the LEP facilities.
10. L.W. Jones, *Cosmic ray issues for accelerator experiments*, presentation at this Symposium.
11. A. Bruhl and J. Strom, *Observation of muon bundles in the ALEPH-detector*, presented at the First Arctic Workshop on Cosmic Ray Muons, Sodankyla, Finland (1999).
12. K. Eggert *et al., CosmoLEP, an underground cosmic ray muon experiment in the LEP ring*, CERN/LEPC 99-5, LEPC/P9 (1999).

ELSEVIER

Nuclear Physics B (Proc. Suppl.) 97 (2001) 16–34

www.elsevier.nl/locate/npe

Study of Nuclear Interactions by Emulsion Chambers at Mt. Chacaltaya

Akinori Ohsawa[a]

[a]Institute for Cosmic Ray Research, University of Tokyo, Kashiwa, Chiba, 277-8582 Japan.

1. Introduction

Study of nuclear interactions by observing cosmic rays permits us to investigate what cannot be done by accelerator experiments. Those are the studies of high energy nuclear interactions in the energy region exceeding 1.5×10^{14} eV (CERN $\bar{p}p$ collider of $\sqrt{s} = 546$ GeV) or 1.6×10^{15} eV (FNAL Tevatron of $\sqrt{s} = 1,800$ GeV), and of exotic events, such as Centauro events [1], etc. which are possibly produced in extremely high energy region and/or by an exotic component among the primary cosmic rays. In this report we discuss high energy nuclear interactions — multiple particle production — based on data by cosmic-ray experiments, leaving important and interesting topics of exotic events elsewhere.

To discuss high energy nuclear interactions it is convenient to start from the energy distribution of produced particles per event, defined as

$$\frac{dN}{dE^*} \equiv \frac{1}{\sigma_{inel}} \frac{d\sigma}{dE^*}$$

because other quantities, such as inelasticity and multiplicity, are derived from it[1]. Feynman scaling law asserts that the distribution, which is expressed by a variable $x^* \equiv 2p^*_{||}/\sqrt{s}$, is independent of the incident energy of collision as $\sqrt{s} \to \infty$ [2]. The law is verified in the energy region of $\sqrt{s} \leq 63$ GeV. And one of the empirical formulae of the energy distribution for *charged* produced particles is

$$\frac{dN}{dx^*} = D\frac{(1-x^*)^d}{x^*} \qquad (1)$$

with $D = 2(d+1)/3$ and $d = 4.0$ [3]. It leads to the energy distribution in the laboratory system

$$\frac{dN}{dx} = D\frac{(1-x)^d}{x}$$

because we have $x^* \simeq E/E_0 \equiv x$ at high energies. Then we have the charged multiplicity and the total inelasticity

$$m(E_0) \equiv \int_{m_\pi/E_0}^{1} \frac{dN}{dx}dx \simeq D\left[\ln\left(\frac{E_0}{m_\pi}\right) - \frac{25}{12}\right]$$

and

$$< K > \equiv \frac{3}{2}\int_0^1 x\frac{dN}{dx}dx = 0.5$$

where the factor 3/2 is due to charge independence of produced pions assuming that all the produced particles are pions. Both of the expressions above reproduce the experimental data in low energy region well (see Figure 12).

In Section 2 we discuss violation of the Feynman scaling law at high energies, based on the experimental data by emulsion chamber experiments at Mt. Chacaltaya. And in Section 3 we formulate the energy distribution of produced particles which shows the violation of the Feynman scaling law, based on data of direct observation by accelerator and cosmic-ray experiments. We discuss the consequences of the formulated distribution at high energies and examine its validity using the data of extremely high energy ($> 10^{18}$ eV) air showers.

2. Emulsion chamber experiments at Mt. Chacaltaya

Emulsion chamber experiments have been carried out at Mt. Chacaltaya (5,200 m, Bolivia) by Brazil-Japan collaboration and by Bolivia-Brazil-Japan collaboration, employing various types of

[1]Quantities with and without an asterisk are those in the center of mass system and in the laboratory system, respectively.

PII S0920-5632(01)01181-1

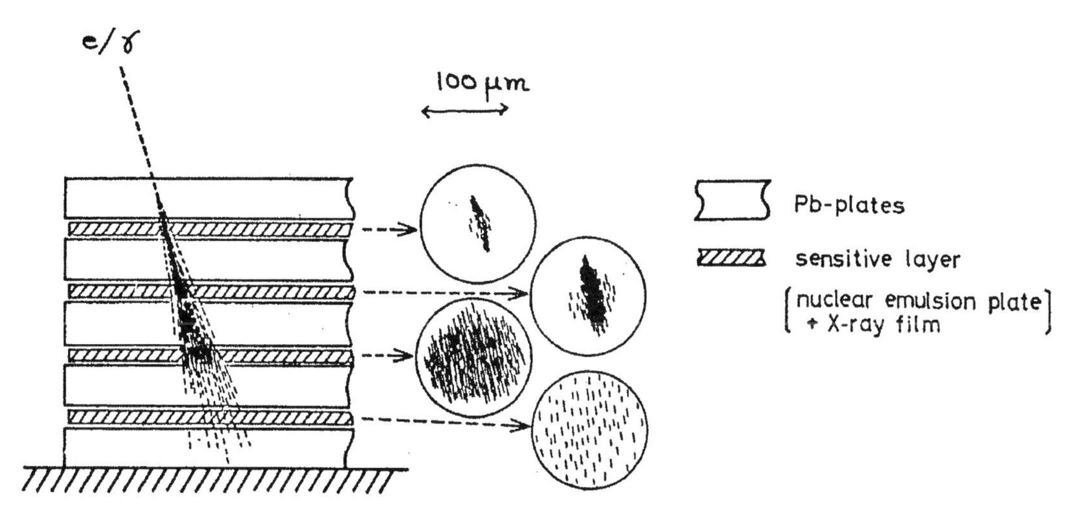

Figure 1. Emulsion chambers consist of lead plates and sensitive layers (X-ray films and/or nuclear emulsion plates), piled up alternately. An electron or a photon, incident upon the chamber, produces a cascade shower in the chamber.

Table 1
Performance of the emulsion chamber

	Nuclear emulsion plate	X-ray film
Shower position	$\Delta x \sim 10\ \mu$m	$\Delta x \sim 100\ \mu$m
Shower energy	electron tracks by microscope absolute value $\Delta E/E \sim 20$ % $E_{th} = 0.1$ TeV	opacity by microphotometer relative value $\Delta E/E \sim 10$ % $E_{th} = 1$ TeV
Shower structure	observable	impossible

chambers according to the purpose of the study. In this section we present some results by the experiments in order to discuss what the experimental data indicate on the nuclear interaction characteristics in high energy region.

2.1. Emulsion chamber

An emulsion chamber is a multiple sandwich of lead plates (1 cm thick each) and sensitive layers (X-ray films and/or nuclear emulsion plates), piled up alternately [4] (see Figure 1). It detects cascade showers which are produced in the chamber by high energy particles incident upon the chamber. That is, an electron or a photon of high energy, incident upon the chamber, produces a number of electrons, positrons and photons — a cascade shower — in the chamber through a chain of electromagnetic interactions with lead. The electron component in the cascade shower is recorded by the sensitive layers, and it appears as a small black spot ($\sim$ 100 μm in radius) on the X-ray film after development.

Since a cascade shower develops over several centimeters of lead, the shower spots are found on X-ray films of several successive sensitive layers in the chamber. The transition of the spot darkness, which is obtained by a microphotometer with a slit of $200 \times 200\ \mu\text{m}^2$, along the depth of the chamber enables us to estimate the energy of the incident particle by comparing it with those which are calculated on the basis of the cascade theory [5] and the sensitivity curve of the X-ray film. In this way we can determine the position and the energy of the incident electron or pho-

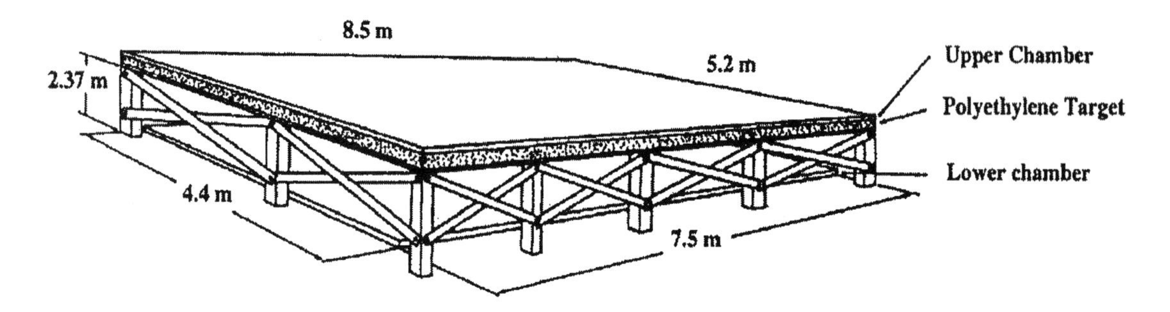

Figure 2. Two-storey chamber, which is used to study C-jets. The chamber consists of upper chamber, target layer, air gap and lower chamber.

Table 2
Structure of two-storey emulsion chambers

	I ('65 - '69)		II ('69 - '83)		III ('83 -)	
	Area	Thickness	Area	Thickness	Area	Thickness
Upper chamber	6.0 m^2	6 cm Pb	44.2 m^2	6 cm Pb	44.2 m^2	6 cm Pb
Target layer	6.0 m^2	70 cm Pitch	44.2 m^2	23 cm Pitch	44.2 m^2	30 cm CH†
Air gap	—	80 cm	—	158 cm	—	237 cm
Lower chamber	6.0 m^2	10 cm Pb	32.0 m^2	10 cm Pb	32.0 m^2	10 cm Pb

† Polyethylene

ton, called (e,γ) hereafter, with high precision (see Table 1). The detection threshold energy of the shower is $E_{th} = 1 \sim 4$ TeV on the X-ray film, depending on the experimental conditions.

On the other hand a high energy hadron, incident upon the chamber, causes multiple particle production through a nuclear collision with lead in the chamber. The π^0's among the produced particles originate a cascade shower through their decays into γ-rays. In this way emulsion chamber is sensitive to both (e,γ)'s and hadrons, incident upon the chamber. It is worth noting,
(1) that hadron detection depends on the total thickness of lead in the chamber, because the mean free path of inelastic N-Pb collisions (N : nucleon) is $\lambda_N = 18.0$ cm, and
(2) that the observed energy E_{ob} of hadron-induced shower is not that of the incident hadron E_h but that released into γ-rays, *i.e.* $E_{ob} = k_\gamma E_h$ where k_γ is the energy fraction of produced γ-rays with $< k_\gamma > \simeq 1/6$. Therefore the detection threshold for hadrons is much higher (~ 6 times) than that for (e,γ)'s.

2.2. Nuclear interactions at 10^{14} eV

Nuclear interactions at 10^{14} eV are observed by the two-storey chamber (Figure 2), which consists of the upper chamber, the target layer, the air gap and the lower chamber [4,6] (see Table 2 for details on the two-storey chamber). The upper chamber is a filter of atmospheric (e,γ)'s, and therefore only the hadrons arrive at the target layer to cause multiple particle production. The γ-rays produced in multiple particle production, which are mainly the decay products of produced π^0's, arrive at the lower chamber with sufficient mutual separation owing to the air gap[2]. The

[2] A γ-ray with the energy $E_\gamma = 1$ TeV has average lateral spread $r = Hp_{T\gamma}/E_\gamma = 100$ (cm) $\times$ 200 (MeV/c)/1 (TeV) = 200 (μm) after traversing the air gap of $H = 1$ (m).

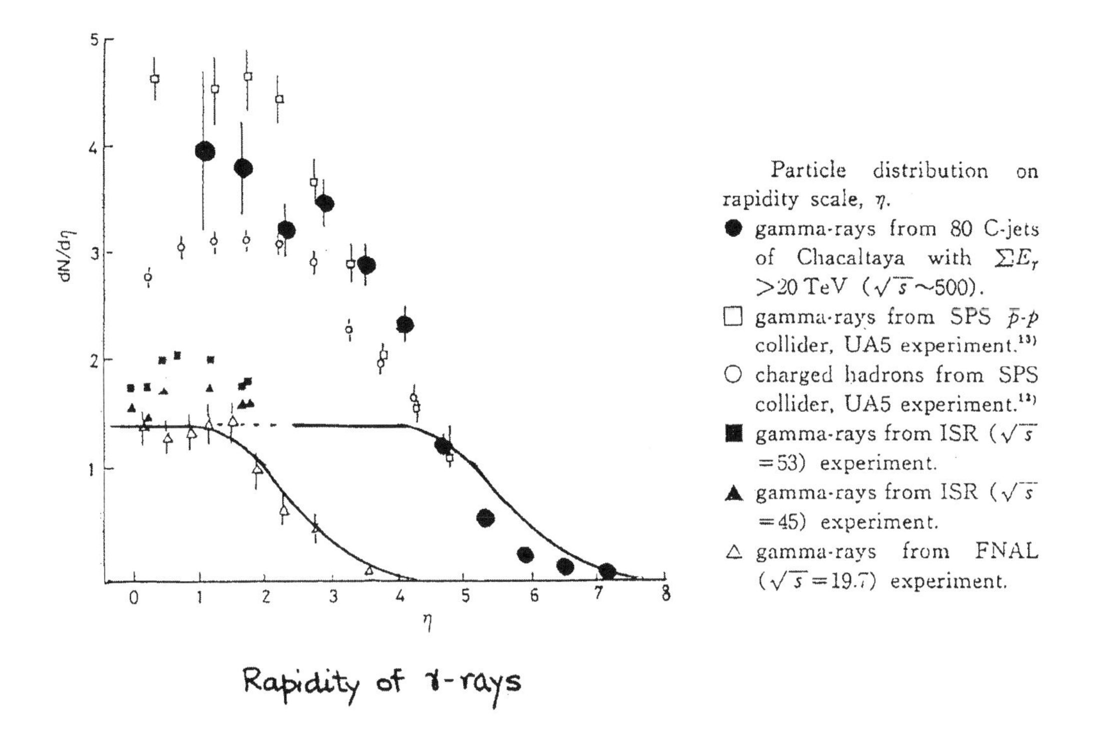

Figure 3. Rapidity density distributions of γ-rays. The solid curves represent the distributions to follow Feynman scaling law at $s^{1/2} = 19.7$ GeV and at $s^{1/2} = 500$ GeV, respectively.

events are called "C-jets", because the events are mainly due to the collisions with carbon in the target layer.

The showers in the lower chamber, originated by these γ-rays, are inspected through a microscope in the nuclear emulsion plates which are employed in the sensitive layers of the lower chamber together with X-ray films. Their energies are determined by counting the number of electron tracks in respective showers. That is, the transition of the electron number in a shower is compared with those which are calculated for various incident energies by cascade theory [5]. The threshold energy for shower detection is $E_{th} = 0.2 \sim 0.5$ TeV when we use nuclear emulsion plates.

The basic idea of a two-storey chamber is to study nuclear collisions by observing π^0's among the produced particles in multiple particle production. One can see that the produced particles in the collision are observed directly in this experiment. It is worth noting that the ambiguity of the interaction point in the target layer is less than the error of energy, determined by electron track counting, and consequently the energy of γ-rays can be calibrated by the kinematical relation of $\pi^0 \to 2\gamma$ decay.

The characteristics of multiple particle production, which are obtained by the experiment, were confirmed by the experiments by CERN $\bar{p}p$ collider [7]. Those are the violation of Feynman scaling law, gradual increase of $< p_{T_\gamma} >$ (the average transverse momentum) with the incident energy, a correlation between dN/dy_γ (the rapidity density of produced γ-rays) and $< p_{T_\gamma} >$, etc.

Figure 3 represents the rapidity density distribution of γ-rays obtained by the experiment [8,6]. The data are from 80 C-jets with the total observed energy $\sum E_\gamma = 20 \sim 80$ TeV and the number of observed γ-rays $n_\gamma \geq 4$. The average incident energy of collision is $< \sqrt{s} >= 500$ GeV. It shows clearly that γ-ray production is suppressed in the forward region and enhanced in the central region, compared with the distribution of the Feynman scaling law.

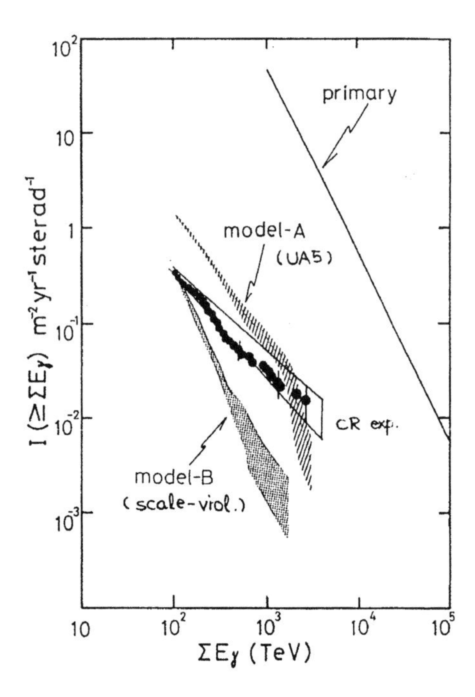

Figure 4. The intensity of families with the energy exceeding $\sum E_\gamma$ (the total observed energy in the family) at Pamirs. Full circles are the experimental data, and "model-A" is by the simulations where UA5 code is assumed for nuclear interactions.

2.3. Nuclear interactions at 10^{15} eV (1)

Direct observation of nuclear collisions is not easy in this energy region because of the scarce intensity of high energy cosmic-rays even at high mountains. Hence nuclear interactions in the atmosphere are observed by the upper chamber of the two-storey chamber or by a simple emulsion chamber of large area.

The event, called "family" or "A-jet", consists of several $\sim$ a few tens showers of (e, γ) and hadron origins, distributed over an area of a few $\sim$ ten centimeters radius. Identification of a family is easy because the constituent showers are parallel one to another on the projection map of all the observed showers. Family is the air shower at its central part or at the early stage of its development, because only the high energy particles are detected by the emulsion chamber.

It is not straightforward to extract characteristics of nuclear collisions from the observed data of families, because most of the events are not due to a single collision but due to complicated cascade processes — nuclear and electromagnetic — in the atmosphere. Therefore usually the data are compared with simulations which follow the atmospheric diffusion of cosmic rays. Main assumptions in the simulations are on the energy distribution of produced particles in multiple particle production and on the mass composition of the primary cosmic rays. We assume UA5 code for the former and proton-dominant composition for the latter. UA5 code is a phenomenological simulation code of multiple particle production, which reproduces what UA5 Collaboration observed at CERN $\bar{p}p$ collider [9]. It predicts an energy distribution of produced particles which strongly violates the Feynman scaling law, both in the central region and in the forward region (see Section III). The assumed proton-dominant composition [10] is the extrapolation of the one in low energy region ($\leq$ several TeV), which is established by the experiments of direct observation[3].

Figure 4 shows the frequency of the families with the energy exceeding $\sum E_\gamma$ (the total observed energy of (e, γ)'s in the family) at Pamir altitude by Chacaltaya-Pamir Collaboration [11], which is one of the basic parameters to discuss

[3] A typical is:

E_0 (eV)	Relative abundance (%)				
	H	He	CNO	Heavy	Fe
10^{15}	42	17	14	14	13
10^{16}	42	13	14	15	16

the families. Results obtained by simulations are presented together. The figure shows that the intensity by the experimental data is lower than that obtained by the simulation where UA5 code and proton-dominant composition are assumed. It indicates that at least either of the assumptions should be revised in the direction to make energy subdivision more rapid through the atmospheric diffusion of cosmic rays. That is, the violation of Feynman scaling law is stronger than that expected by UA5 code and/or the composition of the primary cosmic rays is heavier than a proton-dominant one at 10^{15} eV.

The Fuji-Kambala collaboration obtained the similar results by the emulsion chamber experiment at Mt. Kambala (5,400 m, Tibet), which are shown in Figure 5 [12]. They argue that the composition becomes quite heavier at 10^{15} eV on the assumption of quasi-scaling energy distribution of the produced particles. However, a quasi-scaling energy distribution is not an established one in the energy region of 10^{15} eV.

2.4. Nuclear interactions at 10^{15} eV (2)

We cannot discriminate by emulsion chamber data which characteristics of nuclear interaction or cosmic-ray composition causes the observed low intensity of families. Furthermore it should be pointed out that the above argument needs further assumption of the intensity of the primary cosmic rays, which is obtained by the air shower experiments assuming characteristics of high energy nuclear interactions[4]. To overcome such shortcomings we started an experiment to observe families and air showers simultaneously [13].

The experimental set-up consists of an air shower array, a hadron calorimeter and an emulsion chamber, by which the air shower size (the total number of charged particles in the air shower), hadron data and family data are obtained, respectively. The hadron data which are obtained by the hadron calorimeter, installed beneath the emulsion chamber, provide a clue to link air showers with families. That is, the families without arriving time and the hadron data with arriving time are correlated by the coincidence of their geometrical center, and the hadron data and the air showers by the coincidence of their arriving time.

Figure 6 presents a correlation between $\sum E_\gamma$ and N_e (air shower size) for the air showers which are accompanied by the families [13,14]. Definition of the family is $\sum E_\gamma \geq 10$ TeV and $n_\gamma \geq 4$. Discrepancy between the experimental data and the simulation leads to the same conclusion mentioned in the previous subsection. The similar type of the experiments, which is carried out at Tien Shan (3,300 m, Kazakhstan), obtained the same conclusion, which is shown in Figure 6 [14,15].

Figure 6 shows also that the hypothesis of the heavy composition of the primary cosmic rays, proposed by Fuji-Kambala Collaboration, reduces but cannot describe fully the observed discrepancy. Hence we conclude that violation of the Feynman scaling law is stronger than that assumed in the UA5 code, irrespective of the assumption of the primary cosmic-ray composition.

Figure 7 shows the differential energy spectrum of hadrons in the air shower, which is obtained from the data provided by the hadron calorimeter, together with the results by the simulations [16]. The figure confirms the above conclusion again. KASCADE experiment reaches the same conclusion from the study of hadronic cores of extensive air showers by the large hadron calorimeter [17].

3. Nuclear Interactions at $10^{14} \sim 10^{20}$ eV

As we saw in the previous sections, emulsion chamber experiments show/indicate that the Feynman scaling law of the energy distribution of produced particles, which looks to be valid at $\sqrt{s} \leq 63$ GeV, is violated at $10^{14} \sim 10^{16}$ eV. Then our next question is what kind of energy distribution describes the experimental data. However there is a variety of physically possible energy distributions.

Hence in this section we formulate the energy

[4]It is true that the air shower size at the shower maximum does not depend strongly on the details of nuclear interactions and on the nature of the primary cosmic rays. Hence it is a good measure of the primary energy to initiate the air shower.

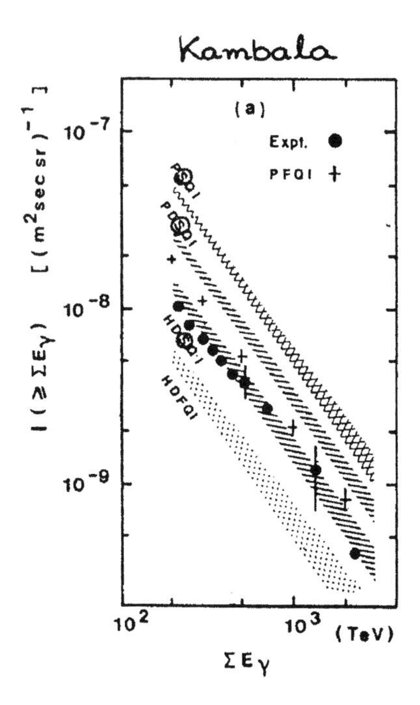

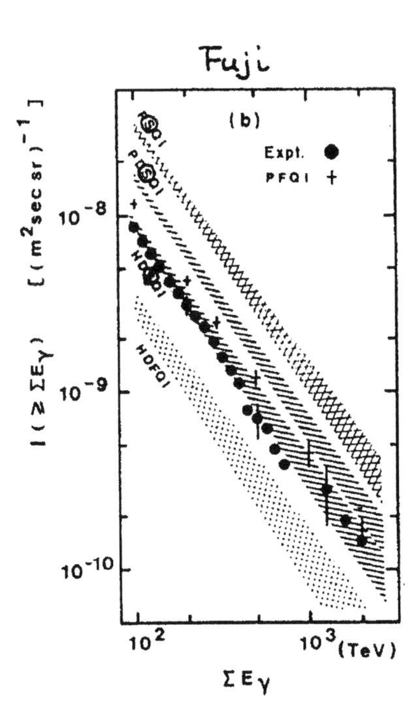

Figure 5. The intensity of the families with the energy exceeding $\sum E_\gamma$ at Mt. Fuji and at Mt. Kambala. The abbreviations, attached to the data by simulations, are as follows:

P (PD) : protons (proton dominant composition) for primary cosmic rays

S (F) : scaling (fire-ball) model for nuclear interactions

Q : QCD-jets included

I : increasing cross section

distribution of produced particles, based on data of direct observation by accelerator experiments and cosmic-ray experiments (C-jets), which show violation of the law. Then the consequences of the formulated distribution are examined and the distribution is compared with those obtained by nuclear interaction models which are used recently in simulations to follow the diffusion of cosmic rays in the atmosphere.

3.1. Formulation of the energy distribution by the data of direct observation

There are several sets of data of direct observation by cosmic-ray and accelerator experiments, which are tabulated in Table 3 in chronological order. One should notice that the experimental data are presented in various quantities as rapidity density, pseudo-rapidity density, etc., owing to the experimental conditions of the respective groups. To discuss these data in relation to the energy distribution of eq.(1), one has to transform dN/dx^* into $dN/d\eta^*$, dN/dy^*, etc.

We assume that the x-distribution is expressed in the following way:

$$\frac{dN}{dx^* dp_T} = aD \frac{(1 - a'x^*)^d}{\sqrt{x^{*2} + \left(\frac{2\mu}{\sqrt{s}}\right)^2}} g(p_T) \qquad (2)$$

$$(\mu \equiv \sqrt{p_T^2 + m_\pi^2})$$

where the parameters a and a' are adjustable ones. This formula reproduces the scaling function of eq.(1) using $a = a' = 1$ and $\sqrt{s} \to \infty$. The parameters a (≥ 1) and a' (≥ 1) express enhancement of eq.(1) in the central region and suppression in the forward region, respectively.

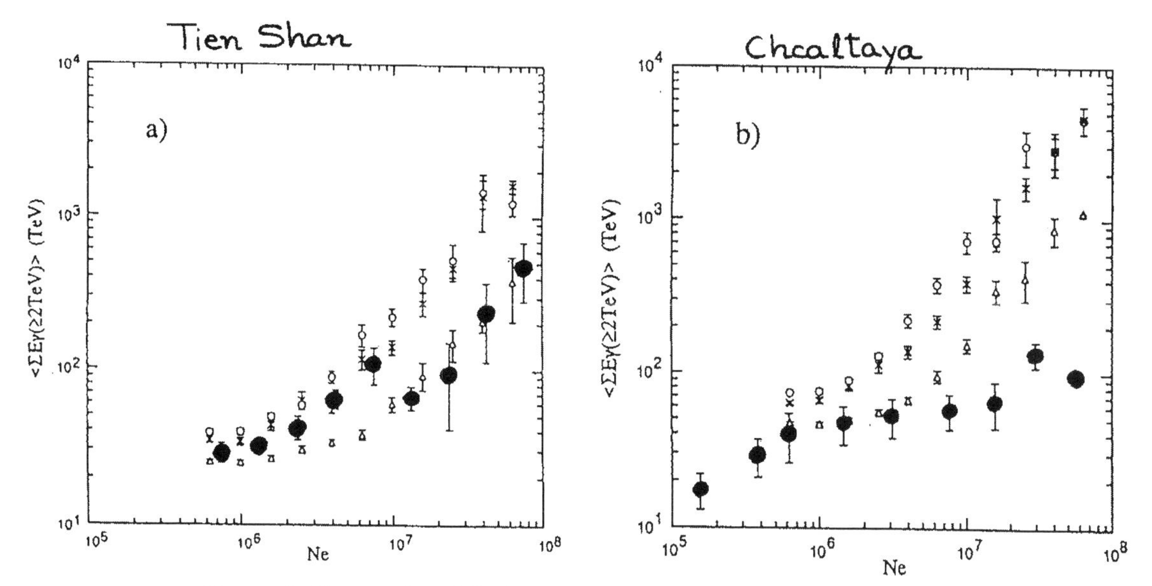

Figure 6. Shower-size (N_e) dependence of average family energy, $\langle \sum E_\gamma \rangle$: (*a*) for the HADRON experiment at Tien-Shan, and (b) for the SYS experiment at Chacaltaya. Symbols are (O) for model-A with 'normal' chemical composition, (×) model-A with 'heavy-dominant' chemical composition, and (Δ) model-B with 'normal' chemical composition and (●) experimental data.

Figure 6. Correlations between $\sum E_\gamma$ (total observed energy in the family) and N_e (air shower size) at Tien Shan (a) and at Chacaltaya (b), for the air showers which are accompanied by families with $\sum E_\gamma \geq 10$ TeV and $n_\gamma \geq 4$. Full circles are for the experimental data; open circles (crosses) are for those by simulations where UA5 code and proton-dominant (heavy-dominant) composition are assumed.

The p_T-distribution is assumed to be

$$g(p_T)dp_T = p_T \exp\left(-\frac{p_T}{p_0}\right)\frac{dp_T}{p_0^2} \qquad (3)$$

with

$$p_0 = \begin{cases} c & (x^* < x_0^*) \\ c\left(\frac{x_0^*}{x^*}\right)^{c'} & (x^* > x_0^*) \end{cases} \qquad (c = 0.2\ \text{GeV}/c)$$

According to eq.(3), the average value of p_T, given by $< p_T >= 2p_0$, is constant (=0.4 GeV/c) in $0 \leq x^* \leq x_0^*$ and becomes smaller in the forward region of $x^* > x_0^*$, which is observed by the experiments [4,18]. The values of c' and x_0^* are determined as

$$c' = 0.57, \qquad x_0^* = 0.08$$

by the rapidity dependence of the average p_T value by UA7 Collaboration [18].

The energy distribution of eq.(2) can be transformed easily to rapidity and pseudo-rapidity distributions:

$$\frac{dN}{dy^* dp_T} = aD\left[1 - a'\frac{\mu}{\sqrt{s}}(e^{y^*} - e^{-y^*})\right]^d g(p_T) \quad (4)$$

$$\frac{dN}{d\eta^* dp_T} = aD\left[1 - a'\frac{p_T}{\sqrt{s}}(e^{\eta^*} - e^{-\eta^*})\right]^d$$

$$\times \frac{(e^{\eta^*} + e^{-\eta^*})}{\sqrt{(e^{\eta^*} - e^{-\eta^*})^2 + \left(\frac{2\mu}{p_T}\right)^2}} g(p_T) \qquad (5)$$

The distributions dN/dy^* and $dN/d\eta^*$ are obtained by numerical integration with respect to the transverse momentum p_T.

Table 3
Particle distribution in multiple particle production

Group	Site	Collision	Energy $\sqrt{s}$ (GeV)	Maximum rapidity	Presented quantity	Observed particles	Observed region
Brazil-Japan Collab. [6]	Mt. Chacaltaya	h–C†	500‡	12.2	dN/dy_γ	γ-rays	$y_\gamma \geq 7.8$
UA5 Collab. [9]	CERN	$\bar{p}p$	53	4.0	$dN/d\eta^*$	charged	$\lvert\eta^*\rvert < 4.5$
			200	5.3	$dN/d\eta^*$	charged	
			546	6.3	$dN/d\eta^*$	charged	
			900	6.8	$dN/d\eta^*$	charged	
UA7 Collab. [18]	CERN	$\bar{p}p$	630	6.4	dN/dy^*	π^0	$y^* = 5.0 \sim 6.6$
Harr et al. [19]	CERN	$\bar{p}p$	630	6.4	$dN/d\eta^*$	charged	$\eta^* = 1.5 \sim 5.5$

† hadron (nucleon, pion)–Carbon collisions
‡ The value is the averaged one, *i.e.* $<\sqrt{s}>= 500$ GeV.

3.2. Scaling violation parameters

Figure 8 shows the pseudo-rapidity distributions of eq.(5) at $\sqrt{s} = 546$ GeV for various values of the scaling violation parameter a'. We assume the energy dependence of parameter a as

$$a = \left(\frac{s}{s_0}\right)^{\alpha} \simeq \left(\frac{E_0}{A}\right)^{\alpha} \qquad (6)$$

where $s_0 = 3.9 \times 10^2$ GeV^2, $A \equiv s_0/2M = 2.0 \times 10^2$ GeV and $\alpha = 0.105$. Then the pseudo-rapidity density at $\eta^* = 0$ is given by

$$\left(\frac{dN}{d\eta^*}\right)_{\eta^*=0} = D\left\langle \frac{p_T}{\sqrt{p_T^2 + m_\pi^2}} \right\rangle a$$

$$= 1.67 \times 0.83 \times \left(\frac{s}{s_0}\right)^{\alpha}$$

which reproduces $\rho(0) = 0.74 s^{0.105}$, given empirically by UA5 Collaboration [9]. It may be worth mentioning that the data of $\rho(0)$, the pseudo-rapidity density at $\eta^* = 0$, is reliable by a collider-type accelerator experiment.

The calculated distributions are compared with the experimental data, listed in Table 3, to obtain the value of the scaling violation parameter a'. The data are those of all inelastic events, but not of NSD (non-single-diffractive) events, because the former are more suitable for discussing cosmic-ray diffusion[5]. The energy dependence of the scaling violation parameters a and a' are in Figure 9. There we observe that the parameter a' has a stronger energy dependence than the parameter a.

Two cases of energy dependence are used in

$$a' = \left(\frac{E_0}{A}\right)^{\alpha'} \qquad (\alpha' = 0.105 \text{ and } 0.210)$$

which are called Model-1 and Model-2 hereafter, respectively. The parameters a and a' in Model-1 have the same energy dependence, and those in Model-2 are the best-fitting ones to the experimental data. The values of a' by Harr *et al.* and by C-jets deviate considerably from other data. Model-0 with $a = a' = 1.0$, which stands for the

[5]Definitions of all inelastic events and NSD events are,
$\sigma_{inel} = \sigma_{\rm NSD} + \sigma_{\rm SD}$ and $\sigma_{\rm NSD} = \sigma_{\rm ND} + \sigma_{\rm DD}$
where ND, SD and DD stand for "non-diffractive", "single-diffractive" and "double-diffractive" events.

According to the data by UA5 Collaboration, the density of the former is smaller by 10 % than that of the latter in the pseudo-rapidity range $0 \leq \eta^* \leq 3.5$ and is almost equal to that of the latter in $\eta^* \geq 3.5$ at $\sqrt{s} = 546$ GeV [9].

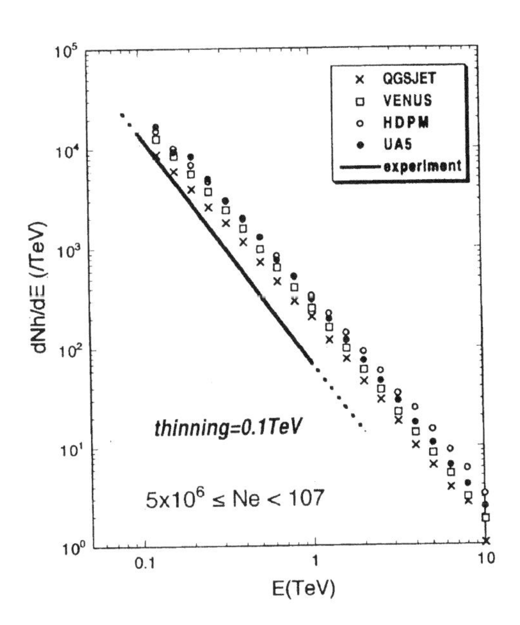

Figure 7. Differential number of hadrons in the air shower, which is obtained by the hadron calorimeter. Full circles are for the data by the simulation where UA5 code and proton-dominant composition are assumed.

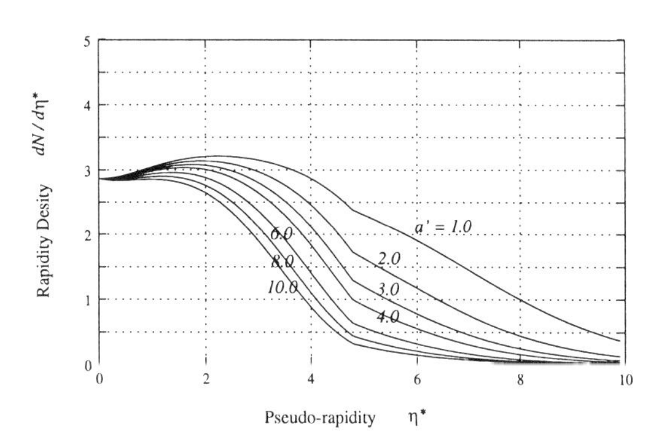

Figure 8. Pseudo-rapidity density distributions at $s^{1/2} = 546$ GeV for various values of the scaling violation parameter a'. The parameter a is assumed as in eq.(6) in the text. The kink of the curve at $\eta^* = 4.8$ is due to the rapidity dependence of the average p_T value.

case of Feynman scaling law, is for the reference (Table 4.)

Figure 10 shows the pseudo-rapidity density distributions (of all inelastic events but not of non-single-diffractive events) by UA5 Collaboration together with those of the parameter values of the best fitting. One can see in the figure that the reproduction is satisfactory by Model-2 and that Model-0 (the Feynman scaling law) cannot reproduce the data both in the central and forward regions.

3.3. x-distribution, multiplicity and inelasticity, predicted by the models

Using an approximate relation $x^* \simeq E/E_0 \equiv x$, valid at high energies, we obtain dN/dx from dN/dx^* of eq.(2):

$$\frac{dN}{dx} = \int_0^\infty aD \frac{(1-a'x)^d}{\sqrt{x^2 + \left(\frac{2\mu}{\sqrt{s}}\right)^2}} g(p_T) dp_T \qquad (7)$$

Figure 11 shows the x-distributions, predicted by Model-2. As can be seen in the figures, the Feynman scaling law is violated strongly both in the central region and in the forward region in Model-2. There is almost no particle with $x \geq 0.01$ at $E_0 = 10^{20}$ eV.

Figure 12 shows the energy dependence of charged multiplicity, predicted by the models, after integration of eq.(7). That is,

$$m(E_0) = \int_0^{1/a'} dx \frac{dN}{dx}$$

One can see in the figure that there is no big difference in multiplicity between Model-1 and Model-2. And no wonder the energy dependence of Model-2 agrees better with the experimental data than that of Model-1.

Figure 12 also shows the energy dependence of the average total inelasticity in the laboratory system, defined by

$$< K >= \frac{3}{2} \int_0^{1/a'} x \frac{dN}{dx} dx$$

It shows that the inelasticity decreases considerably in Model-2 at high energies while it is con-

Table 4
Scaling violation parameters, assumed in the models

Model	Scaling-violation parameters* $a = \left(\frac{E_0}{A}\right)^{\alpha}$	$a' = \left(\frac{E_0}{A}\right)^{\alpha'}$	Average inelasticity	Remarks
Model-0	$\alpha = 0$	$\alpha' = 0$	$< K >= 0.5$	the Feynman scaling law
Model-1	$\alpha = 0.105$	$\alpha' = 0.105$	$< K >= 0.5$	
Model-2	$\alpha = 0.105$	$\alpha' = 0.210$	$< K >= \left(\frac{E_0}{A}\right)^{-0.105}$	consistent with experimental data

* $A = 2.0 \times 10^2$ GeV

stant (= 0.5) in Model-0 and in Model-1[6].

3.4. Comparison with models used in simulations

It is interesting to see how the models assumed here, and those which have been used recently in simulations of atmospheric cosmic-ray diffusion, relate to each other. We compared in Figure 13 the pseudo-rapidity density distributions [21], predicted by UA5 code [22][7], VENUS [23], QGSJET [24], SIBYLL [25], HDPM [26] and DPMJET [27], with those of the present models[8]. The models VENUS, QGSJET and DPMJET are QCD theory inspired ones, while UA5 code, SIBYLL and HDPM are based on the experimental data.

One can see the following points in the figure:
(1) The differences among model predictions are not small one with respect to another, although all models are either QCD-inspired or empirical.
(2) In the central region all model predictions of the pseudo-rapidity density are consistent, except that of HDPM.
(3) In the middle pseudo-rapidity region, *i.e.* $2.0 \le \eta^* \le 6.0$, which is the most important for the atmospheric diffusion of cosmic rays, QCD-inspired models predict higher density than phenomenological models do. The pseudo-rapidity densities by the QCD-inspired models, are between those of Model-1 and Model-2. Those by the phenomenological models are almost consistent with Model-2.
(4) In the most forward region, all model predictions are consistent with those of Model-2.
(5) The experimental data by Harr *et al.* are almost consistent with those by QGSJET.
(6) The distribution by UA5 code is almost consistent with that of Model-2.

[6]It may seem strange that the average inelasticity $< K >$ is 0.5 for Model-1, which has a higher rapidity-density than QGSJET, because it is said that $< K > \simeq 0.6$ by QGSJET. It is due to the difference of sampled events, *i.e.* all inelastic events in the former and NSD events in the latter. In other words the average inelasticity by QGSJET is ~ 0.5 for all inelastic events.

[7]UA5 code is a phenomenological simulation program, made by UA5 Collaboration, which describes the data observed by the collaboration. The code does not necessarily predict the pseudo-rapidity density correctly in the forward region, because the observed pseudo-rapidity region by UA5 Collaboration is limited to $\eta^* \le 4.5$.

[8]One should notice that the pseudo-rapidity density by simulations is for NSD (non-single-diffractive) events while that of the calculation is for all inelastic events. See the footnote[6].

4. Application of the formulated energy distribution to highest energy air showers

We consider highest energy air showers ($> 10^{18}$ eV) as the first step to examine the validity of the formulated energy distribution of produced particles, because the energy dependence of the formulated distribution has larger effect at higher energies.

Our interest is how the nuclear interactions af-

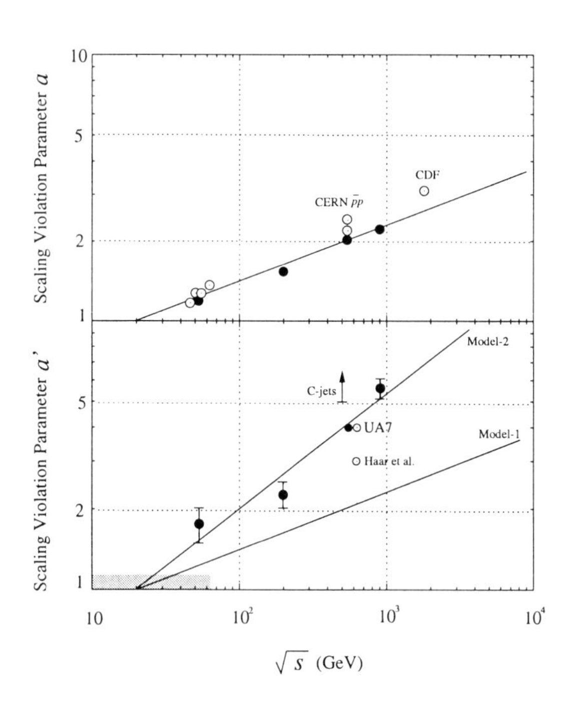

Figure 9. Energy dependence of the scaling violation parameters, a and a', which are obtained by fitting the calculated curves of (pseudo-)rapidity density distribution to those of the experimental data in Table 3. The full circles are the experimental data of UA5 Collaboration and the open circles are the data of other experimental groups. The hatched area indicates the energy region where the Feynman scaling law ($a \simeq 1.0$ and $a' \simeq 1.0$) is verified by the experiments. The lines are the assumed energy dependences of Model-1 and Model-2.

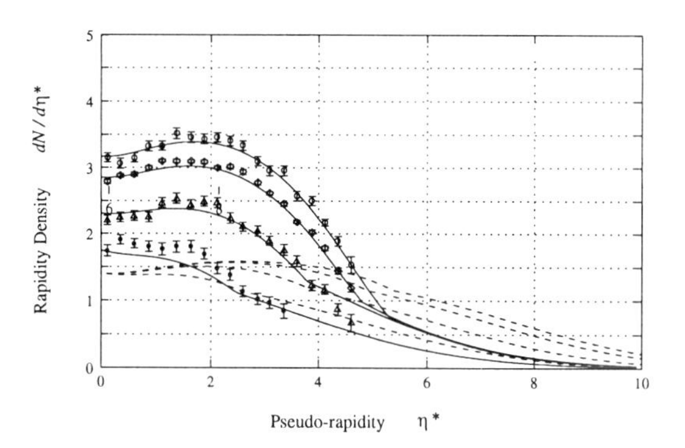

Figure 10. Pseudo-rapidity density distributions. Experimental data are by UA5 Collaboration for all inelastic events at the energies of $s^{1/2} = 53$ GeV (•), 200 GeV (△), 546 GeV (◇), and 900 GeV (○). Solid lines are the eq.(5) in the text, with the parameter values of the best fitting, shown in Figure 2. The chain lines are those of Model-0.

fect the air shower development and how the formulated energy distribution, based on data of direct observation, describes the highest energy air showers.

4.1. Diffusion of Cosmic Rays in the Atmosphere.

A high energy primary cosmic-ray proton[9], incident upon the top of the atmosphere, makes a nuclear collision with an atmospheric nucleus, and many particles one surviving particle and a number of produced particles — are yielded through the collision. The inelastic cross section of the proton-air collision increases with the proton energy. The surviving particle, either a proton or a neutron, repeats inelastic collisions in the atmosphere.

The produced particles are assumed to be pions. The energy distribution of produced pions is substituted by that of $N - N$ collision, for-

[9] We assume that the primary cosmic ray is a proton. And we employ the superposition model for a target nucleus, which assumes that the nucleus of mass number A and energy E_0 is a bundle of A nucleons and energy E_0/A, each. The model is valid when the observation level is sufficiently deep in the atmosphere [28].

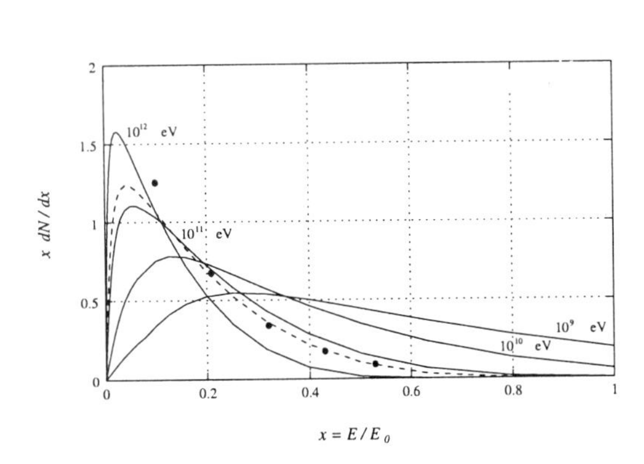

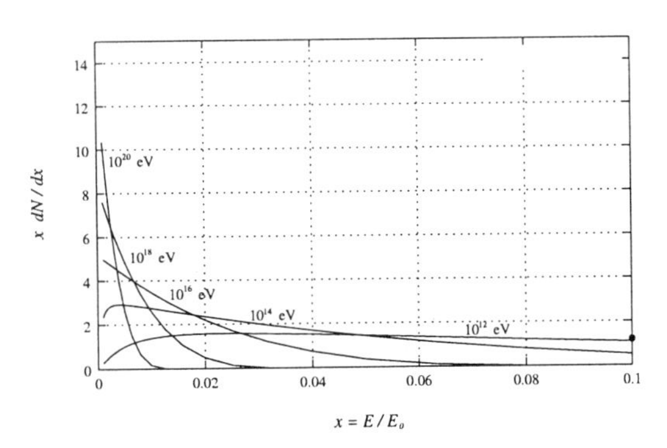

Figure 11. x-distribution of Model-2 for various primary energies E_0. The model describes the strong violation of Feynman scaling law both in the forward region and in the central region. The chain line in the upper figure is the scaling function of eq.(1) in the text. Full circles are the experimental data of $p + p \to \pi^{ch} + X$ at $s^{1/2} = 20$ GeV, where the scaling law is valid [20]. The lower figure shows that the distribution in the forward region shrinks distinctly, in particular, at highest energies.

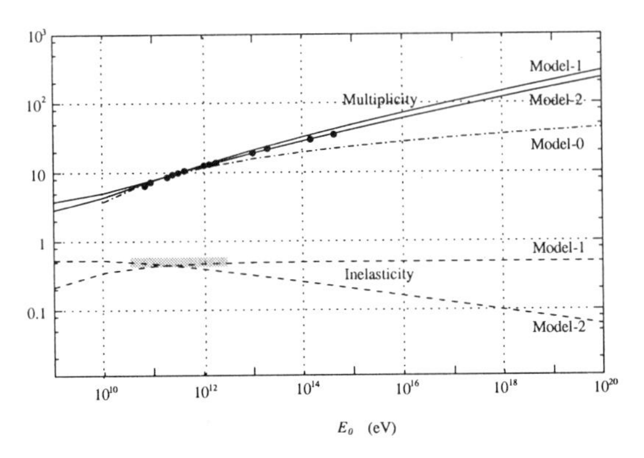

Figure 12. Energy dependence of charged multiplicity and that of total inelasticity, predicted by the models. There is no big difference of the multiplicity between the models. One can see that Model-2 describes the experimental data (full circles) better than Model-1 does, because Model-2 has the best fitted parameters to describe the rapidity density distribution. Inelasticity is decreasing in Model-2, while it is constant (=0.5) for Model-0 and Model-1. The shadowed area indicates the region where the Feynman scaling law is verified by the experiments within the experimental errors.

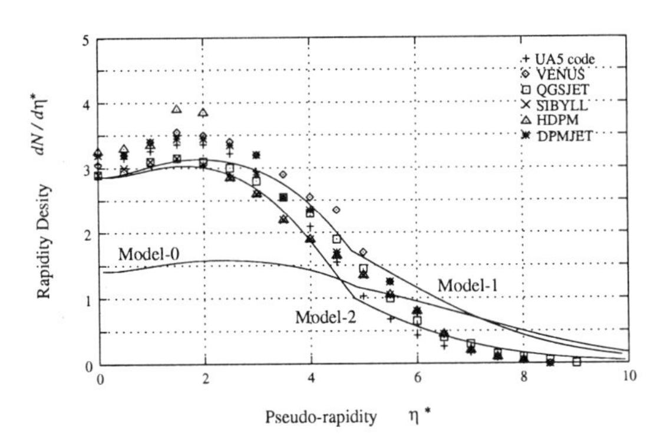

Figure 13. Pseudo-rapidity density distributions at $s^{1/2} = 546$ GeV. Plots are by the models (UA5 code, VENUS, QGSJET, SIBYLL, HDPM and DPMJET) which are employed in the simulations of cosmic ray phenomena. The data by the simulations are based on the NSD (non-single-diffractive) events, while those by Model-0, Model-1 and Model-2 (solid lines) are for all inelastic events. See the footnote[6] for the difference between NSD and all inelastic events.

mulated in Equation 7, Section 3.3, because the effect of the air nucleus target appears only in the backward region[10]. The charged pions among the produced pions make nuclear collisions in the atmosphere again. The collision mean free path of charged pions has the same energy dependence as that of a nucleon. We neglect decays of charged pions into muons.

The multiple particle production, induced by a charged pion, has essentially the same characteristics as the one by a nucleon, which is confirmed in low energy region within the errors of the experimental data. That is, the final state of the collision consists of one surviving pion and produced pions whose energy distribution is the same as the one of proton collision. The differences are that the inelastic cross section of a proton is larger than that of a pion, and that the surviving pion has a probability to be a neutral pion (called the "charge exchange" process) (see Table 5).

The neutral pions, among the produced pions and through the charge exchange process, decay into γ-rays which produce a number of electrons and photons of electromagnetic component, via the electromagnetic cascade process.

Following the above description of the cosmic-ray diffusion in the atmosphere, we formulate the elementary processes in the following way:

(1) Inelastic collision mean free path

The mean free path of the inelastic $N - air$ (N : nucleon) collisions,

$$\lambda_N(E_0) = \lambda_N\left(\frac{E_0}{B}\right)^{-\beta} \qquad (\lambda_N = 80.0 \text{ g/cm}^2)$$

and for $\pi^{\pm} - air$ collisions

$$\lambda_\pi(E_0) = \frac{1}{\xi}\lambda_N(E) \qquad \left(\xi \equiv \frac{\sigma_{\pi N}}{\sigma_{NN}} = 0.71\right)$$

(2) Energy distribution of produced charged pions

The energy distribution of produced *charged* particles through multiple particle production is formulated in Section 3. In the laboratory system it is given by eq.(7). And the scaling violation parameters a and a' are dependent on the incident energy. That is,

$$a = \left(\frac{E_0}{A}\right)^{\alpha} \qquad a' = \left(\frac{E_0}{A}\right)^{\alpha'}$$

where the exponents of the parameters α and α' are tabulated in Table 4.

In other words, we assume that the number of charged pions with the energy between E and $E + dE$ is given by

$$\varphi(E_0, E)dE = D\left(\frac{E_0}{A}\right)^{\alpha}\left[1 - \left(\frac{E_0}{A}\right)^{\alpha'}\frac{E}{E_0}\right]^{d}\frac{dE}{E}$$

where E_0 is the energy of the incident particle. According to the above, the energy of the produced pions is between $E = 0$ and $E = (E_0/A)^{-\alpha'}E_0$.

The average value of the total inelasticity in the laboratory system is constant (= 0.5) in Model-0 and Model-1, while it decreases considerably in Model-2 at high energies (see Figure 12 and Table 4).

(3) Inelasticity distribution

We assume that the inelasticity is distributed uniformly between 0 and $2 < K >$ (≤ 1.0). That is, the inelasticity distribution is

$$u(K)dK = \frac{dK}{2 < K >} \qquad (K = 0 \sim 2 < K >)$$

(4) Energy spectrum of the surviving particle

The energy distribution of the surviving particle is determined by the inelasticity distribution:

$$\int u(K)dK\ \delta(E - (1 - K)E_0)dE$$

which leads to x-distribution of the surviving particle as

$$\frac{dx}{2 < K >} \qquad (1 - 2 < K > \leq x \leq 1)$$

That is, the variable x of the surviving particle is distributed uniformly between $1 - 2 < K >$ and 1, which is consistent with experimental data in low energy region.

There is a probability in pion collisions that the surviving pion is a neutral pion π^0, which decays immediately into γ-rays. This charge exchange process is important from the energy flow point of

[10] The diffusion of cosmic rays in the atmosphere is governed by the high energy particles, produced in the forward region.

view, because the surviving particle has high energy in average, compared with the produced pions. We assume that the charge exchange probability is $b = 0.3$ tentatively, because there is no experimental data of it in the high energy region[11].
(5) Energy distribution of π^0's

We assume that the energy distribution of π^0's is a half of that of charged pions, which is confirmed within the experimental errors in low energy region.

The assumptions, mentioned here, are tabulated in Table 5.

4.2. Size of an air shower

The size of an air shower is defined as the total number of charged particles in the air shower which pass through the plane at the observation level. And the charged particles in an air shower consist of the electron component (electrons and positrons), charged hadrons (mainly charged pions) and muons, among which the electron component is dominant. Hence we calculate only the number of the electron component N_e to express the air shower size.

The calculation is made analytically, because our main interest is to see how the characteristics of nuclear interactions affect the air shower development. That is, the diffusion equations of nucleon, charged pion and electron components are solved analytically for four cases in Table 6. It corresponds to take into account the respective factors of the nuclear interaction characteristics one by one, taking Case A as the zero-th approximation, which is the one of the objectives of the present study.

Figure 14 shows the transition curve of the air shower size for the primary proton with the energies $E_0 = 10^{18}$, 10^{19}, 10^{20} eV for Case A. One can see in the figure that the air showers are at the maximum development at sea level and that the relation $E_0/N_e \simeq 2.0$ (GeV) holds approximately.

Figure 15 shows the ratio of the air shower size between the cases of B, C and D and the case A for the primary energy $E_0 = 10^{20}$ eV. One can see the following in the figure:

[11]The probability may be higher than 0.3, and probably ~ 0.4.

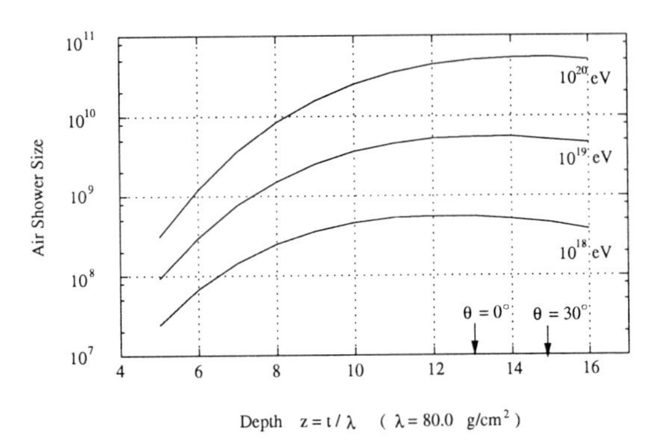

Figure 14. Transition curve of the air shower size for the primary proton with the energy $E_0 = 10^{18}$, 10^{19}, and 10^{20} eV, for Case A (Model-0 and constant cross section). The arrows indicate the depth of the sea level (1,030 g/cm^2) for the air showers with the inclination $\theta = 0°$ and $30°$.

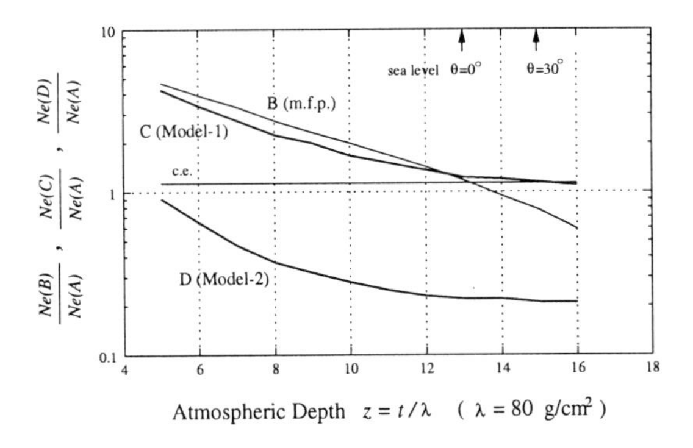

Figure 15. Ratio of air shower size, $N_e(\mathrm{B})/N_e(\mathrm{A})$, $N_e(\mathrm{C})/N_e(\mathrm{A})$ and $N_e(\mathrm{D})/N_e(\mathrm{A})$, along the depth. The cases of A, B, C and D are tabulated in Table 6. The primary energy of a proton is 10^{20} eV.

Table 5
Energy distribution of the produced particles in multiple particle production

Collision	Surviving particle	Produced particles (pions)
$N-N$	nucleon $\delta(E-(1-K)E_0)dE$	charged pions $\varphi(E_0,E)dE$ neutral pions $\frac{1}{2}\varphi(E_0,E)dE$
$\pi^{\pm}-N$	charged pions $(1-b)\,\delta(E-(1-K)E_0)dE^{\dagger}$ neutral pions $b\,\delta(E-(1-K)E_0)dE^{\dagger}$	charged pions $\varphi(E_0,E)dE$ neutral pions $\frac{1}{2}\varphi(E_0,E)dE$

† The charge exchange probability $b=0.3$ is assumed.

Table 6
Cases to solve the diffusion equations

Case	Parameters				Remarks		
	α	α'	β	b	Model	$<K>$	$\sigma(E)$
A	0	0	0	0 and 0.3	Model-0	→	→
B	0	0	0.056	0	Model-0	→	↗
C	0.105	0.105	0	0	Model-1	→	→
D	0.105	0.210	0	0	Model-2	↘	→

Table 7
Constants in the numerical calculation

charge exchange probability (of the surviving pion)	$b=0.3$
collision mean free path	
$N-air$	$\lambda_N=80.0$ (g/cm^2)
$\pi-air$	$\lambda_\pi=113.0$ (g/cm^2)
ratio	$\xi\equiv\lambda_N/\lambda_\pi=0.71$
radiation length in the air	$X_0=37.1$ (g/cm^2)
ratio	$\xi_0=\lambda_N/X_0=2.16$
critical energy in the air	$\epsilon=80.0$ (MeV)

(1) The effect of the charge exchange process of the surviving pion is almost constant over the atmospheric depth, and amounts 13 % (this tendency can be explained by the obtained analytic solutions).
(2) The effect of increasing cross section is large at high altitude, but is small at sea level (this tendency can be explained by the obtained analytic solutions).
(3) The effects of scaling violation, in Model-1 and in Model-2, have similar depth dependence, but the absolute values of them differ by five times.
(4) Model-2 gives smaller air shower size, and the attenuation of the air shower size after the shower maximum is very slow due to the small value of inelasticity.
(5) At sea level the air shower size is affected most by the energy distribution of produced particles, and the effects of the increasing cross section and of the charge exchange are relatively small.

5. Summary and discussion

(i) We formulated three models, Model-0, Model-1 and Model-2, for the energy spectrum of produced particles in multiple particle production (Table 4). Model-2 is based on data of direct observation by cosmic-ray and accelerator experiments in the energy region of $10^{12} \sim 10^{14}$ eV. It indicates the violation of the Feynman scaling law both in the central region and in the forward region. Average inelasticity is $< K >= 0.5$ in Model-0 and in Model-1, but decreases with the incident energy in Model-2. Most of the nuclear interaction models, which are employed in recent simulations to follow the diffusion of cosmic rays in the atmosphere, lie between Model-1 and Model-2, from the view point of the energy distribution of produced particles.

(ii) An analytical expression is given for the size of the air shower, based on the formulated models of nuclear interactions. It enables us to discuss effects of physical processes — the charge exchange of the surviving pion, increasing cross section of *hadron – air* collisions, and the energy distribution of produced particles — to the air shower size. These processes are the major factors to govern the diffusion of high energy cosmic rays in the atmosphere. We obtained the following observations about the size of the extremely high energy air showers of $E_0 = 10^{20}$ eV:
(1) The effect of charge exchange process is almost constant ($\times 1.13$) over the whole depth in the atmosphere.
(2) The effect of increasing cross section is large ($\times\ 2 \sim 3$) at mountain altitudes, but small ($\times 1.18$) at the sea level.
(3) The effect of scaling violation of Model-1 is large ($\times\ 2 \sim 3$) at mountain altitudes, but small ($\times 1.23$) at sea level.
(4) Effect of scaling violation of Model-2 is not negligible at any depth, *i.e.* $\times\ 0.6 \sim 0.4$ at mountain altitudes and $\times 0.22$ at the sea level.

(iii) The air shower size at sea level, expected by the present calculation, is tabulated in Table 8 for the incident proton of $E_0 = 10^{20}$ eV. In the table the effects of the charge exchange process and the increasing cross section are obtained from Figure 15. To calculate the expected air shower size, in which the effects of charge exchange probability and increasing cross section are included, we multiplied all the factors because the factors are near 1.0.

(iv) M. Nagano *et al.* examined the method of energy determination of extremely high energy air showers, employed by AGASA experiment, by the simulation code of CORSIKA [26](with QGSJET model). And they reached the conclusion that the method works well for the highest energy air showers [29]. The simulation gives $N_e = 5.5 \times 10^{10}$ for the proton-induced air showers of $E_0 = 10^{20}$ eV[12]. We can see the following points by comparing the value with those in Table 7:
(1) The value obtained by the simulation is between those of Model-1 and Model-2. In this sense our calculation and the simulation are consistent each other, because we saw in Section 3 that the pseudo-rapidity density distribution by QGSJET model is between those obtained from Model-1

[12] We obtained $N_e = 5.5 \times 10^9$ for the proton-induced air showers of $E_0 = 10^{19}$ eV, from the figure in Reference [29], and multiplied it by 10.

Table 8
Air shower size at sea level, expected from the models, for the incident proton of $E_0 = 10^{20}$ eV

	Model-0	Model-1	Model-2
size*	5.0×10^{10}	6.2×10^{10}	1.1×10^{10}
ratio to Model-0	(×1.0)	(×1.23)	(×0.22)
charge exchange	×1.13	×1.13	×1.13
increasing cross section	×1.18	×1.18	×1.18
size (expected)**	6.7×10^{10}	8.3×10^{10}	1.5×10^{10}

* without the processes of increasing cross section and the charge exchange.
** with the processes of increasing cross section and the charge exchange.

and Model-2.
(2) If we take Model-1, the energy spectrum of the highest energy air showers shifts to the left (toward lower energy) by a factor of 1.5.
(3) If we take Model-2, which is the best-fitted to the experimental data, the energy spectrum shifts to the right (toward higher energy) by a factor of 3.7.

Among the major factors that govern the cosmic-ray diffusion in the atmosphere — the energy distribution of produced particles, the charge exchange probability of the surviving pion and the increasing cross section of hadron-air collisions —, one can see that the first one has the largest effect on the size of extremely high energy air showers. Hence *we have to specify the energy distribution of produced particles in multiple particle production in more detail, in order to confirm the extremely high energy cosmic rays exceeding* 10^{20} *eV.*

(v) The item (3) in the above paragraph (iv) makes the puzzle of extremely high energy cosmic rays more serious. That is, the energy spectrum of the primary cosmic rays, which is estimated from the highest energy air showers, extends beyond the Greisen-Kuzmin-Zatsepin cut-off energy ($\sim 10^{19}$ eV) [30], which is one of the most important and interesting puzzles to be solved at present [31].

Probably it is not irrelevant to conclude that the energy dependences of the scaling violation parameters in Model-2 are not valid at extremely high energy region. In other words, Model-2 does not describe the extremely high energy air showers. This is due to the fact that Model-2 predicts small inelasticity at high energies. For example, the value is as small as 0.2 even at $E_0 = 10^{16}$ eV (see Figure 12).

According to our previous analysis of attenuation mean free paths of hadron and (e, γ) components [32], the inelasticity of $< K >= 0.5$ is compatible but smaller inelasticity is not compatible with the experimental data in the energy region of $10^{14} \sim 10^{16}$ eV. We will discuss in succeeding papers the cosmic-ray data in the energy region of $10^{14} \sim 10^{16}$ eV, presented in section 2, and a model to describe all the available data consistently.

Acknowledgment

The author thanks the local organizing committee of the symposium for offering him an opportunity to give this talk and for holding a very fruitful and enjoyable symposium.

The data, presented here, are obtained by the longstanding (more than 30 years!) Brazil-Japan collaboration and by the Bolivia-Brazil-Japan collaboration which is underway. Universidade Estadual de Campinas, the host institute of the symposium, has been and is a member of the collaboration.

The author is indebted to Prof. Edison H. Shibuya and Prof. Masanobu Tamada for valuable discussions of this work.

REFERENCES

1. C.M.G. Lattes, Y. Fujimoto, S. Hasegawa, Phys. Rep. 65 (1980) 65.
2. R. Feynman, Phys. Rev. Lett. **23** (1969) 1415.

3. T.K. Gaisser, R.J. Protheroe, K.E. Turver, T.J.L. McComb, Rev. Mod. Phys. 50 (1978) 859.
4. C.M.G. Lattes et al. Prog. Theor. Phys. Suppl. No.47 (1971) 1.
5. J. Nishimura, *Handbuch der Physik* Vol. XLVI/2 (Springer-Verlag) (1967) 1.
6. J.A. Chinellato et al., Prog. Theor. Phys. Suppl. No.76 (1983) 1.
7. F. Halzen and N. Yamdagni, Nucl. Phys. **B190** (1987)213.
8. N. Arata, Nucl. Phys. **B211** (1983) 189.
9. G.L. Alner et al. (UA5 Collaboration), Z. Phys. **C33** (1986) 1; Nucl. Phys. **B291** (1987) 445; Phys. Rep. Nos.5 and 6 (1987) 247.
10. S.I. Nikolsky, Proc. 3rd Symposium on Cosmic Rays and Particle Physics, Tokyo (1984) (Inst. for Cosmic Ray Research, Univ. of Tokyo), 507.
11. L.T. Baradzei *et. al.*, Nucl. Phys. **B370** (1992) 365.
12. J.R. Ren *et al.*, Phys. Rev. **D38** (1988)1404.
13. N. Kawasumi et al., Phys. Rev. **D53** (1996) 3534.
14. J. Phys. G **20** (1994) 487.
15. V.V. Arabkin, S.I. Nikolsky, K.V. Cherdyntseva and S.B. Schaulov, FIAN Preprint No.140 (1990) (Lebedev Physical Institute) (in Russian); Proc. 22nd Intern. Cosmic Ray Conf. (Dublin) (1991) Vol.4 pp.141, 269, 273.
16. C. Aguirre *et al.*, Phys. Rev. **D62** (2000) 032003; Talk by N. Kawasumi in this symposium.
17. T. Antoni *et al.*, J. Phys. G **25** (1999) 2161.
18. E. Pare, T. Doke, M. Haugenauer, V. Innocente, K. Kasahara, T. Kashiwagi, J. Kikuchi, S. Lazano, K. Masuda, H. Murakami, Y. Muraki, T. Nakada, A. Nakamoto, T. Yuda, Phys Lett. **B242** (1990) 531.
19. R. Haar, C. Liapis, P. Karchin, C. Biino, S. Erhan, W. Hofmann, P. Kreuzer, D. Lynn, M. Medinnis, S. Palestini, L. Pesando, M. Punturo, P. Schlein, B. Wilkens, J. Zweizig, Phys. Lett. **B401** (1997) 176.
20. M. Adams et al., Z. Phys. **C39** (1988) 257.
21. J. Knapp, D. Heck, G. Schatz, Preprint of Forschungszentrum Karlsruhe, FZKA 5828 (1996).
22. G.J. Alner et al. (UA5 Collaboration), CERN-EP/86-213 (1986).
23. K. Werner, Phys. rep. **232** (1993) 87.
24. N.N. Kalmykov, S.S. Ostapchenko, Yad. Fiz. **56** (1993) 105; N.N. Kalmykov, S.S. Ostapchenko, Phys. At. Nucl. **56** (3) (1993) 346; N.N. Kalmykov, S.S. Ostapchenko, A.I. Pavlov, Bull. Russ. Acad. Sci. (Physics) **58** (1994) 1966.
25. R.S. Fletcher, T.K. Gaisser, P. Lipari, T. Stanev, Phys. Rev. **D50** (1994) 5710; J. Engler, T.K. Gaisser, P. Lipari, T. Stanev, Phys. Rev. **D46** (1992) 5013.
26. J.N. Capdevielle et al., Preprint of Kernforschungszentrum Karlsruhe KfK 4998 (1992).
27. J. Ranft, Phys. Rev. **D51** (1995) 64.
28. G. Schatz, T. Thouw, K. Werner, J. Oehlschläger, K. Bekk, J. Phys. G20 (1994) 1267.
29. M. Nagano, D. Heck, K. Shinozaki, N. Inoue, J. Knapp, Preprint of Forschungszentrum Karlsruhe, FZKA 6191 (1998).
30. S. Yoshida, Rapporteur Talk in 26th Int. Cosmic Ray Conf. (Salt Lake) (1999) AIP Conf. Proc. 516 p.180.
31. J. Cronin, the talk in this symposium.
32. A. Ohsawa, K. Sawayanagi, Phys. Rev. **D49** (1992) 3128-3133.

ELSEVIER

Nuclear Physics B (Proc. Suppl.) 97 (2001) 35–47

www.elsevier.nl/locate/npe

Cosmic ray composition and energy spectrum above 1 TeV: direct and EAS measurements

A.Castellina[a]*

[a]Consiglio Nazionale delle Ricerche, Istituto di Cosmogeofisica, cs.Fiume 4, 10133 Torino, Italy

The most recent experimental results on the cosmic ray composition and energy spectrum above 1 TeV are reviewed and discussed. All data agree on the presence of the so-called "knee" at an energy $E_k \simeq 3\ 10^{15} eV$; the knee is seen in all the components of the Extensive Air Showers. These results support the hypothesis of an astrophysical origin of the knee, while no new features in the hadronic interactions at high energies are envisaged. The cosmic ray composition below and above the knee region is still an open question. According to most experiments, the knee seems to be due to the light component of the primary beam, with a composition getting heavier above the knee. However, results contradicting this conclusion have to be considered and understood.

1. Introduction

The energy spectrum of cosmic rays (CR) spans a very wide energy range, with particle fluxes steeply falling more than 30 orders of magnitude. Above the solar modulation region, the spectrum can be well described by a power law, which steepens around $3\ 10^{15} eV$, a feature called the "knee", discovered in 1958 [1]; it softens again at $\simeq 10^{19} eV$, the "ankle".
Explaining the knee feature would shed light on the CR origin and acceleration mechanism, depending on whether it is a signature of a change in the hadronic interactions at such energies or it reflects a feature of the cosmic ray spectrum, thus concerning mainly astrophysics.
Several arguments involving energetics, composition and secondary γ ray production suggest that cosmic rays at least up to the knee region are confined in the Galaxy.
The most popular theory is that of diffusive shock acceleration in Supernova remnants (SNR), that is particle acceleration by SNRs expanding supersonically in the surrounding medium.
Supernova explosions can easily account for the energy stored in galactic cosmic rays; the spectrum emerging from the SNR is of the type $E^{-2.1}$ up to a maximum energy near 10^{14} eV times the nuclear charge, after which it drops very rapidly. Folding the production spectrum with the effect of diffusion through the Galaxy, and taking a trapping time varying as $E^{-0.6}$ (as found from the proportion of secondary to primary nuclei arriving to Earth), the resulting flux of CR in the Galaxy would be $\propto E^{-2.7}$, in close agreement with expectations. The maximum achievable energy is close to the knee one.
A direct evidence that the nucleonic component of CR is indeed produced in SNRs could be obtained by the observation of γ rays: the accelerated cosmic rays can in fact interact with the local interstellar matter, in this way producing γ rays by either hadronic or leptonic production.
Various experimental groups reported on TeV γ emission from supernova remnants, like SN1006, RJX1713.7-3946, Cassiopea-A (see e.g. [2] and references therein). Unfortunately in all cases the emission can be attributed to electron progenitors and no positive evidence for hadroproduction of TeV γ's has been found yet.
Further information can be obtained from the study of the distribution of CR arrival directions. A recent compilation of the anisotropy measurements can be found in [3]; while the amplitude and phase of anisotropy data below $\simeq 2\ 10^{14} eV$ are consistent and statistically accurate [4], the experimental results at higher energies are not compatible with expectations. This could mean that the diffusion model cannot be simply extrap-

*E-mail: castellina@to.infn.it.

PII S0920-5632(01)01182-3

olated to these energies [5].
Various models have been put forward trying to identify the sites and mechanisms of injection of cosmic rays at higher energies, at and above the knee. If the bend in energy spectrum is related to the maximum achievable energy in the accelerator, then CR at higher energies could be powered by a reacceleration by interstellar turbulence [6], or they could be produced by Supernovae exploding in denser media (their stellar wind cavity) ([7] and references therein). On the other hand, the knee could be attributed to propagation effects. In both cases, one would expect multiple bends due to the different elements bending at fixed rigidity; the composition would become heavier above the knee. An extra-galactic origin for CR above the knee has also been proposed [8], where the accelerator sites are found in Active Galactic Nuclei and the resulting composition is getting lighter above the knee.
A completely different point of view assigns the knee to a new dramatic process of hadronic interaction which takes over around the knee energy. However, even if it is true that in the ΔE of interest we have no direct information about the hadronic interaction cross section for the secondary production relevant to the interpretation of measurements, no experimental data as far show a need for a different interaction mechanism. From the experimental point of view, what is most important in order to test the models is to measure the cosmic ray composition and energy spectrum near the energy limit of the shock models; moreover, measurements of anisotropy and secondary to primary ratio at higher energy are of utmost importance.

2. Direct measurements

Direct measurements of the relative abundances of the cosmic ray nuclei and their distribution in energy are possible only at relatively low energy: they in fact require installation of instrumentation on balloons or space shuttles flying outside the atmosphere at very high altitude. The most recent results still come from two balloon experiments, JACEE and RUNJOB, as summarised in [9,10].
The proton and Helium spectra have been measured by JACEE up to about 800 TeV [11]; no break was found in the proton one, but above 80-90 TeV the experiment does not have enough statistics to either assess or reject its presence [12].
JACEE claim for a flatter He spectrum as compared to the proton one is in agreement with SOKOL result [13], but this is not confirmed by the results by RUNJOB [14]. The JACEE group reported values are $\gamma_p = (-2.80 \pm 0.04)$ and $\gamma_{He} = (-2.68 \pm 0.06)$, while RUNJOB slopes are both $\simeq -2.80$ with an uncertainty between 10 and 20% for both protons and Helium nuclei. It should however be mentioned that the significance of the difference between the slopes for p and He is at the level of only 2σ; on the other hand, the results from RUNJOB are based only on the 1995 and 1996 data ($\simeq$ 30% protons and $\simeq$ 13% He respect to JACEE data). The experimental results on the p and He slopes are of particular importance as regards the models of non linear acceleration of cosmic rays, where the injection rate is an increasing function of the primary particle rigidity [15].
The single component spectra are shown in Fig.1 and in Fig.2. The experimental results agree for what regards the iron spectrum, while RUNJOB gives a factor of 2 lower spectra for the C-N-O and Ne-Si groups. Data are all consistent with an increase of the mean logarithm of the average primary mass $< ln\ A >$ with energy, as shown in Fig.3.

It is clear that more statistics is needed above 100 TeV, and various new projects are under development. ACCESS [16] is estimated to be launched on the International Space Station in 2006. Its primary goal will be the measure of energy spectrum and composition up to $10^{15} eV$, thus testing Supernova shock acceleration models; the charge range at high energy will be $1 \leq Z \leq 28$. Three different detectors are being built for this purpose: a charge identification module, to measures the abundances of all individual elements, a transition radiation detector, to identify and measure the energy of particles with $Z \geq 2$ up to $\simeq 100\ TeV/nucleon$, a calorimeter to measure the particle energies and to identify electrons.

The ATIC [17] project, in its initial design for long duration balloon flights, is devoted to study the energy spectrum of Galactic proton and helium up to $10^{14}eV$, in order give information about the proton/helium ratio, the possible dif-

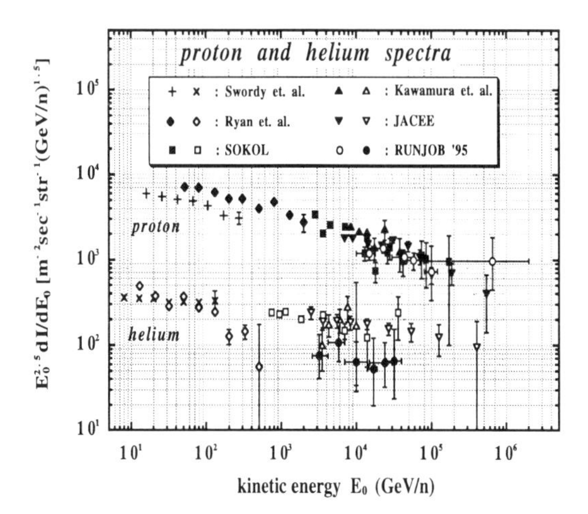

Figure 1. *Differential energy spectra for proton and helium [10].*

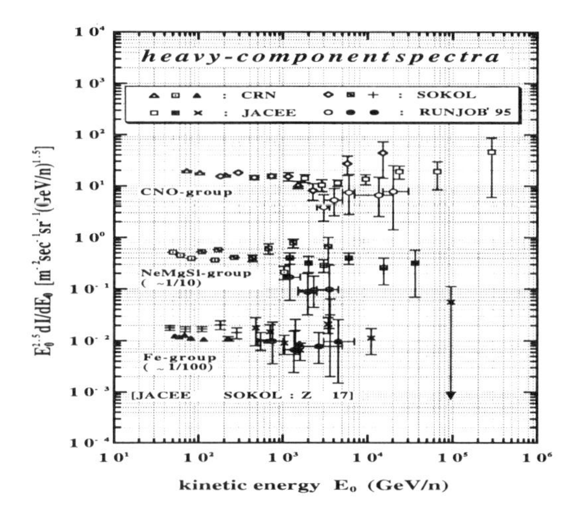

Figure 2. *Differential energy spectra for the CNO, NeMgSi and Iron groups [10].*

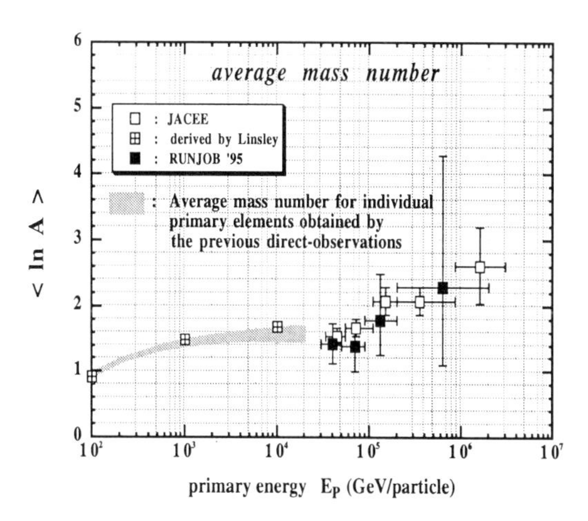

Figure 3. *Average primary mass vs primary energy from direct measurements [10].*

ference in their spectral slopes, the existence of a bend in the proton spectrum.

CREAM [18] plans to explore spectrum and composition up to $\simeq 10^{15}eV$, exploiting ultra long duration balloon flights ($\simeq$ 100 days). With an exposure of $\simeq$ 300 m^2sr *days*, this instrument will collect $\simeq$ 500 proton and helium nuclei above $10^{14}eV$, reaching $\simeq$ 30% statistical accuracy above $10^{15}eV$.

The new Ionization Neutron Calorimeter INCA [19] proposes to study the range 0.1-10 PeV using the well known techniques of ionization and neutron monitor to measure energy and a silicon particle charge detector to determine the charge and coordinates of the primaries.

Combining different detectors, the new projects will have a very powerful tool to overcome the individual technical limitations.

3. The energy spectrum

The experimental observables which are measured in order to extract information about the energy spectrum are the charged components of showers as measured by ground based detectors with scintillator counters, muon and hadron detectors, or the Čerenkov light produced by shower

particles as they propagate through the atmosphere.
The interpretation of these ground level observations in terms of primary particle characteristics is far from straightforward, being strongly dependent on models simulating the production and propagation of particles through the atmosphere. Models in turn depend on extrapolations applied to the data on high energy particle interactions studied at accelerators; the energy region of interest is in fact much higher than that studied at accelerators, the explored kinematic region is the forward one, the collisions among nuclei make the influence of nuclear effects not negligible.
The most recent results of EAS experiments concerning the primary energy spectrum of cosmic rays are described in the following.
The electron size spectra as measured by the EAS-TOP experiment [20] are shown in Fig.4. The knee is clearly visible, and the size corresponding to the knee shifts towards lower values at increasing atmospheric depth, as expected for a knee at given primary energy [21].

The shower size at the knee decreases with increasing atmospheric depth, with an attenuation length $\Lambda_k = (222\pm3)\ g\ cm^{-2}$, in very good agreement with that found for the shower absorption in atmosphere; the integral intensities $I_k(\geq E_k)$ are constant within 20%. The knee in electron size is quite sharp, showing that the change in slope occurs in a limited range of N_e ($\Delta N_e/N_e \leq 25\%$).

The muon size spectra have been measured in 4 different zenith angle intervals, as one can see in Fig.5, where the change of slope is visible at all atmospheric depths despite the large statistical fluctuations. The knee, around $N_\mu^k \simeq 10^{4.65}$, is independent on the number of detected muons, being in fact visible at any core distance [22]. The integral fluxes in electron and muon size are compatible at all atmospheric depths, as expected for a feature occurring at fixed primary energy, also confirming the consistency of the whole procedure.
A further interesting result comes from the relation between the electron and muon size slopes, which can be written as $N_\mu \propto N_e^\alpha$ with $\alpha \simeq 0.75$ in all angular bins: no sudden change in the secondary production when going through the knee

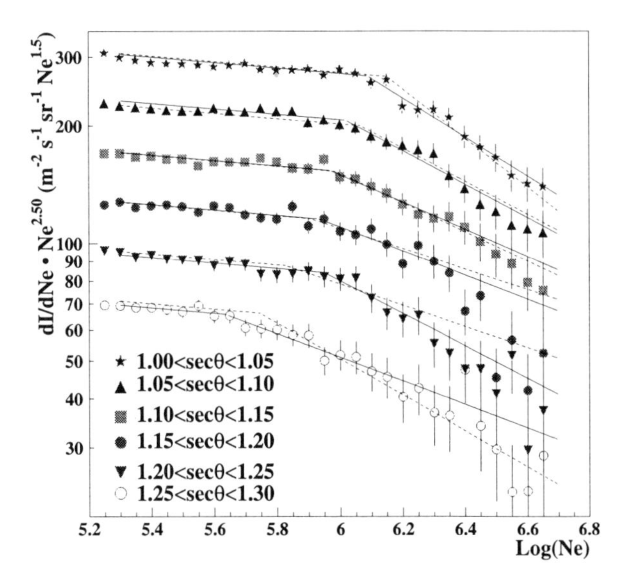

Figure 4. *Differential electron size spectrum at different zenith angles, that is different atmospheric depths, as measured by EAS-TOP. The solid lines show the results of the fitting procedure [21]; the fits represented by dashed lines were obtained requiring a constant integral flux above the knee.*

region is seen, thus showing that, at least from this point of view, no new hadronic effects is needed to explain the knee [23].
A simulation of the shower production and development in atmosphere using the CORSIKA code [24] with the HDPM interaction model allows to find the relation between shower size and primary spectrum $N_e(E_0, A) = \alpha(A_{eff})E_0^{\beta(A_{eff})}$ The effective mass A_{eff} is calculated from the extrapolation of the single nuclear spectra measured at low energies by direct measurements; above the knee, a rigidity dependent cutoff is used.
The final result is shown in Fig.6; the agreement with direct measurements at low energies and with other air shower experimental results at the highest ones is quite good. The systematic uncertainties in the energy spectrum are due to the primary composition and interaction model chosen

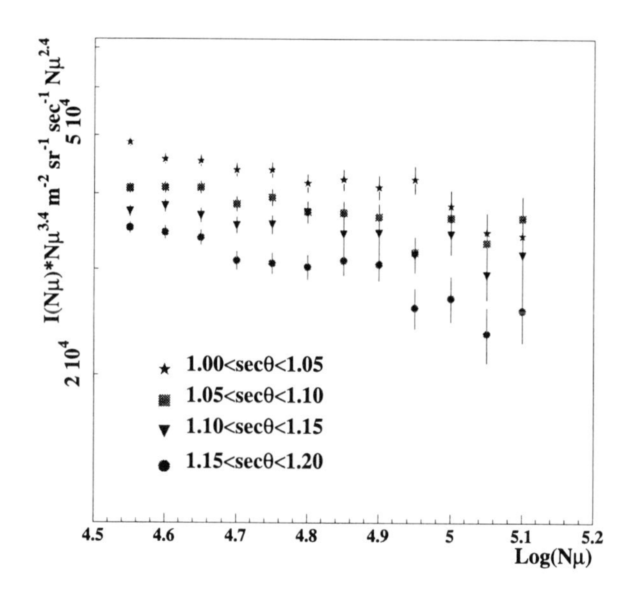

Figure 5. *Differential muon size spectrum at 4 different atmospheric depths as measured by EAS-TOP [22].*

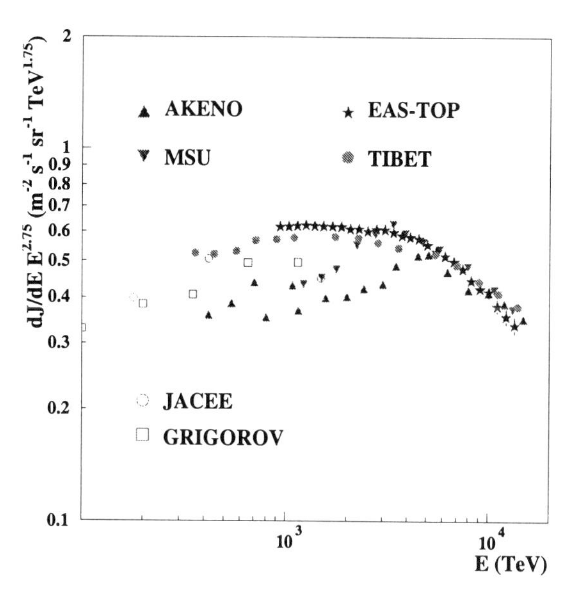

Figure 6. *Primary energy spectrum as obtained by EAS-TOP compared with other experimental results [21].*

in the calculation: the maximum difference in the determination of N_e between different interaction models and HDPM is $\simeq 10\%$; the all-particle flux obtained with "heavy" or "light" limit compositions differs by $\simeq 10\%$ from the above calculated one.

The energy spectrum is determined by CASA-MIA using the muon N_μ and electron N_e^* size measurements [25]. The N_e^* indicates in this case the sum of e^+, e^-, γ at the ground. The sizes combination $F = log_{10}(N_e^* + \psi N_\mu)$ was found to be log-linear in E_0 and, what is most important, independent on the primary mass.

In Fig.7 this relation is shown as found from a simulation of primary protons or iron; the model used was the QGSJET one. The systematic differences in energy assignment for different primary mass A are less than 5%. The average absolute energy reconstruction errors go from $\simeq 25\%$ at $10^{14} eV$ to $\simeq 16\%$ at $\geq 10^{15} eV$.
The parameter ψ, which defines the relative weight of muons and electrons in the showers, is strongly dependent on the model used for

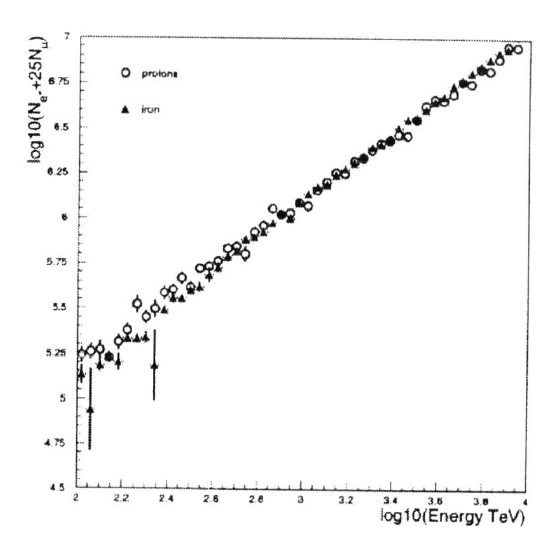

Figure 7. $F = log_{10}(N_e^* + \psi N_\mu)$ *as a function of energy for simulated proton (open circles) and iron (black triangles) primaries [25].*

hadronic interactions, but the change in energy assignment due to this effect is claimed to be $\leq 10\%$. The mass insensitivity allows to determine the energy free of systematic effects (on the contrary, for example, if in some region the energy spectrum changes, N_e vs E also changes). The energy spectrum thus derived is shown in Fig.8; the knee is located at the same primary energy for any atmospheric depths, as expected.

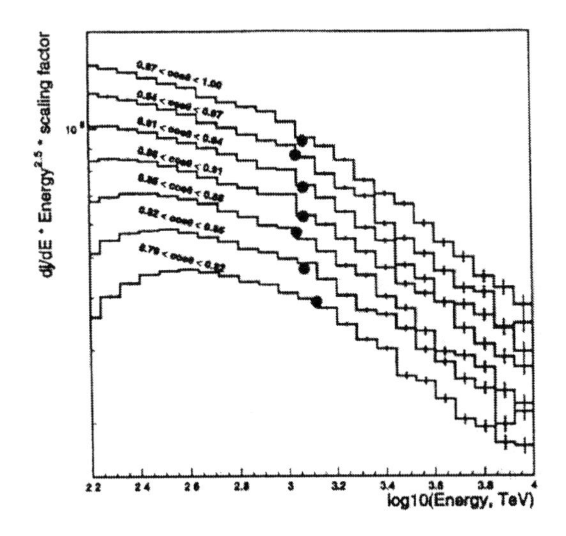

Figure 8. *Primary energy spectrum as found by CASA-MIA [25].*

KASCADE [26] measures all the three charged components of extensive air showers: electrons, muons and hadrons.
The knee is clearly visible in all components; for electrons and muons, the higher statistics allows to study the size at different atmospheric depths, thus finding that the size at the knee decreases at increasing atmospheric depth [27].
The primary energy spectrum can be extracted from the measured size spectra depending on the knowledge of the mass composition as obtained from the observables under investigation and on the relation between size and energy resulting from simulation.
The energy spectrum shown in Fig.9 was found by a combined χ^2 minimisation to fit both the N_e and the N_μ truncated muon size spectra simultaneously (the truncated muon number is that found by fitting the muon lateral distribution within a limited range of 40-200 m) [28]. The evaluated size spectra are in fact the convolution of the energy spectrum and a kernel function describing the probability of a given primary to produce a shower with a certain size and which includes the parametrisations of shower fluctuations for both proton and iron primaries according to Monte Carlo. The knee is found at about 4 PeV.

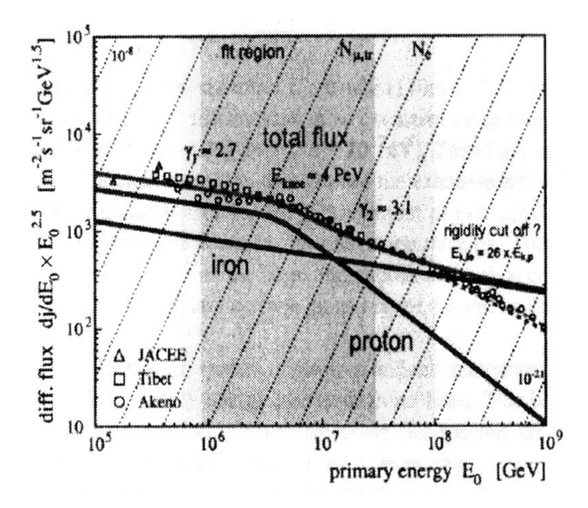

Figure 9. *Primary energy spectrum by KASCADE from electron and muon size spectra [28].*

In Fig.10, the energy spectrum as derived from hadronic data is also shown. The knee is again very clear, and the expectations from pure beams of protons or iron primaries are shown for comparison. Furthermore, KASCADE data show evidence of the knee also in the energy sum of hadrons in the calorimeter [29].
From muon density measurements in the multiwire proportional chambers below the central detector [30], a subdivision of the data in "light" and "heavy" samples (according to the parameter *Log* N_μ/*Log* N_e as described in Sect.4) shows

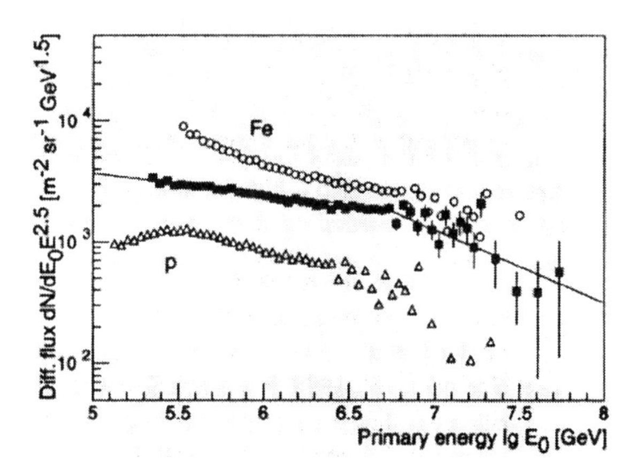

Figure 10. *Primary energy spectrum as found using KASCADE hadrons. Empty circles and triangles are the spectra from simulation of pure iron or proton primaries respectively [29].*

that the knee is strongly dominated by the light component, within a 30% uncertainty due to Monte Carlo statistics.
The broader lateral distribution of the Čerenkov light, due to the smaller absorption of photons in atmosphere, and the high photon number density, that means a better signal-to-noise ratio even for smaller arrays, are the main advantages in using Čerenkov detectors as compared to charge particle counting arrays. The most recent results from apparata based on the detection of Čerenkov light from Extensive Air Showers come from BLANCA [31] and DICE [32], both operating at the same site and sharing some equipment with CASA.
The first one consists of 144 angle-integrating Čerenkov light detectors located in the CASA scintillator array, which provides the trigger and gives core position and shower direction. The Čerenkov lateral distribution function is measured and fitted through the expression $C(r) = C_{120}e^{-sr}$ in the inner part of the distribution ($r \leq 120\ m$). The intensity at a critical radial distance of 120 m, entirely determined by density and scale height of the atmosphere, is proportional to the primary energy and the dependence on the primary mass is fully included in the slope s of the distribution, which is in fact a function of the depth of maximum development X_{max}.
DICE consists of 2 imaging telescopes of 2m diameter. What is measured is the Čerenkov light size N_γ, by summing the total amount of light at each phototube and the depth of maximum development of the shower X_{max} by fitting the shape of the light image in each telescope, with a procedure that is essentially geometrical and not depending on simulations, except for calculations to determine the angular distribution of light around the axis. The core position and shower direction are given by the CASA scintillator array.
The primary energy is estimated through a fit including geometry, N_γ and X_{max} and takes therefore into account the dependence of the lateral distribution and intensity of the Čerenkov light, at fixed primary energy, on the primary mass.
The resulting energy spectra from BLANCA and DICE are shown in Fig.11 (QGSJET model) and Fig.12 in comparison with other experimental results. The knee feature is evident in the BLANCA energy spectrum at $\simeq 3\ PeV$; a 10% shift in the energy scale, which is however less than the instrumental uncertainty, is found by changing the interaction model chosen to interpret the data. The absolute calibration error results in a $\simeq$ 18% systematic error on the energy assignment. According to DICE data, the knee is found around 3 PeV; the systematic uncertainty in the absolute flux is $\simeq$ 30%, due to the intrinsic error in the energy scale of $\simeq$ 15%.

The Čerenkov lateral distribution has also been measured by HEGRA with its AIROBICC detectors [33]. Again, the scintillator array provides the core position and arrival direction of the showers. The total primary energy can be reconstructed by the electromagnetic one, if one assumes a primary composition; however, a determination of the primary energy in a mass independent way can be obtained following the approach of Lindner [34], at the expenses of the energy resolution, which is worse than that found by the mass-dependent method and of a stronger dependence of the result on the fluctuations of X_{max}. The knee is found at about 3 PeV.
In Fig.13, the all-particle primary energy spectrum as obtained with the various experiments here described is shown. A summary of the up-to-date situation is given in Table 1. The differences

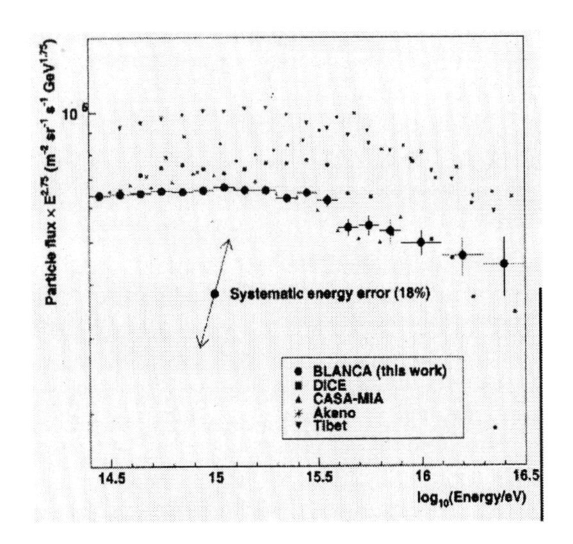

Figure 11. *Primary energy spectrum from BLANCA [31].*

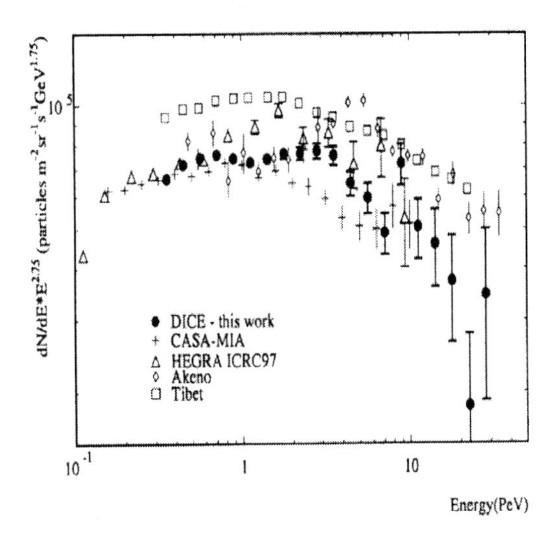

Figure 12. *Primary energy spectrum from DICE events in coincidence with CASA-MIA([32]).*

among the quoted knee energies are mainly due to the assumed composition, which in turn depends on the observables used as will be discussed below, but the existence of the knee is clearly established between 2 and 5 PeV.

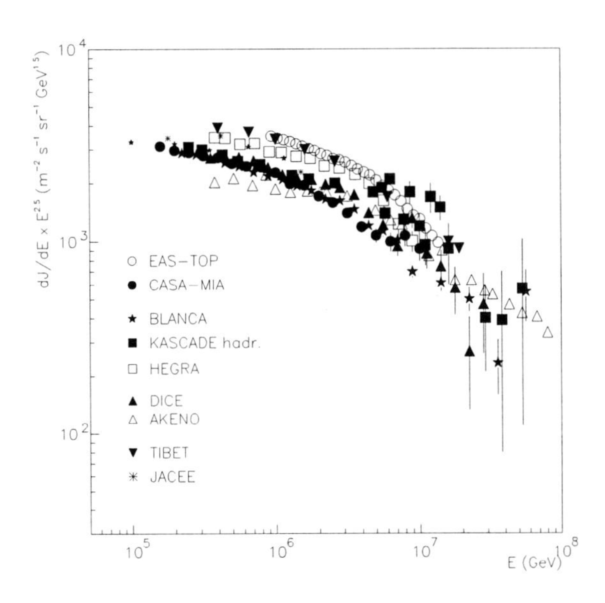

Figure 13. *All-particle primary energy spectrum from the various experiments as described in the text.*

4. Composition

The cosmic ray primary composition measurements around the knee are crucial for the understanding of the mechanisms of acceleration and the source problem.
The experimental observables we are dealing with are: a) the Čerenkov lateral distribution or the image of the Čerenkov light emitted by the shower in atmosphere; they allow to determine the depth of shower maximum X_{max}, which is a logarithmically increasing function of the primary energy. At fixed E_0, heavier primaries are expected to interact earlier, thus giving a smaller value of X_{max} (higher in atmosphere). b) the muon and electron sizes of the showers. For a given primary energy, EAS induced by heavy primaries develop earlier in atmosphere and less energy is released in the electromagnetic component, thus producing a smaller N_e at ground level, as compared to proton showers; on the other hand, muons are produced more copiously in EAS by heavy primaries, because of the higher number

Table 1
Slopes of the energy spectrum below (γ_1) and above (γ_2) the knee. E_k is the energy at which the knee is seen (PeV).

Experiment	γ_1	γ_2	E_k (PeV)
EAS-TOP [21]	-2.76 ± 0.03	-3.19 ± 0.06	$2.7 - 4.9$
KASCADE [35]	-2.70 ± 0.05	-3.10 ± 0.07	$4.0 - 5.0$
KASCADE [29]	-2.66 ± 0.12	-3.03 ± 0.16	5.0 ± 0.5
CASA [25]	-2.66 ± 0.02	-3.00 ± 0.05	smooth
AKENO [36]	-2.62 ± 0.12	-3.02 ± 0.05	$\simeq 4.7$
TIBET [37]	-2.60 ± 0.04	-3.00 ± 0.05	smooth
TUNKA [38]	-2.60 ± 0.02	-3.00 ± 0.06	$\simeq 4.0$
BLANCA [31]	-2.72 ± 0.02	-2.95 ± 0.02	$2.0^{+0.4}_{-0.2}$
DICE [39]	$\simeq -2.7$	$\simeq -3.0$	$\simeq 3.0$
HEGRA [33]	-2.67 ± 0.03	$-3.33^{+0.33}_{-0.41}$	$3.4^{+1.3}_{-0.7}$

of low energy pions.
The EAS-TOP group studied the composition by analysing the behaviour of $\overline{N_\mu}$ as measured in vertical direction in narrow bins of N_e, corresponding to $\Delta N_e/N_e = 12\%$. The result is shown in Fig.14: data are compared with the results of a full simulation including the detector response, where the 1 TeV composition with equal slopes for all components was used, in this way assigning constant composition with energy.

The EAS-TOP data clearly suggest a growth of the mean A with energy, that is a heavier composition above the knee. A change of $\Delta Log(N_e) = 0.5$ results in a $\Delta A/A \simeq 0.4$ [40].
The "K Nearest Neighbour" test was used by CASA-MIA to study the composition [41]. Using the electron and muon densities at different distances from the core and the slope of the electron lateral distribution, samples of event for each different primary mass are generated by Monte Carlo. An experimental event is assigned to the "light primary" or "heavy primary" class by looking at the K nearest neighbours (KNN) in the plane of the used variables: the event will belong to the light primary group if e.g.more than 50% of its KNN are light primaries. Due to fluctuations, which tend to superimpose classes, only broad classes of "p-like" and "Fe-like" events can be used. The proton resemblance, defined as the average fraction of K nearest neighbours which are protons, is shown in Fig.15 for K=5, normalised

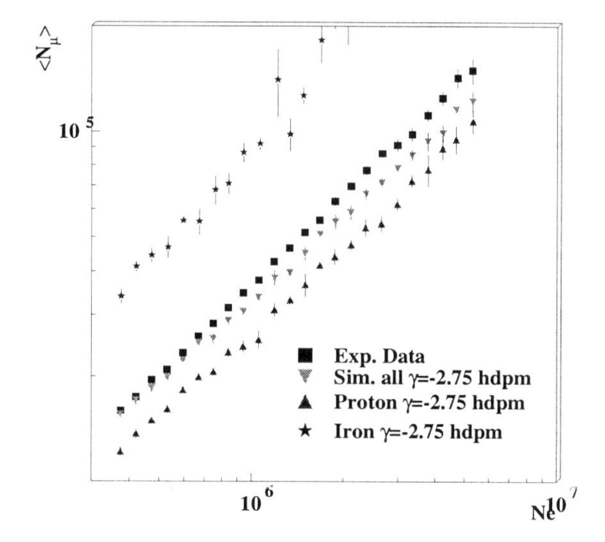

Figure 14. $< N_\mu >$ *vs* N_e *from EAS-TOP experimental data (full squares) as compared to a simulation with mixed composition and all components with the same slope (downward triangles), pure proton (upward triangles) and pure iron (stars) primaries [40].*

such that a pure proton composition would lay along the top of the plot and a pure iron along the bottom border.

The trend towards a heavier composition above the knee is evident; a change in the hadronic in-

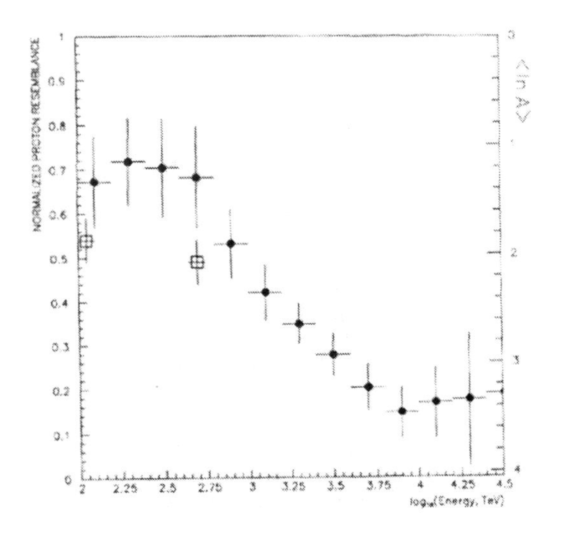

Figure 15. *Normalised proton resemblance plot from CASA-MIA. The open squares give the estimated result if the composition is taken from JACEE direct measurements [41].*

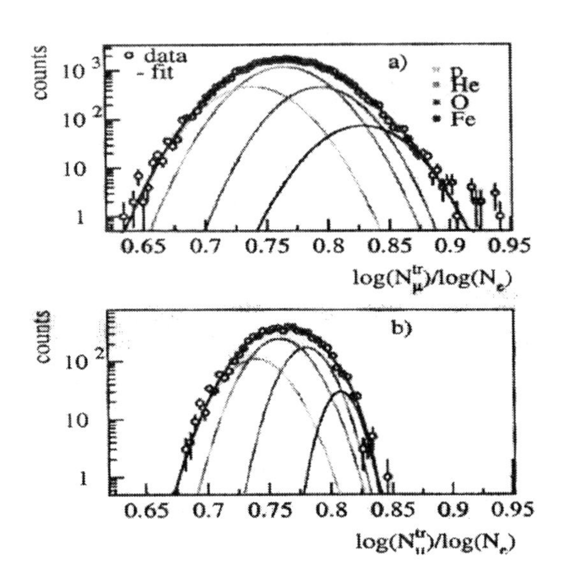

Figure 16. *$log\ N_\mu^{tr}/log\ N_e$ for two of the considered energy bins:* $6.2 \leq log(E/GeV) \leq 6.3$ *and* $6.7 \leq log(E/GeV) \leq 6.8$ *[42].*

teraction model used in the simulation does not change the result. Classifying the events according to their probability of being light or heavy primaries and plotting the energy spectrum for the two classes separately, CASA-MIA data suggest that the knee be due to the light mass group; the spectra are consistent with the idea of cutoffs proportional to the particle rigidities.
The composition problem has been attacked by KASCADE people in a variety of ways, using different observables and analysis methods.
The most sensitive dependence on primary mass was identified in the ratio $log\ N_\mu^{tr}/log\ N_e$, which is found to be Gaussian distributed at fixed A [42]. The experimental ratio is fitted by a superposition of simulated distributions (one for each primary mass group), directly obtaining the fraction of each mass group, as shown in Fig.16 for two energy bins. The composition is dominated by the light component up to about 4 PeV, getting heavier above the knee; the analysis also proves that the composition cannot be described by a single component. The $< ln\ A >$ so obtained is shown in Fig.18.

A number of hadronic observables has also been used, such as the lateral hadron distribution, the hadron energy spectrum, the maximum hadron energy etc. in order to investigate the composition. As one can see in the world survey given in Fig.18, the hadronic data alone give a heavier composition as compared to other data.
An interesting approach was used in [43], where a multivariate analysis using all the measured components of the EAS is performed. The result shows a tendency to a lighter composition approaching the knee, followed by an increase in the average mass above it.
A comparison among the results by KASCADE shows that the absolute scale strongly depends on the observables which are used. It is clear that the balance of energy among the different components of EAS in the simulation does not reproduce the real situation; tests of the high energy interaction models are been performed [44].
In experiments like BLANCA, Spase-VULCAN [45], CACTI [46], Hegra-AIROBICC, X_{max} is measured from the slope of the Čerenkov lateral distribution, which is an almost linear function of the depth of shower maximum. This function is rather independent on the models chosen for the

description of hadronic interactions, while any interpretation of the experimental results in terms of primary composition is not.
In the case of DICE, the imaging technique allows to measure X_{max} in a rather direct way, by fitting the shape of the shower Čerenkov image in each of the 2 telescopes, knowing the arrival direction and the core position of the shower.
A survey of the results is shown in Fig.17, up to the Fly's Eye energies (where air fluorescence is measured); the "direct" point shows the X_{max} that would be expected on the basis of balloon direct measurements [47].

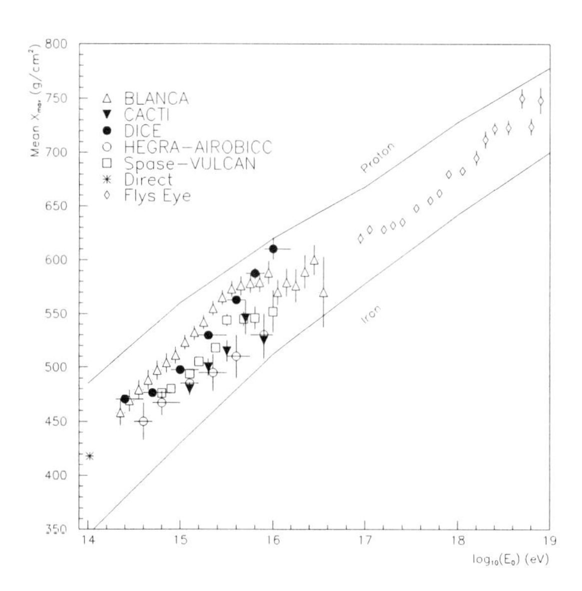

Figure 17. *Mean height of shower maximum vs energy as measured by various devices. HEGRA data from [48], Fly's Eye data from [49]. The lines show the expectations from a pure proton or iron composition using CORSIKA+QGSJET [50].*

BLANCA data suggest a composition getting lighter near the knee and turning to a heavier one after the knee energy. Data from DICE require a composition becoming progressively lighter with increasing energy.
DICE and CASA-MIA groups studied the composition problem also by means of a combination of measured parameters [32]. Two estimates of the mass have been derived, one using the X_{max} as determined by DICE and the other with N_μ and N_e by CASA-MIA, for each detected shower. The combined use of different measurements allows first of all a study of the systematics biasing the composition results; moreover, the requirement of consistency among various measurements allows to limit the range of the parameters used in the models. This analysis suggests a primary composition becoming lighter at and above the knee, not excluding however a constant composition around the knee energy.
A survey of the previously described results on primary composition in terms of $< ln\ A >$ is shown in Fig.18.

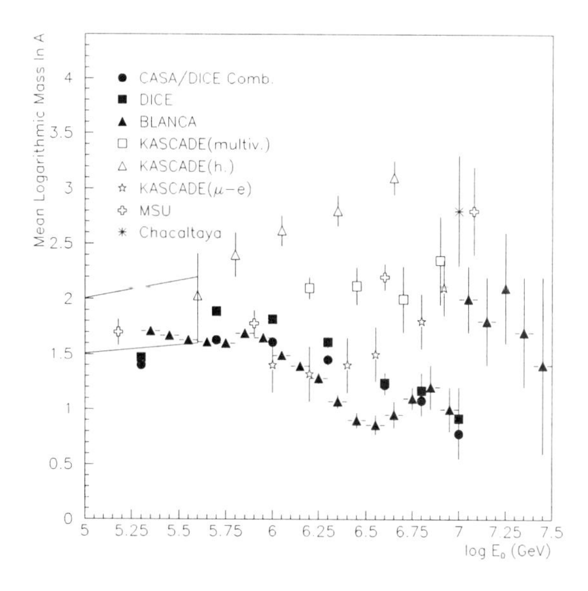

Figure 18. *Mean logarithmic mass vs primary energy. The box represents the region of direct measurements [52]. MSU data from [51], Chacaltaya data from [53].*

5. Conclusion

The cosmic ray energy spectrum and composition are studied with a variety of experimental techniques detecting different air shower components in the energy region above 1 TeV.
Below the knee, no new data are available and the conclusions reached by JACEE and RUNJOB experiments still hold; however new projects, planned to fly on balloons or on the Space Station, are in progress. They will surely extend the explorable energy region and the available statistics on single nuclei.
The energy spectrum has been studied in detail both by charged particles and Čerenkov light ground arrays and some firm conclusions were reached: all data agree on the existence of the knee in the primary energy spectrum of cosmic rays at an energy $\simeq 3-4\ PeV$. The bend has been seen in all shower components, thus supporting an astrophysical interpretation of the knee as opposite to that of a change in the hadronic interaction picture at these energies. All results agree to attribute the knee to the medium-light mass primaries.
Progress has been made as regards the mass composition. Almost all the ground array results show an increase in the primary mean logarithmic mass above the knee, even if the absolute scale can be quite different. There are however contradicting results coming from experiments relying on the Čerenkov light detection from Air Showers; the differences in the measure of X_{max} are however quite big.
It is very important to study in detail the systematics which could bias the results; the combined use of different observables and the comparison among various data sets can help in this task. The different sensitivity to composition of the various used observables, the methods employed to determine the primary energy and the problems found in the models used in the simulations could explain the spread in the results which is apparent in Fig.18.

Acknowledgements

I would like to thank my friends G.Navarra, B.Alessandro and A.Chiavassa for the interesting and useful discussions which were the basis of this paper. Sincere thanks also to all my Brazilian friends, for their warm hospitality during my stay in Campinas.

REFERENCES

1. Kulikov G.V. and Khristiansen G.B., Sov.Phys. JETP **8** (1959) 41.
2. Weekes T., *6th Workshop on GeV-TeV Gamma-Ray Astrophysics : Toward a Major Atmospheric Cherenkov Telescope VI*, Snowbird, UT, USA, 13-16 Aug 1999 Ed. by Dingus, B.L.; ASTRO-PH/9910394.
3. Clay R.W. et al., *Proc.25th Int.Cosmic Ray Conf.*, **4**, Durban (1997) 185.
4. Aglietta M. et al., *Ap.J.*, **470** (1996) 501.
5. Hillas A.M., *Proc.26th Int.Cosmic Ray Conf.*, **4**, Salt Lake City (1999) 225.
6. Seo E.S. and Ptuskin V.S., *Astrophys.J.*, **431** (1994) 705.
7. Biermann P.L. et al., *"The Central Parsecs of the Galaxy"*, ASP Conf.Series, Vol. 186. Ed. by H.Falcke, A.Cotera, W.J. Duschl, F.Melia, M.J.Rieke (1999) 543.
8. Protheroe R.J. et al., *Phys.Rev.Lett.* **69** (1992) 2885.
9. Watson A., *Proc.25th Int.Cosmic Ray Conf.*, Rapporteur Talk, Durban (1997).
10. Shibata T., *Nucl.Phys.B, Proc.Suppl.*, **75A** (1999) 22.
11. Asakimori K. et al., LSU Space Science/Part.Astroph.Prep.11/3/97.
12. Cherry M., *Proc.26th Int.Cosmic Ray Conf.*, **3**, Salt Lake City (1999) 187.
13. Ivanenko I.P. et al., *Proc.23th Int.Cosmic Ray Conf.*, **2**, Calgary (1993) 17.
14. Apanasenko A.V. et al., *Proc.26th Int.Cosmic Ray Conf.*, **3**, Salt Lake City (1999) 163.
15. Berezhko E.G., Ksenofontov L.T., *Proc.26th Int.Cosmic Ray Conf.*, **4**, Salt Lake City (1999) 381.
16. Wefel J.P. et al.,*Proc.26th Int.Cosmic Ray Conf.*, **5**, Salt Lake City (1999) 84. See also http://www701.gsfc.nasa.gov/access/access.htm.
17. Guzik T.G. et al.,*Proc.26th Int.Cosmic Ray Conf.*, **5**, Salt Lake City (1999) 9.
18. Seo E.S. et al.,*Proc.26th Int.Cosmic Ray*

Conf., **3**, Salt Lake City (1999) 207.
19. INCA Coll.,*Proc.26th Int.Cosmic Ray Conf.*, **3**, Salt Lake City (1999) 215.
20. Aglietta M. et al., *Nucl.Instr.Meth.* **A336** (1993) 310.
21. Aglietta M. et al., *Astrop.Phys.* **10** (1999) 1.
22. Aglietta M. et al., *Nucl.Phys.B (proc.Suppl.)*, **85** (2000) 318.
23. Aglietta M. et al., *Nucl.Phys.B, Proc.Suppl.* **75A** (1999) 251.
24. D.Heck et al., Report FZKA 6019, Forschungzentrum Karlsruhe (1998).
25. Glasmacher M.A.K. et al., *Astrop.Phys.* **10** (1999) 291.
26. Klages H.O. et al., *Nucl.Phys.B Proc.Suppl.* **B52** (1997) 92.
27. Kampert K.H. et al., *"Second Meeting on New Worlds in Astroparticle Physics"*, Univ.of Algarve, Faro, Poretugal (1998).
28. Glasstetter R. et al., *Proc.26th Int.Cosmic Ray Conf.*, **1**, Salt Lake City (1999) 222.
29. Hörandel J.H. et al., *Proc.26th Int.Cosmic Ray Conf.*, **1**, Salt Lake City (1999) 337.
30. Haungs A. et al., *Proc.26th Int.Cosmic Ray Conf.*, **1**, Salt Lake City (1999) 218.
31. Fowler J.W. et al.,*ASTRO-PH/0003190.*
32. Swordy S.P., Kieda D.B., *Astrop.Phys.* **13** (2000) 137.
33. Röhring A. et al., *Proc.26th Int.Cosmic Ray Conf.*, **1**, Salt Lake City (1999) 214.
34. Lindner A., *Astrop.Phys.* **8** (1998) 235.
35. Kampert K.H. et al., *Proc.26th Int.Cosmic Ray Conf.*, **3**, Salt Lake City (1999) 159.
36. Nagano M. et al., *J.Phys.C 10*, **9** (1984) 1295.
37. Amenomori M. et al., *Astrophys.J.* **461** (1996) 408.
38. Gress O.A. et al., *Proc.25th Int.Cosmic Ray Conf.*, **4**, Durban (1997) 129.
39. Kieda D.B. and Swordy S.P., *Proc.26th Int.Cosmic Ray Conf.*, **3**, Salt Lake City (1999) 191.
40. Aglietta M. et al., *Proc.26th Int.Cosmic Ray Conf.*, **1**, Salt Lake City (1999) 230.
41. Glasmacher M.A.K. et al., *Astrop.Phys.* **12** (1999) 1.
42. Weber J.H. et al., *Proc.26th Int.Cosmic Ray Conf.*, **1**, Salt Lake City (1999) 341.
43. Roth M. et al., *Proc.26th Int.Cosmic Ray Conf.*, **1**, Salt Lake City (1999) 333.
44. Antoni T. et al., *J.Phys.G:Nucl.Part.Phys.* **25** (1999) 2161.
45. Dickinson J.E. et al., *Nucl.Instr.Meth.*, **A440**, (2000) 114.
46. Paling S. et al., *Proc.25th Int.Cosmic Ray Conf.*, **5**, Durban (1997) 253.
47. Swordy S.P., *Proc.23th Int.Cosmic Ray Conf.*, Invited and Rapporteur, Calgary (1993) 243.
48. Röhring A. et al.,*Proc.23th Int.Cosmic Ray Conf.*, **3**, Salt Lake City (1999) 152.
49. Bird D. et al., *Phys.Rev.Lett.*, **71**, (1993) 3401.
50. Pryke C., *Auger GAP Note* **98-035**, FNAL (1998).
51. Fomin Yu.A. et al., *Proc.16th ECRS*, Alcala (1998) 261.
52. Wiebel-Sooth B., *"All particle Energy Spectrum measured at HEGRA"*, Thesis, **WUB-Dis-98-9**, Wuppertal Univ. (1998).
53. Aguirre C. et al., *SCAN-0009225*, subm.to Phys.Rev.D (2000).

ELSEVIER

Nuclear Physics B (Proc. Suppl.) 97 (2001) 48–65

www.elsevier.nl/locate/npe

Underground experiments for cosmic ray physics: results and future projects

S. Cecchini[a]*

[a]Istituto TeSRE/CNR, via P. Gobetti 101, 4129 Bologna, Italy

A review is made of the main results obtained by large underground detectors on cosmic ray physics, neutrino physics and astrophysics and searches for exotic particles. Future projects and their objectives are also briefly reviewed.

1. INTRODUCTION

Muons and neutrinos are the only known particles capable of reaching deep underground detectors.

The muons detected at great depths arise mainly from the decay of particles produced in the very first collisions of primary cosmic rays (cr) in the atmosphere. Being nearly stable and having a small cross section for interactions the muons are the dominant signal in underground detectors and are traditionally called the "penetrating component" [1]. High energy muons carry specific information concerning the properties of high energy collisions, the chemical composition and, possibly, the origin of high energy cosmic rays.

According to the standard electroweak theory, neutrinos of any energy are capable of penetrating undisturbed large amount of matter and for that reason are known as the most "elusive particle in nature". However their detection would provide us much information about neutrino properties and, being generated in the most energetic events in our Universe, on the deep interior of their far away sources. Though neutrinos are the most abundant cosmic particles at sea level, the neutrino events have to be extracted and identified from an overwhelming background of particles that can either directly or indirectly mimic the genuine neutrino signal. Therefore neutrino detectors have to be located under a very thick shield. Furthermore as a consequence of the small neutrino-nucleon cross-section already many years ago Reines [2] and Greisen [3] pointed out that kiloton or larger sensitive masses are required to study the cosmic and the so called "atmospheric" neutrinos produced in cosmic ray cascades in the atmosphere.

Several examples of cosmic neutrino detection already exist: solar neutrinos (0.1-10 MeV) already observed in the late '60s [4] and neutrinos from SN1987A (10-50 MeV) [5–8]. Cosmic neutrinos with energies > 100 MeV have not yet been observed and they would have profound implications on the search of Ultra High Energy Cosmic Rays (UHECR) origin and sources.

On the other hand the observations by underground experiments of anomalies associated with solar neutrinos [9–12] and atmospheric neutrinos [13–17] seem to challenge standard theoretical expectations.

But large underground detectors operated in low background environment and where secondary cosmic ray flux is greatly reduced can be used to search for not yet observed rare particles, like magnetic monopoles and dark matter candidates.

In this report, after a short history and a description of past and present undeground detectors, the main experimental results of single and multiple muon events, cosmic and atmospheric neutrinos, searches for exotic particles will be reviewed. Perspectives in cosmic ray physics with future detectors will be shortly discussed.

*also INFN, Sezione di Bologna, viale Berti Pichat 6/2, 41027 Bologna, Italy

PII S0920-5632(01)01183-5

2. OVERVIEW OF PAST AND PRESENT UNDERGROUND DETECTORS

The development of non-accelerator physics followed the technical progress in the electronics and computing. The construction of large detectors became possible by easier access to large underground laboratories (Mt. Bianco, Frejus, Kamiokande, Gran Sasso).

By the year 1951 several small detectors were operated at depths of less than 2000 hg of rock [18,19], while only two experiments were carried out at greater depths: Miyazaki et al. [20] in the Shimisu Railway tunnel (1410 and 3000 hg); Bollinger [21] in a salt mine near Ithaca (minimum depth 1510 hg and 1840 hg, and greater depths in non-vertical directions). Both experiments had an area smaller than 1 m^2.

By the end of 1965 three more small experiments were completed and operated: Barton [22] in lake Shore gold mine (min. depth 1690 hg and < 1 m^2), Miyake et al. [23,24] (min. depth 1880 hg; 1.62 m^2) and Achar et al. [25] (min. depth 1880 hg; 1 m^2) in the Kolar Gold Field (KGF) mine, and Castagnoli et al. [26] (min. depth 1319hg; 2 telescopes 0.5 m^2) in the Monte Bianco Tunnel. The main aim of these experiments was to measure muon intensities; the main problem was the very low intensities at those depths.

The years 1965-1980 saw a large increase in detector areas. The emphasis was on the detection of muons at large zenith angles produced by the interactions of high energy atmospheric neutrinos. Moreover, following the advent of the CERN-ISR and the FermiLab accelerators, together with the theoretical developments of limiting fragmentation and scaling, there was a better motivation for cosmic ray measurements sensitive to the detailed features of hadronic interactions. Underground multiple muon events discovered by Wataghin and collaborators in 1941 [27] and independently by Amaldi, Castagnoli and collaborators in 1952 [28] offered a promise in this regard, since they are produced in the early stages of the atmospheric cascades by primary cosmic rays.

During this period Davis started his famous experiment on solar neutrinos in the Homestake mine [29], and already in 1968 a deficit in the neutrino flux was reported [4].

At the beginning of the '70s the proposal for DUMAND (Deep Underwater Muon And Neutrino Detector) appeared and was funded for a feasibility study in 1979.

The beginning of the '80s were a turning point in underground detector developments. One of the main motivations for the birth of new large area detectors came from Grand Unified Theory of the fundamental interactions (GUT). The Standard Model of particle physics was already tested with good accuracy at existing accelerators. To explore sectors of the theory it was necessary to build larger accelerators. But the scale energy of the GUT unification is so high that we will never be able to reach it with accelerators. Fortunately there is another way. Phenomena characterised by a high energy scale do, in fact, occur naturally. But the higher the energy, the more rarely they occur. The proton decay is one of the possible phenomena capable to bring into test the GUT theory. Other possibilities would arise from the phenomena of rare exotic particles (*e. g.* magnetic monopoles) originated at the beginning of our Universe. But to search for such extremely rare events one needs large detectors in an environment with very low background. A background "noise" that is mainly due to natural environmental radioactivity but also by cosmic rays and atmospheric neutrino interactions.

Broadly speaking, the underground detectors built during this period can be classified into three groups following the techniques used: (i) the water Cerenkov detectors (IMB, Kamioka), usually massive (few thousand tons), (ii) liquid scintillator detectors (Baksan, LSD) of modular design (few hundered tons), and (iii) tracking calorimeters (Frejus, NUSEX), tipically of 100-1000 tons, making use of sandwiches of heavy material (*e. g.* iron plates), and proportional, streamer or flash tubes.

These experiments have brought to important results ruling out the minimal SU(5) (from proton decay limits), and making detections of solar and SN1987A neutrinos. The different approches have been rather solid; their main limitations were the collected statistics. The necessity

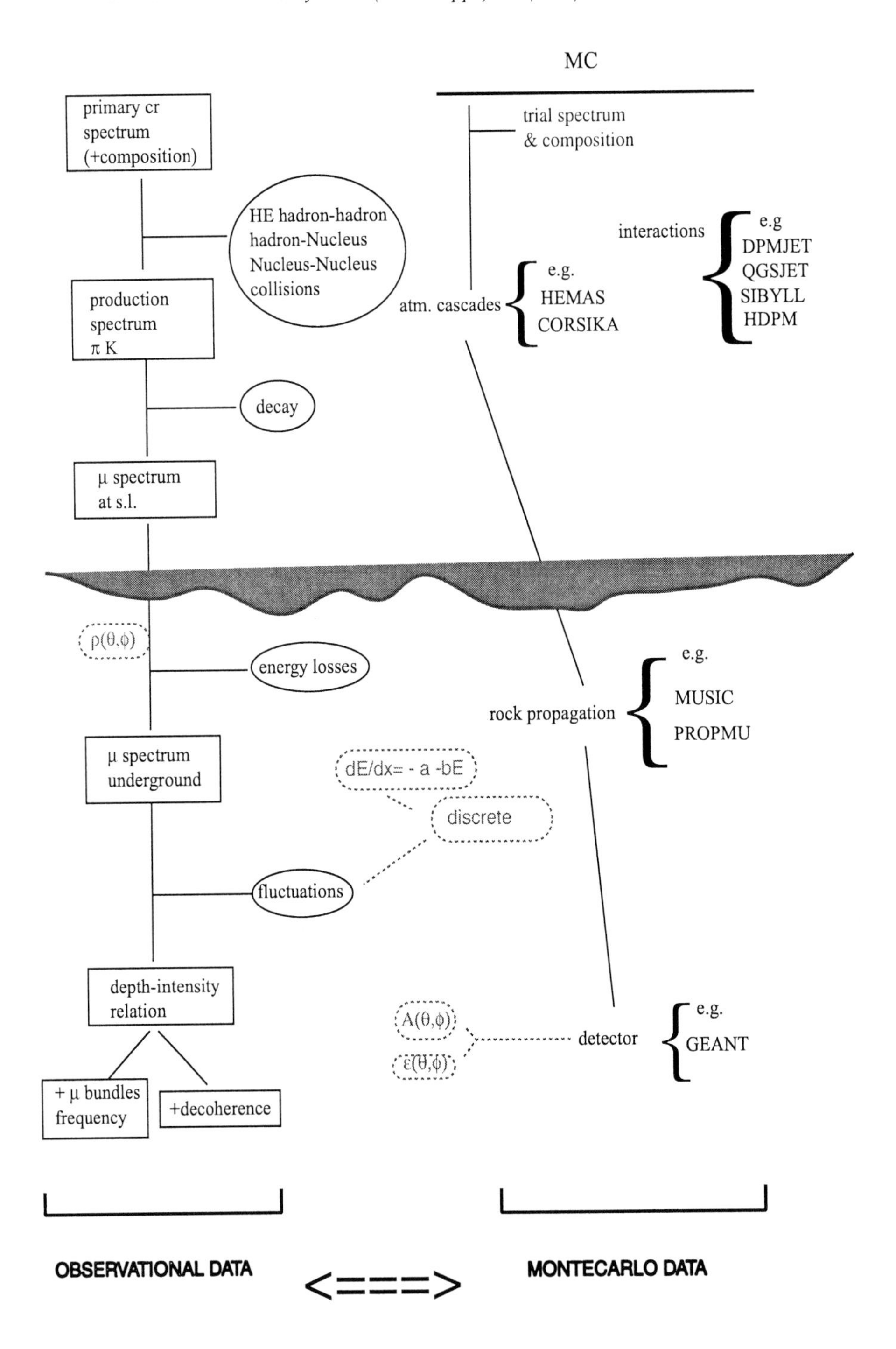

Figure 1. Qualitative scheme of observables (boxes), physical processes (ellipses), origin of systematic-uncertainties (dotted boxes) and relevant steps in MonteCarlo (MC) simulations

to construct similar detectors of larger volumes or areas became clear.

The experience gained with NUSEX was at the base of the MACRO project with an acceptance increased by more than two orders of magnitude. In a similar way the Soudan (I and 2) detector was developed. The successor of the Kamioka and IMB experiments was Super-Kamiokande, with a mass of 50 ktons. LSD inspired LVD at Gran Sasso with the goal of 2ktons of sensitive mass liquid scintillator mass.

The new scale of detectors required easy access to underground laboratories, in order to meet strict safety rules, to provide full instrumented facilities for building and assembling the detetcors, and to accomodate the scientists for long periods of stay. The Gran Sasso Laboratory (LNGS) is the logical outcome of this development. The scientific field covered at Gran Sasso are many: 1) solar neutrinos, 2) dark matter and exotic searches, 3) high energy cosmic ray physics, 4) nuclear stability (double beta decay), 5) low energy cross-section measurements, 6) geophysics, 7) living matter in low radioactivity background, etc.

3. UNDERGROUND MUON PHYSICS AND ASTROPHYSICS

Underground muon studies are relevant for cosmic ray physics: they can give information about their energy spectrum and composition. Moreover the measurements of the muon energy spectra, angular distribution and the depth intensity relation (DIR) allowed us to study the mechanism of charm production in hadron-hadron and hadron-nucleus collisions. The primary energy range covered by the detectors located at depths larger than 2000 m.w.e. for pN and NN interactions goes from 10^{13} eV to almost 10^{17} eV, that is from energies reached at the HERA and Tevatron colliders, and even energies that will be reached by LHC. Furthermore because of the close relationship between muon and atmospheric neutrino production, the knowledge of muon flux can provide an important cross check on the atmospheric neutrino flux. On the other hand the latter constitutes an unavoidable background for future under-water and under-ice neutrino telescopes at energies from about 1TeV to tens of PeV.

Figure 1 summarizes the observable physical quantities that are of prime interest in cosmic ray physics and the sequence of processes at work. In order to make a correct interpretation of underground measurements, and get results on the primary spectrum and composition, one has to rely on Monte Carlo (MC) simulations as a correct description of high energy cosmic ray interactions in the atmosphere, as well as on reliable treatment of the shower development, and carefully consider the muon transport through the overburden. Usually a model for the cosmic ray composition and spectrum is first assumed and then the data obtained by MC simulations are compared with the experimental data.

Table 1 lists the main large underground detectors dedicated to the study of high energy muons, neutrino physics and astrophysics, and the search for exotic particles.

At the energy explored by deep underground detectors one needs to develop models which have a reasonable description of interactions at the energies not yet available at the accelerators. Moreover the muon parent mesons belong to kinematical regions poorly explored at collider energies. Therefore important quantities such as cross-sections, inelasticity and multiplicity must be inferred from extrapolations, often guided by a phenomenological model, e.g. HEMAS [30], or based on physical models, e.g. SYBILL [31], DPMJET [32], QGSJET [33]. Finally other contributions to systematic uncertainties are introduced in the simulation of the muon transport in rock by the inaccurate knowledge of the density and shape of the overburden, of the energy losses and their fluctuations, of the acceptance and efficiency of the detectors.

3.1. Depth-intensity relation

The measurements of the depth intensity relation is important since from it one can study the muon spectrum at sea level and the all-nucleon flux at the top of the atmosphere. Comparison of the all-nucleon spectrum obtained by this technique with direct measurements is limited at high

detector	min. depth	area[m^2]	mass [tons]	technique	purpose	status
KGF	745	4		PS+NF+Pb+FE	CR	> 70
	1500	4		PS+NF+Pb+FE	CR	> 70
	1840	24		PS+NF+Pb+FE	PDK + ν	> 70
	3375	2×4			PDK +ν	> 70
	6045	4×2	10			phase1/75
		$6 \times 6 \times 6.5$	320	PGC+Fe	PDK	phase2/85-92
	7200	$2 \times 6 + 2 \times 8$		PS+NF+Pb	PDK	
CWI/ERPM	8710	170	3.2	LS+NF	ν	67-71
Utah	4400	80-100	2000	SC+WC	ν	69-74
Kamioka	1350	8.5		SC+PS+Pb+Fe	CR	
HPW	4200	120	150+150	WC	PDK	
Artiomovsk	600	30	105	LS	CR	85
Baksan	850	250	120(300)	LS tanks	CR + ν	82
IMB	1570	390	330(6800)	WC	PDK	82
SoudanI	1800	10	31	ST+concrete	PDK	81
Soudan2	2050	130	1100	DT+concrete	PDK	> 91
Kamiokande	2700	120	2100	WC	PDK	83
SK	2700	1200	45000	WC	PDK	> 96
MACRO	3200	900	600	LS+ST+CR39+rock	MM+CR	89-00
LVD	3200	160	1100	LS tank+ST	SN ν	92
Frejus	4850	90	560	ST+Fe	PDK	84-88
NUSEX	5000	10	120	ST+Fe	PDK	82-90
LSD	5200	100	80	LStanks	SN ν	84
Baikal	1350	1300		underwater	HE ν astr	93
AMANDA	1500	10000		underice	HE ν astr	97
RAMAND		10^6		underice	HE ν astr	R&D
ICECUBE				underice	HE ν astr	planned 2008
ANTARES	2400	10^5		underwater	HE ν astr	$> 2001 - 2003$
NESTOR	3800	4.310^5		underwater	HE ν astr	R&D
DUMANDII	4800	20000		underwater	HE ν astr	R&D
NEMO.KM3	3400	2.610^6		underwater	HE ν astr	R&D
ICARUS	3200	150	600	TPC+LArgon	PDK	under construction
MONOLITH	3200	450	34000	RPC+Fe	ν oscill	proposed
PS= plastic scintillator	LS= liquid scintillator	SC= gap spark chamber	NF= neon flash tubes			
WC= water Cerenkov	ST= streamer tubes	DT= drift tubes	PGC= proportional gas counter			
CR39= nuclear track detector						

Table 1: Main large underground detectors which study high-energy muons, neutrino physics and astrophisics, and search for exotic particles

energies mainly by statistics. The DIR also offer a sensitive test of the model of muon production in hadronic showers. A better determination of the muon flux will allow a better normalization of the atmospheric neutrino flux, a vital information for the studies of neutrino oscillations with underground detectors.

Data on the DIR collected by underground detectors from depths from 3 km w.e. to 6-7 km w.e. (corresponding to 3-4 TeV of muon energy at sea level) have reached a good accuracy; data from different experiments appear to be in good agreement among them (see Figure 2 taken from Reference [34]), within 10% of their values. The possible differences are due to systematics in the knowledge of the overburden, to difference in energy losses, and the interaction models.

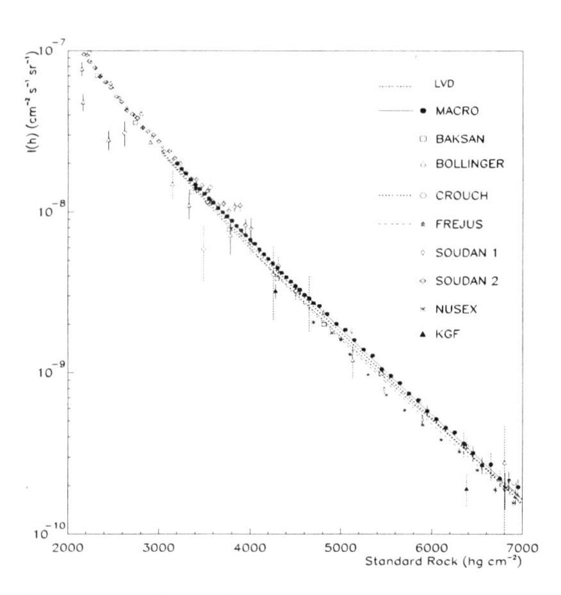

Figure 2. Vertical muon intensity vs. standard rock depth. Data compilation by Crouch and those obtained by other experiments are taken from Reference [34].

The situation is less clear at depths > 7 km w.e. [35] and it is difficult to distinguish between the various model of prompt muon production.

From the DIR it is possible to evaluate the muon spectrum at sea level and compare it to

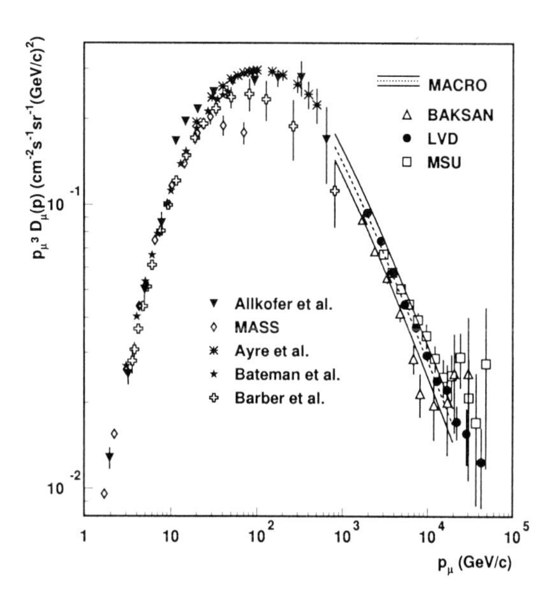

Figure 3. Vertical differential momentum spectrum of muons at sea level. Direct surface measurements extend up to 1 TeV/c. Indirect data are from underground observations for momenta > 1 TeV/c (from [36]).

the measurements made by surface detectors, like magnetic spectrometers and observations of electromagnetic (em) cascades generated by muon *Bremsstrahlung* in calorimeters. In this way it is possible to extend the muon spectrum mesurements at energies >10 TeV. For this it is very important to know: 1) the cross-section of muon interaction, accounting for fluctuations; 2) the composition and density of the rock as well as the slant depth for any azimuthal and zenith angle; 3) the cross-section for prompt muon production for depth > 7 km w.e. The results of surface direct measurements up to 1 TeV/c and of the fit of MACRO together with LVD, MSU and Baksan data are reported in Figure 3 [36]. Data are presented multiplied by p^3 to better observe the variation of the spectrum in the whole energy region and to strengthen a possible flattening in the tail of the spectrum due to charm production. The statistics is too poor to allow any definite assessment on the existence of this effect at energies >10 TeV. The contributions of systematics to the determination of the normalization constant and of the index of the muon spectrum when expressed as in [37] were estimated, for example in the MACRO fit, to be about 5% and 3%, respectively.

An accurate study of the muon flux intensity over different ranges of depth to check for the existence of a prompt muon component has been made by LVD [38], based on its good efficiency in tracking muon events at large zenith angles. As a result of the fitting procedure LVD obtained an upper limit to prompt muon flux (R_c= 2.10^{-3}, 95% C.L.) that favours the model of charm production based on QGSM and the dual parton model [38].

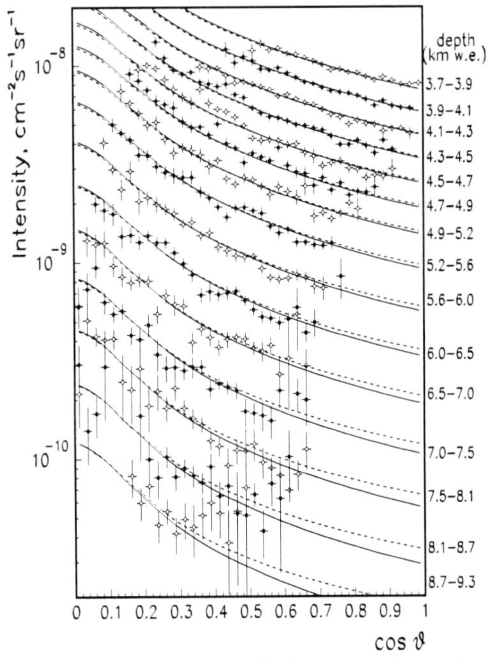

Figure 4. Dependence of the muon intensity on the zenith angle at different depths. Continuous curves are calculations with γ=2.77 and R_c = 0 (best fit of LVD data). The dashed curves are the calculations with γ=2.77 and R_c = 2.10^{-3} (upper limit of LVD). Errors include statistical and systematic uncertainties.

3.2. Composition studies

The elemental composition of primary cosmic rays in the region around the knee can be studied by detectors located at depths larger than about 1000 m. w.e., either in a stand-alone mode or in coincidence with extensive shower arrays at the surface. The most sensitive underground observable is the muon multiplicity distribution. Usually in these studies, an astrophysical model for cosmic ray composition is first assumed and then the muon multiplicity distribution obtained via MC is compared with the experimental one. The first results along this line of research were obtained by the NUSEX group; their observations did not support a proton-dominant composition at the knee region [39]. Soudan 2 [40] adopts three different models: proton rich, identified as "new proton source", a heavy composition and a very heavy composition, called "new source Fe". The three models agree at energies < 3.10^{14} eV/nucleus. Their best fit, however, favours the "new proton" composition model and, in conclusion, their result is in disagreement with a strong enrichment of the composition beyond the knee. The Baksan underground detector, located at shallower depths, is in agreement with a standard primary spectrum in the range 10^{13} – 10^{15} eV. Their statistics however allow only to exclude the hypothesis of a pure proton spectrum composition and the absence of the knee at 3.10^{15} eV [41] (see also [42]).

The KGF underground detector at 6045 hg/cm^2 [43] detected the multiplicity distribution in agreement with a model that gives rise to a mixed composition in the knee region, which becomes richer in heavy elements at higher energies.

A different method, called "direct fit", was developed by MACRO [44]. Such method allows to extract directly the primary spectrum and the chemical composition model preferred by the experimental data. The main results are: both light and heavy components are required in the energy range explored; if one assumes that the knee does not exist, the fit worsens, that is to say, the cosmic ray knee is observable underground. If one looks at the primary cr spectrum, a good agreement is found for CNO, Mg and Fe nuclei in the range where direct data are available. On the contrary, for H and He there is agreement only at energies < 10^{13} eV; tha data suggest a flatter H and He flux at higher energies. The mismatch with the primary spectrum is of 7% at 10^{13} eV and 23% at 10^{14} eV [45]. By representing the results by stand-alone underground detectors as < lnA > vs. energy one can directly compare them to EAS array observations [47–50] and to direct measurements [51] (see Figures 5 and 6). Indeed < lnA > is a useful variable as at a given energy the depth of shower maximum (X_{max}) is known to vary with lnA [46]. We leave the discussion on the results obtained by EAS arrays to [52]. However all the observations appear to indicate that the < lnA > becomes lighter in the knee region before becoming heavier again at higher energies.

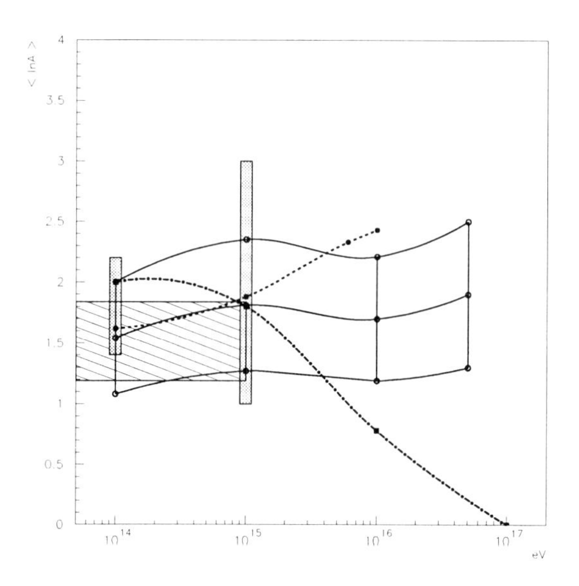

Figure 5. < lnA > vs. energy as deduced by stand-alone underground detectors:(open circles-continuous line) MACRO fit with the band of 1 sigma error, (black squares-dashdotted line) Soudan2 "new proton" source, (black circles-dashed line) KGF, (hatched box) Baksan, (grey boxes) Watson compilation of direct measurements [51].

In a stand-alone research, however, underground experiments can provide a primary composition that is averaged over almost two decades

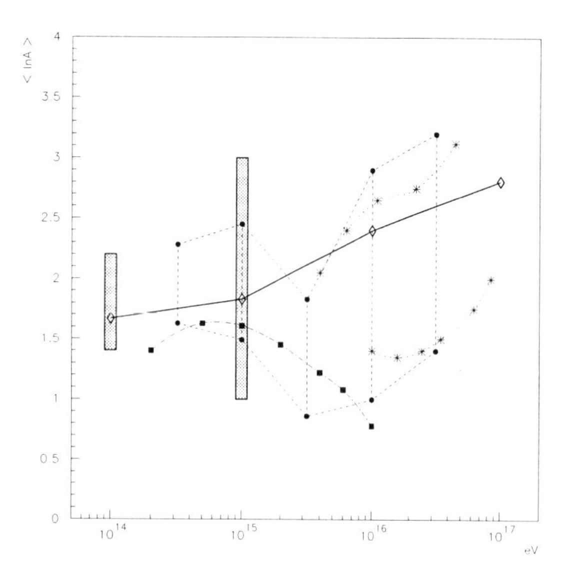

Figure 6. < lnA > vs. energy as deduced by EAS arrays: (black squares) DICE; (black circles) CASA-BLANCA; (asterisks) KASKADE; (open diamonds) EASTOP; (grey boxes) Watson compilation of direct measurements [51]

of primary energy. This is because the energy determination is made by the depth of the overburden. Therefore the sensitivity to variations of the composition with primary energy is lower compared to EAS experiments. A better approach is to use, when possible, coincidences between underground detectors and EAS arrays located directly above them. The LVD and EASTOP experiments at LNGS used the simultaneous measurements of the e.m. size of the shower (N_e) and of the underground mean muon energy losses. For a fixed size (N_e) and a proton initiated shower the mean energy losses of the muon underground should be larger than for heavier primaries. Their result, obtained by analysing about 3500 events collected in 400 days of running time are compatible with a mixed composition up to the knee [53]. The underground muon multiplicity for various shower size windows, was used as a parameter sensitive to composition in the analysis of the 7889 events collected during 348 days of live time by the MACRO and EASTOP experiments [54]. In this sample 240 events were above the knee corresponding to $\log_{10}N_e = 5.85$. The experimental results for data below the knee are reproduced - within 20% error - with a primary spectrum extrapolated from the direct measurements of the various nuclear species at lower energies. Above the knee additional hypotheses must be introduced in order to reproduce the spectrum. There is an indication, however, that the composition becomes heavier above the knee. Regrettably this type of very promising analysis will not continue at LNGS as the EASTOP array is now dismantled. Similar analyses have been made by SOUDAN 1,2 [55,56]. Their results do not support any significant increase in the average primary mass with energy in the range of 10000TeV per nucleus. Some systematic uncertainties remain, however, particularly in the MC modelling of the cosmic ray shower.

3.3. Interactions

In studying cosmic ray physics with underground experiments we have probably reached the point where there is no need of increasing the statistics anymore. The systematic uncertainties are the dominant factors and it is difficult to improve their knowledge to the level required. We need to devise new kinds of analysis to interpret the existing data. Here one example will be reported.

The modelling of high energy interactions is the main problem when using MC simulation to produce atmospheric cascades. In the physics of underground muons two quantities enter at the same time: the composition and the interaction processes. To solve the problem one possibility is to find out observables that depend only on the composition model or on the interaction model; in this way one can hope to disentangle the two effects and reduce the systematic uncertainties of the analysis. The distribution of the muon pair separation in multiple muon events and its shape are sensitive to the transverse structure of the hadronic interaction and are almost independent (in first approximation) [57] of the composition. The agreement found between the HEMAS interaction model used in MC simulations and MACRO data indicates that no anomalous $P_\perp$ components in soft pN and NN collisions are required. By considering the correlation among the muon positions in high multiplicity events one can get information on the composition model whereas it should be less dependent on the hadronic interaction model. The results for energies > 1000 TeV give support to the "direct fit" analysis. Many of the events collected underground by MACRO have a typical structure as the one in Figure 7. The question is: is this kind of structure a statistical fluctuations or does it have a dynamical origin? By a sofisticated analysis [58] based on different mathematical tools new limits on the MC codes used have been placed: in particular, the hadronic interaction code that better reproduces the features of the MACRO analysis is the QGSJET model.

3.4. Muon Astronomy

The search for point sources of underground muons started more than 15 years ago. Few years after the claim that high energy γ rays from Cyg-X3 direction were observed by an EAS experiment [59], NUSEX [60] and Soudan [61] experiments reported an increase in the periodic muon counting rate from the same direction in the sky. Their source could be neutral stable particles (γ, ν). However the signal was too high to be explained by known processes of muon-production in the atmosphere. On the other hand no definite sign of muon excess was found by Kamiokande [62]. New experiments have monitored all the possible or more interesting sources of cosmic γ rays. The capability of underground muon experiments to behave as telescopes has been demonstrated through the method of reconstructing the shadow of the Moon and of the Sun in a similar way to what has been done for the first time by EAS arrays [63–66]. The shadow of the Moon is a useful test of the angular resolution, pointing accuracy and long term stability of a detector. As an example, Figures 8 and 9 show the results obtained by MACRO for the Moon and the Sun, respectively [67]. Similar observations were made by Soudan 2 [69] with a 5σ detection of the Moon shadow, for an angular resolution $< 0.3°$ with an alignment of the detector better than $0.15°$, and by LVD [68] with an observed deficit with a significance of $2.62\,\sigma$.

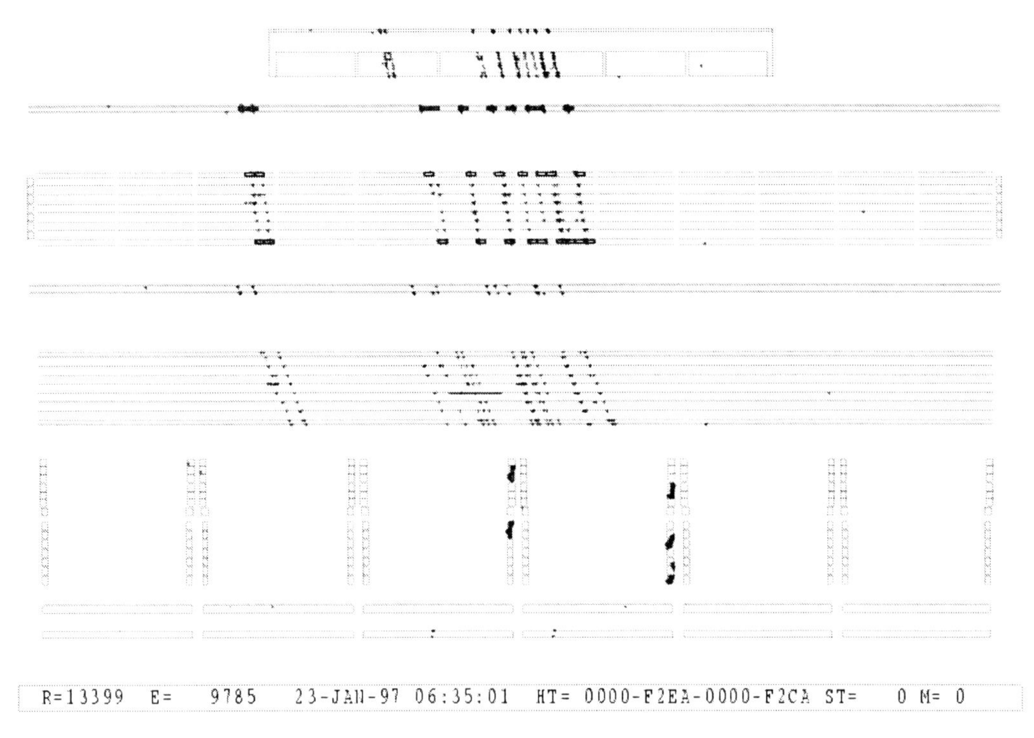

Figure 7. A two cluster event detected by MACRO as seen with the event display of the experiment.

The search for point sources, either steady or variable, has been so far unsuccessful and the sensitivity reached by present underground experiments—which depends on the area, exposure time and resolution—is at the level of few 10^{-13} cm^{-2} s^{-1} or a little lower.

4. NEUTRINO PHYSICS AND ASTROPHYSICS

4.1. Neutrino Oscillations

The interest in atmospheric neutrinos has grown in the last years after the Takayama Neutrino Conference 1998 where the SuperKamiokande collaboration [70] announced the observation of neutrino oscillations. The measured flux of the muons induced by atmospheric ν_μ appeared to be reduced with respect to expectations. Previous observations of an anomaly in the ratio between events with a muon and with an electron were reported by IMB and Kamiokande [71,72], but were not confirmed by NUSEX and Frejus [73,74]. At the conference, however, two other experiments, MACRO [75] and Soudan2 [76], presented results that gave strong support to the same oscillation pattern observed by SK (see Figure 10 from [77]).

The evidence for ν_μ disappearance comes from three experimental effects: 1) the up-down asymmetry (excess of up-going muons) and of an east-west effect in SK observations of muon-like events, 2) a distortion of the zenith angle distribution 3) a reduction of the ν_μ flux compared with the ν_e flux. The SK experiment has been able to give the clearer signal. It is, however, worth noting that the same effect has been observed also by other detectors using completely different techniques. The limitation for a better determination of the oscillation parameters comes from systematics. Sources of uncertainties are related to the spectrum of primary cosmic rays and to the modelling of particle production in hadronic interactions. For the latter the 3-D calculations including geomagnetic effects have been started recently

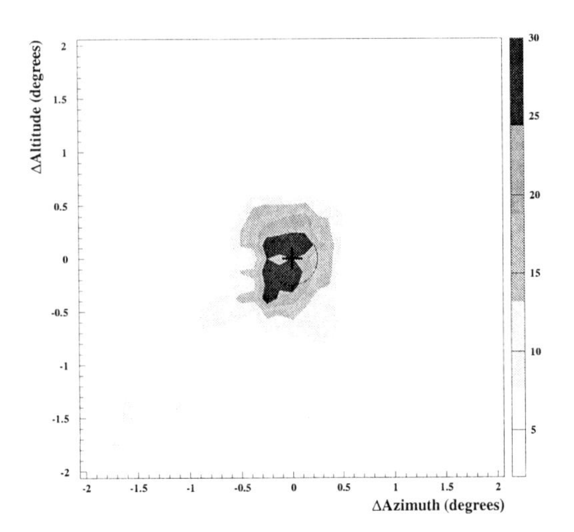

Figure 8. The shadow of Moon observed by MACRO, with a significance of 5.5 s.d.

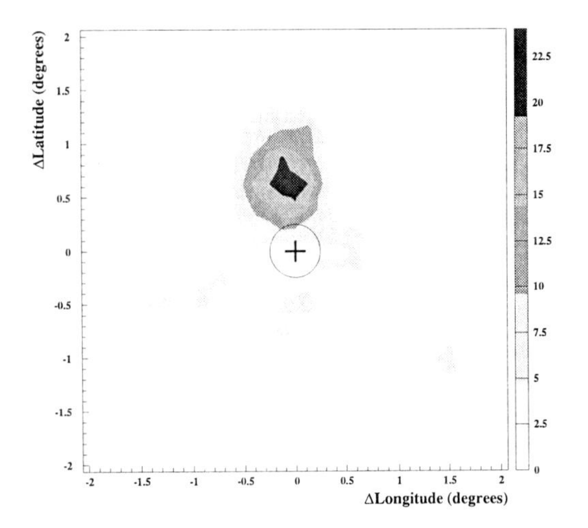

Figure 9. The shadow of the Sun observed by MACRO, with a significance 4.2 s.d.

[78–80]. New precise measurements of the muon flux in the atmosphere or at ground would be important to constrain neutrino flux calculations.

Although the ν_μ to ν_τ oscillation hypothesis provides a simple explanation of the current data, other more complex schemes can fit the data[81]. At present, the ν_μ to ν_s oscillation is disfavored by SK at the 99% C.L. and by MACRO at about the 3σ level; the ν_μ to ν_e is not completely excluded by SK.

In conclusion the ν_μ disappearance in atmosphere is robust represents a great success of underground physics.

A more precise determination of the oscillation parameters of the will be the main goal of accelerator long baseline experiments; They will also try to detect the appearance of the ν_τ which would definitely prove the existence of neutrino oscillations. However the new generation of underground experiments will play an important role, due to the possibility of measuring a sizeable effect for L/E from 10^4 km/GeV to 10^{18} km/GeV.

4.2. Neutrino Astronomy

HE muon neutrinos can be detected via their charged-current interaction inside a detector or in the rock surrounding the detector leading to up-

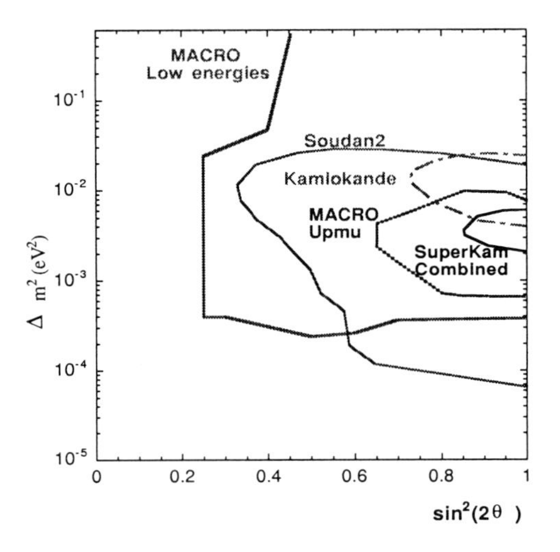

Figure 10. The 90% confidence level regions of the experiments with positive indications for atmospheric neutrino oscillations (from [77]).

going muons. Upward-going muons can be seen directly in Cerenkov detectors and in scintillator dtectors where can be separated by time of flight from down-going muons. At very high energies the $\nu_\mu - \mu$ angle is small and the effective target may be large. A celestial "point" would reveal itself as an excess of events (in a certain direction) above the atmospheric neutrino background. The first attempts to identify neutrinos from discrete point sources were made by [82–84]. No significant signal was observed with respect to the background from atmospheric neutrinos from any event direction by any underground experiment. Recent limits by different underground experiments are reported in Figure 11 (from[85]) for a number of candidate sources. The MACRO values are, at present, the most stringent ones; they are about one order of magnitude higher than the values predicted by plausible neutrino source models. Two high energy telescopes are partially deployed and currently operating: the Lake Baikal detector [86] and AMANDA detector at the South Pole [87,88]; initial results are been presented for 240 and 140 days of life times, respectively. The experiments set upper limits to the diffuse flux of high energy neutrinos ($> 10^{13}$ eV) that extend the limit from the Frejus experiment to much higher energies [89]. These experimental limits are already excluding several models for the origin of such a diffuse flux.

There are also searches for time coincidences with γ ray bursts [90,91], but we need to increase the sensitivity by a factor 100-1000 and to cover the whole sky in order to be able to explore potential sources of high energy neutrinos.

4.3. Neutrinos from Supernovae

Probably the most awaited event in astrophysics is a new supernova in the Milky Way. After SN1987A was successfully detected by various underground experiments, the present-day detectors (Baksan, SK, MACRO, LVD, AMANDA) have been prepared themselves for the event. In the near future the next generation experiments (Borexino, SNO, KamLAND) will be operative. We can foresee to learn a lot about several items of astrophysics and neutrino physics. In fact from the flavor composition, energy spectrum,

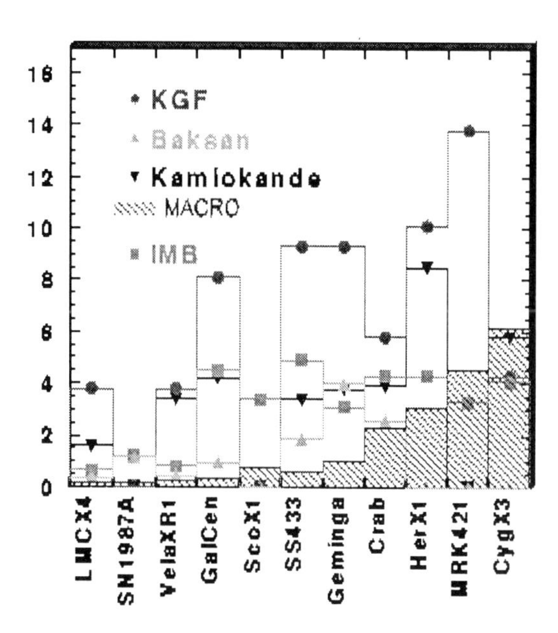

Figure 11. 90% C.L. upper limits for several candidate sources of high energy neutrinos observed by different underground detectors (from [85]).

and time structure of the SN event we will gain new insight in the processes of the stellar collapse and on the nature of the progenitor. By observing the time delay accumulated over galactic distances it should be possible to distinguish between an energy-dependent and flavor-dependent origin of the effect. It is expected that the supernova neutrino burst precedes the optical signal by hours or even days. For this reason it is important to be able to issue a warning to astronomers as early as possible. Since few years an international collaboration - SuperNova Early Warning System - (SNEWS) between MACRO, LVD and SK experiments is active. While MACRO will not any longer take part of it, SNO and AMANDA will join soon. An optimistic estimate of the rate of type-II Supernovae in our galaxy tells us that probably we will have to wait 15-25 years more. So we should keep the detectors alive all the time!

5. EXOTIC PARTICLE SEARCHES

5.1. Magnetic Monopoles and Nuclearites

The search for magnetic monopoles (MMs) is one of the main goals of the MACRO experiment at LNGS; the experiment is able to cover with redundant techniques the widest velocity range. Other experiments (Baikal, AMANDA, Baksan, KGF, Soudan2) are sensitive in more restricted velocity ranges. Supermassive (M_m $=10^{17}$ GeV/c^2) GUT monopoles are expected to have typical velocities of 10^{-3} c, if trapped in the Galaxy; MMs trapped in our solar system or in the supercluster of galaxies may travel with velocities of the order of 10^{-4}c and 10^{-2}c, respectively. In the presence of celestial strong magnetic fields (AGN, black holes), MMs may accelerate and reach higher velocities. MACRO has reached a flux sensitivity below the Parker bound [92], in the MM velocity range 4 10^{-5}c - c. The three subdetectors of MACRO have sensitivities over wide beta ranges, with overlapping regions; thus they allow multiple signatures of the same rare candidate event. No candidates were found in several years of data recorded by any of the subdetectors. Figure 12 shows the present MACRO limit together with those set by other experiments [93].

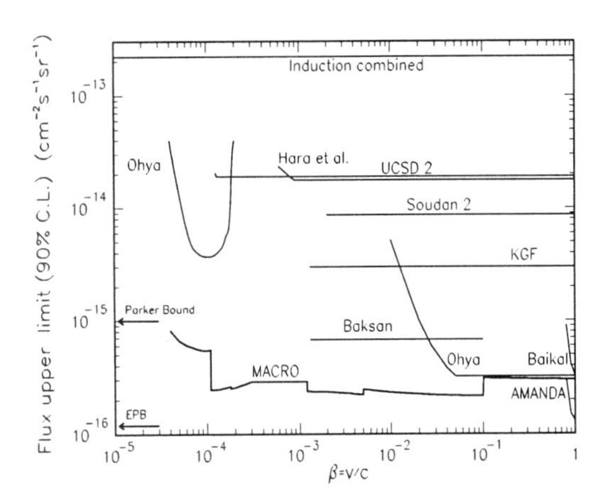

Figure 12. 90% C.L. upper limits for an isotropic flux of bare magnetic monopole (g=g_D) assuming a catalysis cross-section < 10 mb (from [91]).

"Strange Quark Matter" (SQM) should consist of aggregates of comparable amounts of u, d and s quarks; it might be the ground state of QCD [94]. If bags of SQM were produced in a first-order phase transition in the early universe, they

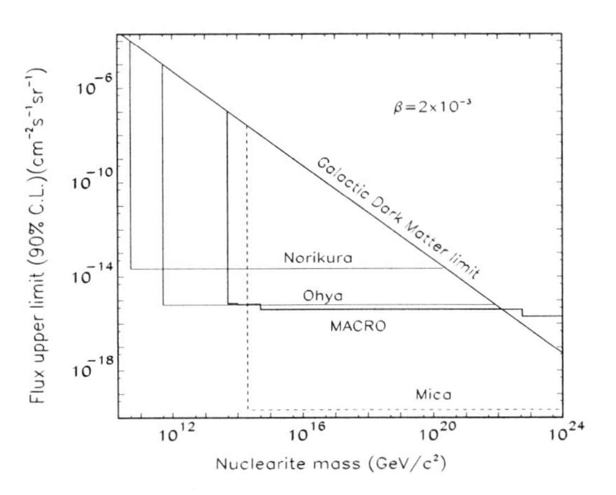

Figure 13. 90% C.L. flux upper limits vs. mass for downgoing nuclearites with velocity 10^{-3}c at ground level. The MACRO limit for nuclearite masses greater than 5.10^{22} GeV/c^2 has been extended and corresponds to an isotropic flux (from [92]).

could be candidates for the Dark Matter, and might be found in the cosmic radiation reaching the Earth. SQM in the cosmic radiation is commonly known as "nuclearites" and "strangelets" [95]. The search based on the scintillator and on the nuclear track subdetectors of MACRO were used to set limits on the flux of cr nuclearites[96]; Figure 13 shows a compilation of limits (see [96] and references therein).

5.2. Indirect searches for WIMPS

Weakly Interactive Massive Particles (WIMPs) could be part of the galactic dark matter; they could be intercepted by celestial bodies, slowed down and trapped in their centers. WIMPs and anti-WIMPs could annihilate in celestial bodies (like the Moon and Sun) and would produce neutrinos of GeV or TeV energies in small angular windows from their centers. The 90% C.L. MACRO limit for the flux from the Earth center is $0.8\,.10^{-14}$ cm^{-2} s^{-1} for a 10° cone around the vertical. From a similar search from the Sun, the limit is $1.4\,.10^{-14}$ cm^{-2} s^{-1}. These limits are presented in Figure 14 together with the values quoted by other underground experiments (from

[85])

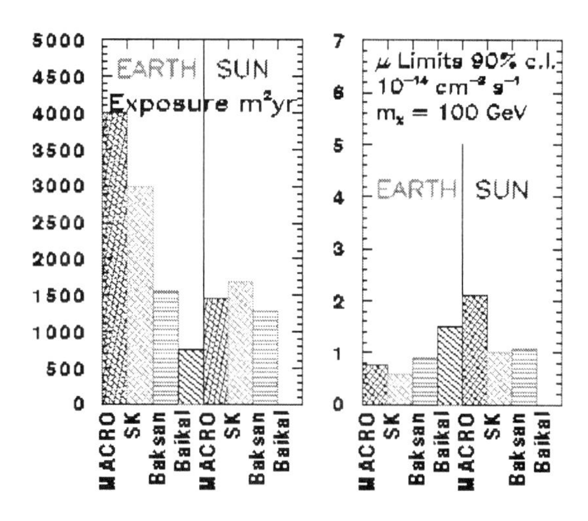

Figure 14. (left) Exposure for indirect WIMPS searches from the Earth center and in the direction of the Sun; (right) 90% C.L. upper limits obtained by different underground detectors(from [85]).

If the WIMPs are identified with the smallest mass neutralino, the MACRO limit may be used to constrain the stable neutralino mass following the SUSY models by Bottino et al. [97]. MACRO data already exclude some of the SUSY models suggested by the annual modulation analysis performed by the DAMA/NaI experiment [98].

5.3. Searches for Proton Decay and for Lightly Ionizing Particles

The quest for the stability of the nucleon, which was one of the main motivations for starting large underground detectors, has now been pursued by SK which has reached a total exposure of 45 kt-yr. No significant signal above background has been found, and the lifetime limits per each corresponding branching rate has been calculated. The limits are reported in Figure 15 together with the best values obtained by IMB and Kamiokande [99].

Despite decades of searching no one has yet observed a free quark. Also the search for leptonic type particles with fractional charges has been in vain. A clear observation of fractional charge would be extremely important since depending on the type of particle seen, it might mean that the confinement breaks down under some circumstances or that an entirely new class of particles exists. The MACRO limit [100] is compared to the LSD [101] and Kamiokande [102] limits in Figure 16 . While LSD has the best scintillator-based limit in the world prior to MACRO, Kamiokande has the lowest limit. Both these experiments claim sensitivities to 1/3e and 2/3e charged particles. To compare the results of the different experiments one should consider specific models of production of the particles, a detailed description of the material above the detectors, and the detector acceptances.

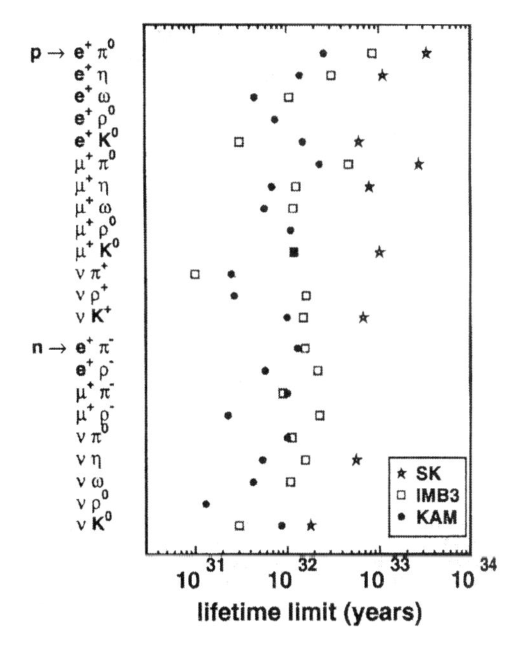

Figure 15. 90% C.L. limits for proton lifetime obtained by SK and the previous IMB and Kamiokande best limits (from [99]).

6. FUTURE DETECTORS

The main goals of the next underground detectors are: 1) the determination of the parameters of neutrino oscillations; 2) the search for high energy neutrinos from the galactic plane, AGN, accelerators of the UHECR and γ-ray bursts from

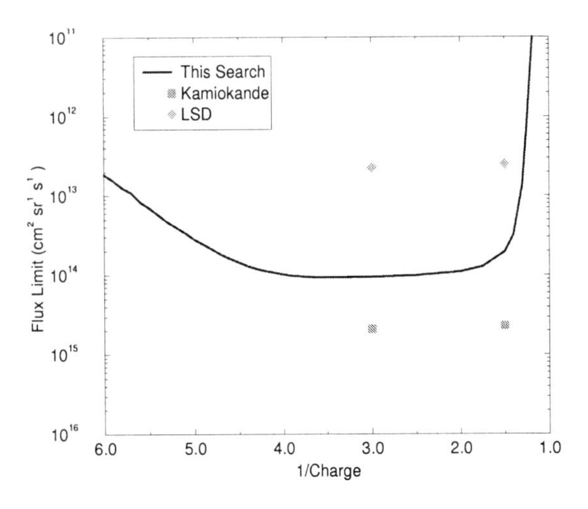

Figure 16. The upper limit on LIP fluxes at 90% C.L.reported by [100] for a continuous range of charges. Also shown are the limits from the searches done by [101,102]

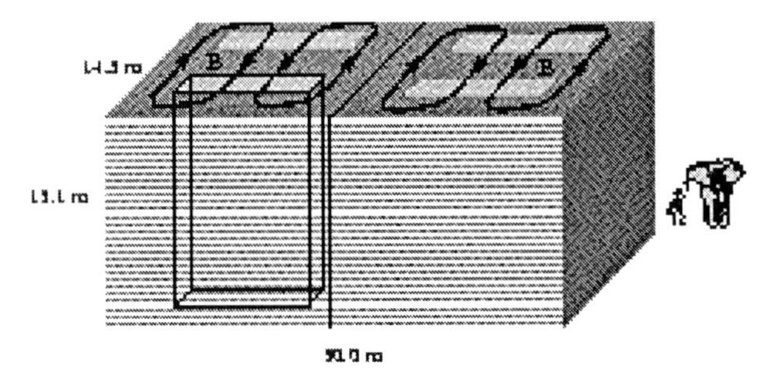

Figure 17. Schematic view of MONOLITH [103]. The embedded magnetic field is also shown.

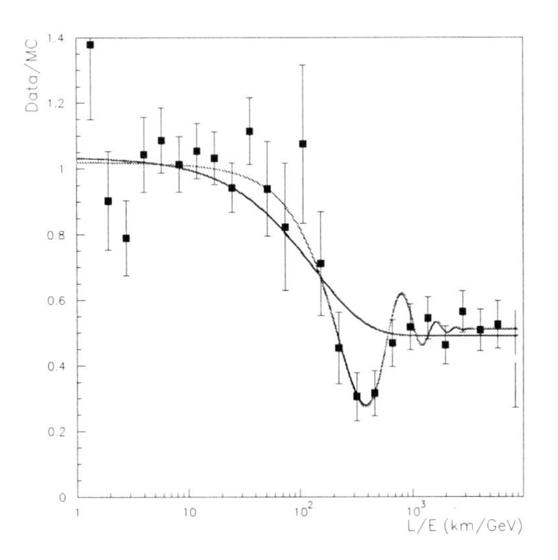

Figure 18. Expected L/E distribution from MONOLITH (point with errors) compared to the best fit of the neutrino decay model of [104](dotted line).

exotic and unexpected sources. The construction of neutrino telescopes of large sizes is under way; probably a km^3 under-water and a under-ice telescope are necessary to reach the required sensitivity. At LNGS the ICARUS 600-ton liquid-argon detector has been approved for development. It is an "electronic bubble chamber" which will provide 3-D images of the recorded events. Its main physical goals are the study of neutrino oscillations (atmospheric and on accelerator beams), ^{8}B neutrinos of the solar spectrum and proton decay in exotic channels. MONOLITH, a magnetised tracking calorimeter (see Figure 17), dedicated to the measurement of atmospheric neutrinos has also been proposed recently [103] for the LNGS. It is designed for the reconstruction of the oscillation pattern with a better L/E resolution with respect to SK (see Figure 18) allowing to distinguish the oscillation hypothesis from non-oscillating new physics (e.g [104]). It will be also a refined spectrometer for measurements of muons of hundreds of TeV.

7. CONCLUSIONS

As we have discussed above, underground detectors increased in area/volume by roughly an order/two orders of magnitude every 10-15 years. Now it is not possible to enlarge their sizes by another order of magnitude. Such mass increase is feasible only with underwater Cerenkov detectors and the proposed scale is 1 km^3. As far as cosmic ray physics studies are concerned the interpretation of the results obtained by many experiments nowadays and new kind of analysis are more urgent than the request for further experimental data. For neutrino physics experiments the proposed refined detectors like MONOLITH and ICARUS with larger sensitivity, larger masses and granularity are still possible and will hopefully give an answer on the nature of the neutrino. Long baseline neutrino experiments may definitively establish the existence of neutrino oscillations. For neutrino astrophysics we have to wait for the next supernova while for high energy neutrino sources only new underwater and underice type detectors will probably be able to uncover some of the misteries of our Universe.

8. ACKNOWLEDGMENTS

I am greately indebted to all the collegues of the MACRO Collaboration, Dr. P. Antonioli, A. Castellina, A. Trinchero for their results and for very useful discussions. I thank all the members of the Organizing Committe for giving the opportunity to present this talk and for their warm hospitality during my stay in Campinas.

REFERENCES

1. Rossi B., Rev. Mod. Phys. 20 (1948) 537.
2. Reines F., Ann. Rev. Nucl. Sci. 10 (1960) 1.
3. Greisen K., Ann. Rev. Nucl. Sci. 10 (1960) 63.
4. Davis R. Jr. et al., Phys. Rev. Lett. 20 (1968) 1205.
5. Hirata K. et al., Phys. Rev. Lett. 58 (1987) 1490 .
6. Bionta R.M. et al., Phys. rev. Lett. 58 (1987) 1494.
7. Alekseev E.N. et al., Phys. Lett. B205 (1988) 209.
8. Aglietta M. et al., Europhys. Lett. 3 (1988) 1315.
9. Y. Suzuki for the Super-Kamiokande Collaboration, Nucl. Phys. B (Proc Suppl.) 77 (1999) 35.
10. Cleveland B.T. et al., Ap. J. 496 (1998) 505.
11. T.A. Kirsten for the GALLEX and GNO Collaborations, Nucl. Phys. B (Proc Suppl.) 77 (1999) 26.
12. Abdurashitov J.N. et al. (SAGE Collaboration) Phys. Rev. Lett. 83 (1999) 4686.
13. Hirata K.S. et al. (Kamiokande Collaboration) Phys. Lett B205 (1988) 416.
14. Casper D. et al. (IMB Collaboration) Phys. Rev. Lett. 66 (1991) 2561.
15. Sobel H. for the Super-Kamiokande Collaboration, Talk at Neutrino2000 Conf.
16. Mann W.A. for the Soudan2 Colaboration, Talk at Neutrino2000 Conf.
17. M. Ambrosio et al. (MACRO Collaboration) Phys.Lett. B434 (1998) 451.
18. Barrett P.H. et al., Rev. Mod. Phys. 24 (1952) 133.
19. George E.P., Prog. Elem. Part. and C.R. Phys. 1 (1952) 393.
20. Miyazaki Y., Phys. Rev. 76 (1949) 1733.
21. Bollinger L.M., Ph.D. Thesis, Cornell University (1951).
22. Barton J.C., Phil. Mag. 6 (1961) 1271.
23. Miyake S. et al., Il Nuovo Cim. 32 (1963) 1505.
24. Miyake S. et al., Il Nuovo Cim. 32 (1963) 1524.
25. Achar C.V.K. et al., Proc. Phys. Soc. 86 (1965) 1305.
26. Castagnoli C. et al., Nuovo Cim. 35 (1965) 969.
27. Damy De Souza Santos M. et al., Phys. Rev. 59 (1941) 902.
28. Amaldi E. et al., Nuovo Cim. 9 (1952) 969.
29. Davis R. Jr., Phys. Rev. Lett. 12 (1964) 200.
30. Forti C. et al., Phys. Rev. D42 (1990) 3668.
31. Fletcher R.S. et al., Phys. Rev. D50 (1994) 5710.
32. Battistoni et al., Astropart. Phys. 3 (1995) 157.
33. Kalmykov N.N. and Ostapchenko S.S., Yad. Fiz. 56 (1993) 105.
34. Ambrosio M. et al. (MACRO Collanoration), Phys. Rev. D52 (1995) 3793.
35. Bugaev E.V. et al., Phys. Rev. D58 (1998) 05401.
36. Cecchini S. and Sioli M., Proc. 5th School on Non-accelerator Particle Astrophysics,(R.A. Carrigan Jr., G. Giacomelli and N. Paver

eds.), EUT (Trieste, 1999).
37. Gaisser T.K., Cosmic Ray and Particle Physics, Cambridge Univ. Press. (Cambridge, 1990).
38. Aglietta M. et al. (LVD Collaboration) Phys. Rev. D60 (1999) 112001.
39. Bologna G. et al. (NUSEX Coll.) Nuovo Cim. 8C (1985) 76.
40. Kasahara S.M. et al., Phys. Rev. D55 (1997) 5282.
41. Bakatanov V.N. et al., Astropart. Phys. 8 (1997) 59.
42. Krenov B.A., Nucl Phys. (Proc. Suppl)33A,B (1998)18.
43. Adarkar H.R. et al., Phys. Rev. D58 (1998) 2635.
44. Ambrosio M. et al. (MACRO Collaboration) Phys. Rev. D56 (1997) 1418.
45. Ambrosio M. et al. (MACRO Collaboration) Nucl. Phys. B (Proc. Suppl.) 75A (1999) 265.
46. Linsley J., Proc. 15th Int. Cosmic Ray Conf. (Plovdiv, 1997) 12, 89.
47. Swordy S.P. and Kieda D.B., Astrop. Phys. 13 (2000) 137.
48. Fowler J.W. et al., astro-ph/0003190.
49. Haungs A. et al., astro-ph/0002025.
50. Aglietta M. et al., Proc. 25th Int. Cosmic Ray Conf. (Durban, 1997) 4, 13.
51. Watson A., Proc. 25th Int. Cosmic Ray Conf. (Durban, 1997) 8, 257.
52. Castellina A., this Conference.
53. Aglietta M. et al. (EASTOP-LVD Collaboration) Astropart. Phys. 9 (1998) 185.
54. EASTOP-MACRO Collaboration, Proc. 25th Int. Cosmic Ray Conf. (Durban, 1997) 4, 41.
55. Das Gupta U. et al., Phys. Rev. D45 (1992) 1459.
56. Longley N.P. et al., Phys. Rev. D52 (1995) 2760.
57. Ambrosio M. et al. (MACRO Collaboration) Phys. Rev. D60 (1999) 032001.
58. Sioli M., PH.D. Thesis, Univ. of Bologna (1999), unpublished.
59. Samorski M. and Stamm W, Ap. J. 268 (1983) l17.
60. Battistoni G. et al., Phys. Lett. B 155(1985) 465.
61. Marshak M.L. (Soudan Collaboration) Phys. Rev. Lett 54 (1985) 2079.
62. Oyama et al., Phys. Rev. Lett. 56 (1986) 991.
63. Aglietta M. et al., Astropart. Phys. 3 (1995) 1.
64. Amenomori M. et al. (Tibet ASγ Collaboration) Phys. Rev. D47 (1993) 2675.
65. Borione A. et al., Phys. Rev. D49 (1994) 1171.
66. Merck M. et al., Astropart. Phys. 5 (1996) 379.
67. Ambrosio M. et al. (MACRO Collaboration) 26th Int. Cosmic Ray Conf. (Salt Lake City, 1999) 7,
68. Aglietta M. et al. (LVD Collaboration) 26th Int. Cosmic Ray Conf. (Salt Lake City, 1999) 7, 222.
69. Cobb J.H. et al., Phys Rev D61 (2000) 092002.
70. Kajita T. for the SuperKamiokande Collaboration, Nucl. Phys. B (Proc. Suppl.) 77 (1999) 123.
71. Casper D. et al., Phys. Rev. Lett. 66 (1991) 2561; Becker-Szendy R. et al., Phys. Rev. D46 (1992) 3720.
72. Hirata K.S. et al., Phys. Lett. B205 (1988); Phys. Lett. B280 (1992) 146.
73. Aglietta M. et al., Europhys. Lett. 8 (1989) 611.
74. Berger Ch. et al., Phys. Lett. B227 (1989) 489; Phys. Lett. B245 (1990) 205.
75. Ronga F. the MACRO Collaboration, Nucl. Phys. B (Proc. Suppl.) 77 (1999) 117.
76. Peterson E. for the Soudan2 Collaboration, Nucl. Phys. B (Proc. Suppl.) 77 (1999) 111.
77. Ronga F., Nucl. Phys. B (Proc. Suppl.) 87 (2000) 135.
78. Battistoni G. et al., Astropart. Phys. 12 (2000) 315.
79. Tserkovnyak Y. et al, hep-ph/9907450.
80. Lipari P., hep-ph/0003013.
81. Fogli G.L. et al., hep-ph/0009269.
82. Krishnaswamy M.R et al., Proc. Asia-Pacific Phys. Conf. (Bangalore, 1986) 1,424.
83. Svoboda R. et al., Astrophys. J. 315 (1987) 420.
84. Berger Ch. et al., Z. Phys. C48 (1990) 221.
85. Montaruli T., Talk at DARK2000 Conf. (Heidelberg, 2000).
86. Balkatanov V.A. et al., Astropart. Phys. 14

(2000) 61.
87. Halzen F. for the AMANDA Collaboration, Nucl. Phys. B (Proc. Suppl.) 77 (1999).
88. Andres. E. et al., astro-ph/0009242.
89. Rhode W et al., Astropart. Phys. 4 (1994) 217.
90. Ambrosio M. et al. (MACRO Collaboration) astro-ph/0002492.
91. The LVD Collaboration, INFN/AE-99/19.
92. Turner M.S. et al., Phys. Rev. D26 (1982) 1926.
93. Ambrosio M. et al., (MACRO Collaboration), hep-ex/0009002.
94. Witten E., Phys. Rev. D30 (1984) 272.
95. De Rujula A. and Glashow S.L., Nature 312 (1984) 734.
96. Ambrosio M. et al. (MACRO Collaboration), Eur. Phys. J. C13 (2000) 453.
97. Bottino A. et al., Astropart. Phys. (1999).
98. Bernabei R. et al., Phys. Lett. B424 (1998) 195; INFN/AE-98/20 (1998).
99. Kielczewska D., AIP Conf. Proc. no. 516 (2000) 225.
100.Ambrosio M. et al., (MACRO Collaboration) Phys. Rev. D62 (2000) 052003.
101.Aglietta M. et al., Astropart. Phys. 2 (1994) 29.
102.Mori M. et al (KamiokandeII Collaboration) Phys. Rev. D43 (1991) 2843.
103.MONOLITH Proposal, LNGS P26/2000, Cern/SPSC 2000-031.
104.Barger V. et al., Phys. Rev. Lett. 82 (1999) 2640.

ELSEVIER

Nuclear Physics B (Proc. Suppl.) 97 (2001) 66–77

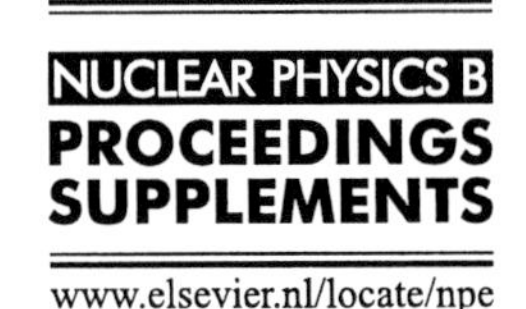

www.elsevier.nl/locate/npe

The Origin of Ultra-High Energy Cosmic Ray: New Physics and Astrophysics.

Angela V. Olinto[a]

[a]Department of Astronomy & Astrophysics,
& Enrico Fermi Institute,
The University of Chicago, Chicago, IL 60637

The lack of a high energy cutoff in the cosmic ray spectrum together with an apparently isotropic distribution of arrival directions for the highest energy events have strongly constrained most models proposed for the generation of these particles. An overview of the theoretical proposals are presented along with their most general signatures. Future experimental tests of the different proposals are discussed.

1. Introduction

The surprising detection of cosmic rays with energies above 10^{20} eV has triggered considerable interest on the origin and nature of these particles. In addition to the ultra-high energy events detected by Fly's Eye, Haverah Park, Yakutsk, and Volcano Ranch, the AGASA experiment has recently reported many hundreds of events (728) accumulated with energies above 10^{19} eV and 8 events above 10^{20} eV [1]. (For a recent review of the observations, see, e.g., [2].)

These observations are surprising because not only the energy requirements for astrophysical sources to accelerate particles to $> 10^{20}$ eV are extraordinary, but the propagation of particles at these energies is prone to large energy losses. Reactions of ultra-high energy proton, nuclei, or photon primaries with cosmic background radiation in intergalactic space suppress the observable flux at the highest energies significantly. In fact, cosmic ray protons of energies above a few 10^{19} eV reach the Δ resonance threshold and produce pions off the cosmic microwave background (CMB) limiting their source to lie not much further than about 50 Mpc away from Earth. This photopion production gives rise to the well-known Greisen-Zatsepin-Kuzmin (GZK) cutoff in the spectrum of cosmic ray photons [3]. Nuclei are photodisintegrated on shorter distances due to the infrared background [4] while the radio background constrains photons to originate from even closer systems [5].

If these UHE particles are protons, they likely originate in extragalactic sources, since at these high energies the Galactic magnetic field cannot confine protons in the Galaxy. If protons are extragalactic, they traverse large intergalactic distances so their spectrum should exhibit the GZK cutoff. The shape of the GZK cutoff depends on the source input spectrum and the distribution of sources in space as well as in the intergalactic magnetic field. In Figure 1, we contrast the observed flux by AGASA with the expected flux for proton sources distributed homogeneously or distributed like galaxies with injection spectrum $J(E) \propto E^{-\gamma}$ and $\gamma = 3$ [6]. We model the distribution of UHECR sources by using the galaxy distribution measured by the recent IRAS redshift survey know as PSCz [7]. As can be seen from the figure, even allowing for the local overdensity the observations are consistently above the theoretical expectation. In fact, when we normalize our simulations by requiring that the number of events with $E \geq 10^{19}$ eV equals the AGASA observations (728), we find that the number of expected events for $E \geq 10^{20}$ eV is only 1.2 ± 1.0 for the PSCz case, i.e., 6 σ away from the observed 8 events.

The gap between observed flux and model predictions narrows as the injection spectrum of the ultra-high energy cosmic ray (UHECR) sources becomes much harder than $\gamma = 3$. The results for

0920-5632/01/$ – see front matter
PII S0920-5632(01)01184-7

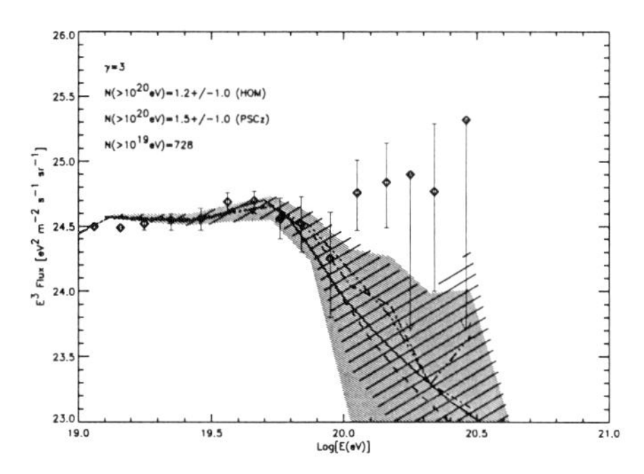

Figure 1. *Simulated fluxes for the AGASA statistics of 728 events above* 10^{19} *eV, and* $\gamma = 3$, *using a homogeneous source distribution (\ hatches) and the PSCz distribution (dense / hatches). The solid and dashed lines are the results of the analytical calculations for the same two cases. The dash-dotted and dash-dot-dot-dotted lines trace the mean simulated fluxes for the homogeneous and the PSCz cases. (see [7]).*

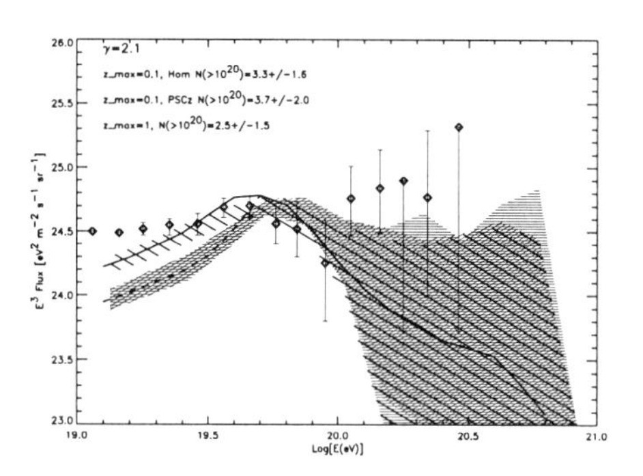

Figure 2. *Simulated fluxes for the AGASA statistics of 728 events above* 10^{19} *eV, and* $\gamma = 2.1$, *using a homogeneous source distribution with* $z_{max} = 0.1$ *(/ hatches), the PSCz distribution with* $z_{max} = 0.1$ *(horizontal hatches), and a homogeneous source distribution with* $z_{max} = 1$ *(\ hatches).*

$\gamma = 2.1$ are shown in Figure 2. For $\gamma = 2.1$, the number of expected events above 10^{20} eV reaches 3.3 ± 1.6 for a homogeneous distribution while for the PSCz catalog it is 3.7 ± 2.0. This trend can be seen also in Figure 3, where analytic solutions for $\gamma = 1.5, 2.1$ and 2.7 are shown.

In addition to the presence of events past the GZK cutoff, there has been no clear counterparts identified in the arrival direction of the highest energy events. If these events are protons or photons, these observations should be astronomical, i.e., their arrival directions should be the angular position of sources. At these high energies the Galactic and extragalactic magnetic fields do not affect proton orbits significantly so that even protons should point back to their sources within a few degrees. Protons at 10^{20} eV propagate mainly in straight lines as they traverse the Galaxy since their gyroradii are $\sim$ 100 kpc in μG fields which is typical in the Galactic disk. Extragalactic fields are expected to be $\ll \mu$G [8,9], and induce at most $\sim 1^o$ deviation from the source. Even if the Local Supercluster has relatively strong fields, the highest energy events may deviate at most $\sim 10^o$ [10,11]. At present, no correlations between arrival directions and plausible optical counterparts such as sources in the Galactic plane, the Local Group, or the Local Supercluster have been clearly identified. Ultra high energy cosmic ray data are consistent with an isotropic distribution of sources in sharp contrast to the anisotropic distribution of light within 50 Mpc from Earth.

The absence of a GZK cutoff and the isotropy of arrival directions are two of the many challenges that models for the origin of UHECRs face. This is an exciting open field, with many scenarios being proposed but no clear front runner. Not only the origin of these particles may be due to physics beyond the standard model of particle physics, but their existence can be used to constrain extensions of the standard model such as violations of Lorentz invariance (see, e.g., [12]).

In the next section, we discuss the issues involved in the propagation of UHECRs from source to Earth. We then summarize the astrophysical Zevatron proposals and followed by the

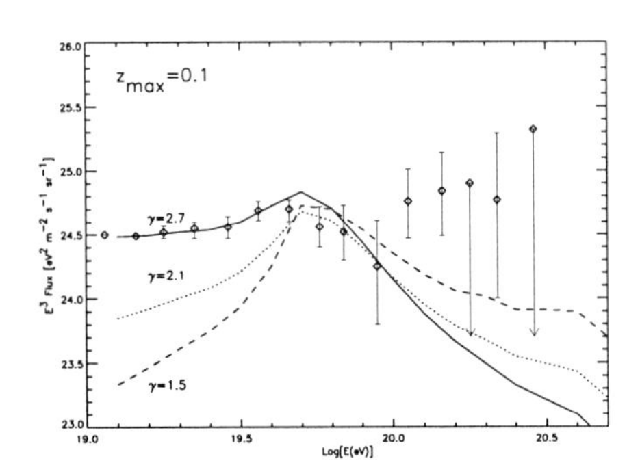

Figure 3. *Propagated spectrum for source spectral index of* $\gamma = 1.5, 2.1, 2.7$

models that involve new physics. To conclude, future observational tests of UHECR models and their implications are discussed. (For recent reviews see [13–16].)

2. Propagation - Losses and Magnetic Fields

Before contrasting plausible candidates for UHECR sources with the observed spectrum and arrival direction distribution, we discuss the effects of propagation from source to Earth. Propagation studies involve both the study of losses along the primaries' path, such as the photopion production responsible for the GZK cutoff, as well as the structure and magnitude of cosmic magnetic fields that determine the trajectories of charged primaries and influence the development of the associated electromagnetic cascade (see, e.g.,[17]).

For primary protons the main loss processes are photopion production off the CMB that gives rise to the GZK cutoff [3] pair production [18]. For straight line propagation, the cutoff should be present at $\sim 5 \times 10^{19}$ eV loss and a significant number of hard sources should be be within ~ 50 Mpc from us. As we discussed in th eprevious section, even with the small number of accumulated events at the highest energies, the AGASA

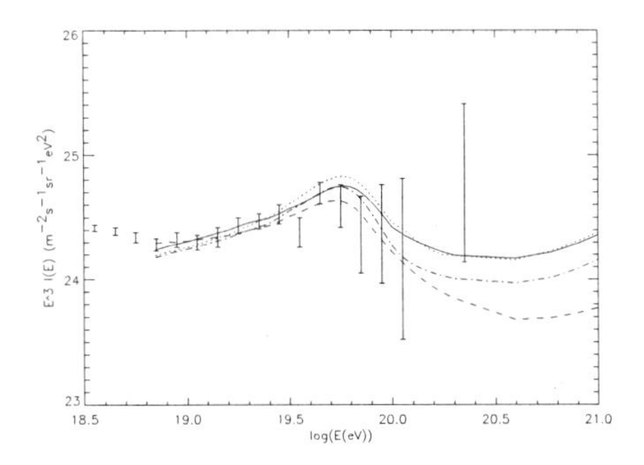

Figure 4. Flux vs. Energy with $E_{max} = 10^{21}$ eV at source. Choices of source distance r(Mpc), spectral index γ, proton luminosity L_p(erg/s), and LSC field B(μG) are: solid line (13 Mpc, 2.1, 2.2×10^{43} erg/s, 0.05 μG); dotted line (10 Mpc, 2.1, 10^{43} erg/s, 0.1 μG); dashed line (10 Mpc, 2.4, 3.2 $\times 10^{43}$ erg/s, 0.1 μG); and dashed-dotted line (17 Mpc, 2.1, 3.3×10^{43} erg/s, 0.05 μG).

spectrum is incompatible with a GZK cutoff for a homogeneous extragalactic source distribution. The shape of the cutoff can be modified if the distribution of sources is not homogeneous [6] and if the particle trajectories are not rectilinear (e.g., the case of sizeable intergalactic magnetic fields) [20–24,11].

Charged particles of energies up to 10^{20} eV can be deflected significantly in cosmic magnetic fields. In a constant magnetic field of strength $B = B_6 \mu$G, particles of energy $E = E_{20} 10^{20}$eV and charge Ze have Larmor radii of $r_L \simeq 110$ kpc $(E_{20}/B_6 Z)$. If the UHECR primaries are protons, only large scale intergalactic magnetic fields affect their propagation significantly [20–24,11] unless the Galactic halo has extended fields [21]. For higher Z, the Galactic magnetic field can strongly affect the trajectories of primaries [25,26].

Whereas Galactic magnetic fields are reasonably well studied, extragalactic fields are still very ill understood [8]. Faraday rotation measures show large magnitude fields ($\sim \mu$G) in the central regions of clusters of galaxies. In regions between clusters, the presence of magnetic fields is evidenced by synchrotron emission but the strength and structure are yet to be determined. On the largest scales, limits can be imposed by the observed isotropy of the CMB and by a statistical interpretation of Faraday rotation measures of light from distant quasars. The isotropy of the CMB can constrain the present horizon scale fields $B_{H_0^{-1}} \lesssim 3 \times 10^{-9}$ G [27]. Although the distribution of Faraday rotation measures have large non-gaussian tails, a reasonable limit can be derived using the median of the distribution in an inhomogeneous universe: for fields assumed to be constant on the present horizon scale, $B_{H_0^{-1}} \lesssim 10^{-9}$ G; for fields with 50 Mpc coherence length, $B_{50\mathrm{Mpc}} \lesssim 6 \times 10^{-9}$ G; while for 1 Mpc coherence length, $B_{Mpc} \lesssim 10^{-8}$ G [9]. These limits apply to a $\Omega_b h^2 = 0.02$ universe and use quasars up to redshift $z = 2.5$. Local structures can have fields above these upper limits as long as they are not common along random lines of site between $z = 0$ and 2.5 [10,9].

Of particular interest is the field in the local 10 Mpc volume around us. If the Local Supercluster has fields of about 10^{-8} G or larger, the propagation of ultra high energy protons becomes diffusive and the spectrum and angular distribution at the highest energies are significantly modified [28,10,24]. As shown in Figure 4 (from [24]), a source with spectral index $\gamma \gtrsim 2$ that can reach $E_{max} \gtrsim 10^{20}$ eV is constrained by the overproduction of lower energy events around 1 to 10 EeV (EeV $\equiv 10^{18}$ eV). Furthermore, the structure and magnitude of magnetic fields in the Galactic halo [21,26] or in a possible Galactic wind can also affect the observed UHECRs. In particular, if our Galaxy has a strong magnetized wind, what appears to be an isotropic distribution in arrival directions may have originated on a small region of the sky such as the Virgo cluster [29]. In the future, as sources of UHECRs are identified, large scale magnetic fields will be better constrained [30].

If cosmic rays are heavier nuclei, the attenuation length is shorter than that for protons due to photodisintegration on the infrared background [4]. However, UHE nuclei may be of Galactic origin. For large enough charge, the trajectories of UHE nuclei are significantly affected by the Galactic magnetic field [26] such that a Galactic origin can appear isotropic [25]. The magnetically induced distortion of the flux map of UHE events can give rise to some higher flux regions where caustics form and some much lower flux regions (blind spots) [26]. Such propagation effects are one of the reasons why full-sky coverage is necessary for resolving the UHECR puzzle.

The trajectories of photon primaries are not affected by magnetic fields, but energy losses due to the radio background constrains photons to originate from systems at $\lesssim 10$ Mpc [31]. If associated with luminous systems, sources of UHE photons should point back to their nearby sources. The lack of counterpart identifications suggests that if the primaries are photons, their origin involves physics beyond the standard model.

3. Astrophysical Zevatrons

The puzzle presented by the observations of cosmic rays above 10^{20} eV have generated a number of proposals that we divide here as *Astrophysical Zevatrons* and *New Physics* models. Astrophysical Zevatrons are also referred to as bottom-up models and involve searching for acceleration sites in known astrophysical objects that can reach ZeV energies. New Physics proposals can be either hybrid or pure top-down models. Hybrid models involve Zevatrons and extensions of the particle physics standard model while top-down models involve the decay of very high mass relics from the early universe and physics way beyond the standard model. Here we discuss astrophysical Zevatrons while new physics models are discussed in the next section.

Cosmic rays can be accelerated in astrophysical plasmas when large-scale macroscopic motions, such as shocks, winds, and turbulent flows, are transferred to individual particles. The maximum energy of accelerated particles, $E_{\max}$, can be estimated by requiring that the gyroradius of the particle be contained in the acceleration region: $E_{\max} = Ze\,B\,L$, where Ze is the charge of the particle, B is the strength and L the coherence length of the magnetic field embedded in the plasma. For $E_{\max} \gtrsim 10^{20}$ eV and $Z \sim 1$, the only known astrophysical sources with reasonable BL products are neutron stars ($B \sim 10^{13}$ G, $L \sim 10$ km), active galactic nuclei (AGNs) ($B \sim 10^4$ G, $L \sim 10$ AU), radio lobes of AGNs ($B \sim 0.1\mu$G, $L \sim 10$ kpc), and clusters of galaxies ($B \sim \mu$G, $L \sim 100$ kpc). In Figure 5, we highlight the B vs. L for objects that can reach $E_{max} = 10^{20}$ eV with $Z = 1$ (dashed line) and $Z = 26$ (solid line). We discuss each of these candidates below.

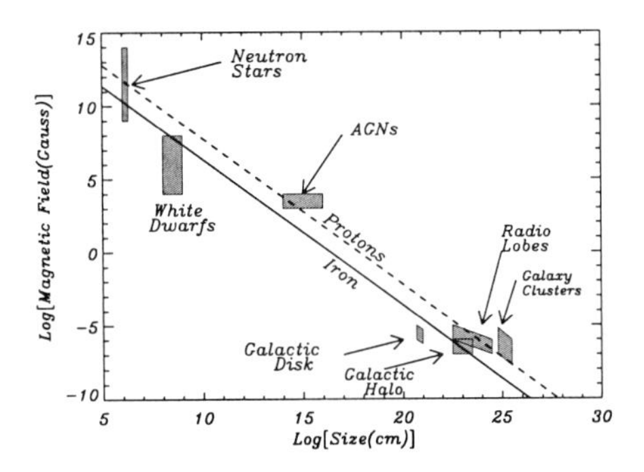

Figure 5. *B vs. L, for* $E_{max} = 10^{20}$ *eV,* $Z = 1$ *(dashed line) and* $Z = 26$ *(solid line) from [14].*

Clusters of Galaxies: Cluster shocks are reasonable sites to consider for ultra-high energy cosmic ray (UHECR) acceleration, since particles with energy up to $E_{\rm max}$ can be contained by cluster fields. However, efficient losses due to photo-pion production off the CMB during the propagation inside the cluster limit UHECRs in cluster shocks to reach at most ~ 10 EeV [32].

AGN Radio Lobes: Next on the list of plausible Zevatrons are extremely powerful radio galaxies [33]. Jets from the central black-hole of an active galaxy end at a termination shock where the interaction of the jet with the intergalactic medium forms radio lobes and 'hot spots'. Of special interest are the most powerful AGNs where shocks can accelerate particles to energies well above an EeV via the first-order Fermi mechanism. These sources may be responsible for the flux of UHECRs up to the GZK cutoff [34].

A nearby specially powerful source may be able to reach energies past the cutoff. However, extremely powerful AGNs with radio lobes and hot spots are rare and far apart. The closest known object is M87 in the Virgo cluster (~ 18 Mpc away) and could be a main source of UHECRs. Although a single nearby source with especially hard spectra may fit the spectrum for a given strength and structure of the intergalactic magnetic field [24], it is unlikely to match the observed arrival direction distribution. If M87 is the primary source of UHECRs a concentration of events in the direction of M87 or the Virgo cluster should be seen in the arrival direction distribution. No such hot spot is observed (*Hot spot* in Table 1). The next known nearby source after M87 is NGC315 which is already too far at a distance of ~ 80 Mpc. Any unknown source between M87 and NGC315 would likely contribute a second hot spot, not a isotropic distribution. The very distant radio lobes will contribute a GZK cut spectrum (*dist RLs* in Table 1).

The lack of a clear hot spot in the direction of M87 from the arrival direction distribution has encouraged the idea that a strong Galactic magnetic wind may exist that could help isotropize the arrival directions of UHECRs. A Galactic wind with a strongly magnetized azimuthal component [29] (B_{GW} in Table 1) can significantly alter the paths of UHECRs such that the observed arrival directions of events above 10^{20} eV would trace back to the North Galactic pole which is close to the Virgo where M87 resides. If our Galaxy has a such a wind is yet to be determined. The proposed wind would focus most observed events into the northern Galactic pole and render point source identification fruitless [35]. Future observations of UHECRs from the Southern Hemisphere by the Southern Auger Site will provide precious data on previously unobserved parts of the sky and help distinguish plausible proposals for the effect of local magnetic fields on arrival directions. Full sky coverage is a key discriminator of such proposals.

AGN - Central Regions: The powerful engines that give rise to the observed jets and radio lobes are located in the central regions of active galaxies and are powered by the accretion of matter onto supermassive black holes. It is reasonable to consider the central engines themselves as the likely accelerators [36]. In principle, the nuclei of generic active galaxies (not only the ones with radio lobes) can accelerate particles via a unipolar inductor not unlike the one operating in pulsars. In the case of AGNs, the magnetic field (B_{source} in Table 1) may be provided by the infalling matter and the spinning black hole horizon provides the imperfect conductor for the unipolar induction.

The problem with AGNs as UHECR sources is two-fold: first, UHE particles face debilitating losses in the acceleration region due to the intense radiation field present in AGNs, and second, the spatial distribution of objects should give rise to a GZK cutoff of the observed spectrum. In the central regions of AGNs, loss processes are expected to downgrade particle energies well below the maximum achievable energy. This limitation has led to the proposal that quasar remnants, supermassive black holes in centers of inactive galaxies, are more effective UHECR accelerators [37]. In this case, losses at the source are not as significant but the propagation from source to us should still lead to a clear GZK cutoff since sources would be associated with the large scale structure of the galaxy distribution (*LSS* in Table 1). From Figure 1–3, these models can only succeed if the source spectrum is fairly hard ($\gamma \lesssim 2$) [6].

Neutron Stars: Another astrophysical system capable of accelerating UHECRs is a neutron star. In addition to having the ability to confine $10^{20} eV$ protons (Figure 5), the rotation energy of young neutron stars is more than sufficient to match the observed UHECR fluxes [38]. However, acceleration processes inside the neutron star light cylinder are bound to fail much like the AGN central region case: ambient magnetic and radiation fields induce significant losses. However, the plasma that expands beyond the light cylinder is free from the main loss processes and may be accelerated to ultra high energies.

One possible source of UHECR past the GZK cutoff is the early evolution of neutron stars. In particular, newly formed, rapidly rotating neutron stars may accelerate iron nuclei to UHEs through relativistic MHD winds beyond their light cylinders [39]. This mechanism naturally leads to vary hard injection spectra ($\gamma \simeq 1$) (see Table 1). As seen in Figure 3, $\gamma \sim 1$ improves the agreement between predicted flux and observations for energies above 10^{20} eV. In this case, UHECRs originate mostly in the Galaxy and the arrival directions require that the primaries be heavier nuclei. Depending on the structure of Galactic magnetic fields, the trajectories of iron nuclei from Galactic neutron stars may be consistent with the observed arrival directions of the highest energy events [25]. Moreover, if cosmic rays of a few times 10^{18} eV are protons of Galactic origin, the isotropic distribution observed at these energies is indicative of the diffusive effect of the Galactic magnetic fields on iron at $\sim 10^{20}$ eV. This proposal should be constrained once the primary composition is clearly determined (see *Iron* in Table 1).

It has also been suggested that young extragalactic highly magnetized neutron stars (magnetars) may be sources of UHE protons which are accelerated by reconnection events [40]. These would be prone to a GZK cut spectrum and would need a very hard injection spectrum to become viable explanations.

Gamma-Ray Bursts: Transient high energy phenomena such as gamma-ray bursts (GRBs) may also be a source of ultra-high energies protons [41]. In addition to both phenomena having unknown origins, GRBs and UHECRs have other similarities that may argue for a common source. Like UHECRs, GRBs are distributed isotropically in the sky, and the average rate of γ-ray energy emitted by GRBs is comparable to the energy generation rate of UHECRs of energy $> 10^{19}$ eV in a redshift independent cosmological distribution of sources, both have $\approx 10^{44}$erg /Mpc^3/yr.

However, recent GRB counterpart identifications argue for a strong cosmological evolution for GRBs. The redshift dependence of GRB distribution is such that the flux of UHECR associated with nearby GRBs would be too small to fit the UHECR observations [42]. In addition, the distribution of UHECR arrival directions and arrival times argues against the GRB–UHECR common

Table 1
Zevatrons

	Composition	Source γ	Sky Distrib.	B Needs	Best Tests
Radio Lobes	Proton	2–3	M87 +dist RLs	B_{GW}	Hot spot & γ
AGN Center	Proton	2–3	LSS	B_{source}	GZK feature
YNSWs	Iron	1	Gal. Disk	B_{gal}	Iron & Disk
GRBs	Proton	2–3	Hot spot or	large B_{IGM}	Hot spot & Flux

origin. Events past the GZK cutoff require that only GRBs from $\lesssim$50 Mpc contribute. Since less than about *one* burst is expected to have occurred within this region over a period of 100 yr, the unique source would appear as a concentration of UHECR events in a small part of the sky (a *Hot spot* in Table 1). In addition, the signal would be very narrow in energy $\Delta E/E \sim 1$. Again, a strong intergalactic magnetic field can ease the some of these difficulties giving a very large dispersion in the arrival time and direction of protons produced in a single burst (*large* B_{IGM} in Table 1) [41]. Finally, if the observed small scale clustering of arrival directions is confirmed by future experiments with clusters having some lower energy events clearly precede higher energy ones, bursts would be invalidated [23].

4. New Physics Models

The UHECR puzzle has inspired a number of different models that involve physics beyond the standard model of particle physics. New Physics proposals can be top-down models or a hybrid of astrophysical Zevatrons with new particles. Top-down models involve the decay of very high mass relics that could have been formed in the early universe.

The most economical among hybrid proposals involves a familiar extension of the standard model, namely, neutrino masses. If some flavor of neutrinos have mass (e.g., $\sim 0.1eV$), the relic neutrino background is a target for extremely high energy neutrinos to interact and generate other particles by forming a Z-boson that subsequently decays [43] (see *ν Z burst* in Table 2). If the universe has very luminous sources (Zevatrons) of extremely high energy neutrinos ($\gg 10^{21}$ eV), these neutrinos would traverse very large distances before annihilating with neutrinos in the smooth cosmic neutrino background. The UHE neutrino Zevatrons can be much further than the GZK limited volume, since neutrinos do not suffer the GZK losses. But if the interaction occurs throughout a large volume, the GZK feature should also be observed. For plausible neutrino masses $\sim 0.1eV$, the neutrino background is very unclustered, so the arrival direction for events should be isotropic and small scale clustering may be a strong challenge for this proposal. The weakest link in this proposal is the nature of a Zevatron powerful enough to accelerate protons above tens of ZeVs that can produce ZeV neutrinos as secondaries. This Zevatron is quite spectacular, requiring an energy generation in excess of presently known highest energy sources (referred to as *νZevatron* in Table 2).

Another suggestion is that the UHECR primary is a new hadronic particle that is also accelerated in Zevatrons. The mass of a hypothetical hadronic primary can be limited by the shower development of the Fly's Eye highest energy event to be below $\lesssim$50 GeV [44]. As in the Z-burst proposal, a neutral particle is usually harder to accelerate and are usually created as secondaries of even higher energy charged primaries. But once formed these can traverse large distances without being affected by cosmic magnetic fields. Thus, a signature for future experiments of hybrid models that invoke new particles as primaries is a clear correlation between the position of powerful Zevatrons in the sky such as distant compact radio quasars and the arrival direction of UHE events [45]. Preliminary evidence for such a correlation has been recently reported [46].

Another exotic primary that can be accelerated to ultra high energies by astrophysical systems is the vorton. Vortons are small loops of superconducting cosmic string stabilized by the angular momentum of charge carriers [47]. Vortons can be a component of the dark matter in galactic halos and be accelerated by astrophysical magnetic fields [48]. Vortons as primaries can be constrained by the observed shower development profile.

It is possible that none of the astrophysical scenarios or the hybrid new physics models are able to explain present and future UHECR data. In that case, the alternative is to consider top-down models. Top-down models involve the decay of monopole-antimonoploe pairs [49], ordinary and superconducting cosmic strings, cosmic necklaces, vortons, and superheavy long-lived relic particles. The idea behind these models is that relics of the very early universe, topological defects (TDs) or superheavy relic (SHR) particles, produced after or at the end of inflation, can decay today and generate UHECRs. Defects, such as cosmic strings, domain walls, and magnetic monopoles, can be generated through the Kibble mechanism as symmetries are broken with the expansion and cooling of the universe. Topologically stable defects can survive to the present and decompose into their constituent fields as they collapse, annihilate, or reach critical current in the case of superconducting cosmic strings. The decay products, superheavy gauge and higgs bosons, decay into jets of hadrons, mostly pions. Pions in the jets subsequently decay into γ-rays, electrons, and neutrinos. Only a few percent of the hadrons are expected to be nucleons. Typical features of these scenarios are a predominant release of γ-rays and neutrinos and a QCD fragmentation spectrum which is considerably harder than the case of Zevatron shock acceleration.

ZeV energies are not a challenge for top-down models since symmetry breaking scales at the end of inflation typically are $\gg 10^{21}$ eV (typical X-particle masses vary between $\sim 10^{22-25}$ eV). Fitting the observed flux of UHECRs is the real challenge since the typical distances between TDs is the Horizon scale about several Gpc. The low flux hurts proposals based on ordinary and superconducting cosmic strings which are distributed throughout space (*Distant TD* in Table 2). Monopoles usually suffer the opposite problem, they would in general be too numerous. Inflation succeeds in diluting the number density of monopoles and makes them too rare for UHECR production. To reach the observed UHECR flux, monopole models usually involve some degree of fine tuning. If enough monopoles and anti-monopoles survive from the early universe, they may form a bound state, named monopolonium, that can decay generating UHECRs. The lifetime of monopolonia may be too short for this scenario to succeed unless they are connected by strings [50].

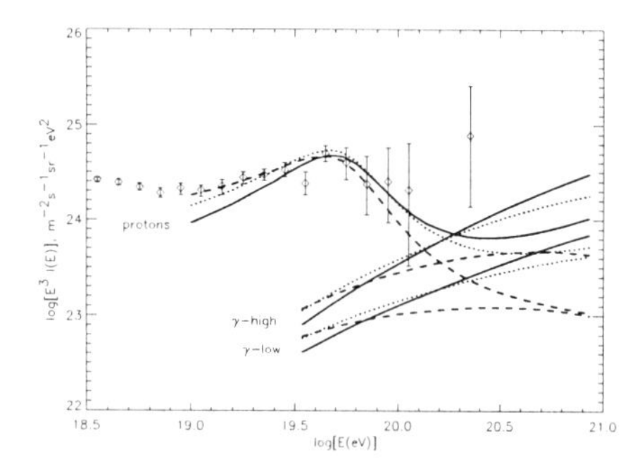

Figure 6. *Proton and γ-ray fluxes from necklaces for $m_X = 10^{14}$ GeV (dashed lines), 10^{15} GeV (dotted lines), and 10^{16} GeV (solid lines) normalized to the observed data. γ-high and γ-low correspond to two extreme cases of γ-ray absorption (see, [35]).*

Once two symmetry breaking scales are invoked, a combination of horizon scales gives room to reasonable number densities. This can be arranged for cosmic strings that end in monopoles making a monopole string network or even more clearly for cosmic necklaces [51]. Cosmic necklaces are hybrid defects where each monopole is connected to two strings resembling beads on a cosmic string necklace. Necklace networks may evolve to configurations that can fit the UHECR flux which is ultimately generated by the annihilation of monopoles with antimonopoles trapped in the string [51,52]. In these scenarios, protons dominate the flux in the lower energy side of the GZK cutoff while photons tend to dominate at higher energies depending on the radio background (see Figure 6 and *Distant TD* in Table 2). If future data can settle the composition of UHECRs from 0.01 to 1 ZeV, these models can be well constrained. In addition to fitting the UHECR flux, topological defect models are constrained by limits on the flux of high energy photons, from 10 MeV to 100 GeV, observed by EGRET.

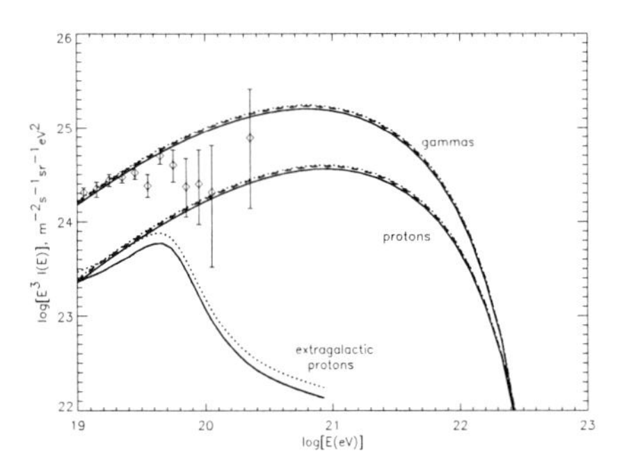

Figure 7. *SHRs or monopolia decay fluxes (for $m_X = 10^{14}$ GeV): nucleons from the halo (protons), γ-rays from the halo (gammas) and extragalactic protons. Solid, dotted and dashed curves correspond to different model parameters (see [35]).*

Another interesting possibility is the recent proposal that UHECRs are produced by the decay of unstable superheavy relics that live much longer than the age of the universe [53]. SHRs may be produced at the end of inflation by non-thermal effects such as a varying gravitational field, parametric resonances during preheating, instant preheating, or the decay of topological defects. These models need to invoke special symmetries to insure unusually long lifetimes for SHRs and that a sufficiently small percentage decays today producing UHECRs [53,54]. As in the topological defects case, the decay of these relics also generates jets of hadrons. These particles behave like cold dark matter and could constitute a fair fraction of the halo of our Galaxy. Therefore, their halo decay products would not be limited by the GZK cutoff allowing for a large flux at UHEs (see Figure 7 and *SHRs* in Table 2). Similar signatures can occur if topological defects are microscopic, such as monopolonia and vortons, and decay in the Halo of our Galaxy (*Local TD* in Table 2). In both cases (*SHRs* and *Local TD*) the composition of the primary would be a good discriminant since the decay products are usually dominated by photons.

Future experiments should be able to probe these hypotheses. For instance, in the case of SHR and monopolonium decays, the arrival direction distribution should be close to isotropic but show an asymmetry due to the position of the Earth in the Galactic Halo [52] and the clustering due to small scale dark matter inhomogeneities [55]. Studying plausible halo models for their expected asymmetry and inhomogeneitis will help constrain halo distributions especially when larger data sets are available in the future. High energy gamma ray experiments such as GLAST will also help constrain SHR models via the electromagnetic decay products [56].

5. Conclusion

Next generation experiments such as the High Resolution Fly's Eye which recently started operating, the Pierre Auger Project which is now under construction, the proposed Telescope Array, and the EUSO and OWL satellites will significantly improve the data at the extremely-high end of the cosmic ray spectrum [2]. With these observatories a clear determination of the spectrum and spatial distribution of UHECR sources is within reach.

The lack of a GZK cutoff should become clear with HiRes and Auger and most extragalactic Zevatrons may be ruled out. The observed spectrum will distinguish Zevatrons from new physics models by testing the hardness of the spectrum

Table 2
NEW PHYSICS

	Composition	Source γ	Sky Distrib.	Th. Challenge	Best Tests
ν Z burst	photons	νZevatron	Isotropic	νZevatron	photon, Isotropy
Distant TD	phot/GZK p	QCD frag	Isotropic	flux	photon, GZK p
Local TD	photons	QCD frag	Gal Halo	origin	photon, Halo
SHRs	photons	QCD frag	Gal Halo	lifetime	photon, Halo

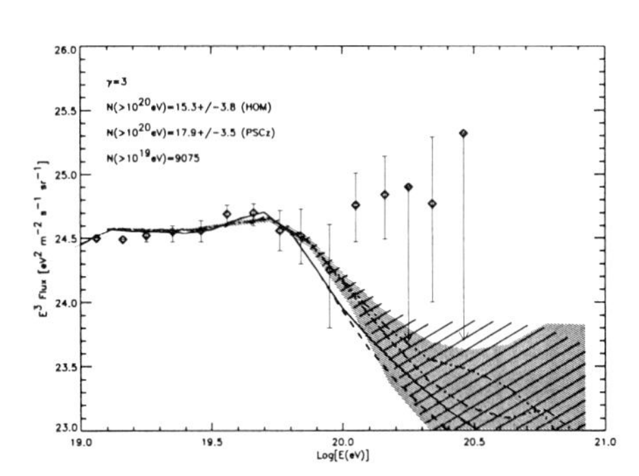

Figure 8. *Simulated fluxes for the Auger projected statistics of 9075 events above* 10^{19} *eV, and* $\gamma = 3$, *using a homogeneous source distribution (\ hatches) and the PSCz distribution (/ hatches). The solid and dashed lines are the results of the analytical calculations for the same two cases. The dash-dotted and dash-dot-dot-dotted lines trace the mean simulated fluxes for the homogeneous and the PSCz cases. (see [7]).*

and the effect of propagation. Figure 8 shows how clearly Auger will test the spectrum in spite of clustering properties. The cosmography of sources should also become clear and able to discriminate between plausible populations for UHECR sources. The correlation of arrival directions for events with energies above 10^{20} eV with some known structure such as the Galaxy, the Galactic halo, the Local Group or the Local Supercluster would be key in differentiating between different models. For instance, a correlation with the Galactic center and disk should become apparent at extremely high energies for the case of young neutron star winds, while a correlation with the large scale galaxy distribution should become clear for the case of quasar remnants. If SHRs or monopolonia are responsible for UHECR production, the arrival directions should correlate with the dark matter distribution and show the halo asymmetry. For these signatures to be tested, full sky coverage is essential. Finally, an excellent discriminator would be an unambiguous composition determination of the primaries. In general, Galactic disk models invoke iron nuclei to be consistent with the isotropic distribution, extragalactic Zevatrons tend to favor proton primaries, while photon primaries are more common for early universe relics. The hybrid detector of the Auger Project should help determine the composition by measuring the depth of shower maximum and the muon content of the same shower. The prospect of testing extremely high energy physics as well as solving the UHECR mystery awaits improved observations that should be coming in the next decade with experiments under construction such as Auger [57] or in the planning stages such as the Telescope Array [58], EUSO [59], and OWL [60].

6. Acknowledgment

I am grateful to the organizers of this excellent meeting. This work was supported by NSF through grant AST-0071235 and DOE grant DE-FG0291 ER40606.

REFERENCES

1. N. Hayashida et al., astro-ph/0008102, appendix for *Astrophys. J.* 522 (1999) 225.
2. A. A. Watson, Phys. Rept. 333-334 (2000) 309; X. Bertou, M. Boratav, and A. Letessier-

Selvon, *Int.J.Mod.Phys.* A15 (2000) 2181-2224.
3. K. Greisen, *Phys. Rev. Lett.* 16 (1966) 748; G. T. Zatsepin and V. A. Kuzmin, *Sov. Phys. JETP Lett.* 4 (1966) 78.
4. J. L. Puget, F. W. Stecker, and J. H. Bredekamp, *Astrophys. J.* 205 (1976) 638; F. W. Stecker and M. H. Salamon, *Astrophys. J.* 512 (1999) 521.
5. V. S. Berezinsky, Yad. Fiz. 11 (1970) 339; R. J. Protheroe and P. L. Biermann, *Astropart. Phys.* 6 (1996) 45; erratum ibid. 7 (1997) 181.
6. M. Blanton, P. Blasi, & A. V. Olinto, astro-ph/0009466.
7. W. Saunders et al., (astro-ph/0001117).
8. P. P. Kronberg, Rep. Prog. Phys., 57 (1994) 325; J. P. Vallée, *Fundamentals of Cosmic Physics*, 19 (1997) 1.
9. P. Blasi, S. Burles, and A. V. Olinto, *Astrophys. J.* 514 (1999) L79.
10. D. Ryu, H. Kang and P. L. Bierman, *Astron. Astrophys.* 335 (1998) 19.
11. G. Sigl, M. Lemoine, and P. Biermann, *Astropart. Phys.* 10 (1999) 141.
12. L. Gonzalez-Mestres, *Nucl. Phys. B (Proc. Suppl.)* 48 (1996) 131; S. Coleman and S. L. Glashow, *Phys. Rev. D* 59 (1999) 116008; R. Aloisio, P. Blasi, P Ghia, and A. Grillo, preprint INFN/AE-99/24.
13. A. V. Olinto, *Phys. Rept.* 333-334 (2000) 329.
14. P. Bhattacharjee and G. Sigl, *Phys. Rept.* 327 (2000) 109.
15. V. S. Berezinsky, *Nucl.Phys. (Proc.Suppl.)* 70 (1999) 41.
16. R. D. Blandford, *Particle Physics and the Universe*, eds. Bergstrom, Carlson and Fransson (World Scientific, 1999).
17. S. Lee, A. V. Olinto, and G. Sigl, Astrophys. J. 455 (1995) L21; R. J. Protheroe and P. A. Johnson, Astropart. Phys. 4 (1996) 253; erratum, ibid 5 (1996) 215.
18. G. B. Blumenthal, Phys. Rev. D 1 (1970) 1596.
19. V. Berezinsky and S. Grigorieva, Astron. Astroph. 199 (1988) 1.
20. E. Waxman and J. Miralda-Escude, Astrophys. J. 472 (1996) L89.
21. T. Stanev, Astrophys. J. 479 (1997) 290.
22. G. A. Medina Tanco, E. M. de Gouveia Dal Pino, and J. E. Horvath, Astropart. Phys. 6 (1997) 337.
23. G. Sigl, M. Lemoine, and A. V. Olinto, Phys. Rev. 56 (1997) 4470.
24. P. Blasi and A. V. Olinto, *Phys. Rev.* D **59** (1999) 023001.
25. V. N. Zirakashvili, D. N. Pochepkin, V. S. Ptuskin, and S. I. Rogovaya, Astron. Lett. 24 (1998) 139.
26. D. Harari, S. Mollerach, and E. Roulet, (1999) astro-ph/9906309
27. J. D. Barrow, P. G. Ferreira, and J. Silk, Phys. Rev. Lett. 78 (1997) 3610.
28. J. Wdowczyk and A. W. Wofendale, Nature 281 (1979) 356; M. Giler, J. Wdowczyk, and A. W. Wolfendale, J. Phys. G.: Nucl. Phys. 6 (1980) 1561; V.S. Berezinsky, S.I. Grigorieva, and V.A. Dogiel 1989, Sov. Phys. JETP 69, 453; ibid. Astronomy and Astrophysics 232 (1990) 582.
29. E. J. Ahn, G. Medina-Tanco, P. Biermann, T. Stanev, astro-ph/9911123 (1999).
30. M. Lemoine, G. Sigl, A. V. Olinto, and D. Schramm, Astropart. Phys. 486 (1997) L115.
31. F. W. Stecker, Astrophys. J. 157 (1969) 507; G. G. Fazio and F. W. Stecker, Nature 226 (1970) 135; V. S. Berezinsky, Yad. Fiz. 11 (1970) 339; R. J. Protheroe and P. L. Biermann, Astropart. Phys. 6 (1996) 45; erratum ibid. 7 (1997) 181.
32. H. Kang, D. Ryu, T.W. Jones, *Astropart. Phys.* 456 (1996) 422.
33. P.L. Biermann and P. Strittmatter, *Astropart. Phys.* **322** (1987) 643.
34. J. P. Rachen and P. L. Biermann, *Astron. Astrophys.* **272** (1993) 161.
35. P. Billoir and A Letessier-Selvon, astroph/000142 (2000).
36. K.S. Thorne, R. Price, & D. MacDonals, *Black Holes: The Membrane Paradigm* (New Haven: Yale Press) (1986).
37. E. Boldt and P. Ghosh, *Mon. Not. R. Astron. Soc.*, in press (1999).
38. A. Venkatesan, M. C. Miller, and A. V. Olinto, *Astrophys. J.* 484 (1997) 323.
39. A. V. Olinto, R. I. Epstein, and P. Blasi, *Proceedings of 26th ICRC*, Salt Lake City, **4**,

361 (1999); P. Blasi, R. I. Epstein, and A. V. Olinto, *Astrophys. J. Letters* 533 (2000) L123.
40. E.M. de Gouveia Dal Pino, A. Lazarian, *Astrophys. J.* 536 (2000) L31-L34.
41. E. Waxman, *Phys. Rev. Lett.* 75 (1995) 386; M. Vietri, *Ap. J.* 453 (1995) 883.
42. F. W. Stecker, *Astropart. Phys.* 14 (2000) 207.
43. D. Fargion, B. Mele, and A. Salis, Astrophys.J. 517 (1999) 725-733; T. Weiler, *Astropar. Phys.* **11** (1999) 303.
44. I. F. Albuquerque, G. Farrar, and E. Kolb, *Phys. Rev.* D **59** (1999) 015021.
45. G. R. Farrar and P. L. Biermann, *Phys. Rev. Lett.* **81** (1998) 3579.
46. A. Virmani, et al., astro-ph/0010235
47. R. L. Davis and E. P. S. Shellard, *Nucl. Phys.* B 323 (1989) 209.
48. S. Bonazzola and P. Peter, *Astropart. Phys.* **7**, 161 (1997).
49. C. T. Hill, *Nucl. Phys.* B **224** (1983) 469; D. N. Schramm and C.T. Hill, Proc. 18th ICRC (Bangalore) **2** (1983) 393.
50. J. J. Blanco-Pillado and K. D. Olum, Phys.Rev. D60 (1999) 083001.
51. V. Berezinsky and A. Vilenkin, *Phys. Rev. Lett.* **79** (1997) 5202.
52. V. Berezinsky, P. Blasi, and A. Vilenkin, *Phys. Rev.* D **58** (1998) 103515-1.
53. V. Berezinsky, M. Kachelrieß and A. Vilenkin, *Phys. Rev. Lett.* **79** (1997) 4302; V. Kuzmin and V. Rubakov, Yad. Fisika **61** (1998) 1122.
54. D. J. H. Chung, E. W. Kolb, and A. Riotto, *Phys. Rev. Lett.* **81** (1998) 4048; V. Kuzmin and I. Tkachev, *Phys. Rev.* D **59** (1999) 123006.
55. P. Blasi & R. K. Seth, *Phys. Lett.* B 486 (2000) 233-238.
56. P. Blasi, *Phys. Rev.* D **60** (1999) 023514.
57. J. W. Cronin, *Nucl. Phys. B. (Proc. Suppl.)* 28 (1992) 213.
58. M. Teshima et al., *Nucl. Phys. B (Proc. Suppl.)* 28B (1992) 169.
59. L. Scarsi, in the proceedings of the International Workshop on Observing Ultra High Energy Cosmic Rays From Space and Earth, Metepec, Puebla, Mexico (2000).
60. R. E. Streitmatter, Proc. of *Workshop on Observing Giant Cosmic Ray Air Showers from* $> 10^{20}$ *eV Particles from Space*, eds. J. F. Krizmanic, J. F. Ormes, and R. E. Streitmatter (AIP Conference Proceedings 433, 1997).

Contributed Papers

ELSEVIER

Nuclear Physics B (Proc. Suppl.) 97 (2001) 81–84

www.elsevier.nl/locate/npe

Self-organized criticality in atmospheric cascades

M.Rybczyński, Z.Włodarczyk [a] and G. Wilk[b]

[a]Institute of Physics, Pedagogical University, Kielce, Poland
emails: mryb@pu.kielce.pl and wlod@pu.kielce.pl

[b]The Andrzej Soltan Institute for Nuclear Studies, Nuclear Theory Department, Warsaw, Poland
email: wilk@fuw.edu.pl

We argue that atmospheric cascades can be regarded as an example of the self-organized criticality and studied by using Lévy flights and a nonextensive approach. It allows us to understand the scale-invariant energy fluctuations inside cascades in a natural way.

1. INTRODUCTION

It is well known that energy spectra of particles from atmospheric family events which are observed by the emulsion chambers at mountain altitudes are essentially following power-like dependence. On the other hand the occurrence of power-law distributions is a very common feature in nature and it is usually connected with such notions as criticality, fractals and chaotic dynamics and studied using generalized (nonextensive) maximum entropy formalism characterized by a nonextensivity parameter q. Such formalism leads in a natural way to power-laws in the frame of equilibrium processes [1]. For nonequilibrium phenomena the sources of power-law distributions are self-organized criticality [2] and stochastic multiplicative processes [3]. In the former, nonequilibrium systems are continuously driven by their own internal dynamic to a critical state with power-laws omnipresent. In the later, power-law is generated by the presence of underlying replication of events. All these approaches can be unified in terms of generalized nonextensive statistics mentioned above (widely known as Tsallis statistics) [1].

We shall look therefore at the development of cascades from this point of view. To be more specific, we shall use the notion of Lévy flights as representatives of power-laws emerging from generalized statistics [1].

2. MONTE-CARLO SIMULATION OF MULTIPLICATIVE PROCESSES

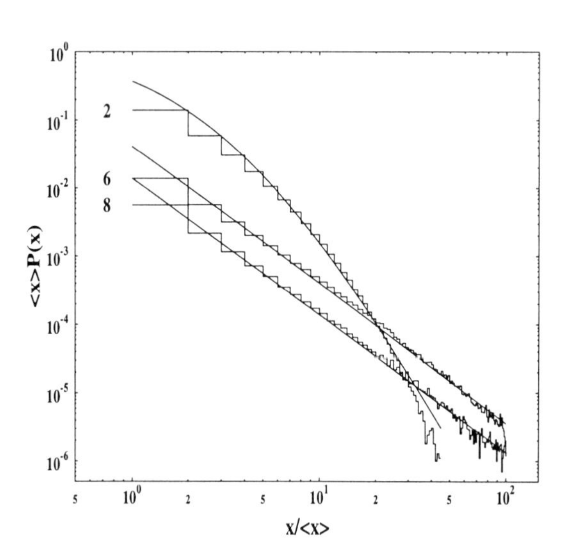

Fig.1 Distributions $P(x)$ for multiplicative process, as given by eq. (1) (histograms), for different generation numbers $N = 2, 6$ and 8 are fitted by Lévy distribution (eq. (3)) (solid lines) with parameters $\alpha = 4.81, 2.05$ and 2.015, respectively.

Let us start with a numerical (Monte-Carlo) consideration of the following model of multiplicative process:

$$x_{N+1} = \xi x_N, \qquad x_0 = 1, \tag{1}$$

0920-5632/01/$ – see front matter
PII S0920-5632(01)01185-9

where ξ is random number taken from the exponential distribution (to account for the essentially exponential scaled energy dependence of the single elementary interaction):

$$f(\xi) = \frac{1}{\xi_0} \exp\left(-\frac{\xi}{\xi_0}\right). \tag{2}$$

As was shown in [4], every exponential distribution with a fluctuating parameter results in the Lévy distribution (cf. Fig. 1), which in our case has the following form

$$P(x_N) = \frac{1}{\langle x_N \rangle} \cdot \frac{\alpha - 1}{\alpha - 2} \cdot \left[1 + \frac{1}{\alpha - 2} \frac{x_N}{\langle x_N \rangle}\right]^{-\alpha}. \tag{3}$$

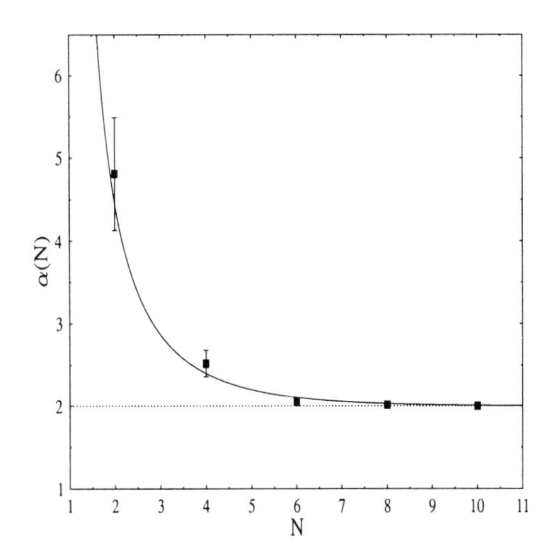

Fig.2 Dependence of the parameter α of the Lévy distribution (3) on the generation number N. Solid line represents fit by the formula (4) with $c = 0.55$. Asymptotically $\alpha = 2$ (dotted line).

In our present case, the parameter α depends on the number of generations N and on the parameter ξ_0 which defines the exponential distribution of the variable ξ (in the limit of large N, the dependence on ξ_0 effectively vanishes). As is illustrated in Fig. 2, the dependence $\alpha(N)$ can be described by a simple formula

$$\alpha = \alpha(N) = \frac{2}{1 - c^{N-1}} \tag{4}$$

where c is the generation independent parameter which should be fitted (in our case $c = 0.55$).

If we use the canonical form of the Lévy distribution as emerging from Tsallis statistics [1],

$$P(x) = \frac{2 - q}{\chi} \left[1 - (1 - q)\frac{x}{\chi}\right]^{\frac{1}{1-q}}, \tag{5}$$

where mean $\langle x \rangle$ is given by the parameter $\chi = (3 - 2q)\langle x \rangle$ and q is a nonextensive parameter mentioned before such that $q = 1 + 1/\alpha$ [4], then

$$q = \frac{3}{2} - \frac{c^{N-1}}{2}. \tag{6}$$

Notice that for $N \to 1$, the parameter $q \to 1$ (or, respectively, $\alpha \to \infty$), i.e., we are recovering the initial exponential distribution. In the limit $N \to \infty$ the parameter q approaches the value $q = 3/2$, which is a limiting value available for q emerging from the normalization condition imposed on the probability distribution $P(x_N)$. In this limit, $\alpha \to 2$. It is interesting to note that for $[x_N/\langle x_N \rangle] \cdot [1/(\alpha - 2)] >> 1$, i.e., for x sufficiently large, one gets a power-like behavior of $P(x_N)$:

$$P(x_N) \propto \left(\frac{x_N}{\langle x_N \rangle}\right)^{-\alpha}. \tag{7}$$

Actually such situation is reached reasonably fast, because $\langle x_N \rangle = \xi_0^N$ and α tends to its limiting value $\alpha = 2$ rather quickly with increasing number of generations N. In practice the equilibrium distribution $P(x_N) \propto x_N^{-2}$ is reached (for $x_N >> \langle x_N \rangle$) already for $N > 6$.

3. COMPARISON WITH EXPERIMENTAL DATA

There exists a number of experimental data from emulsion chambers exposed at mountain altitudes, which are relevant for our approach and which we shall compare with our results [5,6]. Although, as we have already mentioned, the energy spectra of particles from atmospheric family events they represent are roughly expressed by power-law distributions, so far there was no analysis showing whether it is true and to what extend. Our work is the first attempt in this direction as we show that they all can be described by the Lévy type spectra.

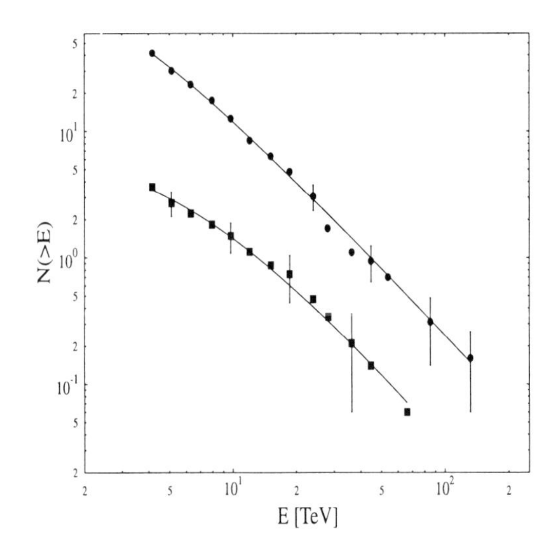

Fig.3 Integral energy spectra for gamma quanta (circles) and hadrons (squares) in families registered at Mt. Kambala are fitted by Lévy distributions (solid lines).

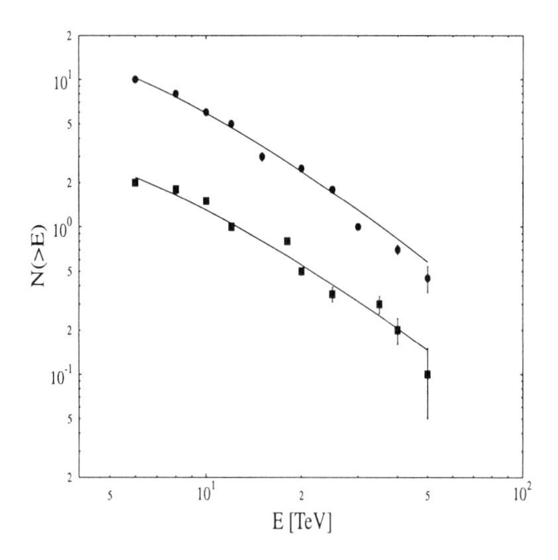

Fig.4 Integral energy spectra for gamma quanta (circles) and hadrons (squares) in families registered at Mt. Pamir are fitted by Lévy distributions (solid lines).

Integral spectra $N(>E)$ from Kambala experiment (at 520 g/cm^2) [5] are shown in Fig. 3 where they are fitted using parameters $\alpha - 1 = 1.8$ (for gamma families) and $\alpha - 1 = 2.0$ (for hadronic families). Notice that hadronic component is little "younger" (higher α means smaller N, cf. Fig. 2) than the electromagnetic component.

Families observed deeper in the atmosphere (what means higher N) have smaller α, as expected, cf. Fig. 4, where $N(>E)$ from Pamir experiment (at 600 g/cm^2) [6] are presented. Here the corresponding parameters $\alpha - 1$ are equal to 1.8 and 1.7 for gamma and hadronic components, respectively.

It is evident from Figs. 3 and 4 that the observed gamma-hadron families are already reaching their quasi-equilibrium states. It should be noticed that most of the families registered come not from a single nuclear interaction but they usually contain particles which are decascadents of several (~ 7 at $\Sigma E > 100$ TeV) genetically connected nuclear interactions [7]. Our analysis indicates that the average number of cascade generations leading to the observed distributions is about $N \simeq 3$.

4. SUMMARY AND DISCUSSION

We have provided a description of cascade processes encountered in cosmic ray emulsion experiments at high altitudes, which is based on the notion of Lévy distributions (which, in turn, originate from the nonextensive statistics described by parameter q [1,4]). In this way, we have found that the observed distributions are to a high accuracy power-like, but at the same time we are able to account for the small deviations from the exact power-like behavior. They are, in our approach, directly connected with the finite number N of the cascade generations in the way provided by formulas (4) or (6).

The results presented before should be confronted with the result of pure nuclear origin, which is the case of the so called "halo" events with strong concentration of particles in the central region. Fig. 5 shows such a case exposing the integral energy distribution of shower cores (recorded at distance $R < 0.65$ cm from the energy-weighted center) obtained in a "halo" event P06 registered at Chacaltaya [8]. In this case we observe almost exponential distribution ($\alpha - 1 = 16.0$ in this case, what translates to $q = 1.06$) with no influence of multiplicative processes.

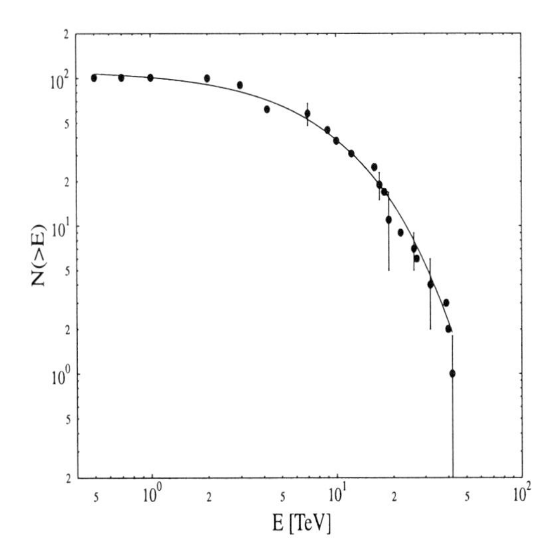

Fig.5 Integral energy spectra for shower cores from halo event (circles) registered at Chacaltaya is fitted by Lévy distribution (solid lines).

Finally, it should be noticed that there exists a number of special cases more-or-less directly connected with our results:
(a) Energy transport equation [9] for cascade processes gives

$$N(E,t) = N_0 E^{-(s+1)} \exp\left(-\frac{s-1}{s+1} \cdot \frac{t}{t_0}\right) \quad (8)$$

as number of particles with energy E at depth t. Notice that for the age parameter $s \to 1$ one approaches equilibrium spectrum $N(E) = N_0 E^{-2}$ with the slope $\alpha = 2$.
(b) One observes strong intermittent behavior in the families [10], which results from the fluctuations in the atmospheric cascades alone and is not sensitive to details of the elementary interactions.
(c) One observes multifractal behavior in the cascades with fractal dimension D_N at each cascade stage N where $P(x) \propto x^{D_N}$ [11]. For each successive emission the distribution should be more inhomogeneous, i.e., $D_N > D_{N+1}$.
(d) Self-organized criticality [2] to which all cascades considered here bear close resemblance because of the Zipf's law [12], which predicts power-law behavior of the type seen in eq.(7) (with $\alpha = 2$).

Acknowledgments: The partial support of Polish Committee for Scientific Research (grants 2P03B 011 18 and 621/E-78/SPUB/CERN/P-03/DZ4/99) is acknowledged.

REFERENCES

1. See, for example, G.Wilk and Z.Włodarczyk, *Nucl. Phys.* **B** (Proc. Suppl.) **A75** (1999) 191 and references therein. Cf. *Braz. J. Phys.* **29** No.1 (1999) (also at the URL: http://sbf.if.usp.br/ WWW_pages/Journals/ BJP/Vol29/Num1/ index.htm) for review.
2. P.Bak, C.Tang and K.Wiesenfeld, *Phys. Rev. Lett.* **59** (1987) 381; M.Paczuski, S.Maslow and B.Pak, *Phys. Rev.* **E53** (1996) 414.
3. S.C.Manrubia and D.H.Zanette, *Phys. Rev.* **E59** (1999) 4945.
4. G.Wilk and Z.Włodarczyk, *Phys. Rev. Lett.* **84** (2000) 2770.
5. J.R.Ren et al., (Kambala Coll.), *Proc.* 19th *Int. Cosmic Ray Conf.*, La Jolla, **6** (1985) 429.
6. A.S.Borisov et al. (Pamir Coll.), *Bull. Soc. Sci. Lett. de Łódź*, **XLII-5** (1992) 71.
7. J.Nowicka, A.Tomaszewski and Z.Włodarczyk, *J. Phys.* **G11** (1985) 1365.
8. N.M.Amato, N.Arata and R.H.C.Maldonato, Proc. Int. Symp. Cosmic Ray Superhigh Energy Int., Beijing (1986) 4-37.
9. S.Hayakawa, *Cosmic Ray Physics*, John Willey & Sons, New York, 1969.
10. G.Wilk and Z.Włodarczyk, *J. Phys.* **G19** (1993) 761.
11. I.Sarcevic and H.Satz, *Phys. Lett.* **B233** (1989) 252; cf. also: S.V.Chekanov, *Eur. Phys. J.* **C6** (1999) 331.
12. Cf., for example, R.V.Sole, S.C.Manrubia, M.J.Benton and P.Bak, *Nature* **388** (1997) 764 and references therein.

ELSEVIER

Nuclear Physics B (Proc. Suppl.) 97 (2001) 85–88

www.elsevier.nl/locate/npe

Muons from strangelets

M.Rybczyński, Z.Włodarczyk [a] and G. Wilk[b]

[a]Institute of Physics, Pedagogical University, Kielce, Poland
emails: mryb@pu.kielce.pl and wlod@pu.kielce.pl

[b]The Andrzej Soltan Institute for Nuclear Studies, Nuclear Theory Department, Warsaw, Poland
email: wilk@fuw.edu.pl

The hypothesis is discussed that muon bundles of extremely high multiplicity observed recently by ALEPH detector (in the dedicated cosmic-ray run) can originate from the strangelets colliding with the atmosphere.

1. INTRODUCTION

In the astrophysical literature [1] one can find a number of phenomena which can be regarded as a possible manifestation of the existence of the so called *Strange Quark Matter* (SQM) (in the form of lumps called strangelets), extremely interesting possibility of a possible new stable form of matter. They include, among others, anomalous cosmic ray burst from *Cygnus X-3*, extraordinary high luminosity gamma-ray bursts from the *supernova remnant N49* in the Large Magellanic Cloud or *Centauro* type events. There are also several reports suggesting direct candidates for the SQM. In particular, anomalous massive particles, which can be interpreted as strangelets, have been apparently observed in cosmic ray experiments [1]. All this makes a search for other possible candidates or signals for SQM extremely interesting topic.

Proceeding along this line we would like to bring ones attention to the recent (still unpublished, however) data from the cosmic ray run of the ALEPH detector at CERN-LEP experiment. The hypothesis which we shall discuss in what follows is that, if confirmed, the muon bundles of extremely high multiplicity observed recently by ALEPH in its dedicated cosmic-ray run [2] can originate from the strangelets propagating through the atmosphere and interacting with the air nuclei.

2. MUON BUNDLES FROM CosmoLEP

Why the CosmoLEP data are potentially so important? The reason is twofold. First, the studies of high multiplicity cosmic muon events (called muon bundles) is potentially a very important source of information about the composition of primary cosmic rays. It is because muons transport (in essentially undisturbed way) significant information on the first interaction of the cosmic ray particle with atmosphere. In comparison electromagnetic cascades are more calorimetric in nature and less sensitive to any model uncertainties, which could be important for establishing the primary spectrum. The second point has to do with the fact that multi-muon bundles have never been studied with such precise detectors as provided by LEP program at CERN, nor have they been studied at such depth as at CERN [3]. The underground location of the LEP detectors (between 30 and 140 meters) is ideal for the muon based experiments because the corresponding muon momentum cut-off is then between 15 and 70 GeV, i.e., in the most sensitive range from the point of view of the primary interaction, where interaction and decay probabilities are equal at the starting point of the cascade.

The present situation is the following. Data archives from the ALEPH runs have revealed a substantial collection of cosmic ray muon events. More than $3.7 \cdot 10^5$ muon events have been recorded in the effective run time 10^6 seconds. Multi-muon events observed in the 16 m^2

0920-5632/01/$ – see front matter
PII S0920-5632(01)01186-0

time-projection chamber with momentum cut-off $70GeV/c$ have been analysed and good agreement with the Monte Carlo simulations (performed using *CORSIKA* code [4]) is obtained for multiplicities N_μ between 2 and 40. However, there are 5 events with unexpectedly large multiplicities N_μ (up to 150) whose rate cannot be explained, even assuming pure iron primaries. They will be our central point of interest here.

3. SOME FEATURES OF STRANGELETS

For completeness we shall summarize now features of strangelets and their propagation through the atmosphere, which will be relevant to our further discussion. More detailed information can be found in [5]. Typical SQM consists of roughly equal number of up (u), down (d) and strange (s) quarks and it has been argued to be the true ground state of QCD [6,7]. For example, it is absolutely stable at high mass number A (excluding weak interaction decays of strange quarks, of course) and it would be more stable than the most tightly bound nucleus as iron (because the energy per barion in SQM could be smaller than that in ordinary nuclear matter). On the other hand it becomes unstable below some critical mass number A_{crit}, which is of the order of $A_{crit} = 300 - 400$, depending on the various choices of relevant parameters [7]. At this value of A the separation energy, i.e., the energy which is required to remove a single barion from a strangelets starts to be negative and strangelet decays rapidly by evaporating neutrons.

In [5] we have demonstrated that the geometrical radii of strangelets $R = r_0 A^{1/3}$ are comparable to those of ordinary nuclei of the corresponding mass number A (i.e., in both cases r_0 are essentially the same). We have shown at the same place how it is possible that such big objects can apparently propagate very deep into atmosphere. The scenario proposed and tested in [5] was that after each collision with a atmosphere nucleus, a strangelet of mass number A_0 becomes a new one with mass number approximately equal $A_0 - A_{air}$ and this procedure continues unless either strangelet reaches Earth or (most probably) disintegrates at some depth h of atmosphere reaching $A(h) = A_{crit}$.

This results, in a first approximation (in which $A_{air} << A_{crit} < A_0$), in the total penetration depth of the order of

$$\Lambda \simeq \frac{4}{3}\lambda_{N-air}\left(\frac{A_0}{A_{air}}\right)^{1/3} \qquad (1)$$

where λ_{N-air} is the usual mean free path of the nucleon in the atmosphere.

4. RESULTS

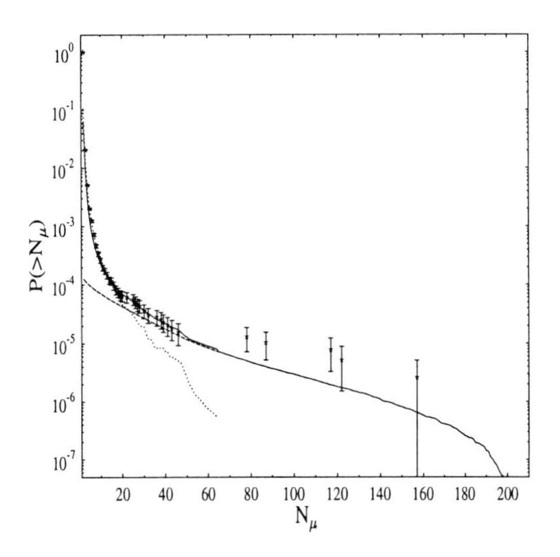

Fig.1 Integral multiplicity distribution of muons for the CosmoLEP data (stars), Monte Carlo simulations for primary nuclei with "normal" composition (dotted line) and for primary strangelets with $A = 400$ (broken line). Full line shows the summary (calculated) distribution.

This is the picture we shall use to estimate the production of muon bundles resulting from the interaction of strangelets with atmospheric nuclei. We use for this purpose the SHOWERSIM [8] modular software system specifically modified for our present purpose. This Monte Carlo program describes the interaction of the primary particles at the top of atmosphere and follows the resulting electromagnetic and hadronic cascades through the atmosphere down to the observation level. Muons with momenta exceeding $70 GeV/c$ are then registered with the sensitive area of 16 m^2 (randomly scattered with respect to the shower axes). Primaries initiated showers were sampled from the usual power spectrum $P(E) \propto E^{-\gamma}$ with the slope index equal to $\gamma = 2.7$ and with energies above $10 \cdot A$ TeV.

flux can accommodate experimental data. Taking into account the registration efficiency for different types of primaries one can estimate the amount of strangelets in the primary cosmic flux. To describe the observed rate of high multiplicity events one needs the relative flux of strangelets $F_S/F_{total} \simeq 2.4 \cdot 10^{-5}$ (at the same energy per particle).

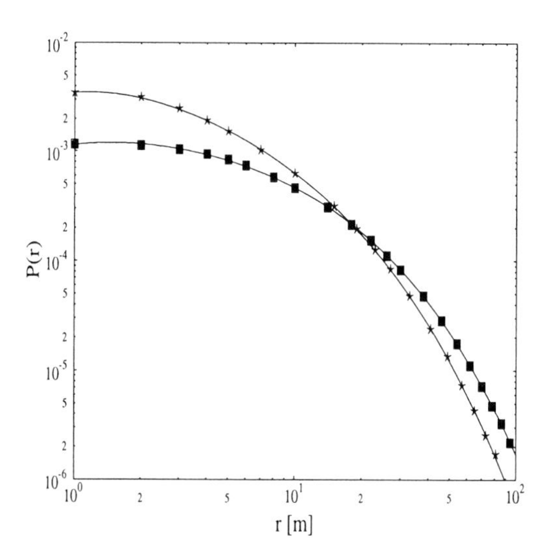

Fig.2 Lateral distribution of muons in the bundles with multiplicities $90 < N_\mu < 110$, which originated from primary proton (stars) and primary strangelet with mass number $A = 400$ (squares) (both with energy 10^4 TeV per particle).

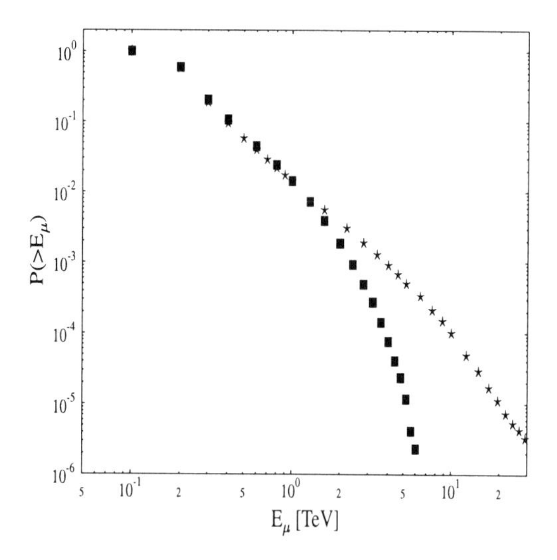

Fig.3 Energy distribution of muons in the bundles with multiplicities $90 < N_\mu < 110$, which originated from primary proton (stars) and primary strangelet with mass number $A = 400$ (squares) (both with energy 10^4 TeV per particle).

The integral multiplicity distribution of muons from ALEPH data are compared with our simulations in Fig. 1. We have used here the so called "normal" chemical composition of primaries [9] with 40 % of protons, 20 % of helium, 20 % of C-N-O mixture, 10 % of Ne-S mixture and 10 % of Fe. It can describe low multiplicity ($N_\mu \leq 20$) region only. On the other hand, muon multiplicity from strangelet induced showers are very broad. As can be seen, the small amount of strangelets (with the smallest possible mass number $A = 400$, i.e., the one being just above the critical one estimated to be $A_{crit} \sim 320$ here) in the primary

Fig. 2 shows the lateral distribution and Fig. 3 the energy distribution of muons in the bundles. They allow us to test the origin of high multiplicity events. Note that, in order to obtain high N_μ tail from normal nuclei only, one needs much higher primary energies per nucleon than in the case where strangelets were also added. Distribution of muons from strangelets is broader and their energy spectrum softer in comparison to events with the same N_μ induced by protons. It is interesting to observe that the high multiplicity events discussed here (with $N_\mu \simeq 110$ recorded on 16 m^2) correspond to ~ 5600 muons with $E_\mu \geq 70$ GeV (or 1000 muons with energies above 220 GeV). These numbers are in surprisingly good agreement with results from other experiments like Baksan Valley, where 7 events with more than 3000 muons of energies exceeding 220 GeV were observed [10].

5. CONCLUSIONS

To recollect: we have demonstrated that the recently observed extremely high multiplicity of muons can be most adequately described by relatively minute (of the order of $\sim 2.4 \cdot 10^{-5}$ of total primary flux) admixture of strangelets of the same total energy. This is precisely the flux we have estimated some time ago [5] when interpreting direct candidates for strangelets and is fully consistent with existing experimental estimations provided by [11]. It accommodates also roughly the observed flux of Centauro events as was shown in [12]. The CosmoLEP studies of multi-muon bundles will therefore significantly improve our understanding of the nature and importance of the SQM candidates.

Acknowledgments: The partial support of Polish Committee for Scientific Research (grants 2P03B 011 18 and 621/E-78/SPUB/CERN/P-03/DZ4/99) is acknowledged.

REFERENCES

1. Cf., for example, R.Klingerberg, *J. Phys.* **G25** (1999) R273 and references therein.
2. C.Taylor et al., (CosmoLEP Coll.), *CosmoLEP and underground cosmic ray muon experiment in the LEP ring* CERN/LEP 99-5 (1999) LEPC/P9 and *Cosmic multi-muon events in ALEPH as part of the CosmoLEP project*, CosmoLEP Report 1 (1999); cf. also: CERN Courier, Vol. 39-8, October (1999) 29.
3. C.Timmermans (L3+C Coll.), 26th *Int. Conf. Cosmic Ray Conf.*, Salt Lake City (1999), Contributed Papers, Vol. 2 (1999) 9.
4. D.Heck, J.Knapp, J.G.Capdevielle, G.Schatz and T.Thouw, *Corsika: A Monte Carlo Code to Simulate Extensive Air Showers*, Wissenschaftliche Berichte FZKA 6019 (1998), Forschungszentrum Karlsruhe.
5. G.Wilk and Z.Włodarczyk, *J. Phys.* **G22** (1996) L105 and *Heavy Ion Phys.* **4** (1996) 395.
6. E.Witten, *Phys. Rev.* **D30** (1984) 272.
7. E.Fahri and R.L.Jaffe, *Phys. Rev.* **D30** (1984) 2379; M.S.Berger and R.L.Jaffe, *Phys. Rev.* **C35** (1987) 213.
8. A.Wrotniak, Report No. 85-195, Univ. of Maryland (1985).
9. S.Nikolskii, *Izv. Akad. Nauk SSSR* (Ser. Fiz.) **39** (1975) 1160; cf. also: T.Shibata, *Nuovo Cim.* **19C** (1996) 713.
10. V.N.Bakatanov et al., 24th *Int. Cosmic Ray Conf.*, Roma, Vol. 1 (1995) 561 and 23th *Int. Cosmic Ray Conf.*, Calgary (1993) 399.
11. T.Saito, Y.Hatano and Y.Fukada, *Phys. Rev. Lett.* **65** (1990) 2094.
12. G.Wilk and Z.Włodarczyk, *Nucl. Phys.* **B** (Proc. Suppl.) **52B** (1997) 215.

ELSEVIER

Nuclear Physics B (Proc. Suppl.) 97 (2001) 89–92

www.elsevier.nl/locate/npe

Cosmic Ray Issues for Accelerator Experiments

Lawrence W. Jones[a]

[a]University of Michigan, Department of Physics, Ann Arbor, MI 48109-1120

Cosmic ray experimentalists seek to understand the physics of the first interaction of primary cosmic rays with air nuclei a) to comprehend the character of particle interactions at very high energies, and b) to correctly simulate that interaction in order to properly interpret data collected (in most cases) under an overburden of atmosphere. With regard to b), John Ellis (CERN) has suggested that the cosmic ray community develop a compendium of specific questions to address to the accelerator experiment community, in the context of current and planned accelerator experiments at LHC, RHIC, and the Tevatron. This paper is an attempt to begin such a discussion.

1. Background

Historically, cosmic ray experiments have focused on discoveries concerning the nature of particle interactions at the highest accessible energies, and have indeed produced much physics of unquestioned importance; the discoveries of the positron, muon, pion, kaons, hyperons; the increase of total cross section with energy, etc. More recently the Centauros, Chirons, the Long-Flying Component, aligned events, and other phenomena have been reported by cosmic ray physicists, but largely ignored by experimentalists working at accelerators and by particle theorists. This has understandably frustrated members of the cosmic ray community, who are certainly stimulated by these claims of "new physics". However the accelerator community is much more interested in pursuing the searches for theoretically-predicted particles, such as the Higgs boson and evidence for particles predicted by Super-Symmetry theories. This focus of the accelerator community is very well summarized in the talk of L. Foa at the X ISVHECRI at Gran Sasso two years ago [1].

2. Recent Accelerator-Based Activities and Discussions

The primary focus of the large "4π" detectors at CERN and Fermilab is on the discovery of the Higgs boson, the searches for evidence of Super-Symmetry (SUSY), and b-quark physics. However, as has been frequently noted, the angular coverage of these very large detectors misses the small-angle forward energy flow, although including over 99% of the 4π solid angle. And the interesting, relevant cosmic ray events generally involve specifically these most forward secondaries which are missed by the 4π detectors, as they carry the majority of the final-state energy.

In fact these cosmic ray reports were not *totally* ignored. The Centauro phenomenon, in part, stimulated Bjorken, Kowalski, and Taylor in their development of the concept of Disoriented Chiral Condensates (DCC), a dynamic consequence of which would be a distribution of pion charges in the final states of hard quark-quark collisions different from the usual (and expected) binomial distribution [2], so that events with an anomalously large or small ratio of charged-to-neutral pions in the final state would not be unusual. This discussion stimulated a modest, parasitic experiment, MiniMax, at the Fermilab collider to search for evidence of such anomalous charge ratio events. As reported earlier, the result of this experiment was negative; no evidence for DCC was detected [3]. It should be noted, however, that a cosmic ray experiment has reported indirect evidence of DCC, using the analysis methods developed for MiniMax [4].

However proposals to extend these investigations through more ambitious experiments, at both the Fermilab Tevatron and at the LHC have been met with unwelcome reactions by the respective program committees. Thus, at CERN, the FELIX Letter of Intent (Karsten Eggert and

PII S0920-5632(01)01187-2

Cyrus Taylor, co-spokesmen) was discussed by the LHC Committee in late 1997 and subsequently was not approved [5]. On the balance, the detector would have represented an investment on a scale of about 10% of the large, 4π detectors, ATLAS and CMS.

Subsequently, a more modest proposal was developed for the Fermilab Tevatron; P-899 (M. J. Longo, spokesman) [6]. This would have utilized the C-0 intersection region (wherein MiniMax was located) for observing secondaries produced through zero degrees in the forward direction, although with less complete and simulations coverage than the FELIX proposal at CERN. This proposal also was not approved.

It is clear that the decision makers at these laboratories do not take the cosmic ray claims of "New Physics" seriously, and that they believe that available resources should be focused on known (i.e., theoretically predicted) problems (such as the Higgs and SUSY) rather than 'fortune hunts'.

3. Models and Monte Carlo Programs

Meanwhile, the cosmic ray community has been studying air showers for decades, seeking to learn the primary composition at energies above those accessible to direct balloon and satellite observation; about 10^{14}eV. The problem is that the primary composition is inextricably linked to the dynamics of the first interaction, as far as air shower observables are concerned. This has been a recurring topic of discussion at past meetings of this symposium series, and has recently been underscored by papers from the Karlsruhe group [7], where they compare observed hadrons in air showers with Monte Carlo predictions. Specifically, they compared the predictions of the Monte Carlo models VENUS, QGS-Jet, and SIBYLL with their data, and found that, at energies above about 10^{16}eV, none of the models matched the data, although near 10^{14}eV, QGS-Jet agreed reasonably well with data.

An effort is now under way to merge QGS-Jet and VENUS and build a model that will give better agreement between data and the model. However, the problem is daunting - as the meager progress over the past decade has demonstrated. One must have a single self-consistent model which, with the correct energy-dependant composition, will agree with *all* reliable air-shower data, including the energy and angular distributions of muons and hadrons as well as the electromagnetic component. Another disagreement with the Monte Carlo models is apparent in the data for muon multiplicities in studies with the ALEPH detector at CERN, discussed at the Sodankyla Workshop and reported there by Karsten Eggert's group [8].

Engel, Gaisser and Stanev report in their recent publication on the low-energy "atmospheric" neutrino data that the greatest uncertainly in calculating the neutrino spectrum, knowing the primary cosmic ray spectrum and composition at these relevant energies, is in the uncertainty in inclusive pion production by protons of 6 to 30 GeV [9]. It is surprizing, and a bit sad, that these production data do not exist, after 40 years of physics experiments at the CERN, Brookhaven, and Fermilab accelerators. However a recent issue of the CERN Courier contains a brief report on a planned experiment, HARP (Hadron Production Experiment at the CERN PS), which is being designed to collect just these data [10].

4. Utah ICRC Discussions

At the XXVI International Cosmic Ray Conference at Salt Lake City, Utah in August, 1999, John Ellis of the CERN particle physics theory group took an interest in the discussions of the high energy particle interactions as observed and inferred by cosmic ray studies. While he was less persuaded of the significance of the emulsion chamber claims of "new physics" and of the case for exploratory experiments such as FELIX, he was much more sympathetic toward the need for solid experimental data as input to Monte Carlo simulations. Specifically, he suggested that the cosmic ray community develop a listing of experimental data which would be useful, desirable, or necessary as inputs to Monte Carlo simulations for air showers at energies above those where direct particle interactions of cosmic rays are practical; e.g., above about 10^{14}eV. This kind of physics has been referred to by Leon Lederman as "engineering physics", i.e., physics whose objectives are to facilitate the interpretation of other physics experiments, rather than to produce new physics *per se*.

This conversation with Ellis suggested that, indeed, it would be useful to draw up such an agenda of desired experimental measurements. This should ideally involve the architects of the Monte Carlo programs in use and/or being developed, however the following section is an attempt at a beginning of such a list.

5. Specific Shopping List

As inputs to Monte Carlo simulations, it seems that what is most needed are inclusive production data, particularly in the forward direction, both for proton-proton interactions, but also for heavier targets, specifically air nuclei (nitrogen and oxygen) and heavier beams. The HARP experiment at CERN, noted above, is an appropriate starting point, with incident proton beams of up to 15 GeV. It would seem appropriate to extend those measurements to incident proton beams of 100 and 400 GeV, to explore the validity of any scaling rules, etc. The CERN SPS is also an attractive accelerator since it has had an active program accelerating heavy nuclei (e.g. lead). Inclusive production could therefore be studied with incident beams of different atomic masses, up through iron, with 150 - 200 GeV per nucleon.

The emphasis should, of course, be on the most energetic particles; secondaries with $x > 0.2 or 0.25$, for example. And, of course, neutral secondaries should be included as well as charged secondaries, recalling again the fact that, in high-energy interactions initiated by a proton beam, the leading final-state hadron is a neutron about 1/4 of the time.

For higher energies, specifically for production data relevant to Monte Carlos studying the "knee" region (1 to 10 PeV primaries), data from hadron colliders will be essential. Because of the capability - and plans - to employ beams of heavy nuclei, the Brookhaven RHIC and CERN LHC colliders are obvious foci of attention.

It may be useful, in considering experiments at RHIC and the LHC, to refer to the following tables, where the equivalent maximum energy of various cosmic ray nuclei incident on air (nitrogen) target nuclei are noted. "Beams" of protons, alphas, and of nitrogen, silicon, and iron nuclei are included. Of course, lower-energy collisions at either machine are readily accessible.

6. Conclusions

It would certainly be useful to the cosmic ray community to present a "wish list" of desired production data to the program committees of the large accelerator laboratories, with those active in the development of Monte Carlo programs as major contributors. Of course, members of the traditional cosmic ray community should be willing to assist in building and operating any resulting experiments and analyzing the data. Ideally, it should then be possible to collect the information needed to construct the appropriate Monte Carlo models for use in interpreting surface cosmic ray data. At the same time, these data could be studied for evidence of the unusual phenomena reported from the cosmic ray community down through the years as well as for indications of other "new physics".

REFERENCES

1. L. Foa, Nuclear Physics B (Proc. Suppl.) **75A** (1999) 28-36.
2. J.D. Bjorken, SLAC-PUB-5545, Int. J. Mod. Phys. **A7**, (1992) 489; K.L. Kowalski and C.C. Taylor, CWRUTH-92-6, heh-pph/9211282; J.D. Bjorken, K.L. Kowalski, and C.C. Taylor, SLAC-PUB-6109, Proc. of Les Recontres de la Vallee D-Aoste, La Thuile, 1993, ed. M. Greco, Editions Frontier (1993) 507.
3. T.C. Brooks, et al. Phys. Rev **D 61** 032003 (2000).
4. C.E. Navia, C.R.A. Augusto, F.A. Pinto, S.L. Barrosa, and E.H. Shibuya, Phys. Rev. **D 55** (1997) 5384; C.R.A. Augusto, S.L.C. Barroso, V. Kopenkin, M. Moriya, C.E. Navia, and E.H. Shibuya, Nuclear Physics B (Proc. Suppl.) **75A** (1999) 206-208.

Table 1
RHIC Nucleus-Nucleus Collision Energies

"Projectile" beam		"Target" beam		Nucleon-nucleon	Equivalent E of projectile
Particle	E per nucleon (GeV)	Particle	E per nucleon (GeV)	E (center of mass) (GeV)	nucleus on a stationary tgt. (TeV)
p	250	p	250	500	133
p	250	p	125	354	67
p	125	p	125	250	33
p	250	^{14}N	125	354	67
^{4}He	125	^{14}N	125	250	133
^{14}N	125	^{14}N	125	250	462
^{28}Si	125	^{14}N	125	250	912
^{56}Fe	125	^{14}N	125	241	1720

Table 2
LHC Nucleus-Nucleus Collision Energies

"Projectile" Beam		"Target" Beam"		Nucleon - nucleon	Equivalent E of projectile
Particle	E per nucleon (TeV)	Particle	E per nucleon (TeV)	E (center of mass) (TeV)	nucleus on a stationary tgt. (PeV)
p	7	p	7	14	104
p	7	p	3.5	9,9	52
p	3.5	p	3.5	7	26
p	7	^{14}N	3.5	9.9	52
^{4}He	3.5	^{14}N	3.5	7	104
^{14}N	3.5	^{14}N	3.5	7	364
^{28}Si	3.5	^{14}N	3.5	7	728
^{56}Fe	3.25	^{14}N	3.5	6.75	1355

5. FELIX Collaboration, "FELIX, a full acceptance detector for the LHC" Report No. CERN/LHCC 97-45, LHCC/110, 1997.
6. M.J. Longo, et al. "Particle Production at 0° from the Collider" Fermilab proposal 899 (1997), L.W. Jones, Vulcano Workshop 1998, Italian Physical Society Conference Proceedings **Vol 65** (1999) 343;
7. T.Antoni, et al. J. Phys. G; Nucl. Part. Phys. **25** (1999) 2161-2175, H. Rebel "Cosmic Rays Interfacing Astrophysics and Particle Physics" Karlsruhe preprint (2000).
8. A. Bruhl and J. Strom, "Observation of muon bundles in the Aleph detector", presented at the First Arctic Workshop on Cosmic Ray Muons, Sodankyla, Finland; April, 1999.
9. R. Engel, T.K. Gaisser, and T. Stanev Physics Letters **B 472** (2000) 113-118.
10. CERN Courier **40** (2000) p.8 "Harping on about hadrons".

ELSEVIER

Nuclear Physics B (Proc. Suppl.) 97 (2001) 93–96

www.elsevier.nl/locate/npe

Recent Results of KASCADE
Phenomenology of Extensive Air Showers

G. Schatz[a*], T. Antoni[a], W.D. Apel[a], F. Badea[b], K. Bekk[a], K. Bernlöhr[a,†] H. Blümer[a,d], E. Bollmann[a], H. Bozdog[b], I.M. Brancus[b], C. Büttner[a], A. Chilingarian[c], K. Daumiller[d], P. Doll[a], J. Engler[a], F. Feßler[a], H.J. Gils[a], R. Glasstetter[d], R. Haeusler[a], W. Hafemann[a‡], A. Haungs[a], D. Heck[a], J.R. Hörandel[d§], T. Holst[a], K.-H. Kampert[a,d], J. Kempa[e], H.O. Klages[a], J. Knapp[d¶], D. Martello[d], H.J. Mathes[a], P. Matussek[a], H.J. Mayer[a], J. Milke[a], D. Mühlenberg[a], J. Oehlschläger[a], M. Petcu[b], H. Rebel[a], M. Risse[a], M. Roth[a], T. Thouw[a], H. Ulrich[a], A. Vardanyan[c], B. Vulpescu[b], J.H. Weber[a], J. Wentz[a], T. Wiegert[a], J. Wochele[a], J. Zabierowski[f], S. Zagromski[a]
The KASCADE Collaboration

[a]Institut für Kernphysik, Forschungszentrum Karlsruhe, P.O. Box 3640, D–76021 Karlsruhe, Germany
[b]Institute of Physics and Nuclear Engineering, RO–7690 Bucharest, Romania
[c]Cosmic Ray Division, Yerevan Physics Institute, Yerevan 36, Armenia
[d]Institut für Experimentelle Kernphysik, Universität Karlsruhe, D–76021 Karlsruhe, Germany
[e]Department of Experimental Physics, University of Lodz, PL–90236 Lodz, Poland
[f]Soltan Institute for Nuclear Studies, PL–90950 Lodz, Poland

KASCADE (KArlsruhe Shower Core and Array DEtector) is a multi-detector setup to observe the electromagnetic, muonic and hadronic air shower components simultaneously in the energy region around the "knee" of the primary spectrum. Its main aim is to determine energy spectrum and composition of hadrons in primary cosmic rays. This is attempted by registring a large number of observables for each EAS including measurements of electrons, muons and hadrons. This contribution gives a short description of the experiment and then presents some results on the lateral distributions of various particle types and on the spectrum of hadrons. The status of our analyses to determine mass composition is presented in an accompanying contribution by A. Haungs.

1. INTRODUCTION

The Karlsruhe extensive air shower (EAS) experiment KASCADE aims to investigate the composition of primary cosmic rays in the knee region of the energy spectrum. The experiment is a multi-detector system which registers simultaneously a large number of observables for each individual EAS. Particles of all three main shower components (electromagnetic, muonic and hadronic) are measured. The basic idea behind KASCADE is to use this multi-parameter feature to reduce, or hopefully avoid, the ambiguities of interpretation which are known to result from our incomplete knowledge of high energy strong interactions. Building of KASCADE started in 1990 and first measurements, with a yet incomplete detector system, started in 1996.

This contribution gives an up-date of the hardware of the experiment and discusses results concerning the measured properties of EAS. The analysis of the data to determine energy and mass of primary cosmic ray particles from EAS measurements rely heavily on the comparison with simulations. It is therefore essential to verify that EAS are phenomenologically well described by such calculations. This is the idea behind the discussions of this contribution. The present status of our attempts to determine energy spectrum and composition of primary cosmic rays are described in an accompanying paper presented by A. Haungs [1].

The experimental activities of the KASCADE

*corresponding author; present address: Habichtweg 4, D-76646 Bruchsal, Germany, e-mail: bgschatz@t-online.de
†now at University of Hamburg, Hamburg, Germany
‡now at: University of Heidelberg, Heidelberg, Germany
§now at: University of Chicago, Chicago, IL 60637
¶now at: University of Leeds, Leeds LS2 9JT, U.K.

PII S0920-5632(01)01188-4

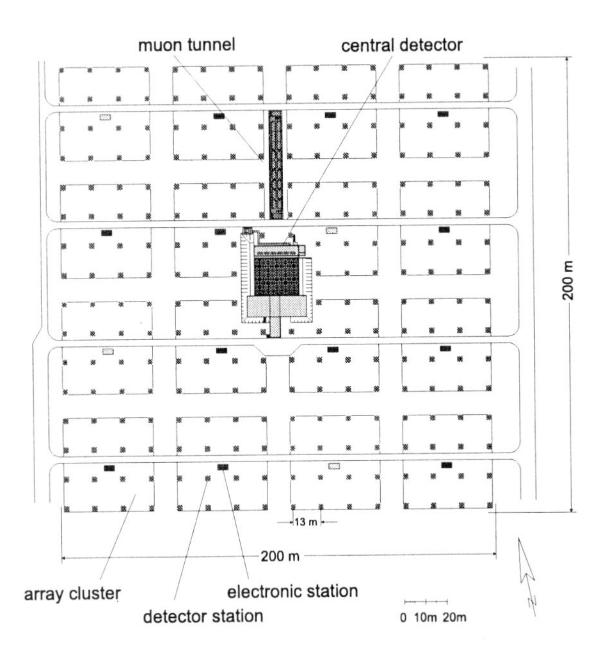

Figure 1. *Lay-out of the KASCADE experiment.*

collaboration are complemented by the development and continual improvement of the Monte Carlo code CORSIKA [2]. This program simulates the shower development for all particle components in three dimensions down to observation level. Several high-energy interaction models have been implemented to allow an estimate of systematic uncertainties resulting from different treatments of high energy interactions.

2. STATUS OF HARDWARE AND ANALYSIS

The KASCADE experiment has been described in detail previously [3] and the reader is referred to this reference for more details. A lay-out is shown in Fig.1. Its main parts are a detector array, a central detector and an underground tunnel. The latter part is a new addition to the experiment. It houses 600 m^2 of streamer tube detectors arranged as a three-layer tracking detector of 150 m^2 effective area. This part of the experiment allows to measure the direction of muons in EAS[4]. The muon tunnel became operational stepwise between fall 1999 and spring 2000. Therefore no data obtained with these detectors are included in the present paper.

A nontrivial problem of EAS experiments is the reconstruction of shower observables such as the number of electrons, muons etc. from the various detector outputs. The signals of the muon detectors in the array, e.g., will contain contributions from e.m. particles (owing to punchthrough) and vice versa (since muons pass through the e.m. detectors above). A discussion of this important question is beyond the scope of this contribution. Details can be found in [5–7]. The development of analysis procedures was based on CORSIKA and a complete Monte Carlo program to simulate the response of the various detector components to all kinds of shower particles.

3. LATERAL DISTRIBUTIONS OF ELECTRONS AND MUONS

In a study completed recently the average lateral distibution functions (LDFs) of electrons, muons and hadrons were investigated [8]. The somewhat unexpected result was that, within the radial range of KASCADE, all LDFs are best described by the usual NKG function albeit sometimes with other than the usual parameters. Measured electron densities were averaged for showers within a certain bin of total electron number N_e. Fig.2 shows the results. The fits have been restricted to densities below $200\,\mathrm{m}^{-2}$ to avoid saturation. As can be seen NKG functions fit the LDFs very well. For these fits the age parameter was kept fixed at s=1.65 and the Molière radius (which we therefore prefer to call r_e) was allowed to vary. The optimum values of r_e then are in the range of 20 to 30 m, i.e. much smaller than the usual Molière radius of c. 80 m. When other age values are preselected the optimum value of r_e changes revealing the well-known strong nonlinear correlation between s and r_e. The fact that a good fit can be obtained with a fixed age implies a simple radial scaling of LDFs within the range of radii and sizes considered in this paper. The characteristic radius r_e increases with zenith distance in proportion to its secant and decreases with N_e, not unexpectedly.

KASCADE measures muons at three different thresholds: 230 MeV (array), 490 MeV (trigger

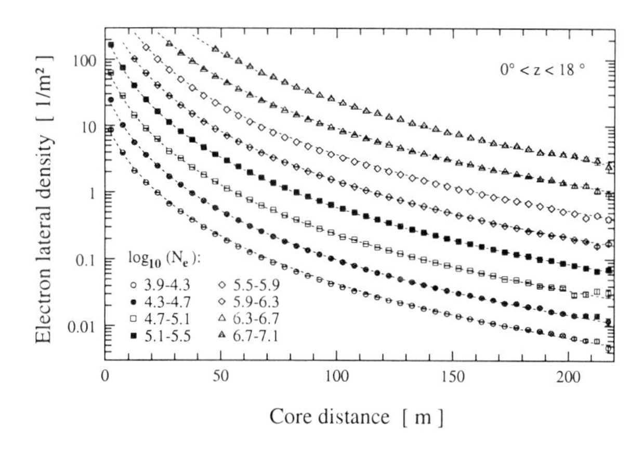

Figure 2. *Observed LDFs for electrons above 5 MeV and their NKG fits (from ref. [8]).*

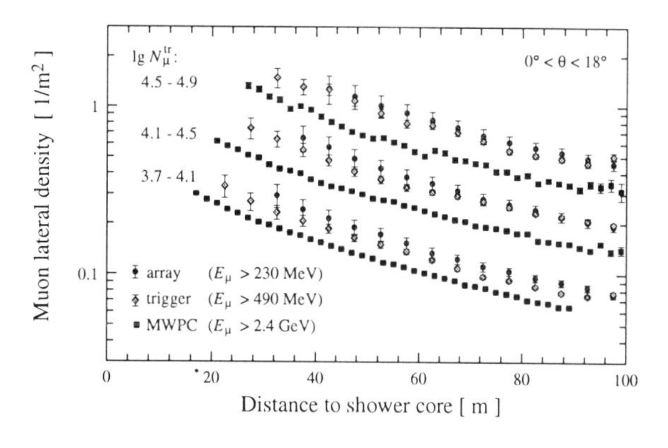

Figure 3. *Measured muon LDFs for three different energy thresholds (from ref. [8]) .*

plane) and 2.4 GeV (multiwire counters; all values refer to vertical incidence). The fit ranges are 40 to 220 m, 20 to 100 m and $\leq$ 100 m, respectively. The lower limits are imposed in order to avoid excessive punchthrough by hadrons and e.m. particles. The upper limit is of course much lower than the diameter of the muon lateral distribution of an EAS. We therefore prefer to quote, as our experimental result, the number of muons in the radial range 40 to 200 m ('truncated muon number') rather than a total number because the latter would require extrapolation into a range not covered by our experiment. The observed LDFs are described well by NKG functions with fixed radial parameter r_{μ}=420 m and variable age parameter s. Alternative functions proposed by Greisen [9], Linsley [10] and Hillas et al. [11] did not yield better fits. Again a strong correlation between the optimum s and r_{μ} is observed when the latter is varied. It should be mentioned that the choice of r_{μ} does not effect the value of the truncated muon number but would result in very different values of the total muon number. Fig.3 shows some results of our measurements. The optimum age parameters decrease with increasing muon number and decreasing zenith distance reflecting steeper LDFs for larger and for vertical showers.

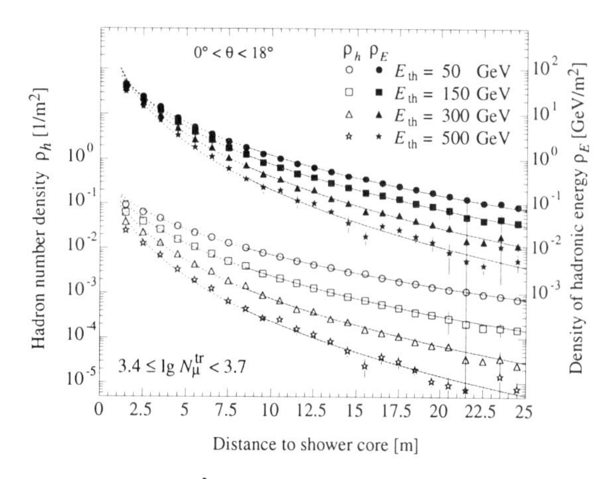

Figure 4. *Measured hadron LDFs for different energy thresholds (from ref. [8]).*

4. HADRONS: LATERAL DISTRIBUTIONS AND SPECTRA

In this section we only discuss hadrons in the cores of EAS. Spectra of single hadrons not associated with EASs have been studied previously [12]. It is worth mentioning, though, that data on single hadrons extend to above 50 TeV by now. Hadrons in EASs are the more concentrated in the shower core the higher their energy. Hence the LDFs depend strongly on energy threshold. This is borne out by the results shown in Fig.4. The lines represent NKG fits with a radius pa-

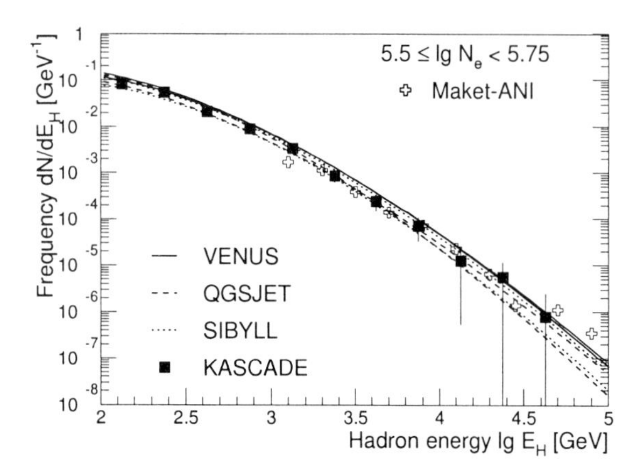

Figure 5. *Measured hadron spectrum in EAS with a limited range of shower sizes (from ref. [13]).*

rameter fixed at 10 m. In addition to the hadron densities the energy densities of hadrons are displayed. The fits describe the data well for radii above c. 2m. At smaller radii saturation effects may come into play especially for the number densities.

Energy spectra of hadrons in shower cores have been studied previously [13]. An example of a measured spectrum, corresponding to a primary energy of c. 6 PeV, is shown in Fig.5 together with data from the MAKET ANI experiment [14] and simulations with the CORSIKA program. Our main conclusion from more refined analyses is that the QGSJET model developed by the Moscow group [15] reproduces the experimental data satisfactorily at low energies (i.e. below approximately the "knee" energy). At higher energies all interaction models fail to describe the data in every respect. For more details we refer to the extensive publication [13].

5. CONCLUSIONS

The KASCADE experiment is in full operation by now and yields high quality data on electrons, muons and hadrons in EASs in a primary energy range from a few hundred TeV to c. 10 PeV. The lateral distributions of all particles can be well described by the NKG function. The derived parameters exhibit a number of expected features: steepening of lateral decline with increasing size (i.e. primary energy) and, for hadrons, with increasing particle energy, flattening with increasing zenith distance. Most observed features, though clearly not all, are well described by simulations with the CORSIKA program. We feel we have collected a high quality data base for approaching the really important problems: energy distribution and composition of primary cosmic rays in the region of the "knee" [1].

REFERENCES

1. A. Haungs et al., KASCADE collaboration, these proceedings
2. D. Heck et al., FZKA Report 6019, Forschungszentrum Karlsruhe (1998).
3. H.O. Klages et al., KASCADE collaboration, Nucl. Phys. B (Proc. Suppl.) 52B (1997) 92.
4. I. Atanasov et al., Report FZKA 6474, Forschungszentrum Karlsruhe (2000)
5. J. Weber, PhD thesis, University of Karlsruhe, 1999; Report FZKA 6339, Forschungszentrum Karlsruhe (in German)
6. J. Hörandel, PhD thesis, University of Karlsruhe, 1998; Report FZKA 6015, Forschungszentrum Karlsruhe (in German)
7. A. Haungs, PhD thesis, University of Heidelberg, 1996; Report FZKA 5845, Forschungszentrum Karlsruhe (in German)
8. T. Antoni et al. - KASCADE collaboration, Astropart. Phys., in press
9. K. Greisen, Ann. Rev. Nucl. Sci. 10 (1960) 63
10. J. Linsley, Proc. 8th ICRC, Jaipur 1963, vol. 4, p. 77
11. A. M. Hillas and J. Lapikens, Proc. 15th ICRC, Plovdiv 1977, vol.8, p.460
12. H. H. Mielke et al., J. Phys. G: Nucl. Part. Phys. 20 (1994) 637
13. T. Antoni et al., KASCADE collaboration, J. Phys. G: Nucl. Part. Phys. 25 (1999) 2161
14. S. V. Ter-Antonian et al., Proc. 24th ICRC, Rome 1995, vol. 1, p. 369
15. N.N. Kalmykov and S.S. Ostapchenko, Yad. Fiz. 56 (1993) 105.

ELSEVIER

Nuclear Physics B (Proc. Suppl.) 97 (2001) 97–100

www.elsevier.nl/locate/npe

Towards the Energy Spectrum and Composition of Primary Cosmic Rays in the Knee Region: Methods and Results at KASCADE

A. Haungs[a*], T. Antoni[a], W.D. Apel[a], F. Badea[b], K. Bekk[a], K. Bernlöhr[a†], H. Blümer[a,c], E. Bollmann[a], H. Bozdog[b], I.M. Brancus[b], C. Büttner[a], A. Chilingarian[d], K. Daumiller[c], P. Doll[a], J. Engler[a], F. Feßler[a], H.J. Gils[a], R. Glasstetter[c], R. Haeusler[a], W. Hafemann[a‡], D. Heck[a], J.R. Hörandel[c§], T. Holst[a], K.–H. Kampert[a,c], J. Kempa[e], H.O. Klages[a], J. Knapp[c¶], D. Martello[c], H.J. Mathes[a], P. Matussek[a], H.J. Mayer[a], J. Milke[a], D. Mühlenberg[a], J. Oehlschläger[a], M. Petcu[b], H. Rebel[a], M. Risse[a], M. Roth[a], G. Schatz[a‖], T. Thouw[a], H. Ulrich[a], A. Vardanyan[d], B. Vulpescu[b], J.H. Weber[a], J. Wentz[a], T. Wiegert[a], J. Wochele[a], J. Zabierowski[f], S. Zagromski[a] :
The KASCADE Collaboration

[a]Institut für Kernphysik, Forschungszentrum Karlsruhe, P.O. Box 3640, D–76021 Karlsruhe, Germany
[b]Institute of Physics and Nuclear Engineering, RO–7690 Bucharest, Romania
[c]Institut für Experimentelle Kernphysik, Universität Karlsruhe, D–76021 Karlsruhe, Germany
[d]Cosmic Ray Division, Yerevan Physics Institute, Yerevan 36, Armenia
[e]Department of Experimental Physics, University of Lodz, PL–90236 Lodz, Poland
[f]Soltan Institute for Nuclear Studies, PL–90950 Lodz, Poland

KASCADE (KArlsruhe Shower Core and Array DEtector) is a multi-detector setup to observe the electromagnetic, muonic and hadronic air shower components simultaneously at primary energies in the region of the "knee". A large number of observables per single shower are registered. The main aims of the experiment are the determination of the primary energy spectrum around the "knee" and the energy variation of the chemical composition. The measurements reveal an increasing mean mass of the primary cosmic rays above the observed kink, and a sharper knee for the light primary component than for the all-particle spectrum, and the absence of a knee for the heavy component between 1 and 10 PeV.

1. INTRODUCTION

In the 40 years since the first observation of a kink in the size spectrum of extensive air showers (EAS) between 1 and 10 PeV primary energy [1], in a lot of experiments the so-called "knee" has been confirmed. The proof of the kink was mainly shown in EAS observables like the shower size, i.e. the total number of particles of the EAS. But to clarify its origin, a precise knowledge about the energy variation of the mass composition and the shape of the primary energy spectrum is required. The Karlsruhe air shower experiment KASCADE [2] aims to investigate the knee region of the charged cosmic rays by precise measurements together with sophisticated methods of data analyses. An energy and mass estimation of the primaries are aspired on an event-by-event basis. KASCADE is a multi-detector setup for measuring simultaneously a large number of observables in the different EAS particle (electromagnetic, muonic and hadronic) components (see also the contribution of G. Schatz to these proceedings [3]). This feature of KASCADE enables to perform a multivariate analysis of the registered EAS on an event-by-event basis to account for the stochastic process of the EAS development. In parallel the KASCADE collaboration tries to improve the tools for the Monte Carlo simulations of the relevant physics. The program code CORSIKA [4] allows the detailed three dimensional simulation of the shower development of all particle components down to observation level,

*corresponding author, e-mail: haungs@ik3.fzk.de
†now at: MPI Heidelberg, Heidelberg, Germany
‡now at: University Heidelberg, Heidelberg, Germany
§presently at: University of Chicago, Chicago, IL 60637
¶now at: University of Leeds, Leeds LS2 9JT, U.K.
‖present address: Habichtweg 4, 76646 Bruchsal, Germany

PII S0920-5632(01)01203-8

and has implemented several high-energy interaction models to facilitate comparisons [5].

2. ENERGY AND MASS ESTIMATION

The KASCADE array consists of 252 detector stations in a $200 \times 200\,\mathrm{m}^2$ rectangular grid containing unshielded liquid scintillation detectors (e/γ-detectors) and below 10 cm lead and 4 cm steel plastic scintillators as muon-detectors. In the center of the array a hadron calorimeter ($16 \times 20\,\mathrm{m}^2$) is built up, consisting of more than 40,000 liquid ionisation chambers in 8 layers. Below the calorimeter a setup of position sensitive multiwire proportional chambers in two layers measures EAS muons with an energy larger than 2.4 GeV, used for the estimation of a local muon density $\rho_\mu^\star(R_{core})$ for each single event.

In general, for each single shower a large number of observables are reconstructed with small uncertainties. For example, the errors are 10-20% for the shower sizes, i.e. total numbers of electrons N_e and the number of muons in the range of the core distance $40 - 200\,\mathrm{m}$ N_μ^{tr}, both reconstructed from the array data. The array data are also used for determining core position and shower direction. Especially for central showers additional EAS parameters can be reconstructed from data of the Central Detector. Examples of such observables are the number of reconstructed hadrons in the calorimeter ($E_h^{\mathrm{thres}} = 50GeV$), their energy sum, the energy of the most energetic hadron ("leading particle" in the EAS), the number of muons in the shower center $N_\mu^\star$, or parameters deduced by a fractal analysis of the hit pattern of muons and secondaries produced in the calorimeter by high-energy hadrons. The latter ones are sensitive to the structure of the shower core which is mass sensitive due to different shower developments of light and heavy primaries in the atmosphere.

The principle way for the estimation of energy and mass composition of the cosmic rays is based on the comparison of the measurements with the output of Monte Carlo simulations. At KASCADE these comparisons are followed in a non parametric way on an event-by-event basis as well as by parametrisations of the measured and simulated distributions.

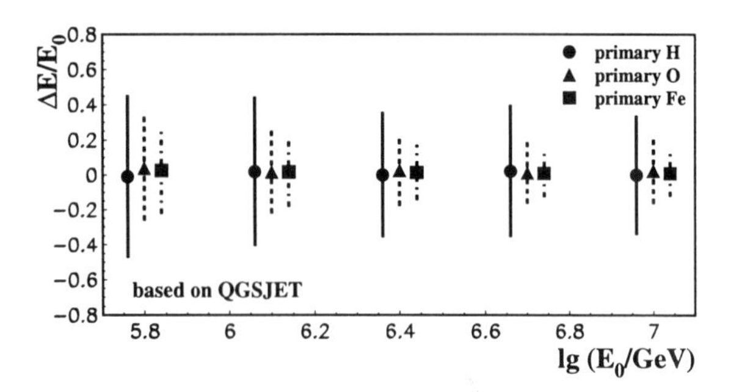

Figure 1. *Energy resolution for different primaries estimated by a neural net analysis.*

The estimate of the primary energy of each single shower is performed by a non-parametric multivariate analysis [6]. Fig.1 shows the energy resolution determined by a neural net analysis for different primaries in the relevant energy range. A large number of Monte Carlo simulations allows the well trained net to calculate the energy of each measured shower in a relatively mass independent way. As multivariate methods are based on a-priori knowledge gained by Monte Carlo simulations, the model dependence effects the largest systematical uncertainty (Fig.4). But it allows to compare different high-energy interaction models. The results of the analysis shown in Fig.4 are cross-checked by using various non-parametric techniques (Bayesian Classifier, Neural Network, k-Nearest-Neighbors), different samples of EAS (central showers with more different observables and showers with core inside the array with large statistics) and by using different sets of observables [7].

Similar multivariate analysis methods are applied for the estimation of the chemical composition of the primary cosmic rays around the knee. This is especially done for showers with their cores in the Central Detector where a larger number of observables are simultaneously available. As an example, the result of a neural net analysis using the shower size N_e, $N_\mu^\star$ and the fractal parameters as input is shown in Fig.2 [8]. The obtained classification of the EAS is corrected with a misclassification matrix (estimated including detailed de-

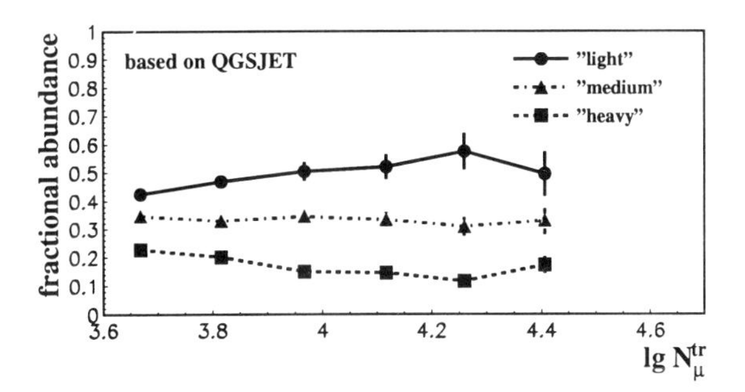

Figure 2. *Relative abundances of different mass groups obtained by a neural net analysis using N_e, $N_\mu^\star$ and fractal parameters as observables.*

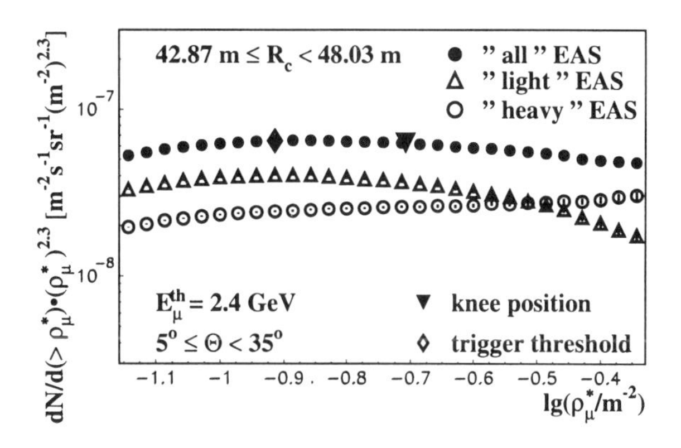

Figure 3. *Local muon density spectra for a certain core distance range. The division into light and heavy induced EAS are performed by analysing the shower size ratio of the EAS.*

tector simulations) leading to relative abundances (e.g. of three primary mass groups), from which the mean logarithmic mass is calculated (Fig.5). The number of primary mass groups considered is limited by the number of Monte Carlo events and by intrinsic fluctuations of the EAS development reducing the power of mass discrimination. In Fig. 5 results of Bayesian analyses are also shown [9]. The limited number of Monte Carlo simulations reduces not only the reasonable number of mass groups, but also the number of observables which can be used for one multivariate analysis. Hence a set of approaches using different observables are averaged in case of the actual result of the Bayesian analyses, but performed independently on basis of two models (QGSJET [10] and VENUS [11]).

Parametric approaches for the analyses of the KASCADE data lead to the results also shown in Fig.4: A simultaneous fit to the N_e and N_μ^{tr} size spectra is performed for the reconstruction of the primary energy spectrum [12]. The kernel function of this fit contains the size-energy correlations for two primary masses (proton and iron) obtained by Monte Carlo simulations. This approach leads to the all-particle energy spectrum as a superposition of the spectra of light and heavy particles. For the light particle spectra a steep kink is revealed, whereas for the heavy particle component a knee is missing between 1 and 10 PeV, thus leading to an increase of the average mass above the knee. An analysis of the $lg(N_\mu^{tr})/lg(N_e)$ size-ratios of the same data sample results in the elemental composition also shown in Fig.5. The measured distribution of these ratios in a certain energy range is assumed to be a superposition of parametrised Gaussian distributions for different primary masses with mean values and widths expected from the simulations [13], leading to relative abundances of the primaries. The results show a tendency of an increasing mean mass above the knee energy in agreement with the results of the non-parametric methods. All these results are obtained by considering a certain range of zenith angles, only. Extensions of the analyses to different EAS angles of incidence are presently in work.

A further approach to the primary energy spectrum is the analyses of local muon densities. For a fixed core distance the local muon density (measured by the Central Detector) reflects the primary energy spectrum and the features of the knee region [14]. A classification of all events in light and heavy induced samples by the ratio of the muon to electron number (registered with the array detectors) allows to analyse the spectra for the different mass groups separately (Fig.3). The analyses results in a clear kink for the spectra of light particle induced showers, whereas the heavy particle induced EAS do not display a corresponding kink in the energy region of 1-10 PeV.

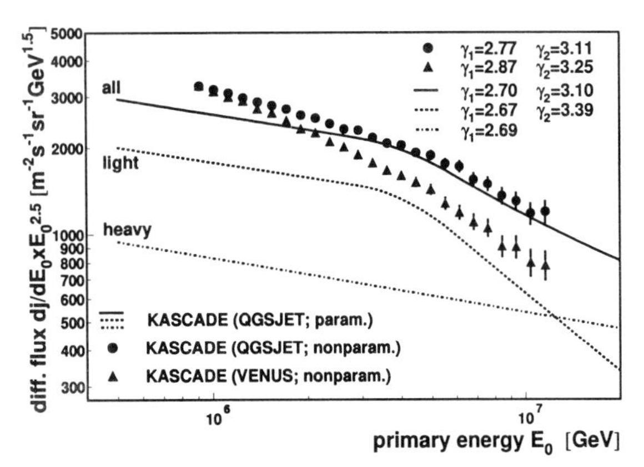

Figure 4. *The primary cosmic ray energy spectrum from KASCADE reconstructed by two different methods.*

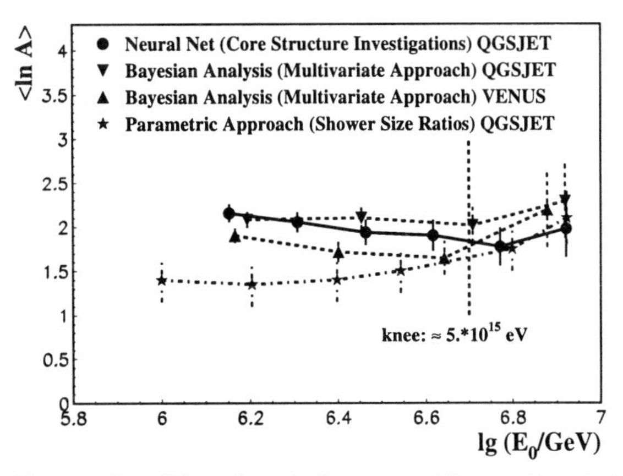

Figure 5. *The chemical composition estimated with the KASCADE data, using different methods and different sets of observables from different particle components.*

Conversion of the spectra to the primary energy spectrum is performed with help of Monte Carlo simulations (QGSJET), resulting in comparable slopes and knee positions as shown in Fig.4. An analogous analysis of densities of the hadronic component is presently in work.

3. CONCLUSIONS

The first results of KASCADE concerning the primary energy spectrum and mass composition between 1 and 10 PeV confirm the knee in the all-particle spectrum around $5 \cdot 10^{15}$ eV with a change of the power law index of $\Delta\gamma \approx 0.3 - 0.4$. When classifying the measured EAS in light and heavy induced showers, the kink is obviously more pronounced for the "light" spectrum, whereas a knee is missing in the spectrum of the heavy particle induced EAS. Consistent with that result, estimates of the elemental composition of the considered energy region show an increase of the heavy component above the knee.

But it should be mentioned, that the quantitative values of knee position, slopes of the spectrum, mean mass etc., depend on the high-energy interaction model underlying the interpretation and, in addition, on the observables used. This shows that our results are still subject to systematic uncertainties which require further investigations.

REFERENCES

1. G.V. Kulikov and G.B. Khristiansen, Soviet Physics JETP 35 (1959) 441.
2. H.O. Klages et al.-KASCADE collaboration, Nucl. Phys. B (Proc. Suppl.) 52B (1997) 92.
3. G. Schatz et al.-KASCADE collaboration, these proceedings.
4. D. Heck et al., FZKA report 6019, Forschungszentrum Karlsruhe (1998).
5. M. Risse et al.-KASCADE collaboration, these proceedings.
6. A. Chilingarian et al.-KASCADE collaboration, Proc.26th ICRC 1999, HE 2.2.04.
7. M. Roth, PhD thesis, University of Tübingen 1999, FZKA report 6262, Forschungszentrum Karlsruhe (in German).
8. A. Haungs et al.-KASCADE collaboration, Proc.26th ICRC 1999, HE 2.2.39.
9. M. Roth et al.-KASCADE collaboration, Proc.26th ICRC 1999, HE 2.2.40.
10. N.N. Kalmykov and S.S. Ostapchenko, Yad. Fiz. 56 (1993) 105.
11. K. Werner, Physics Reports 232 (1993) 87.
12. R. Glasstetter et al.-KASCADE collaboration, Proc.26th ICRC 1999, HE 2.2.03.
13. J.H. Weber et al.-KASCADE collaboration, Proc.26th ICRC 1999, HE 2.2.42.
14. A. Haungs et al.-KASCADE collaboration, Proc.26th ICRC 1999, HE 2.2.02.

ELSEVIER

Nuclear Physics B (Proc. Suppl.) 97 (2001) 101–104

www.elsevier.nl/locate/npe

Test and Analysis of Hadronic Interaction Models with KASCADE Event Rates

M. Risse[a*], T. Antoni[a], W.D. Apel[a], F. Badea[b], K. Bekk[a], K. Bernlöhr[a†], H. Blümer[a,c], E. Bollmann[a], H. Bozdog[b], I.M. Brancus[b], C. Büttner[a], A. Chilingarian[d], K. Daumiller[c], P. Doll[a], J. Engler[a], F. Feßler[a], H.J. Gils[a], R. Glasstetter[c], R. Haeusler[a], W. Hafemann[a‡], A. Haungs[a], D. Heck[a], J.R. Hörandel[c§], T. Holst[a], K.-H. Kampert[a,c], J. Kempa[e], H.O. Klages[a], J. Knapp[c¶], D. Martello[c], H.J. Mathes[a], P. Matussek[a], H.J. Mayer[a], J. Milke[a], D. Mühlenberg[a], J. Oehlschläger[a], M. Petcu[b], H. Rebel[a], M. Roth[a], G. Schatz[a||], T. Thouw[a], H. Ulrich[a], A. Vardanyan[d], B. Vulpescu[b], J.H. Weber[a], J. Wentz[a], T. Wiegert[a], J. Wochele[a], J. Zabierowski[f], S. Zagromski[a]
(KASCADE Collaboration)

[a]Institut für Kernphysik, Forschungszentrum Karlsruhe, P.O. Box 3640, D–76021 Karlsruhe, Germany
[b]Institute of Physics and Nuclear Engineering, RO–7690 Bucharest, Romania
[c]Institut für Experimentelle Kernphysik, Universität Karlsruhe, D–76021 Karlsruhe, Germany
[d]Cosmic Ray Division, Yerevan Physics Institute, Yerevan 36, Armenia
[e]Department of Experimental Physics, University of Lodz, PL–90236 Lodz, Poland
[f]Soltan Institute for Nuclear Studies, PL–90950 Lodz, Poland

Based on the KASCADE multi-detector system with its large hadron calorimeter and using the CORSIKA simulation program with the implemented high-energy hadronic interaction models QGSJET, VENUS, DPMJET, SIBYLL, and HDPM, a method for the test of models by comparing event rates is described. Preliminary results show differences of the model predictions both among each other and when confronted with measurements. The rates are strongly influenced by the inelastic cross sections and the elasticity, especially by the contribution of diffractive dissociation. The discrepancy to measurements at primary energies below $\simeq 3 \cdot 10^{13}$ eV can be reduced by increasing the non-diffractive inelastic cross section.

1. INTRODUCTION

The interpretation of extensive air shower (EAS) data often is closely connected to the use of high-energy hadronic interaction models. However, for the EAS development hadronic interaction processes which are beyond the kinematical and energy region of accelerators, such as diffraction, are of crucial importance. Perturbative QCD calculations are not applicable for particle production with low transverse momenta, and various model concepts and realizations exist for the necessary extrapolations. Apart from different extrapolations, an additional source of uncertainty for all models originates from the experimental situation especially concerning cross section measurements. For example, inelastic nucleon-carbon cross sections at energies of about $E_{lab} = 200-280$ GeV have been measured as 225±7 mb [1] and 237±2 mb [2], the errors being mainly systematical. At the highest energies of $\sqrt{s} = 1.8$ TeV ($E_{lab} \simeq 1.7$ PeV), results obtained for the total proton-antiproton cross section are 72.8±3.1 mb (E710) [3], 80.03±2.24 mb (CDF) [4], and 71.71±2.02 mb (E811) [5] with a probability of the values being consistent of 1.6 % [5]. These systematic uncertainties of about 5–10 % are propagated when constructing hadron-air and nucleus-air cross sections. Effects on EAS predictions are large: An increase of 10 % of the inelastic cross section, e.g., reduces the number of high-energy hadrons (> 100 GeV) by 40–50 % and the electron number (> 3 MeV) by $\simeq$ 15 % (values for

*corresponding author, e-mail: risse@ik1.fzk.de
†now at: MPI Heidelberg, Heidelberg, Germany
‡now at: University Heidelberg, Heidelberg, Germany
§presently at: University of Chicago, Chicago, IL 60637
¶now at: University of Leeds, Leeds LS2 9JT, U.K.
||present address: Habichtweg 4, 76646 Bruchsal, Germany

PII S0920-5632(01)01204-X

proton primaries of 10^{14}–10^{15} eV, $\Theta = 0°$, and KASCADE observation level, i.e., 110 m a.s.l.). The total muon number (> 300 MeV) changes only little ($\simeq +4$ %) but the lateral distribution becomes considerably flatter (reduction of up to 15 % of $\rho_\mu(r < 100\text{m})$).
Therefore, a test of the model predictions is not only important in terms of astrophysical interpretations, e.g., for the determination of the mass composition of cosmic rays, but also interesting information on particle physics might be revealed.

2. CONCEPT AND REALIZATION

A recent analysis [6] of the hadronic structure in EAS cores at energies around the knee exhibited differences between the models QGSJET [7], VENUS [8], and SIBYLL 1.6 [9]. Especially in the SIBYLL version an imbalance between the hadronic and muonic component pointed to an underestimation of the muon number, and new developments of the SIBYLL model have been started [10]. In order to test the simulations at primary energies where results of direct flux measurements are given, in this analysis event rates measured with the KASCADE experiment [11] are used. We start with a coincidence trigger of ≥ 9 out of 456 scintillators of the trigger layer in the central detector (Fig.1). After a trigger, it is searched for at least one hadron in the calorimeter [12] with a reconstructed energy of more than 90 GeV. The frequency of these events per time unit defines the trigger rate and the hadron rate.

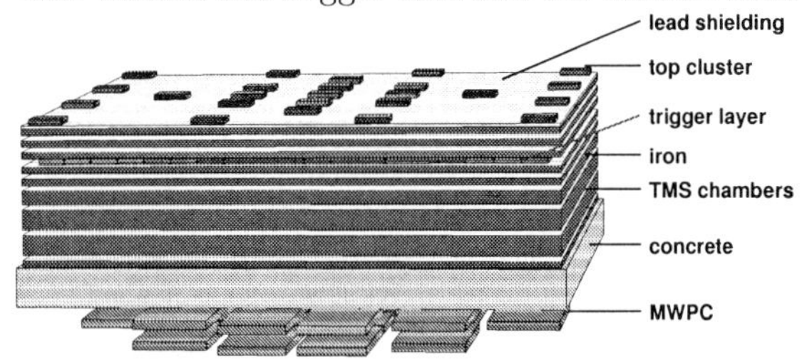

Figure 1. The KASCADE central detector (size $16 \times 20 \times 4$ m^3).

The measured rates show long-term stability on the percent level after being corrected for dead time and air pressure effects (each ≤ 10 %). Already with data of a few days statistical errors become negligible (< 1 %). No significant detector malfunctions have been found.
The air shower simulations are performed using the CORSIKA code [13]. The detector response is calculated with the GEANT package [14]. The five primary particle groups p, He, O, Mg, and Fe are simulated in accordance with the spectra given by direct measurements [15]. For extrapolations to higher energies ($> 5 \cdot 10^{14}$ eV), the individual spectra are assumed to drop with constant spectral indices up to a knee energy of $\lg(E/\text{GeV}) = 6.5$ and to steepen by about 0.3 after the knee. The complete acceptance in terms of primary energy, zenith angle, and distance of shower core to the central detector is considered. In addition to QGSJET (CORSIKA 5.62), VENUS 4.12, and SIBYLL 1.6, the models DPMJET 2.4 [16] and HDPM (CORSIKA 5.62) [13,17] have been used. For each model, the simulation statistics corresponds to a real time flux of about 20 min (HDPM: 10 min).

3. RESULTS

Figure 2 displays for VENUS and SIBYLL 1.6 the contribution of primary energies to the rates. The trigger rate of SIBYLL 1.6 is lower when compared with VENUS for energies above $\lg(E/\text{GeV})=5.0$. Simulations show that triggers at these energies mainly stem from muons while at smaller energies also cascading hadrons in the central detector contribute. Hence, the lowered trigger rate for SIBYLL agrees with the above mentioned results in reference [6]. The hadron rate originates from energies for which the flux is quite well determined by direct measurements. Thus, this observable does not suffer from the composition and flux uncertainties at high primary energies.
The integral values of all models are compiled in Figure 3 and compared with the measured KASCADE value. Differences of about a factor of 1.7 in the predicted trigger and of 2 in the hadron rates occur. No model prediction agrees well to another. The main contribution to the systematic errors of the predictions, which are mostly correlated in both rates, results from the uncertainty in the absolute fluxes and, for the trigger

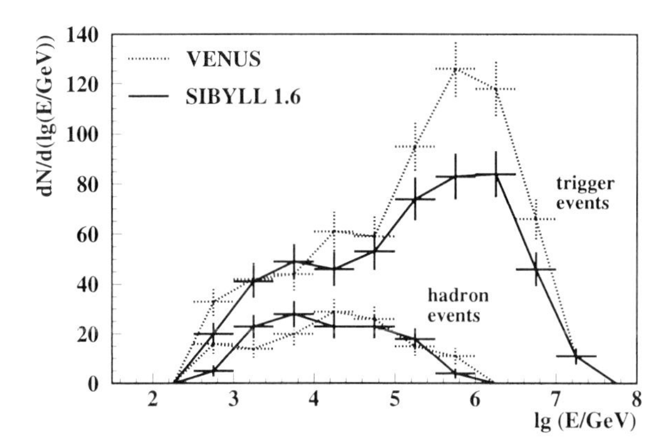

Figure 2. Contribution of primary energies to the trigger and hadron rate for the models VENUS and SIBYLL 1.6 (with statistical errors).

rate, its extrapolation to high primary energies (see reference [18] for details). However, to a first approximation all models would be effected in the same manner, and discrepancies between the predictions would persist. The trend of the rates can be understood in terms of the predicted muon lateral distribution and the number of high-energy hadrons which are closely connected to the inelastic cross section and the elasticity. For example, DPMJET adopts large cross sections and low elasticity values and produces early developing air showers. The hadron number on observation level is small. The muon lateral distribution is flat with low muon densities at distances which contribute to the trigger rate. Thus, small values for the rates emerge (vice versa for HDPM).
Compared to measurement, QGSJET, DPMJET, and SIBYLL 1.6 show reasonable agreement with respect to the systematic uncertainty. All models overestimate the hadron rate. In the modified version SIBYLL 2.0, the effect of an increased cross section and a reduced elasticity (corresponding to a different x_f-distribution of the leading particle) becomes apparent. Although the total muon number is increased in SIBYLL 2.0 compared to SIBYLL 1.6 ($\simeq$ 10 % at 10^{15} eV for a proton primary, the difference becoming larger with increasing energy), the early shower development causes a much flatter muon lateral distribution and therefore a strong suppression of trigger events in the region of $10^{13}-10^{15}$ eV.

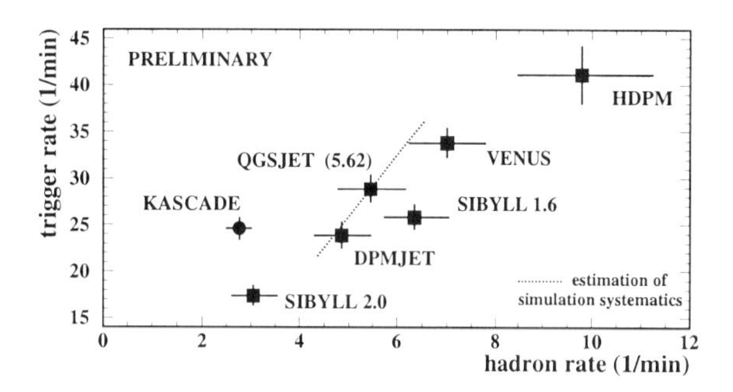

Figure 3. Measured (with total error) and predicted (with statistical error) trigger vs. hadron rates. An estimation of the systematic uncertainty for the simulations is indicated for QGSJET by the dashed line.

Since hadrons also contribute to the trigger, less hadrons reaching the detector will additionally reduce the trigger rate. The authors of the SIBYLL code are currently implementing further developments which will change the predictions [19]. The large decrease of the rates can be understood quantitatively when modifying cross section or elasticity by changing the diffraction dissociation in a sensitivity analysis, as has been performed with QGSJET calculations [18].
In order to classify the rates into different primary energy bins, the muon detectors of the KASCADE array are included in the analysis. Using them as a veto, low primary energies ($< 3 \cdot 10^{13}$ eV) are selected. Comparing with the measurements, the overestimation of the hadron rate can be adressed to these primary energies. The simulations show that these hadrons are mostly the highest energetic in the shower and have suffered either very few and/or mainly diffractive collisions. This is plausible, since otherwise the energy loss from the primary particle of a few TeV to the hadron energy cut in the analysis of 90 GeV would be too large. This selection effect decreases with primary energy. At energies above 10^{14} eV, also secondary hadrons contribute to hadron rate events. With higher simulation statistics, this high primary energy region can be investigated in more detail.
Consequently, at low primary energies the mea-

surements show that the leading particles lose more energy while traversing the atmosphere. This can be accounted for in the simulations either by increasing the inelastic cross section (more interactions) or by decreasing the elasticity (enhanced energy loss per interaction). When keeping the inelastic cross section fixed, this leads to a reduction of the contribution of diffraction dissociation. In total, an increase of the non-diffractive inelastic cross section for energies below $\simeq 3 \cdot 10^{13}$ eV will yield a better agreement between the predicted and measured hadron rates.

4. CONCLUSIONS

The KASCADE detector system combined with the CORSIKA code allows to test high-energy hadronic interaction models. Preliminary results of the investigation of the models QGSJET, VENUS, DPMJET, SIBYLL 1.6, HDPM, and SIBYLL 2.0 reveal differences in the predictions which can be understood in terms of different shower development of the muonic and hadronic component. These are mainly related to the adopted inelastic cross sections and elasticities. Discrepancies to the measured rates are found in both rates and point out possible modifications. The overestimation of high-energy hadrons from small primary energies can be reduced by increasing the non-diffractive inelastic cross section. In close connection with the authors of the models, further studies which have not been discussed here concerning, e.g., the hadron multiplicity in hadron rate events, the inclusion of the electromagnetic component in the analyses, or selecting high primary energies will help to improve our understanding of the interaction processes in air showers.

Acknowledgements

We are very grateful to R. Engel and S.S. Ostapchenko for illuminating discussions and appreciate the possibility to show preliminary SIBYLL results with a non-finalized version.

REFERENCES

1. A.S. Carrol et al., *Phys. Lett.* B80 (1979) 319
2. T.J. Roberts et al., *Nucl. Phys.* B159 (1979) 56
3. N.A. Amos et al., E710 Collaboration, *Phys. Rev. Lett.* 68 (1992) 2433
4. F. Abe et al., CDF Collaboration, *Phys. Rev.* D50 (1994) 5550
5. C. Avila et al., E811 Collaboration, *Phys. Lett.* B445 (1999) 419
6. T. Antoni et al., KASCADE Collaboration, *J. Phys. G: Nucl. Part. Phys.* 25 (1999) 2161
7. N.N. Kalmykov, S.S. Ostapchenko & A.I. Pavlov, *Nucl. Phys. B (Proc. Suppl.)* 52B (1997) 17
8. K. Werner, *Phys. Rev.* 232 (1993) 87
9. R.S. Fletcher et al., *Phys. Rev.* D50 (1994) 5710;
   J. Engel et al., *Phys. Rev.* D46 (1992) 5013
10. R. Engel et al., *Proc. 26th ICRC (Salt Lake City)* 1 (1999) 415
11. H.O. Klages et al., KASCADE Collaboration, *Nucl. Phys. B (Proc. Suppl.)* 52B (1997) 92; see also contributions of A. Haungs and G. Schatz, KASCADE Collaboration, these proceedings
12. J. Engler et al., *Nucl. Instr. Meth.* A427 (1999) 528
13. D. Heck et al., *CORSIKA: A Monte Carlo Code to Simulate Extensive Air Showers*, FZKA 6019, Forschungszentrum Karlsruhe (1998).
14. *GEANT, Detector Description and Simulation Tool*, CERN Program Library Long Writeups W5013 (1993)
15. B. Wiebel, *Chemical composition in high energy cosmic rays*, WUB 94-08, Bergische Universität – Gesamthochschule Wuppertal (1994); B. Wiebel-Sooth, P.L. Biermann & H. Meyer, *Astron. Astrophys.* 330 (1998) 389
16. J. Ranft, *Phys. Rev.* D51 (1995) 64
17. J.N. Capdevielle, *J. Phys. G: Nucl. Part. Phys.* 15 (1989) 909
18. M. Risse, *Test und Analyse hadronischer Wechselwirkungsmodelle mit KASCADE-Ereignisraten*, PhD Thesis, Universität Karlsruhe (2000)
19. R. Engel, *personal communication* (2000)

ELSEVIER

Nuclear Physics B (Proc. Suppl.) 97 (2001) 105–108

NUCLEAR PHYSICS B
PROCEEDINGS
SUPPLEMENTS

www.elsevier.nl/locate/npe

NESTOR Neutrino Telescope Status Report

Peter K.F. Grieder for the NESTOR Collaboration [a]

[a]Physikalisches Institut
Sidlerstrasse 5
CH-3012 Bern, Switzerland

We present a short status report of the NESTOR deep sea high energy muon and neutrino telescope now under construction in Pylos, Greece.

1. Introduction

NESTOR is a deep sea high energy muon and neutrino telescope now under construction at the NESTOR Institute and Laboratory in Pylos, southern Greece. It consists of a *tower* of 12 hexagonal *floors* of 32 m diameter positioned at vertical intervals of 20 m. It is equipped with two highly sensitive 15 inch photomultiplier tubes (PMT) at the corner points of the hexagonal floor structures and in the center, totaling 168 units for phase 1.

The PMTs are arranged in pairs, one above the other, one facing up the other down, to insure uniform spherical response [1] [2]. The basic tower concept can be extended arbitrarily to an array of towers forming a large 3-dimensional matrix of optical sensors. Figure 1 shows a photograph of a fully equipped NESTOR floor suspended from a crane during a deployment operation for a system test.

2. Mode of Operation

NESTOR will detect the Cherenkov radiation produced by muons (and electrons for electron neutrino induced reactions) in a large volume of very clear water. The muons recorded by the tower are either of atmospheric origin produced by cosmic rays, i.e., downward going muons, or the result of neutrino reactions in the water within the detector matrix, its immediate surroundings or in the sea floor below the tower.

Neutrino induced muons manifest an omnidirectional distribution. However, they can only be identified as such within the zenith angular range that is inaccessible for atmospheric muons. The latter depends on the overburden of water and therefore on the depth below the surface at which the detector matrix is being operated. Relatively shallow detectors are therefore restricted to upward going muons only as a safe signature for neutrino induced muons.

On the other hand, deep detectors such as NESTOR whose prospective site is located at a depth of $\sim$4000 m, approximately 20 km off shore from the town of Pylos in southern Greece, will be able to use almost 3π steradian for the neutrino work, excluding only a zenith angular cone about the vertical upward direction of $\simeq 70°$ to eliminate atmospheric muons subintending zenith angles of $\lesssim 70°$.

3. Scientific Goals

The scientific goals of NESTOR had been outlined in several papers [1] and [3]. They are essentially the same as those discussed initially and very extensively in numerous papers of the pioneering but discontinued DUMAND telescope, that serve as guideline for similar proposals. They can be summarized briefly as follows:

3.1. High Energy ν-Astronomy.

Search for high energy (TeV) neutrino point sources in an attempt to locate potential sources of high energy cosmic rays, and search for high energy diffuse neutrinos from unresolved AGNs and similar objects.

PII S0920-5632(01)01205-1

Figure 1. NESTOR floor consisting of the hexagonal support structure with the optical detector modules attached at the end points. The central housing holds the data handling and transmission electronics.

3.2. Muon and Neutrino Physics.

Explore the muon spectrum, muon multiplicities and decoherence. Investigate the atmospheric neutrino flux, study neutrino oscillations using either atmospheric or accelerator neutrinos.

3.3. Search for Dark Matter

Nestor can also search for certain forms of dark matter, such as neutralinos which, if captured by the Sun (or less likely by the Earth) can annihilate in pairs, producing a detectable neutrino source in the host body.

4. Effective Area and Minimum Detectable Flux

The effective area and volume of a NESTOR type detector is energy dependent because of the energy dependence of the muon range. The effective area is approximately 20.000 m^2 for TeV muons, for a single tower as described above. The minimum detectable flux (MDF), too, is energy dependent. In addition it is angular resolution dependent because it is an interplay between signal and noise. The signal to noise ratio increases with increasing angular resolution for point sources. For a single NESTOR tower it amounts to $\sim 10^{-10}$ cm^{-2}s^{-1} for muon neutrinos from point sources with energy > 1TeV and a one degree square resolution. Detectors of this kind are in general signal and not background limited.

5. Project Status

- The laboratories in the new NESTOR institute building in Pylos are now fully equipped and operational and the construction of the NESTOR telescope is well under

Figure 2. The floating NESTOR deployment and service platform. It measured 60 m on the side and carries a workshop container.

way.

- The site studies including water transparency, water current and a variety of other environmental parameter measurements have been completed. A summary of the optical measurements is shown in Fig. 3 and of the environmental parameters in Fig. 4.

- The cable laying operation of the 31 km, 18 fiber electro-optical cable is currently in progress. 25 km of this cable that follow the sea floor are armored, the remaining 6 km of the riser cable are torque balanced to avoid mechanical problems during deployment of the floors when assembling the tower.

- The shore station where the cable terminates is situated in the Methoni Meteorological Station 10 km away from Pylos and

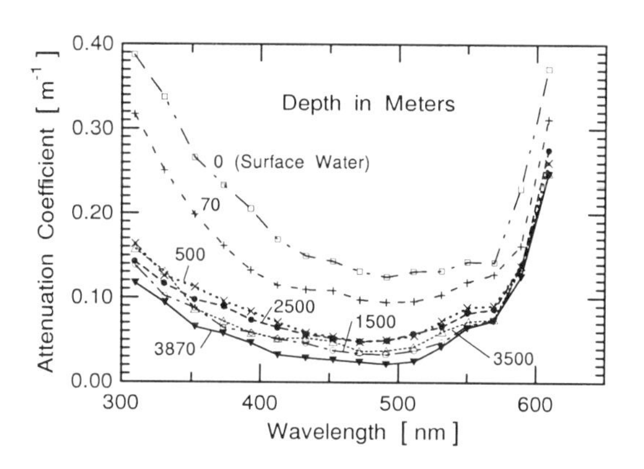

Figure 3. Typical optical attenuation coefficients versus wavelength of water samples taken at different depth in the Mediterranean at the NESTOR site [4].

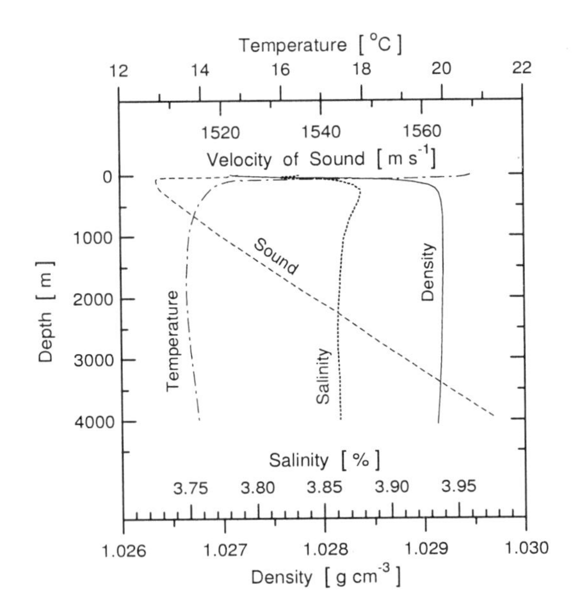

Figure 4. Environmental parameter profiles of the NESTOR detector site in the Mediterranean [5].

is operational.

- Toward the end of June a small size test floor of 12 m diameter equipped with only 12 PMTs (non at the center) and the necessary support equipment, including environmental monitoring units, will be deployed, installed 85 m above the sea floor at the NESTOR site and connected to the shore cable. The purpose of this operation is to test and explore the performance of a number of subsystems such as module control, data acquisition, data handling and transmission, as well as environmental parameter monitoring under realistic conditions for some time while assembling of the full size floor units continues.

- A special floating deployment platform of triangular shape measuring 60 m on the side as shown in figure 2 is under construction. It will significantly facilitate deployment and installation of the NESTOR floors when assembling the tower and will also serve as servicing platform later on.

REFERENCES

1. Anassontzis, S., et al., NESTOR Collaboration: A Neutrino Astroparticle Physics Laboratory for the Mediterranean, Proposal Vol. 1 and 2, May 31 (1995).
2. Bottai, S., for the NESTOR Collaboration: Proc. Internat. Cosmic Ray Conf., Salt Lake City, 2, p. 456 (1999).
3. Resvanis, L., et al., Europhys. News, 23, p. 172 (1992).
4. Khanaev, S.A., and A.F. Kuleshov, Proc. 3rd NESTOR Workshop, ed. L. Resvanis, University of Athens, p. 253 (1993).
5. Anassontzis, S., et al., Proc. 3rd NESTOR Workshop, ed., L. Resvanis, University of Athens, p. 614 (1993).

ELSEVIER

Nuclear Physics B (Proc. Suppl.) 97 (2001) 109–112

www.elsevier.nl/locate/npe

The project of comprehensive cosmic ray investigations with mountain elevation experimental setups in the energy range 10^{15}–10^{18} eV

S.A.Slavatinsky[a], on behalf of the Pamir and Tien Shan Collaboration

[a]Lebedev Physical Institute, Leninskii pr. 53, Moscow 117924, Russia

The project of upgrading of two currently operating experimental setups, i.e., the *Pamir* and the *Hadron* setups, located at high mountain altitudes is proposed for comprehensive investigations of the primary cosmic rays (PCR) in the energy range 1-1000 PeV. The main goals of the project are the research of primary energy spectrum peculiarities, namely, the direct and feasible reverse bends of the spectrum and the study of the PCR mass composition as well as revealing the nature of some new phenomena in cosmic rays recently observed at mountain elevations.

1. Introduction

There are several unresolved problems concerning the energy spectrum shape and the PCR mass composition. The main questions that need to be answered are concerning the reason of a "knee" in the PCR energy spectrum at energy $\sim 3 \div 4$ PeV [1] and the existence of local irregularities ("fine" structure) in the extensive air shower (EAS) spectrum [2].

Besides, a series of new unusual phenomena and events is observed in various cosmic ray experiments just in the same energy range where the "knee" exists [3], namely, Centauro-type events, a phenomenon of coplanar emission of particles [3], an existence of large diffuse darkened spots ("halos") on X-ray films, penetrating hadrons under the lead depth of 50 cm, a flux of neutrons in the EAS cores with abnormal delaying time [4].

One may suppose that unusual particles, e.g., strangelets or even more exotic ones, may be present in the PCR flux at energies above several units of PeV and their existence is the origin of the observed anomalous phenomena in experiments with cosmic rays.

Therefore the problem of a comprehensive study of processes, constituting the basis for unusual phenomena in the "knee" region is very important.

With this object in view, we intend to upgrade experimental setups which are now in operation at the Pamirs and the Tien Shan Mountain Station. The main point of this upgrading which will result in assembling a new *HADRON-M* setup is the employment of a radically new detector, namely, an ionization neutron calorimeter (INCA) of large area (Fig.2).

2. The present status of the Hadron Setup and its upgrading

The *Hadron* setup consists of the following detectors and instruments (Fig.1):

- in the center, there is a X-ray emulsion chamber (XREC) with ionization chambers of 160 m^2 in sensitive area. A conjunction with EAS is established by coincidence of space coordinates and angular characteristics of electron- photon cascades in XREC with those of signals detected by ionization chambers positioned in a cross-way manner. The ionization chambers measure also the sum of the energy of the electron-photon component distributed over the whole area of the XREC. The spatial resolution of ionization channels is 0.11 by 0.11 m^2;

- the ground surface shower array of 120 scintillation detectors for recording EAS electron-photon component. The arrangement of the detectors makes it possible to record EAS inside a circle of 200 m in radius positioned in the center of the setup and thus to refine lateral distribution functions (LDF) at large distances;

- the *Chronotron* system for determination of EAS arrival angles by means of measuring the

0920-5632/01/$ – see front matter
PII S0920-5632(01)01206-3

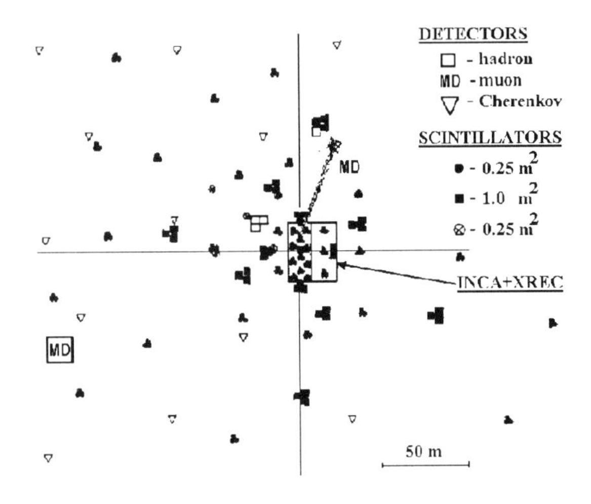

Figure 1. A layout of detector location of the *HADRON-M* setup at Tien Shan (3340 m above sea level).

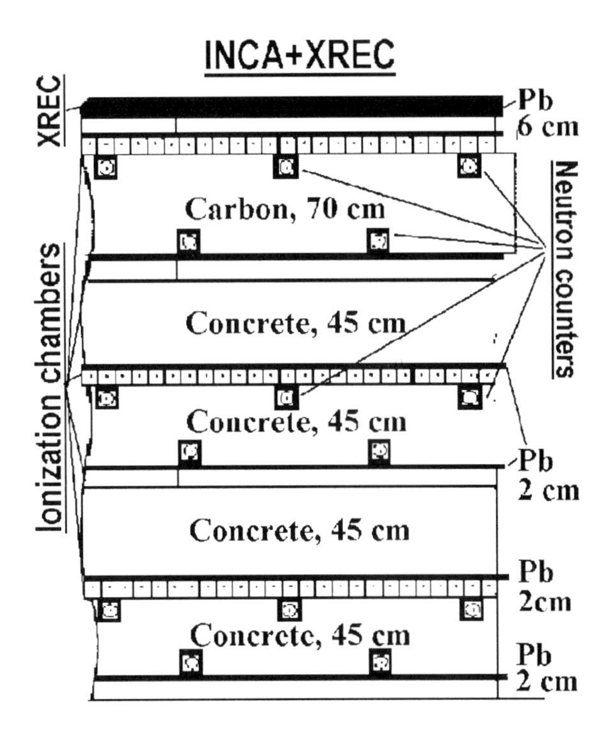

Figure 2. Dissection of central detector of the *HADRON-M* setup, i.e., INCA+XREC.

delay of the EAS front arrival in various points of the shower array;

- the underground central muon detector of 100 m^2 in sensitive area for recording muons with energy $E_m > 5$ GeV;

- for measuring direction and time of arrival of singular muons with respect to the EAS front, there is a *MUON-T* central underground setup consisting of 18 wide-gap spark chambers with nanosecond electronic recording channels;

- the neutron supermonitor 18 NM-64 located at 34 m from the center and connected with the EAS recording system for measuring the energy flux of the hadron component with $E_h > 50$ MeV;

- the *VEGA* system consisting of 13 photomultipliers without mirrors for recording EAS Cherenkov radiation in the atmosphere.

The main trends of upgrading the *HADRON-M* central part are as follows:

- replacing of existent old-type ionization impact recording facility in the central part of the *Hadron* setup with the INCA (Fig.2);

- replacing of old-type electronics by a modern one.

3. Statistics within three year exposition of the *HADRON-M* setup and expected results

The problems which could be solved as a result of three year exploitation of the *HADRON – M* setup are listed beneath with a short description of the applied techniques for their solution:

1. The conclusion of the diffusion model on the nature of the PCR energy spectrum bend at energy of 3-4 PeV will be confirmed or rejected. For these purpose, the partial energy spectra for the main PCR nuclei groups (α-particles, middle nuclei with $Z = 6 \div 8$, heavy nuclei with $Z = 10 \div 25$ and iron nuclei) in the energy range 0.5-100 PeV, i.e., in the range below and above the first energy bend, will be determined for the first time.

The total number of recorded EAS with axes

within the INCA borders will be as high as $\sim 10^5$ events after three year setup operation term. For all such events, the EAS energy will be determined by the most reliable method as an energy sum of the main EAS components, i.e., hadron, muon and electron-photon ones. The nature of the primary particle for each particular event can be established from the collection of several EAS characteristics and, in some cases, it can be determined using characteristics of the correspondent gamma-family recorded by XREC [5]. To reconstruct the mass composition of PCR particles from characteristics of EAS induced by them we shall apply the technique of a reverse problem solution and multivariate analysis, exploited for image recognizing. We might expect that reliability of attribution of recorded events to the specified five groups of PCR particles will be high enough and the error in the determination of nucleus nature will be within 10-20%.

The expected numbers of events for each specified group of PCR particles will be near 10^4.

Thus the information on existence of PCR partial spectrum bends will be originally obtained in the experiment. The comparison of these data with predictions of various diffusion models will make it possible to draw conclusion on applicability of a model under consideration. In any case such an information will clarify not only the nature of the spectrum bend but perhaps the actuality of existence of some cosmic ray source nearby the Galaxy which can produce irregularities in the PCR composition around the Earth.

2. The search for local gamma-sources and the study of diffuse gamma-radiation in the weakly explored energy range of 0.2-10 PeV will be performed. For this purpose, intensity and shape of energy spectrum of muon-less, hadron-less and, especially, neutron-less EAS will be measured in the specified energy range.

In the three year term of *HADRON-M* operation, it is expected to observe 1200 EAS in the 0.2-10 PeV energy range. It will be possible not only to discover the source of primary gamma-quanta but also to determine their energy spectrum.

Besides, the flux value for the primary diffuse gamma-quanta with energy higher than 200 TeV will be established.

3. A unique information about unusual phenomena and events previously observed in cosmic rays will be obtained due to EAS recording both in the calorimeter and by scintillation and neutron counters that makes it possible not only to measure the primary energy E_0 reliably and accurately enough but also to determine effectively the nature of a primary particle producing each event.

Such an information may be crucial for interpretation of the following unusual events formerly observed :

a) Events with coplanar divergence of secondary particles possessing highest energies. At EAS energies above 10 PeV, the total number of four core coplanar events will be about 10. It is significant that, for all these events, a complete information will be obtained both on the primary particle generating the EAS and on properties of the EAS itself. Besides, a few dozens of three core coplanar events in multi-core EAS will be observed.

b) Events related to the production of "halos", i.e., large diffuse darkened spots on X-ray films produced by high-energy hadrons. The expected number of these events will amount 25 ± 5. Note that information on the primary particle nature and its energy is of particular importance for the interpretation of these events. More over, it is hard to exclude that the nature of these primary particles is unusual, namely, it may be related to production of strangelets. In this case, the corresponding EAS will be also unusual in what concerns their characteristics though it is difficult now to predict their peculiarities.

c) In the *HADRON-M* operation term we expect to detect several Centauro and Anti-Centauro events, but their recording will be supplied with information of principle importance, i.e., primary particle nature and its energy. Perhaps the energy threshold for their observation is higher than available at accelerators.

d) Knowledge of the nature of the primary particle will make it possible to perform detailed analysis of EAS characteristics in order to discover the signature of quark-gluon plasma in heavy nuclei interactions (e.g., Fe-Air).

e) Physics processes responsible for appearance of

an abnormal intense neutron component in EAS cores delayed in 500 νs or more, originated from primary particles with energy higher than 3 PeV will be established.
4. The problem of existence of the second ("reverse") bend in the PCR energy spectrum at energies higher than 50 PeV will be solved. The expected number of EAS with energies higher than 50 PeV drastically depends on the presence or absence of the second bend and will be around 600 and 200, respectively, after three years of the experiment.

4. XRECs of the *Pamir* Experiment: status, upgrading and expected statistics

An extensive experimental site destined for exposition of large-area XREC at elevation of 4370 m above sea level at the Pamirs was constructed by the *Pamir Experiment* Collaboration in the past years.

At present a steel frame 36 m^2 in area is constructed at Pamirs. The frame makes it possible to assemble an XREC consisting of two lead blocks separated from each other in the vertical direction by a 2.5 m air gap. The total depth of the lead absorber amounts 50 cm. If the effect of weak absorption arises from charmed Λ_C and D-meson production, the corresponding absorption curve which is expected after a three year exposition must have the clear bump.

The absence of the bump will manifest a new type of hadrons with production cross section about one order of magnitude less than the nuclear one. It is quite evident that the discovery of such particles, probably consisting of quarks of higher color symmetries, would be of fundamental importance for physics.

In conclusion, we are glad to invite all cosmic-ray physicists to take part in these experiments with the *Pamir* and *Tien Shan* setups.

This work is supported by Russian Foundation for Basic Research, project N 00 - 02 -17271.

REFERENCES

1. Berezinsky V.S., Ginsburg V.L., *et al.*, Astrophysics of Cosmic Radiation. Moscow, Nauka, Fizmatgiz, 1990.
2. Nesterova N.M., Proc. 25th ICRC, Durban, v.4, 173, (1997).
3. Slavatinsky S.A., Nucl.Phys. B (1997.), 52 B, 56-70.
4. Antonova V.A., Chubenko A.P. *et al.*, Nucl.Phys. B, 75 A (1999), p.333.
5. Roinishvili N.N. *et al.* Nucl.Phys. B75A (1999), 180.

ELSEVIER

Nuclear Physics B (Proc. Suppl.) 97 (2001) 113–115

www.elsevier.nl/locate/npe

High energy gamma-families with halo and mass composition of primary cosmic rays in energy region above 10 PeV

A.S.Borisov[a], Z.M.Guseva[a], S.A.Karpova[a], J.Kempa[b], A.Krys[b], V.M.Maximenko[a], V.S.Puchkov[a], S.A.Slavatinsky[a]

[a]Lebedev Physical Institute
Leninskii pr. 53, Moscow 117924, Russia

[b]Lodz University
90-236 Lodz ZFWE, Pomorska 149/153, Poland

The experimental spectrum of halo areas for gamma-families with $\sum E_\gamma \geq 500$ TeV is obtained and compared with simulations in quark-gluon string model. The comparison suggests a $\sim$ 2 times decrease of the proton fraction in primary cosmic ray (PCR) composition in the energy region $10^{16} \div 10^{17}$ eV. However at an energy above 10^{17} eV, the experimental flux of halo events with area exceeding $300 \div 400$ mm^2 is $5 \div 10$ times larger than the simulated one. Such a difference may be explained by the assumption that in this region the fraction of protons increases as much as 3 times, while the slope of the spectrum decreases ($\gamma = 2.6 \div 2.7$).

1. Introduction

The aim of the present work is the further analysis of Pamir experimental data with respect to a possible decrease of the light components of PCR at the energy region $10 \div 100$ PeV [1,2]. The Pamir experimental data on spectrum of halo areas are compared with calculations based on the quark-gluon string model MQ [3], which reproduces the main features of gamma-hadron families with $\sum E_\gamma = 100 \div 400$ TeV (that corresponds to $E_0 = 10^{15} \div 10^{16}$ eV) under the assumption that mass composition of PCR is normal (p+α: 50% and subiron+iron: 20%). This nuclear composition is near to the extrapolation of direct measurements from the lower energy region.

The halo is a large diffuse dark spot, recorded in X-ray emulsion chamber along with a high-energy gamma-hadron family. The area of a halo, i.e., the region where optical density of darkness exceeds overall background of the X-ray film, amounts to several centimeters squared.

The halo formation accompanying superfamilies with $\sum E_\gamma \geq 500$ TeV is connected usually with a narrow bundle of high energy particles from the atmosphere. A halo may arise from a pure electromagnetic cascade in the atmosphere initiated by a neutral pion or may be related to a nuclear interaction produced not so far from the observation level if the created flow of energy density is ≥ 20 TeV$\cdot$mm^{-2}. Therefore halo events are most efficiently produced by primary protons due to their high penetrating power.

2. Experimental Data

The experimental data used in the present analysis were recorded by a thin X-ray emulsion chamber with lead absorber (Γ-block) exposed at the Pamir's (4370 m asl or 600 g$\cdot$cm^{-2}) either as the separate unit or as a part of thick hadron chamber with carbon absorber. The total thickness of the Γ-block is 6 cm of Pb with 2 or 3 recording layers of X-ray film. The halo measurements were performed employing an automatic scanning densitometer. Densitograms of halo events were measured on X-ray films, exposed at the depth 9 - 11 c.u. ($4 \div 6$ cm of Pb). The following phenomenological quantitative criterion of halo creation is accepted : the area S bounded by the isodense with an optical darkness density of $D = 0.5$ at a depth of 10 c.u. for

PII S0920-5632(01)01207-5

one side emulsion plate is

$$S = \int_{D \geq 0.5} dS \geq 4 \text{ mm}^2 \quad (1)$$

or, for multi-core "halo",

$$\sum S_i = \sum_i \int_{D \geq 0.5} dS \geq 4 \text{ mm}^2, \ S_i \geq 1 \text{ mm}^2 \ .(2)$$

The halo darkness $D = 0.5$ corresponds to the particle density 0.04 μm^{-2}. The total exposure of Γ-block chambers and the total number of superfamilies with halo areas larger than 4 mm^2 are 2600 m^2·year and 64, respectively. Besides, 8 events with halo area $S \geq 300$ mm^2 obtained during total exposure ST = 10000 m^2·year were used in the present analysis.

3. Model Calculations

Model calculations of the halo spectrum were carried out for four mass compositions of PCR from normal to heavy one, for the primary particles in the energy region $2 \cdot 10^{15} \div 5 \cdot 10^{18}$ eV. Monte-Carlo sampling procedure [4] taking into account the isotropic angular distribution of the primary particles was used. The calculations show that the events with halo area $4 \div 100$ mm^2 are produced by the primary particles in the energy region $2 \cdot 10^{15} \div 10^{17}$ eV and superfamilies with halo size larger than 300 mm^2 correspond to the highest energy region ($> 10^{17}$ eV).

4. Results

Figure 1 presents the experimental integral spectrum of halo areas along with calculated spectra for two mass compositions specified as normal (MQ1) = (p - 40%, α - 10%, M+H - 30%, VH - 20%) and heavy (MQ2) = (p - 10%, α - 10%, M+H - 20%, VH - 60%). As is seen from Fig. 1, the gradual decreasing of the fraction of protons and α−particles in the energy range $2 \cdot 10^{15} \div 10^{17}$ eV will make it possible to fit experimental data with simulated ones in the halo area range 4 − 100 mm^2. The best fit is attained when the fraction of p+α is ≈ 30%. But at the halo area larger than 300 mm^2, the integral spectrum of simulated halo

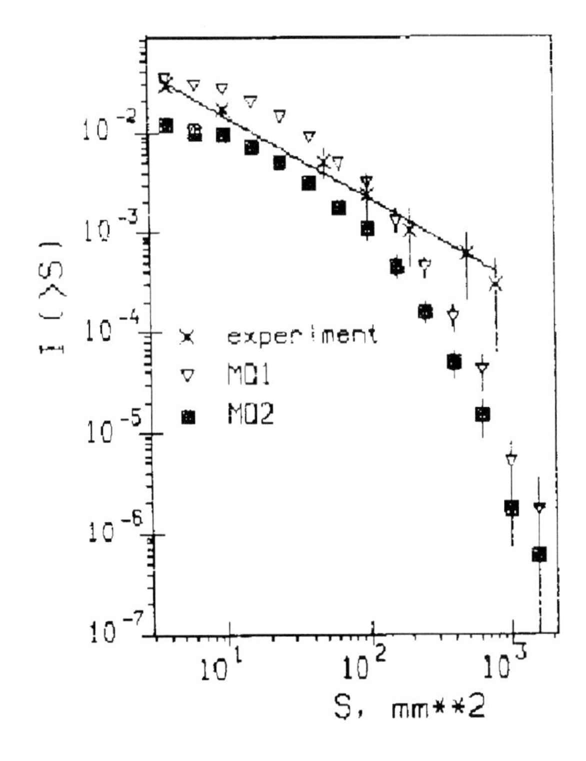

Figure 1. Intensity of Halo events in terms of geometric halo size S ($D \geq 0.5$).

areas S strongly deviates from the experimental values. The difference between calculated data and the experimental values at $S \geq 1000$ mm^2 is about 20 times.

Up to now 5 events are known in the world statistics with visible energy $(E_\gamma, E_h, E_{\text{halo}}) \geq 10^{16}$ eV and halo areas $S \gtrsim 1000$ mm^2 (total effective exposure for the Pamir level is about 13000 m^2·year). The primary energies of these events are estimated as 10^{17} eV and more.

In order to analyse which primary particles are responsible for the halo with areas greater than 300 mm^2, the probability of halo creation was calculated for protons, CNO nuclei and Fe at the energy $E = 10^{18}$ eV. As is seen in Fig.2, protons and may be light nuclei can produce halo areas $S \geq 1000$ mm^2 but the probability is not more than $5 \div 10\%$. Nearly the same result was

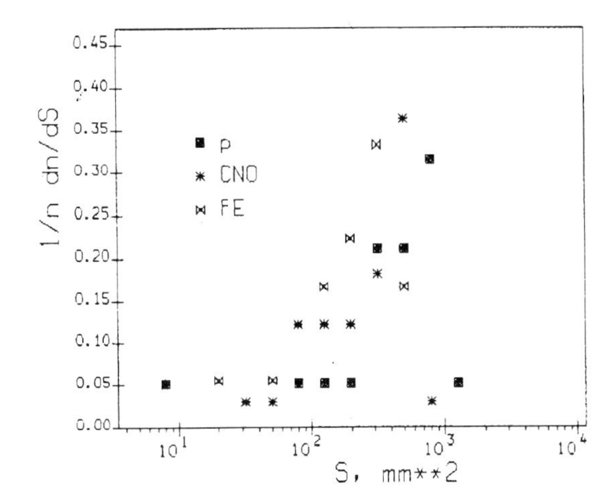

Figure 2. The probability of the halo formation for different primary particles (p, CNO, Fe) at the energy of 10^{18} TeV.

obtained in calculations [5] based on CORSIKA code, employing the QGSJet model for nuclear interaction.

The deviation of our experimental results from calculated ones strongly suggests the change of primary mass composition around 10^{17} eV. An agreement between the experiment and the simulations may be attained if the contribution of the proton primaries is close to 100% and the real air-shower intensity around 10^{18} eV is 2 : 3 times higher than generally accepted [6]. But this conclusion is valid if the efficiency of the halo formation depends only on primary energy and mass composition. An alternative interpretation is also possible, namely that nuclear composition remains unchanged but the properties of nuclear interaction at this energy change. The inelasticity coefficient in MQ-model strongly increases up to 0.8 at 10^{17} eV. It seems that the decreasing of this value may reduce the discrepancy between the experiment and calculations in the energy region $10^{17} \div 10^{18}$ eV, but in the energy region $10^{16} \div 10^{17}$ eV the disagreement will increase and, therefore, primary mass composition will be heavier. In any case the assumption of essential softening of the secondary particle spectra in fragmentation region [7] does not favor the halo formation. The third hypothesis of halo creation is connected with long-flying particles which interact or decay not far from the observation level. This idea can explain many unusual phenomena observed in cosmic rays at the energy around 10^{16} eV (for example, the phenomenon of coplanar emission of high energy secondary particles).

This work was supported by Russian Foundation for Basic Research, project N 00 - 02 -17851.

REFERENCES

1. Borisov A.S. *et al.*, Proc. 24th Int. Cosmic Ray Conference, Roma, V. 1, (1995), p. 182.
2. Borisov A.S., Guseva Z.M., Karpova S.A. *et al.*, Bull.Russ.Acad.Sci. (Physics), V. 61,N. 3, p. 449.
3. Dunaevsky A.M. *et al.*, Proc. 5th ISVHECRI, Lodz, (1988) p. 59.
4. Borisov A.S., Guseva Z.M., Karpova S.A. *et al.*, Nuclear Physics B, (Proc. Suppl.), 52B (1997) p. 185.
5. Puchkov V.S., Guseva Z.M., Karpova S.A.*et al.*,Proc. 26*th* Int.Cosmic Ray Conference, Salt Lake City, USA, (1999), V. 1, p. 104.
6. Nesterova N.M. *et al.*, Proc. 24th Int. Cosmic Ray Conference, Roma, (1995), V. 2, p. 748.
7. Nikolsky S.I., Proc. 26th Int. Cosmic Ray Conference, Salt Lake City, USA, (1999), V. 1, p. 159.

Nuclear Physics B (Proc. Suppl.) 97 (2001) 116–117

www.elsevier.nl/locate/npe

Absorption of gamma-families in the atmosphere at the level of the Pamirs

A.S.Borisov, Z.M.Guseva, V.G.Denisova, E.A.Kanevskaya, S.A.Karpova, V.M.Maximenko, R.A.Mukhamedshin, V.S.Puchkov, and S.A.Slavatinsky [a]

[a]Lebedev Physical Institute, Moscow, Russia

The attenuation length of the γ-families with $\sum E_\gamma \geq 100$ TeV ($N_{fam} = 1003$), $\lambda_{att} = (77 \pm 3)$ g/cm^2 and their vertical intensity $I_{\text{vert}} = 0.63 \pm 0.04$ m^{-2}·year^{-1}·sr^{-1} are determined from the angular distribution. The experimental data are compared with model simulations.

Over many years the collaboration of the Experiment PAMIR [1] investigated high-energy nuclear interactions in cosmic rays with X-ray emulsion chambers (XREC) exposed at the Pamirs at an altitude of 4370 m above sea level (600 g/cm^2). Various components of nuclear-electromagnetic cascades (NEC) induced by protons and nuclei of primary cosmic rays (PCR) are recorded. The investigations are concerned mainly with "families", i.e. bundles of the most energetic secondary particles (threshold for particle detection in the XREC is 1-2 TeV) in the core of just the same NEC. The secondaries are hadrons and particles of electromagnetic nature (γ-rays and electrons) which, for brevity, hereafter are referred to as γ-rays.

In this paper we consider families recorded in a thin XREC with lead absorber (Γ-block); its structure schematically is as follows: 4 cm Pb + X-ray film + 1 cm Pb + X-ray film + 1 cm Pb + X-ray film. The XREC of this type were exposed either as a single Γ-block or as an upper part of a compound experimental set-up with hadron blocks. Particles recorded in Γ-block are mostly γ-rays, and the corresponding families are called γ-families.

Attenuation of the flux of γ-families in the atmosphere depends both on the properties of the PCR (mass composition, energy spectrum) and characteristics of strong interaction at extremely high energy. Therefore exploration of the attenuation length λ_{att} and the intensity of γ-family flux provides good means for experimental estimation of inelastic cross section $\sigma_{in}(E)$ and inelasticity $K_{tot}(E)$ of hadron interactions at energies above those attainable in accelerators. The dependence of λ_{att} on $\sigma_{in}(E)$ and $K_{tot}(E)$ in general case is rather complicated, therefore the estimation of these values based on experimental data may be inferred by comparison of experimental values with those simulated by different models based on different assumptions on properties of the strong interaction and the PCR [2,3].

One possible approach to experimental determination of λ_{att} consists in the measurement of the differential distribution of zenith incidence angle ϑ of γ-families with total energy $\sum E_\gamma$ at a fixed depth H in the atmosphere, $I(> E, H, \vartheta)$. Under assumption of an exponential dependence of absorption on the thickness of absorber, the angular distribution is

$$\frac{dI(\sum E, H, \vartheta)}{d\Omega} = A_1 \cos\vartheta \exp\left(-\frac{H}{\lambda_{abs}}\frac{1}{\cos\vartheta}\right) \quad (1)$$

or

$$\ln\left[\frac{dI(\sum E, H, \vartheta)}{d\cos^2\vartheta}\right] = A_2 - \frac{H}{\lambda_{abs}}\sec\vartheta \quad (2)$$

(the extra multiplier $\cos\vartheta$ arrives from the expression for the effective area in the case of a flat detector).

The other possible approach consists in measuring vertical intensity of family number at two different depths in the atmosphere (H_1, H_2), and then relating them in

$$\lambda_{abs} = (H_2 - H_1)/ln\left[\frac{I(> E, H_1, 0)}{I(> E, H_2, 0)}\right] \quad (3)$$

0920-5632/01/$ – see front matter
PII S0920-5632(01)01208-7

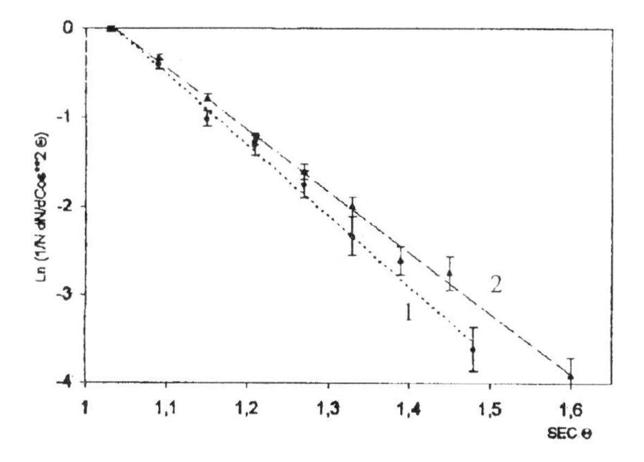

Figure 1. Angular distribution of families with $\sum E_\gamma \geq 100$ TeV. 1 - experiment. 2 - simulations by code MC0.

However it should be noted that this approach is not quite independent from the previous one, because for the calculation of the vertical intensity one has to use the angular distribution.

The present analysis is based on γ-families recorded during a total XREC exposition $ST = 2635\ \mathrm{m}^2\cdot$year and selected by following criteria: a. total energy of γ-rays in the family $\sum E_\gamma \geq 100$ TeV; b. energy of γ-rays and their incidence angle (with respect to vertical) $E_\gamma \geq 4$ TeV and $\vartheta \leq 45^o$, respectively; c. deviation of separate γ-rays from the energy-weighted center of a family in the target diagram plane at the observation level $R \leq 15$ cm; d. total number of γ-rays in the family satisfying above criteria $N \geq 3$. The total number of selected and analyzed γ-families is $N_{fam} = 1003$ ($ST = 2635$ m^{2}year).

The angular distribution obtained (Fig.1) is well described by a straight line in accordance with Eq.(2). Its parameters determined by least-squares method give a value of $\lambda_{att} = (77 \pm 3)$ g/cm^2. Figure 1 presents also simulations based on a version of the quark-gluon string model (code MC0). Evidently experimental angular distribution and simulations are in good agreement. The attenuation length for simulated events is $\lambda_{att}^{MC0} = 88 \pm 2$ g/cm^2.

In an earlier publication of the Experiment PAMIR [4] $\lambda_{att} = 102 \pm 16$ g/cm^2 was obtained by the same method of zenith angular distribution. Later possible distortions of the experimental data were analyzed more thoroughly which result in a refined, appreciably smaller, value of the attenuation length [5]. This is confirmed now by enlarged statistics.

According to (2), the vertical flux may be presented as

$$I(\sum E_\gamma, H, 0) = \frac{N_{fam}}{ST\Omega_{eff}} \quad (4)$$

where

$$\Omega_{eff} = 2\pi \int_0^{\vartheta_0} e^{-m(\frac{1}{\cos\theta}-1)} \cos\vartheta \sin\vartheta d\vartheta \quad (5)$$

where ϑ_0 is the bound value of incidence angle and $m = \frac{H}{\lambda_{att}}$. At $\vartheta = 45^o$ and with the obtained value of λ_{att}, $\Omega_{eff} = 0.60 \pm 0.02$ sr. Consequently, the global intensity of γ-families is $I_{\text{glob}} = N_{fam}/ST = 0.38 \pm 0.02\ \mathrm{m}^{-2}\cdot\mathrm{year}^{-1}$ and the vertical intensity is $I_{\text{vert}} = I(\sum E_\gamma > 100$ TeV, 600 g$\cdot$cm^{-2}, 0$) = 0.63 \pm 0.04\ \mathrm{m}^{-2}\cdot\mathrm{year}^{-1}\cdot\mathrm{sr}^{-1}$.

REFERENCES

1. S.G.Bayburina et al. Trudy FIAN (Transactions of Lebedev Physical Institute), V.154, pp. 1 - 141 (1984).
2. A.M.Dunaevsky, Thesisses, Lebedev Physical Institute (19).
3. G.F.Fedorova and R.A.Mukhamedshin, Bulletin de la Societe des Sciences et des Lettres de Lodz, V.16, pp.43 - 47 (1984).
4. Collaboration of Experiment "Pamir", Izv. AN SSSR (Bul. of USSR Ac. of Sc.), ser. fiz. V. 54(4), pp. 43-47.
5. R.Aliaga *et al.*, Izv.RAN (Bul. of RAS), V.61(3), pp. 441 - 447 (1997) .

ELSEVIER

Nuclear Physics B (Proc. Suppl.) 97 (2001) 118–121

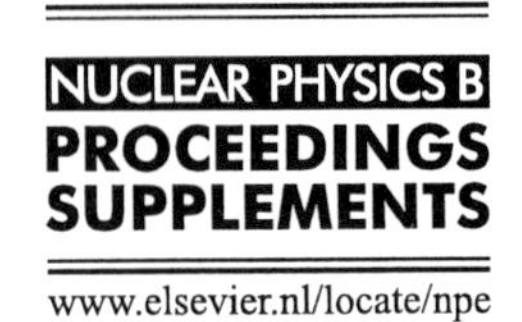

www.elsevier.nl/locate/npe

On the nature of gamma-hadron family alignment

A.S.Borisov[a], R.A.Mukhamedshin[b], V.S.Puchkov[a], S.A.Slavatinsky[a] and G.B.Zdanov[a]

[a]Lebedev Physical Institute, Leninskii pr. 53, Moscow 117924, Russia

[b]Institute for Nuclear Research, Prosp. 60-letiya Oktyabrya 7a, Moscow 117312, Russia

Analysis of experimental and simulated γ−families calculated in QGS model revealed that two different mechanisms of strong interactions are responsible for the observed alignment of high energy γ−rays in superfamilies. One of them relates to conventional QCD-jet production. Another one has a clear energy threshold behavior ($E_0^{th} \gtrsim 8$ PeV) and manifests itself via alignment of 4-core cluster structures observed in superfamilies with $\sum E_\gamma \geq 700$ TeV. Both mechanisms are characterized by large transverse momenta. A phenomenological model with energy threshold production of coplanar hadrons makes it possible to reproduce experimental fraction of aligned events.

1. Introduction

The phenomenon of alignment of high-energy particles in superfamilies has been permanently studied by the *Pamir* Experiment Collaboration for many years since its first observation in multi-core halo events [1]. Due to large scales of the experimental setup consisting of various types of X-ray emulsion chambers (XREC) with lead absorber (total exposition is $\sim$ 7000 m^2) and its high mountain elevation (4400 m a.s.l), the *Pamir* Collaboration possesses a unique statistics of superfamilies, i.e., about one hundred events with $\sum E_\gamma \geq$700 TeV. However, the fraction of aligned events among those is not more than 20-30 %, essentially hampering the analysis of the phenomena. A few but decisive events were observed in some other experiments with XREC [2,3].

Recently, in order to observe the signature of coplanar hadron production, the Chinese physicists have treated superhigh-energy events recorded some years ago in *Mt.Kambala* Experiment. Their conclusions are in a good agreement with those of the *Pamir* Collaboration though the lack of statistics disabled them to confirm the results of the *Pamir* Experiment to the full [4]. Nevertheless many specific features of the phenomenon are quite clear already and the most remarkable among them is its energy threshold behavior ($E_0^{th} \gtrsim 8$ PeV) [5]. But still there is no adequate theoretical explanation of the alignment phenomenon until now.

2. Family anisotropy and QCD-jets

To check the possibility to explain the aligned family generation by QCD-jets, we treat simulated events calculated with the MC0 code, which is based on QGS model of strong interactions supplemented with a QCD-jet production mechanism [6]. MC0 code provides a good fit to most experimental family characteristics, especially, in the $\sum E_\gamma$= 100-400 TeV energy range. Besides we use also the MC0_ NJ data base with simulated events calculated using the same QGS model but without introducing the QCD-jet generation.

As usual, for quantitative description of the alignment phenomenon, we employ λ_N−criterion [5] which characterizes the degree of alignment of N most energetic cores on the target diagram. Recall that λ_N varies from $-1/(N-1)$ to 1 and structures with $\lambda_4 \geq 0.8$ are considered to be aligned. Previously we reported [5] that examination of the so-called nuclear clusters increases sensitivity to the alignment manifestation due to effective reconstruction of hadrons (h^*) from the last but one generation of nuclear cascade in the atmosphere by means of clusterization procedure. To extract clusters in families, a parameter χ_{ij} is used with χ_c = 48 TeV$\cdot$ cm for critical value.

Formerly it was established that existence of so-called binocular-type events can be accounted for by QCD-jets [7,8]. Binocular events are a very narrow class of γ−families ($\leq$ 5%) and its contribution to family anisotropy is rather small.

PII S0920-5632(01)01209-9

Table 1
Mean values of α_3-criterion $< \alpha_3 >$ calculated for 3-core cluster structures

$\sum E_\gamma$, TeV	100 – 400		400 – 700	700 – 2000	$N_c \geq 6$, $E_c^{th} = 50$ TeV
$< ER >$ $TeV \cdot cm$	> 0	> 40	> 0	> 0	> 0
EXP	$0.07 \pm .02$	$0.21 \pm .05$	$0.03 \pm .04$	$0.17 \pm .07$	$0.25 \pm .11$
MC0	$0.04 \pm .02$	$0.25 \pm .05$	$0.00 \pm .04$	$0.05 \pm .02$	$-0.03 \pm .03$
MC0_ NJ	$0.01 \pm .01$	$0.08 \pm .05$	$-0.06 \pm .04$	$0.06 \pm .04$	$-0.07 \pm .06$

Table 2
Fraction of families with aligned ($\lambda_4^c \geq 0.8$) 4-core cluster structures

$\sum E_\gamma$, TeV	100 – 400	400 – 700	700 – 2000	$N_c \geq 6$, $E_c^{th} = 50$ TeV
EXP	$0.03 \pm .01$	$0.05 \pm .02$	$0.15 \pm .05$	$0.27 \pm .09$
MC0	$0.04 \pm .01$	$0.04 \pm .02$	$0.05 \pm .01$	$0.05 \pm .02$
MC0_ NJ	$0.05 \pm .01$	$0.04 \pm .02$	$0.04 \pm .02$	

Nevertheless large p_T jet production results in appreciable $\gamma-$family asymmetry which can be described in terms of α_N-criterion [7]. Table 1 presents mean values of α_3-criterion calculated for 3-core cluster structures as a function of released energy. It is seen that QCD-jets increase anisotropy of clusterized $\gamma-$families in the low-energy region $\sum E_\gamma$= 100-400 TeV.

To enhance contribution of this mechanism we need to select families with large lateral sizes. Table 1 displays also that asymmetry of wide $\gamma-$families with $< ER >_c \geq 40$ TeV·cm is considerably larger as compared with usual ones.

The QCD-jet production can result also in appearance of aligned cluster structures due to trivial fluctuations. However this mechanism fails to account for both the anisotropy (see Table 1) and the alignment of families in the energy range $\sum E_\gamma \geq 700$ TeV. The last conclusion comes from Table 2 which contains fraction of aligned families with 4-core structures subject to energy range.

Especially high discrepancy between experimental and simulated values of anisotropy and alignment arises when families are selected by the number N_c of high-energy clusters ($E_c \geq 50$ TeV) in a family. It was revealed from analysis of simulated events that there is a strong dependence of primary particle energy E_0 on the number of high-energy clusters N_c not to be compared with that between E_0 and the family energy $\sum E_\gamma$. Particularly, the selection criterion $N_c \geq 6$, $E_c^{th} = 50$ TeV corresponds to a primary energy of $E_0 \gtrsim 8$ PeV which is considered to be the threshold energy for coplanar production of high energy hadrons [5].

3. The aligned event features

As seen from Table 3, the integral characteristics of aligned and not-aligned families are practically the same. The difference appears when we treat the highest energy elements in the central part of a family, namely, the lateral spread of 4 most energetic clusters. It means that a subsequent nuclear-electromagnetic cascade, which takes place after the coplanar emission of high-energy hadrons occurs, almost completely destroys the initial coplanar pattern. This conclusion is confirmed also by treating various 4-core high-energy structures composed of reconstructed ("elder") particles, i.e., initial quanta γ^* resulting from the decay of neutral pions produced in the last nuclear interaction just above the chamber, reconstructed neutral pions π^{0*} themselves and reconstructed hadrons h^* from the last but one nuclear interaction which refer to extracted nuclear clusters. These types of structures inside families are especially sensitive to the coplanar emission of high-energy particles [5].

An important result which favored the same

Table 3
Characteristics of aligned and not-aligned families with $N_c \geq 6$ ($E_c^{th} = 50$ TeV, $E_\gamma^{th} = 6$ TeV)

	$<< R_\gamma >>$ /cm/	$<< R_c >>$ /cm/	$<< R_c >_4>$ /cm/	$<< E_c R_c >_4>$ $/GeV \cdot km/$	$cos(\theta)$
EXP:$\lambda_4^c \geq .8$	$2.4 \pm .6$	$3.9 \pm .6$	$1.8 \pm .5$	$2.2 \pm .4$	$0.97 \pm .01$
EXP:$\lambda_4^c < .8$	$2.3 \pm .3$	$3.6 \pm .3$	$0.7 \pm .1$	$1.2 \pm .1$	$0.92 \pm .02$
MC0:$\lambda_4^c \geq .8$	$2.3 \pm .2$	$3.9 \pm .2$	$0.6 \pm .1$	$0.9 \pm .1$	

Table 4
Fraction of aligned 4-core structures composed of reconstructed particles ($N_c \geq 6$, $E_c^{th} = 50$ TeV)

Type of reconstructed particle	γ^* ($z_c = 1.2$ TeV·cm)	π^{0*} ($z_c = 3.4$ TeV·cm)	h^* ($\chi_c = 48.$ TeV·cm)
EXP F($\lambda_4^c \geq .8$)	$0.13 \pm .04$	$0.22 \pm .05$	$0.27 \pm .09$
$< R_c >_4$ /cm/	1.5 ± 0.5	1.3 ± 0.3	1.8 ± 0.5
CPM F($\lambda_4^c \geq .8$)	$0.16 \pm .01$	$0.24 \pm .01$	$0.20 \pm .02$
$< R_c >_4$ /cm/	1.2 ± 0.1	1.4 ± 0.1	1.2 ± 0.1
MC0 F($\lambda_4^c \geq .8$)	$0.06 \pm .01$	$0.07 \pm .02$	$0.05 \pm .02$
$< R_c >_4$ /cm/	0.6 ± 0.1	0.8 ± 0.2	0.6 ± 0.1

scenario of aligned family production is that the incidence angle θ of the primary particle originating the aligned event is close to vertical. This should take place because inclined showers can not preserve the coplanar frame due to greater number of contributing successive interactions.

4. A phenomenological model of coplanar particle production

To constitute a mechanism of aligned event production it is useful to determine the destruction rate of a coplanar pattern of secondary hadrons produced at the top of the atmosphere by their successive interactions with air nuclei above the chamber.

For this purpose, a phenomenological model of coplanar particle production was incorporated in the MC0 code. The CPM model [9] assumes a step-like threshold for coplanar particle production in the proton-air nucleus interaction at an energy of $E_0^{th} = 8$ PeV. At energies above the threshold, the strong interactions drastically change since the probability for the process of coplanar emission of secondaries at these energies was taken to be equal to 1. The coplanarity of the particle emission means that the longitudinal component of the average transverse momentum of secondaries in reference to the coplanarity plane $< p_t >^{\parallel}$ is as high as 2.3 GeV/c while its transverse component $< p_t >^{\perp} = 0.22$ GeV/c, corresponding to the mean transverse momentum of hadrons inside QCD jets. Thus the resultant ratio $< p_t >^{\parallel} / < p_t >^{\perp}$ is ~ 10, which is consistent with the correspondent scattering of aligned cores in experimental families about the line of their alignment $< R_c >_4^{\parallel} / < R_c >_4^{\perp} = 12 \pm 3$.

As seen from Table 4, which contains fractions of families with aligned 4-core structures composed of reconstructed particles and their mean sizes, the model provides a good fit to experimental data on alignment phenomenon.

It is quite evident that conventional hard interaction processes could not completely vanish. To make inelastic cross section of coplanar particle production realistic, one needs to introduce also penetrating particles in order to decrease the destruction of coplanar frames by nuclear cascade. The different behavior of $< R_c >$ and $< R_c >_4$ for aligned and not aligned events favors the same conclusion. A mean free path of $200 - 300$ g/cm^2 for a penetrating particle, producing coplanar hadrons in nuclear interactions, seems to be optimal to accord all variety of observable characteristics. The consequences of such a hypothesis is discussed elsewhere [9].

5. Conclusions

a) The alignment of superfamilies can not be explained by the QCD-jet production but results from some new physical mechanism of coplanar emission of particles in strong interactions.

b) The coplanar emission of high-energy particles is characterized by a sharp energy threshold at $E_0^{th} \gtrsim 8$ PeV and a large transverse momentum of about several GeV/c.

c) A phenomenological model under discussion provides a good fit to the main characteristics of aligned experimental events.

This work is supported by the Russian Foundation for Basic Research, project N 00-02-17271.

REFERENCES

1. L.T.Baradzei *et al.*, Proc. 3rd ISVHECRI, Tokyo (1984), 136.
2. N.M.Amato *et al.*, Preprint CBPF-NF-056/86, Rio de Janeiro (1986).
3. J.-N.Capdevielle *et al.*, Proc. 25th ICRC, Durban, 6 (1997), 57.
4. L.Xue *et al.*, Proc. 26th ICRC, Salt Lake City, 1 (1999), 127.
5. A.S.Borisov *et al.*, Nucl. Phys. B (Proc. Suppl.), 75A (1999), 144.
6. R.A.Mukhamedshin, Bull. Soc. Sci. Lettr. Lodz, Ser. Rech. Def., XVI (1994), 137.
7. A.M.Dunaevsky *et al.*, Proc. 3rd ISVHECRI, Tokyo (1984), 178.
8. L.K.Ding *et al.*, Proc. 3rd ISVHECRI, Tokyo (1984), 142.
9. R.A.Mukhamedshin, Nucl. Phys. B (Proc. Suppl.), to be published in this volume.

ELSEVIER

Nuclear Physics B (Proc. Suppl.) 97 (2001) 122–125

www.elsevier.nl/locate/npe

Phenomenological approach to the problem of alignment

R.A.Mukhamedshin [a], *

[a]Institute for Nuclear Research, Moscow 117312 Russia

Phenomenological features of models of coplanar particles generation are considered.

Phenomenologically, all the models proposed to explain the alignment of the most energetic cores of $\gamma - h$ families [1] can be distinguished by a specific coplanar interaction, mean free path (m.f.p.) for this interaction, characteristic transverse momentum, and so on. Results of the analysis of alignment dependence on these parameters made to estimate model features are given below. Families are assumed to be aligned, if the condition $\lambda_4 \geq 0.8$ is satisfied for their Energetically Distinguished Centers (EDCs), the most energetic objects in gamma-hadron families (particles, clusters, haloes).

1. INTERACTION FEATURES

1.1. Correlation of longitudinal and transverse momenta

From two leading jets (1 and 2) and two QCD jets (3 and 4) generated in the CMS of colliding hadrons, jets 1 and 3 can be really observed in Lab system (Fig 1*a*). The i–th particle's total transverse momentum is proportional to the longitudinal one in both systems: $q^*_{t\,i} \propto p^*_{L\,i} \propto p_{L\,i}$ and $\sin\theta_i \propto const$. As a result, jet 3 can form a separate cluster on the target plane, but the alignment cannot be produced by QCD jets.

Within the framework of the SHDID model [2] used in Sects. 1.2 and 2, which assumes the alignment to be a result of rupture of a quark-gluon string between scattered quarks of the projectile, one can expect the necessary $q_{t\,i} - p_{L\,i}$ correlation in Lab system that produces an observable alignment (Fig. 1*b*).

The favorable $q_{t\,i} - p_{L\,i}$ correlation also takes place in the case of disintegration of a flat (due to a high spin, e.g.) hadron system (Fig. 1*c*).

1.2. Scenario of momentum transfer

Number and coplanarity of particles are determined by the scenario of distributing the transferred momentum $\overrightarrow{Q}_t$ between them as well as by the behavior of $p_t^{\perp}$ –transverse momentum (normal to the coplanarity plane). One can consider the following cases: (*a*) the i–th particle gets a transverse momentum, $q_t^i = q_t^{i-1} + \Delta q_t$, where $\Delta q_t = const$; (*b*) Δq_t decreases with increasing q_t; (*c*) Δq_t and $\langle p_t^{\perp}\rangle$ decrease with increasing q_t.

Fig. 2 demonstrates target diagrams of three interaction versions, which are primarily identical but differ in scenario of $\overrightarrow{Q}_t$ transfer. In case (*b*) the particles are closer one to another, and the particle number increases. In case (*c*) the alignment degree is most high.

2. ROLE OF DECASCADING

Fig. 3 shows the dependence of the fraction of aligned families, $F(\lambda_4 \geq 0.8) \equiv F_{0.8}$, on distance between the interaction point and observation level at varying Z_c for all particles and separately for γ–rays ($Z_c = 0, 1, 20$ TeV·cm). It is seen that (a) $F_{0.8}$ strongly depends on distance and Z_c; (b) at small distances ($x \lesssim 200$ g/cm^2) $F_{0.8}$ depends on the particle type; (c) if $F_{0.8}$ depends on particle type, the distance is not large.

Figure 4 shows the dependence of $F_{0.8}$ on $Z'_c = Z_c/H \simeq q_t$ at a small distance between the interaction point and observation level for γ–rays and all particles. For γ–rays $F_{0.8}(Z'_c \sim 0)$ is not large due to γ–ray emission by π^0–mesons. At $Z'_c \gtrsim 50$ MeV, γ–rays from the parent π^0–mesons begin to be recollected; the maximum is at $Z'_c \approx m_{\pi^0}$. For all particles, $F_{0.8}$ is large at once.

*This work is supported by the Russian Foundation for Basic Research, project no. 98-02-17157.

PII S0920-5632(01)01210-5

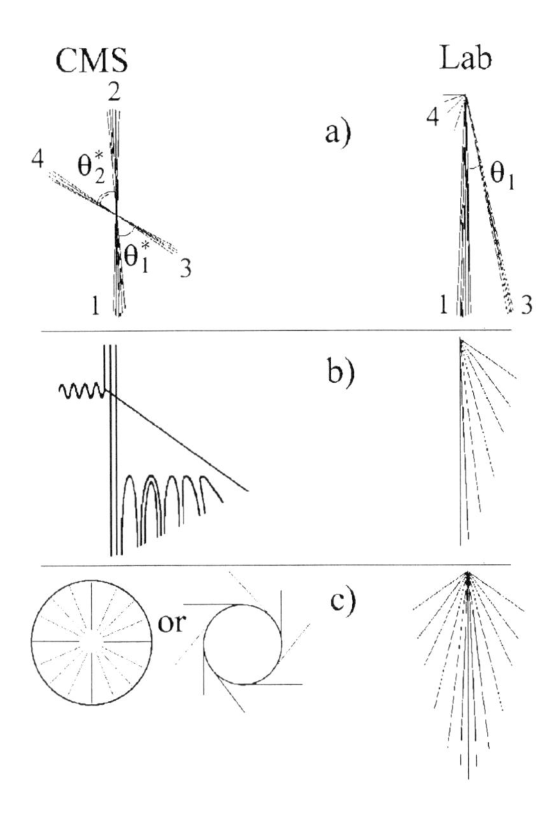

Figure 1. Correlations between longitudinal and transverse momenta of particles in CMS and Lab system.

With increasing the distance from the point of coplanar interaction, cascade processes change the picture. Fig. 5 shows that the Z_c dependence of γ–families is a function of distance as well. So, it could give some information on effective height of coplanar interactions.

3. ANALYSIS OF OBSERVABLES

A simplified model of coplanar particle production (CPM) with the following parameters was used as a heuristic tool: $\langle n_{ch} \rangle \approx 20$; $\langle q_t \rangle = 2.34$ GeV/c; $\langle p_t^{\perp} \rangle = 0.22$ GeV/c; $E - \langle q_t \rangle$ correlation (Fig. 1*c*) is assumed in coplanar interactions produced by a particle with m.f.p. λ_{copl} (for primary protons $\lambda_{copl} = 0$) and decay path λ_{dec}; primary spectrum's index $\beta = 2.05$ at $E_0 \geq 5$ PeV.

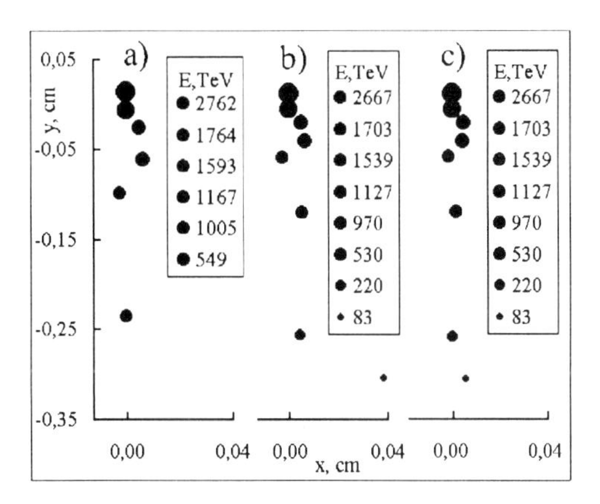

Figure 2. Dependence of alignment on mechanism of distribution of the transferred transverse momentum between particles.

Characteristics of four EDCs in simulated $\gamma - h$ families with energy $\sum E_{\gamma} = 500 - 2000$ TeV are considered below. To select EDCs, a two-stage decascading procedure ($Z_c = 1$ and 20 TeV·cm) is applied. Direct lines designate results found with the QGSM-type code MC0 [3].

Below the dependence of γ–family's characteristics of 4 EDCs on λ_{copl} and λ_{dec} (km, $E = 10$ PeV) is analyzed. Lines are drawn by eye.

Fig. 6 shows that $F_{0.8}$ depends rather weakly on λ_{copl} and λ_{dec}.

Figs. 7 and 8 show rather weak dependence of $\langle R_4 \rangle$ and $\langle (ER)_4 \rangle$ on λ_{copl} and λ_{dec}. For $\langle R_4 \rangle$, virtually only the presence of penetrating components is of importance because this shifts the distribution of interactions deeper into the atmosphere.

Fig. 9 demonstrates rather strong dependence of average zenithal angle $\langle \theta \rangle$ on λ_{copl}, which is suppressed, in part, with reduction of λ_{dec}.

Analysis of the aligned events shows that $\langle R_4 \rangle$ and $\langle \theta \rangle$ do not virtually vary, whereas $\langle (ER)_4 \rangle$ increases slightly ($\sim 20\%$).

To estimate the influence of $\langle q_t \rangle$, the simulation was carried out at $\langle q_t \rangle = 1.28$ GeV/c. Values of $F_{0.8}$ (Fig. 10), $\langle R_4 \rangle$ and $\langle (ER)_4 \rangle$ decrease by a factor of ~ 1.5.

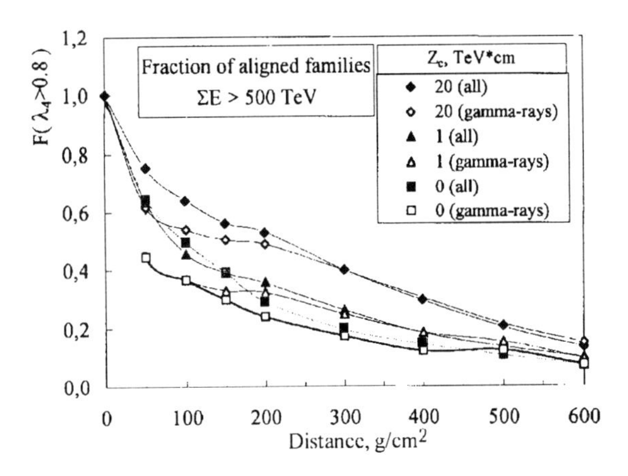

Figure 3. Dependence $F_{0.8}$ on distance between the interaction point and observation level.

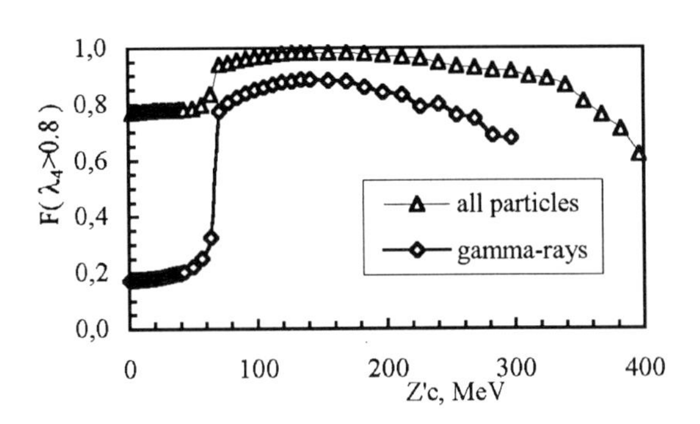

Figure 4. Dependence of $F_{0.8}$ on Z_c' at a small distance from the interaction point.

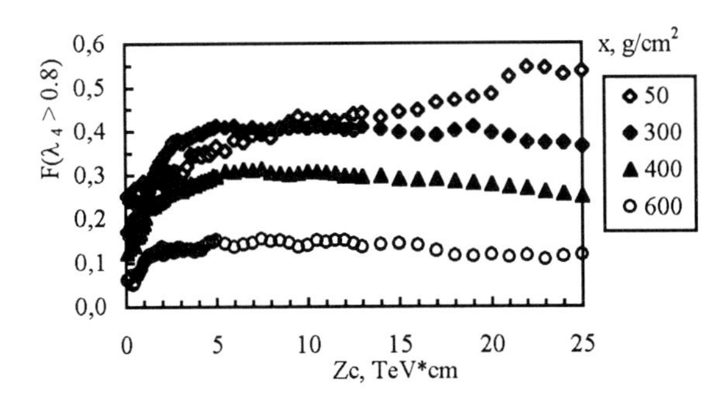

Figure 5. Dependence of $F_{0.8}$ on Z_c and distance between the interaction point and observation level for γ–families.

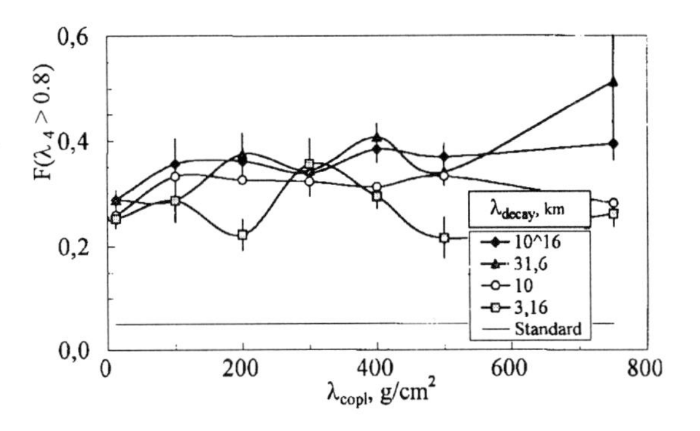

Figure 6. Dependence of $F_{0.8}$ on λ_{copl} and λ_{dec}.

4. CONCLUSION

The simulation and comparison with experimental data [4,5] show that

- if the fraction of aligned events $F_{0.8} \gtrsim 0.2$ at a mountain level then $\sigma_{copl}^{p-air} \simeq \sigma_{inel}^{p-air}$ (if the effect is produced by protons).
- $F_{0.8}$ rather poorly depends on λ_{copl} and λ_{dec};
- characteristic transverse momentum $\langle q_t \rangle$ in the alignment direction is $\simeq 1-2$ GeV/c.
- specific correlation between longitudinal and transverse momenta is preferable;
- the appearance of a component with $\lambda_{copl} \lesssim 200\,\mathrm{g/cm^2}$ is not excluded.

REFERENCES

1. Pamir Collaboration, Proc. 4th ISVHECRI, Beijing (1986) p.4
2. Royzen I.I., Mod. Phys. Lett. A (1994) v.9, no.38, p.3517.
3. Mukhamedshin R.A. *et al.*, Bull. Soc. Sci. Lettr. Lodz, Ser.Rech.Def. (1994), XVI 137.
4. Borisov A.S. *et al.*, Nucl. Phys. B (Proc. Suppl.) **75A** (1999), p. 144.
5. Managadze A.K. Private communication.

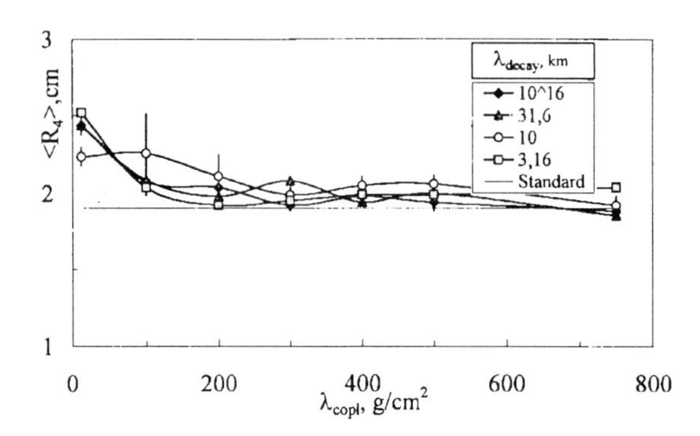

Figure 7. Dependence of $\langle R_4 \rangle$ on λ_{copl} and λ_{dec}.

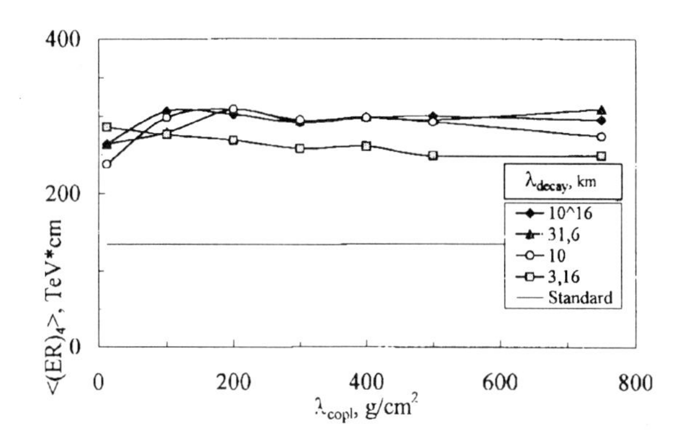

Figure 8. Dependence of $\langle (ER)_4 \rangle$ on λ_{copl} and λ_{dec}.

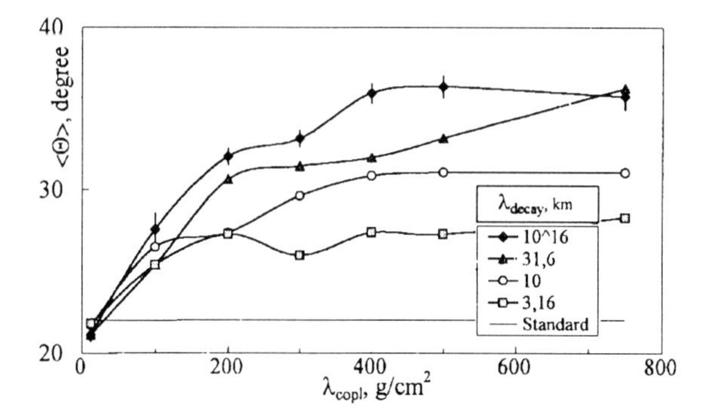

Figure 9. Dependence of $\langle \theta \rangle$ on λ_{copl} and λ_{dec}.

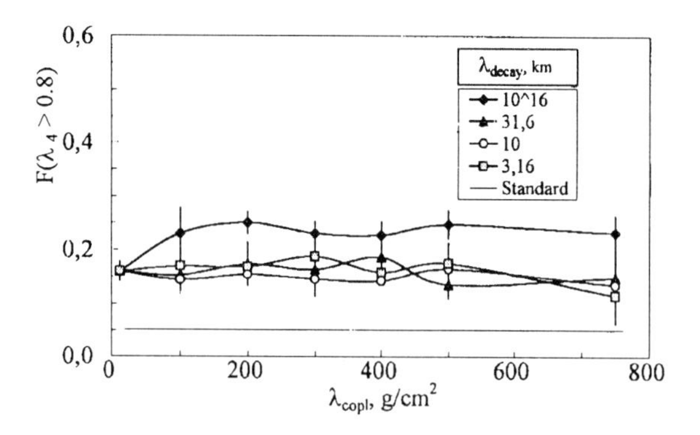

Figure 10. Dependence of $F_{0.8}$ on λ_{copl} and λ_{dec} at $\langle q_t \rangle = 1.28$ GeV/c.

ELSEVIER

Nuclear Physics B (Proc. Suppl.) 97 (2001) 126–129

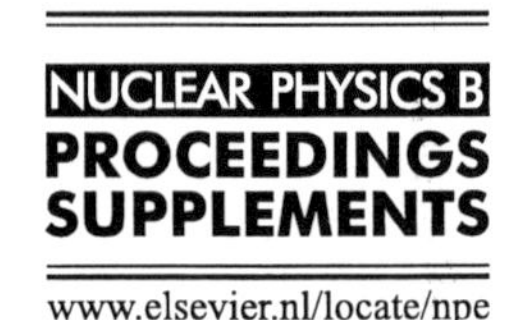

www.elsevier.nl/locate/npe

MULTIPLE PRODUCTION AND COPLANAR EMISSION

Jean-Noel CAPDEVIELLE *
PCC, College de France, Paris

Some samples of stratospheric gamma ray families have been simulated with CORSIKA and compared to the remarkable event JF2af2 obtained during an exposure of emulsion chambers on the Concorde near 10^7 GeV.

CORSIKA has been especially adapted according to X-ray emulsion chamber measurements and aligned events are selected by a least square fit method to one straight line. After comparing the characteristics of alignments and gamma's energetic distributions, for different situations, (diffractive component or on opposite non diffractive component with large multiplicities), it is shown that the coincidence of particular topology of the primary interaction and favorable geometric incidence gives a probability of $1 - 3\%$ to observe alignments from normal multiple production.

In such case,there is a characteristic gap between one of the most energetic gammas and the others and the probability to observe co-planar emission would be increasing with zenith angle. Events with such common features from simulation and experiment are shown.

1. INTRODUCTION

During the last 20 years, 8 emulsion chambers have been flown on the Concorde for different measurements, very high energy jets, stratospheric γ - ray families, γ - ray flux (Iwai et al. 1982, Capdevielle et al. 1979, 1987, 1988) , hyperstrange baryonic matter (Capdevielle et al. 1996) and emulsions for dosimetry. All the detectors, but the last one, were enough thick to measure, at least, the energy of secondary γ - rays up to 1000 TeV, allowing to initiate the collection of informations on multiproduction in the energy range between the limit of the present colliders and the LHC. The present work is devoted to the most energetic event showing a coplanar emission.

2. THE EMULSION CHAMBER CONCORDE EXPERIMENTS

Regular supersonic Atlantic flights provide one "plateau exposure" of more than 2 hours. The same X - ray emulsion chamber carried one hundred times above the Atlantic is then exposed during 200 hours at least at an average altitude of 17km. This corresponds, as indicated by the Concorde flight curve to an atmospheric depth of 100 - 105 $g.cm^{-2}$.The exceptional event described here was obtained in such typical run with a classical structure (2 sheets of Sakura N type X - ray film and one emulsion ET7B plate below 0.25cm lead plates). The X - ray films are used to build the map of the chamber and have a first estimation of the energy. The target diagrams (in case of contained jets) are constructed with the emulsion plates and the development of each individual cascade is followed by counting the number of electron tracks within circles of radii from 10 up to 100 μm to measure the shower energy with an accuracy of 17%. Events occurring above the chamber or in the upper part of the chamber are then measured from their e.m. component.

The photography of this event as it appeared at naked eyes from the X ray film is reproduced on fig.1 with a clear alignment of the most energetic gamma rays.

3. THE EVENT OF 10^7 GeV

The 211 γ's (above 200 GeV) of this event were identified with their respective coordinates and energies; the analysis was then focused on the typical multicluster structure and the planarity, ascertained by naked eyes on the X - ray film, was confirmed, suggesting a multijet structure (Capdevielle 1988).

*capdev@cdf.in2p3.fr

PII S0920-5632(01)01212-9

Figure 1. Coplanar emission visible in the event JF2af2 from the X-ray film.

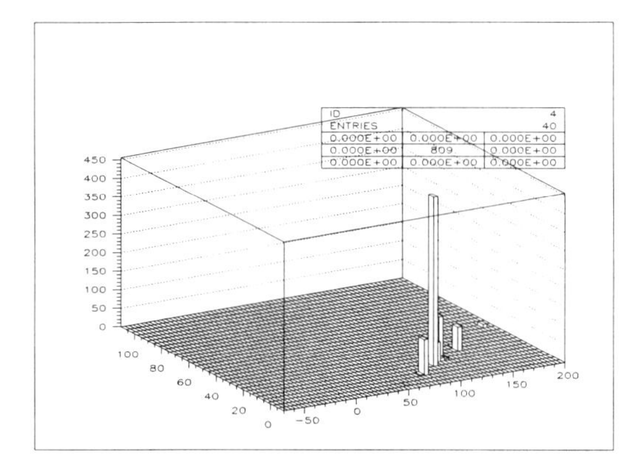

Figure 2. Central part of the event JF2af2.

3.1. ALIGNMENTS

The analysis of individual γ - rays revealed, that, when taken in the order of decreasing energies, the most energetic γ's are situated on a perfect geometrical straight line.

The general aspect of the event is shown on the Lego-plot of fig. 2 (central part) where the coordinates are plotted in mm and the visible energy is plotted on the vertical axis in TeV.

This Lego-plot displays the topology obtained for the 4 most energetic γ rays (above 50 TeV) listed on table 3 and the 38 γ - ray aligned extracted from the 211 γ's of the event.

Table 1
The 4 most energetic γ - rays in JF2af2

E_γ(TeV)	x(mm)	y(mm)
300	100.410	11.077
105	117.468	18.109
75	84.24	5.022
53	110.428	15.551

The regression coefficient of the straight line determined by the coordinates of of those 4 γ's is 0.9993 and it remains equal to 0.992 for the 38 γ's aligned containing 808 TeV, i.e., 51% of the visible energy.

The atmospheric layer, the cabin atmosphere and the cabin wall represent less than $30g.cm^{-2}$ at this altitude and the alignment structure has been pointed out without decascading procedure. When compared to the events recorded in the Pamir experiment (Slavatinski 1997) suggesting a more frequent coplanar emission above 10^7 GeV, i.e. $\sqrt{s} \sim 5$ TeV, this event confirms this tendency (which was not observed for the events recorded at lower energies). The very large multiplicity of this event, for such energy, would imply a larger isotropy for standard multiproduction models: on the opposite, the near perfect alignment observed remains one serious interpellation of the pre - LHC physics in terms of unidimensional properties of hadronic matter at ultra high energy.

4. MULTIPRODUCTION AND COPLANAR EMISSION

From some simulations with CORSIKA, we noticed firstly that the energy distribution of the photons of JF2af2 are easy to reproduce and that the superposition of 3 successive collisions up to the level of Concorde was a common circumstance. A special set of simulations has been carried out with same zenith angle than for JF2af2 of 52^0 for 100 events. This was done with the model HDPM2 taking into account recent features of collider physics such asp_t versus central rapidity density (UA1-MIMI exp.) and recent re-

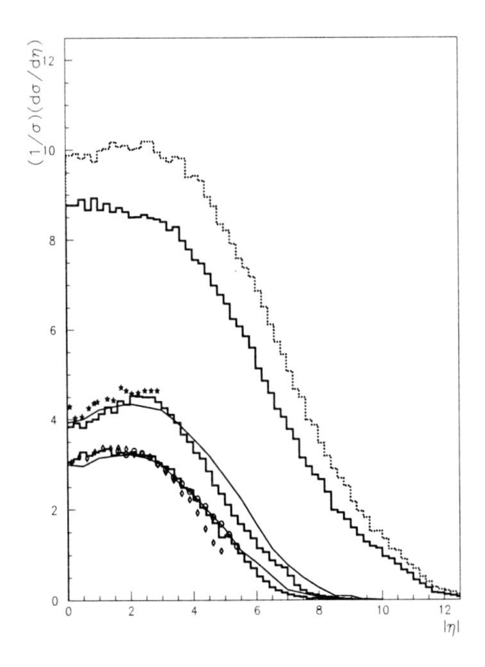

Figure 3. Pseudo-rapidity distributions with HDPM2.

sults of Fermi-lab for pseudo-rapidity up to 5.5 (Capdevielle 99). The pseudo-rapidity distributions obtained with HDPM2 for 2000 collisions NSD are shown on fig.3 (Capdevielle et al.00).

As it was not possible to scan each individual event simulated, it was requested to calculate the linear regression line from the coordinates of the gammas and to plot only the events with a regression coefficient larger than 0.9. The event $n^0 47$ of this serial selected by this method shows a nice alignment of 32 gammas (fig.4).

The similarity of the clusters (gammas above 10 TeV are plotted) with those of fig.1 is especially interesting. Tracing back the genetics of this event, it belongs to the normal NSD multiple production with high multiplicity.

The first collision occurred at an altitude of $24km$ and the regression coefficient of the 32 gammas (above 10 TeV) is 0.968 with a total energy

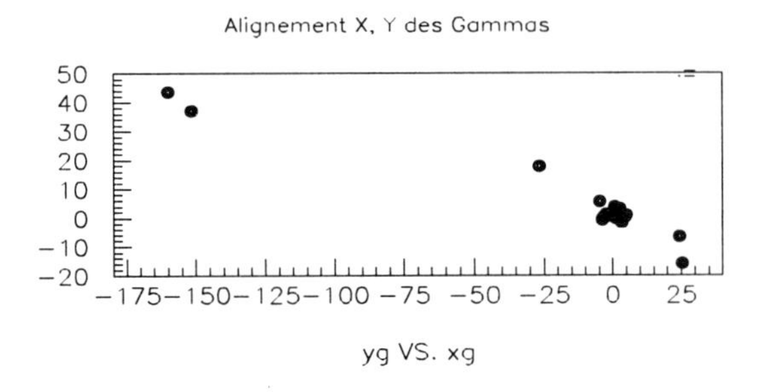

Figure 4. 32 gammas aligned in the simulated event 47.

deposited of 815 TeV. The main features generating the alignment are here :

- high multiplicity
- important zenith angle

Both circumstances combine as follows ; the probability to get a large impulsion transverse is enhanced in high multiplicity events and this large p_t can be devoted to a high energy gamma. This emission can be close, as here, of the vertical plane. The rest of the cluster is displaced in the opposite direction (p_t conservation and the maximal separation between gammas appears in the horizontal plane (the emulsion or X sheets) with a characteristic gap visible in fig.1, fig.4 and in some events of Pamir. On our sample of 100 events, we have counted 2% of events with $r \geq 0.95$, 6% events with $r \geq 0.9$, 10% and 18% with r exceeding respectively 0.8 and 0.7.

The minimal and major displacements corresponding to bremsstrahlung or pair production near the chamber and original emission near the first collision can be seen on table 2 reproducing the event 47 as seen on an horizontal film in the emulsion, following the CORSIKA particle code (1 for gamma, 2 for electron, and 3 for positron) and the indication of generation number (1 for particle produced at 1st interaction..).

Table 2
The 32 most energetic $\gamma's$ in simulated evt 47

particle code	E_γ(TeV)	x(mm)	y(mm)
11	15.6	4.6	0.3
11	58.2	3.1	-1.4
31	29.3	3.4	-1.5
11	56.6	3.2	-1.4
21	23.1	0.5	1.9
31	56.1	4.9	1.0
11	11.9	0.6	3.9
11	14.8	-0.6	1.3
31	33.5	-0.8	1.3
11	19.9	1.1	0.7
11	40.8	0.9	0.2
21	12.8	-160.0	43.6
21	11.2	-151.6	37.1
11	56.5	-151.5	37.1
21	33.6	-0.6	1.7
14	13.2	-2.3	0.8
14	35.0	1.0	-0.2
13	16.3	25.3	-15.8
12	19.6	24.3	-6.5
22	23.0	-4.6	5.6
14	32.3	2.7	3.2
24	22.0	-3.0	0.5
34	28.3	-3.0	0.5
14	25.5	-3.6	-0.7
15	14.8	-2.3	1.3
12	17.3	-2.6	0.1
22	10.1	-26.6	17.8
22	21.5	-3.8	-0.5
12	18.9	-3.7	-0.5
12	17.6	-3.7	-0.3
13	10.7	2.8	-0.7
13	14.2	2.7	-0.6

5. CONCLUSION

The combination of a large transverse momentum generated in the vertical plane (or in the neighborhood) for an energetic secondary at 1st collision can be at the origin of the alignments observed in emulsion chambers. In that case the coplanar emission appears near 10^7 GeV where the ratio primary energy - energy threshold of the chamber is the most favorable. More simulations will be needed to confirm the increase of the probability of alignment observation at large zenith angle with a characteristic gap. The zenith angle distribution of coplanar events, associated to the frequency versus observed energy are probably the best criteria to understand if we have here a pure geometrical- high multiplicity fluctuation in NSD component artifact or if this is a footprint of new physics.

REFERENCES

1. J. N. Capdevielle et al., Proc. 16th ICRC (Kyoto), 6, 324 (1979).
2. J. N. Capdevielle et al., Proc. 20th ICRC (Moscow), 5, 182 (1987).
3. J. N. Capdevielle et al., J. Phys. G, 14, 503, 1988.
4. J. N. Capdevielle et al., Proc. 24th ICRC (Roma), 1, 910 (1995).
5. J. N. Capdevielle, Proc. 26th ICRC (Salt Lake City), 1, 111 (1999)
6. J. N. Capdevielle et al., Astrop.Phys. , 13, 259, 2000
7. J. Iwai et al., Nuovo Cimento A69, 295 (1982).
8. S. A. Slavatinski, Nucl. Phys. 52B, 56 (1997).

ELSEVIER

Nuclear Physics B (Proc. Suppl.) 97 (2001) 130–133

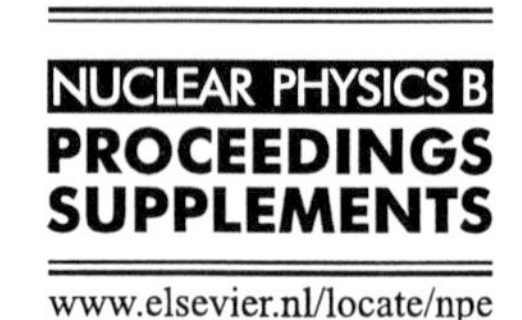

www.elsevier.nl/locate/npe

The distribution of $E \cdot R$ in e-γ families

A.Haungs[a], A.Iwan[b], J.Kempa[b] * †

[a]Institut für Kernphysik, Forschungszentrum Karlsruhe, Karlsruhe, Germany

[b]Department of Experimental Physics, University of Lodz, Lodz, Poland

Various comparisons are made of average characteristics of e-γ families measured in Pamir, with different theoretical models. Among these parameters the mean value of $\overline{E \cdot R}$ (E is energy of gamma quantum and R is the distance from the centre of family) and the distributions of $E \cdot R$ of individual families seem not to agree with theoretical predictions. In the present paper an attempt was made to explain the cause of this disagreement. The reasons which may lead to misinterpretation of primary mass composition of primary cosmic radiation will be discussed based on e-γ families data.

1. INTRODUCTION

Emulsion calorimeters constructed in Pamir consist of multiple layers of lead (or carbon) plates interposed with layers of X-ray films. Having entered the calorimeter, a high-energy electron and/or gamma ray initiates an electron- photon cascade in it. Such a cascade, when passing through the X-ray film layer, produces a black-spot, which can be detect by naked-eye scanning. The shower spots can be noticed when the energies of the entering electrons and photons are greater than 1–2 TeV. It has been assumed that in the upper block of the calorimeter (the so-called Γ-block) γ- quanta and electrons with energy exceeding the threshold value of 4 TeV produce a black-spot, which can be measured with almost 100% efficiency. However, a detailed investigation of fluctuations in the development of electromagnetic cascades shows [1] that it is very difficult to fix such a 100% threshold energy for all the angles and energies. Moreover the fluctuations are very wide. In these calculations energy was reconstructed from the obtained darknesses D. Like in the experiment, only such events were taken for further analysis, in which the energy after reconstruction was higher than 4 TeV.

Groups of spots observed in the calorimeter, caused by electrons and photons, are habitually called "e-γ families". As the energy of the primary nucleus which created a fixed individual family is not known, experimental data are grouped according to the value of the sum of energies of all electrons and photons included in the family (so-called $\Sigma E_{e,\gamma}$). The threshold value of this energy is assumed as $\Sigma E_{e,\gamma} = 100$ TeV in Pamir experiment.

A serious problem and still not completely solved, apart from the wide fluctuations mentioned above, are the so-called overlapping spots. They are cascades, which have its centers at distances smaller than several tens of μm. They may be created by two photons resulting from a π^0 decay just above the calorimeter. In the experiment, the place is found where the darkness in such a dim spot is the greatest and the measurements are carried out in this position with the use of photometers with diaphragms of 48 μm, 84 μm, and 140 μm in diameter. In the final elaboration such two photons are treated as a single photon. This effect gets particularly annoying as we come to greater families $\Sigma E_{e,\gamma} > 500$ TeV. In such families the contribution of such spots is so significant that it requires a separate analysis.

This is the reason why in the present paper we are limited to only two groups of families: $\sum E_{e,\gamma} = (100\text{–}200$ TeV) and $\sum E_{e,\gamma} = (200\text{–}400$ TeV).

*corresp. author, e-mail: kempa@fizwe3.fic.uni.lodz.pl

†partly supported by the Rector of the Lodz University (grant 2000/505/704)

0920-5632/01/$ – see front matter

PII S0920-5632(01)01215-4

Table 1
The average characteristics of "e-γ families" for various ranges of family energies for the string–like model and different primary nuclei [3]. The fluctuations in the cascade development in the calorimeter were not taken into account. The energy threshold for individual photons and electrons is 4 TeV.

Mean value of radius $\overline{R}$, mm

$\Sigma E_{e,\gamma}$, TeV	proton	He	N	Fe	Pamir exp.
100–200	19 ± 1	25 ± 2	34 ± 2	46 ± 3	23 ± 1
200–400	15 ± 1	25 ± 2	33 ± 3	40 ± 5	22 ± 1
400–700	15 ± 1	20 ± 3	33 ± 2	35 ± 2	18 ± 2

Mean value of $\overline{E \cdot R}$, TeV mm

$\Sigma E_{e,\gamma}$, TeV	proton	He	N	Fe	Pamir exp.
100–200	167 ± 6	207 ± 10	271 ± 13	358 ± 21	210 ± 10
200–400	144 ± 7	210 ± 10	250 ± 15	390 ± 52	220 ± 10
400–700	133 ± 10	168 ± 13	230 ± 15	281 ± 21	210 ± 20

2. RESULTS OF MONTE CARLO SIMULATIONS

The QGSJET model and the CORSIKA [2] program were applied in the simulation of the development of nuclear – electromagnetic cascades for different primary particles from proton to iron nuclei. Three series involving 10000 nuclei with the energy sampled from the differential spectrum with the slope $\gamma = -3.0$ above the breaking were taken into account in our calculations. Each series was analyzed separately. Such "e-γ families" were searched for, in which $\Sigma E_{e,\gamma}$ of electrons and photons with the energy above 4 TeV exceeded 100 TeV. All the possible electrons and photons were registered for the theoretical families obtained in such a way. Then, for each electron and each photon, the simulation of the development of individual cascades was made. The darkness D and the reconstructed energy $(E_{e,\gamma})_{\rm rec}$ of the electron (photon) which had initiated the cascade were estimated in the "working" layer under 5 cm of Pb, exactly in the same way as it is done in the experimental approach. The same parameters were calculated for the second time, with the energy threshold $(E_{e,\gamma})_{\rm rec}$ 4 TeV. In each case, we had $E_{e,\gamma}$ above the calorimeter at our disposal, already known and theoretically derived from the calculations, as well as $(E_{e,\gamma})_{\rm rec}$ obtained after the reconstruction. These were the only tracks which were taken into account in the final part of our elaboration for each family above the threshold 4 TeV. Other parameters such as: the number of tracks n_γ and $\bar{R}$ and $\overline{E \cdot R}$ distributions were calculated for each family as well. The families were grouped in the intervals $\Sigma E_{e,\gamma} = (100 - 200)$ TeV and $\Sigma E_{e,\gamma} = (200 - 400)$ TeV twice: before and after reconstructions.

3. MONTE CARLO SIMULATIONS WITH NO RECONSTRUCTION

In the paper [3] the results of Monte Carlo simulations obtained for different string models QGSJET, MC0, MQ as well as for the dual parton model and VENUS were presented. It was shown that the predictions of the string models are in agreement within the error limit and differ from the VENUS productions. This fact makes us believe that when investigating internal parameters of the families we will be able to distinguish between these two types of models in near future.

Whenever the fluctuations in the cascade development are not being taken into account (Table 1) the theoretical results seem to support the primary composition with helium nuclei as those

Table 2
Average characteristics of "e-γ families" for hydrogen and helium primary nuclei with the fluctuations in the cascade development in the calorimeter. The energy threshold for individual photons and electrons is 4 TeV.

Mean value of radius $(\overline{R})_{rec}$, mm

$\Sigma E_{e,\gamma}$, TeV	p	He	Pamir exp.
100–200	21.8 ± 2.2	27.9 ± 3.0	23 ± 1
200–400	17.0 ± 1.8	27.9 ± 3.1	22 ± 1

Mean value of $(\overline{E \cdot R})_{rec}$, TeV mm

$\Sigma E_{e,\gamma}$, TeV	p	He	Pamir exp.
100–200	184 ± 29	228 ± 34	210 ± 10
200–400	158 ± 26	231 ± 34	220 ± 10

which mainly contribute to the interactions in which the "e-γ families" observed in the Pamir experiment are initiated.

4. THE RESULTS OF CORSIKA – GEANT CALCULATIONS

In the present paper, as mentioned above, the fluctuations in the cascade development in the experiments were considered with the use of the GEANT program. The results concerning the parameters of the family after the reconstruction are presented in Table 2 for hydrogen and helium nuclei. The predictions for heavier nuclei are in evident disagreement with experimental results.

5. RESULTS

The obtained results are preliminary. The same tendency may be observed for the mean values in the individual series of calculations, in spite of the fact that the fluctuations had been neglected. If fluctuations had been taken into consideration in the calculation of the development of the cascade in the calorimeter, then the increase of all the parameters characteristic to electron-gamma families is observed in the both groups: for average n_γ by 1, for $\bar{R}$ by 14% and $\overline{E \cdot R}$ by about 10% for both types of nuclei. It is seen from Table 2 that the mean values of the parameters as observed in the experiment are placed between the predictions for proton and for helium. If the fluctuations are not taken into account we can only conclude that heavier nuclei are responsible for the production of families.

It is possible that such an effect was observed in the paper [4]. The fact that characteristics of the families with $\Sigma E_{e,\gamma} > 1000$ TeV differ from the theoretical predictions may be caused by the three factors:

1. fluctuations in the cascade development were not taken into consideration,
2. overlapping spots, i.e. very large families, were not taken into consideration,
3. an old UA5 random generator was used in the Monte Carlo simulation.

All these factors mean that a meticulous care should be taken when dealing with the information [4] about the observed change of the iron contribution from (2 – 3 %) to (27 ± 10 %) in the primary spectrum at the highest energies.

REFERENCES

1. A.Haungs, J.Kempa, 1997, Proc. 25th ICRC, Durban, **6**, 101

2. D.Heck et al., 1998, FZKA Report 6019, Forschungszentrum Karlsruhe
3. A. Borisov, V. Denisova, Z. Guseva, A. Iwan, S. Kasparova, J. Kempa, A. Krys, V. Maximenko, V. Puchkov and S. Slavatinsky, 1999, Proc. 26th ICRC, Salt Lake City, **1**, 84
4. C.E. Navia, C.R.A. Augusto, V. Kopenkin, Y. Fujimoto, M. Moriya, A. Ohsawa, K.A. Managadze, I.V. Rakobolskaya, T.M. Roganova, L.G. Sveshnikova, E.H. Shibuya, and S.L.C. Barroso, 1999, Proc. 26th ICRC, Salt Lake City, **3**, 284

ELSEVIER

Nuclear Physics B (Proc. Suppl.) 97 (2001) 134–137

www.elsevier.nl/locate/npe

Test of High-Energy Interaction Models with Data of the Emulsion Chamber Experiment PAMIR

A. Haungs[a*], J. Kempa[b†], J. Malinowski[b†]

[a]Institut für Kernphysik, Forschungszentrum Karlsruhe, P.O. Box 3640, D–76021 Karlsruhe, Germany

[b]Department of Experimental Physics, University of Lodz, PL–90236 Lodz, Poland

The emulsion chamber experiment PAMIR at high mountain altitude (4370m a.s.l.) registers hadronic and electromagnetic secondary particles of cosmic rays with a high threshold energy in the TeV region. These particles are produced preferentially in the extreme forward direction of interactions of cosmic ray protons or Helium nuclei in the atmosphere with primary energies mainly below 1 PeV. Hence the measurements are sensitive to the physics of high-energy hadronic interaction models as implemented in the Monte Carlo air shower simulation program CORSIKA. By use of detailed detector simulations we compare data of the PAMIR experiment with predictions of different models for the energy spectrum of the electromagnetic component.

1. INTRODUCTION

Monte Carlo simulations of extensive air showers based on different high-energy interaction models are used for the reconstruction and interpretation of indirect cosmic ray measurements. The validity question of these models increases with the improved precision of the measurements by experiments like KASCADE [1,2]. The experimental results are meanwhile sensitive to the various theoretical approaches for the interactions of high-energy protons or nuclei with air-molecules. Alas, the unknown elemental composition and primary energy spectrum in the PeV region smear out the information content of the data concerning the interactions, or vice versa the unknown interaction features hinder to follow the main aims of these experiments to determine composition and energy of the PeV cosmic rays.

Experiments at high mountain altitudes on Pamir or Mt. Chacaltaya measure electrons, gammas and hadrons with high particle energy thresholds in the TeV region by emulsions or X-ray films [3]. Besides the reconstruction of the cores of extensive air-showers in the knee region by so called particle families, integral measurements of single hadrons and electrons/gammas are performed. At high altitudes these particles stem mainly from primary cosmic rays of energies below 1 PeV and were produced in extremely forward direction. This explains the special suitability of these measurements for the test of high-energy hadronic interaction models: In the primary energy region of 10-100 TeV the elemental composition and flux of the cosmic rays are approximately known from direct measurements on balloons or satellites. The sensitivity of the emulsion experiments to the extreme forward direction of the interaction has complementary information to the data of accelerator experiments to which the cross sections of the interaction models are adjusted.

In the present paper we compare the expected inclusive flux of e/γ and hadrons for the PAMIR experiment for seven different high-energy interaction models, all embedded in the air-shower simulation program CORSIKA [4]. The measured distribution of the electromagnetic component with the e/γ-block of the PAMIR experiment is compared with the expectations of the simulations including detailed detector simulations. The considered observable is the so-called optical density of spots displayed in X-ray films.

*corresponding author, e-mail: haungs@ik3.fzk.de

†partly supported by the Polish State Committee for Scientific Research (KNB grant No PB 929/P3/95/7) and partly by the Rector of the Lodz University (grant 2000/505/704)

PII S0920-5632(01)01219-1

2. CORSIKA SIMULATIONS

For the following analyses air-shower events are generated for three different nuclei (H, He, Fe), using seven different interaction model codes (VENUS vers.4.12 [5], QGSJET version of CORSIKA 5.62 [6], SIBYLL vers.1.6 [7], HDPM [4], DPMJET vers.2.4 [8], SIBYLL vers.2.1 [9], and neXus vers.2β [10]) implemented in CORSIKA (basically vers. 5.62). For each primary and model 500,000 events were generated, except for the models DPMJET and neXus, where somewhat fewer events were simulated in view of the long computing time required for these codes. In the case of primary protons the simulations cover the energy range of 10^{13} eV - 10^{16} eV with slope $\gamma_H = 2.75$ and isotropic incidence up to 40°. In the case of primary helium (iron) the used slope is $\gamma_{He} = 2.62$ ($\gamma_{Fe} = 2.60$) in the energy range $2 \cdot 10^{13}$ eV - 10^{16} eV (10^{14} eV - 10^{16} eV). The slopes were taken from the compilation of direct measurements [11]. All secondary particles with energies larger than 1 TeV at the observation level of the Pamir experiment (4370 m) are taken into account.

Figure 1 compares the energy spectra of hadrons and electromagnetic particles produced by primary protons for all interaction models. Obviously for both particle types the spectra reflect somehow the primary energy slope, and differences of the slopes are within statistical uncertainties. But the absolute scales of the predictions differ. The SIBYLL 1.6 and HDPM produce significantly more and DPMJET significantly fewer particles than the other models. QGSJET and neXus give quite similar predictions for both components, VENUS agrees with them in the electromagnetic part, but produces some more hadrons. SIBYLL 2.1 behaves in the opposite way: It is in agreement with QGSJET and neXus for the hadronic part, but predicts a higher number of electromagnetic particles. Roughly, the particle numbers are inversely proportional to the proton-air total cross sections used in the respective simulations [4].

The measurable spectra at the experiment Pamir are mainly due to primary protons. This is demonstrated in Figure 2, where the contributions of the different primaries to the total flux of electromagnetic particles is shown. The contribution of heavy primaries is not only reduced due to the assumed fluxes, but also by the faster development by showers induced by heavy particles.

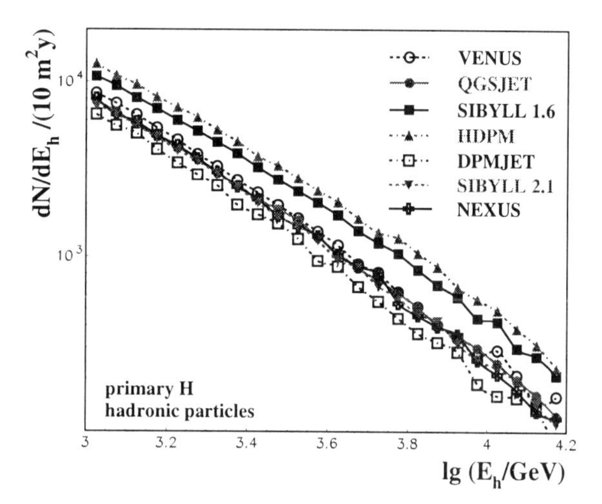

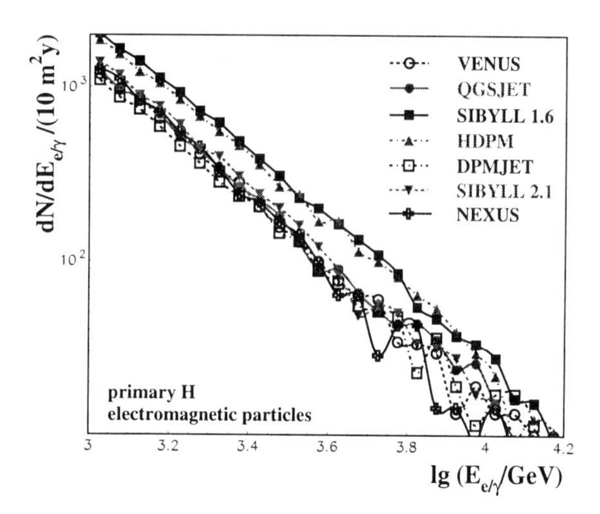

Figure 1. *The energy spectra of hadronic and electromagnetic secondary shower particles above 1 TeV produced by primary protons as expected for the PAMIR observation level for different interaction models. The statistical uncertainties of the simulations are not plotted, but affect the tail of the spectra. The lines are for guiding the eyes, only.*

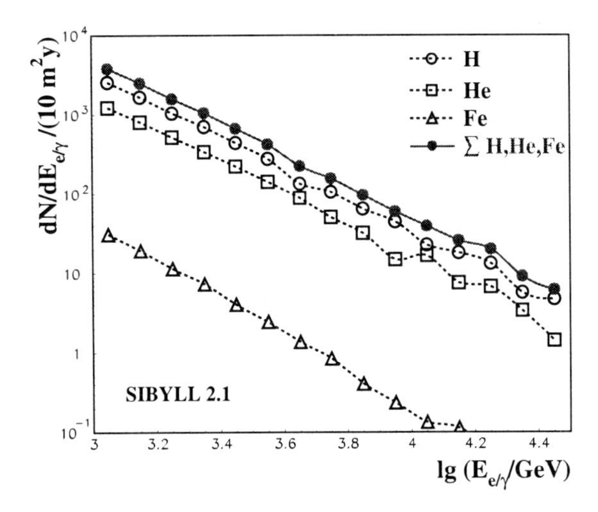

Figure 2. *Contributions of different primary nuclei to the expected electromagnetic particle spectrum for the PAMIR observation level.*

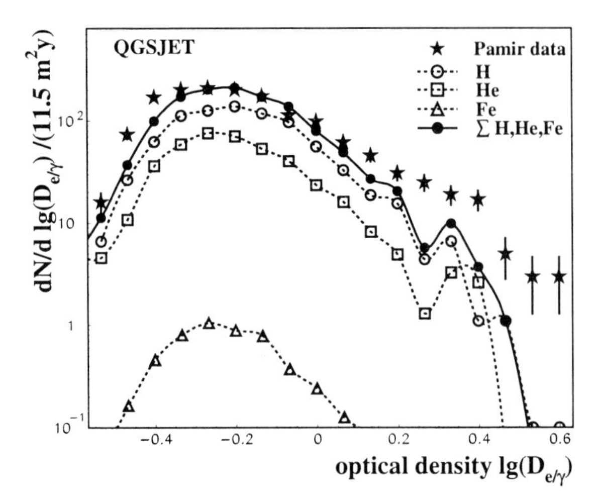

Figure 3. *Measured optical density spectrum ($ST = 11.5\, m^2 yr$) of single particles in the working layer (12 c.u.) of the Pamir experiment ($N_{e,\gamma} = 1469$) compared with the simulated spectra (QGSJET) of different primaries normalized to the exposure time of the measurement.*

Primaries with larger energies do not compensate this effect, due to the steeply decreasing primary spectrum.

3. DETECTOR SIMULATIONS

For electromagnetic particles the optical densities in the X-ray films were calculated for photons and electrons of different energies and angles of incidence by a detailed detector simulation using the GEANT tool [12]. The results were then parameterized. An optical density is then calculated for each e.m. particle (of the CORSIKA simulations) with given energy and angle of incidence by interpolating the density distributions [13]. These procedures account for the response of the detector, including its fluctuations, and for the efficiency and threshold effects of the PAMIR experiment. Figure 3 shows a measured optical density distribution [14] compared with distributions (QGSJET) simulated including the detector response as described above. The absolute values result from the exposure time of the measurements, measured fluxes and chemical composition [11]. The comparison shows a good agreement in both, the slope of the spectrum and the total number of particles. Even the threshold effects (detection threshold is around 4 TeV, but with large fluctuations [15]) seem to be reproduced in a correct way. Deviations between simulation and measurement at large optical densities (large energies) may be attributed to the limited maximum primary energy in the simulations. Please note the negligible contribution of particles from primary iron nuclei and the only slight differences in the slopes for different primaries. In general the QGSJET simulations match the data satisfactorily. This is confirmed by using various measured data samples of the same observable for the comparisons.

In Figure 4 the predictions of all regarded interaction models are compared with the measurements for the medium region of the optical density on a linear scale. The differences in particle numbers as discussed in Fig. 1 are still visible after the detector simulation and conversion into the measured observable. That means, the SIBYLL (version 1.6) and HDPM models predict the largest numbers, and their particle fluxes

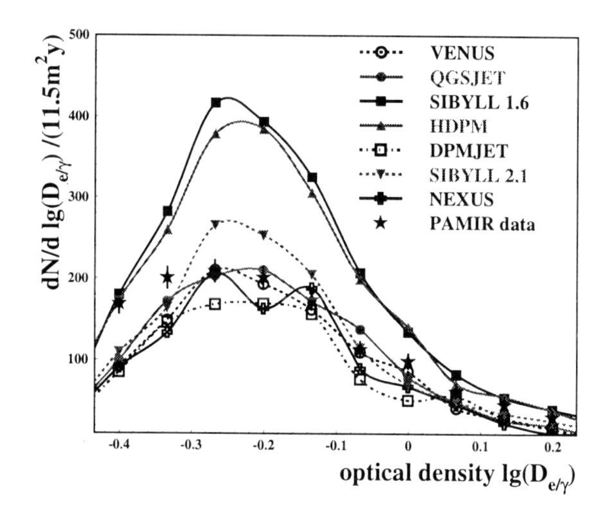

Figure 4. *Comparison of the predictions of the optical density spectrum from different high-energy interaction models with the measurements. For the simulated spectra expectations of different primaries are normalized to the exposure time of the measurement and summed up.*

are far off the measurements. Even SIBYLL 2.1 seems to produce too many electromagnetic particles above TeV energies at the observation level of PAMIR. All the other models agree with the data within the statistical uncertainties of simulations and measurements. The DPMJET model tends to underestimate the total number of particles.

4. CONCLUSIONS

Advanced tools for the simulation of the detector response of cosmic ray experiments improve the quality of the interpretation of the measurements. The combination of the air shower simulation in the atmosphere (CORSIKA with different high-energy interaction models) and the simulation of the detector response (GEANT) leads to a reasonable reconstruction of measured parameters. The present results can be used as hints for the interpretation of the various interaction features in the models, especially for the extremely forward direction at relatively low primary energies of 50-1000 TeV. In future the present comparisons of the electromagnetic part including the detector simulation will be extended to the hadronic secondaries. This is expected to increase the sensitivity of the measurements to the various interaction models.

ACKNOWLEDGEMENT

We would like to thank D. Heck for preparing the CORSIKA program, especially for the embedding of the newly developed codes SIBYLL 2.1 and neXus.

REFERENCES

1. G. Schatz et al.-KASCADE collaboration, these proceedings.
2. A. Haungs et al.-KASCADE collaboration, these proceedings.
3. L.T. Baradzei et al.-Chacaltaya and Pamir collaboration, Nucl. Phys. B 370 (1992) 365.
4. D. Heck et al., FZKA Report 6019, Forschungszentrum Karlsruhe (1998).
5. K. Werner, Phys. Rep. 232 (1993) 87.
6. N.N. Kalmykov, S.S. Ostapchenko, and A.I. Pavlov, Nucl. Phys. B (Proc. Suppl.) 52B (1997) 17.
7. J. Engel et al., Phys. Rev. D46 (1992) 5013.
8. J. Ranft, Phys. Rev. D51 (1995) 64.
9. R. Engel, Proc.26th ICRC 1999, HE 2.5.03.
10. H.J. Drescher et al., preprint hep-ph/9903296, March 1999.
11. B. Wiebel-Sooth et al., Astron. Astroph. 330 (1998) 389.
12. A. Haungs and J. Kempa, Proc.25th ICRC 1 (1997) 101.
13. A. Haungs and J. Kempa, Proc. 16th ECRS 1998, Alcala, Spain, Ed. J.Medina, p.583.
14. H. Bialobrzeska et al., Nucl. Phys. B (Proc. Suppl.) 75A (1999) 162.
15. A. Haungs et al., Proc.26th ICRC 1999, HE 1.2.23.

ELSEVIER

Nuclear Physics B (Proc. Suppl.) 97 (2001) 138–141

www.elsevier.nl/locate/npe

Inelasticity Fluctuations and the Efficiency of Emulsion Chambers for Hadrons

G. Schatz[a]*, J. Oehlschläger[a]

[a]Institut für Kernphysik, Forschungszentrum Karlsruhe, P.O. Box 3640, D–76021 Karlsruhe, Germany

Emulsion chamber experiments register high energy hadrons in cosmic rays indirectly, via gamma rays produced by nuclear reactions in the chamber materials (usually carbon and lead). If hadrons fail to produce any gamma rays above the detection threshold of typically a few TeV they will escape detection even in an infinitely thick chamber. The hadron energy is determined by dividing the energy sum of the secondary gamma rays by a mean inelasticity. Here the questions arise whether this conversion factor is the same for mesons and baryons and how large the fluctuations of this inelasticity are. These problems are investigated by simulations of the interactions of charged pions and protons with carbon and lead nuclei using the VENUS interaction model. Our calculations indicate considerable deviations of the detection probability from 1 in the energy region one to two orders of magnitude above the detection threshold for gamma rays, and clear differences between mesons and baryons, also as far as inelasticities and their fluctuations are concerned.

1. INTRODUCTION

Emulsion chamber experiments register high energy hadrons and electromagnetic particles in cosmic rays, with a threshold of typically a few TeV. These particles are usually concentrated in the cores of extensive air showers (EAS). For gamma rays, electrons and positrons this threshold is comparatively sharp and their energy can be determined to a reasonable accuracy. (Since these types of particles cannot be distinguished experimentally we will refer to them collectively as gammas in the following text.) Hadrons however are only detected via secondary gamma rays produced in interactions with the detector material, mainly carbon and lead. The energy sum of these gamma rays, usually referred to as 'visible energy' or E_γ^h, is clearly smaller than that of the interacting hadrons. The energy of the latter is then estimated by dividing E_γ^h by a mean (in)elasticity determined theoretically. This inelasticity is known to fluctuate considerably. In addition, it may be expected to depend on the nature of the hadron, whether meson or baryon, and on its energy. The latter effect results from the the fixed threshold of the detector for gamma rays and the fact that the energy of secondary gammas will increase with the hadron energy. These effects have been known for a long time but, to the best of the authors' knowledge, do not seem to have been studied in detail. It is the purpose of this contribution to investigate these aspects on the basis of simulations with the VENUS interaction model [1].

*corresponding author; present address: Habichtweg 4, D-76646 Bruchsal, Germany, e-mail: bgschatz@t-online.de

2. SIMULATIONS AND RESULTS

Interactions of charged pions and protons with the nuclei ^{12}C and ^{208}Pb were simulated in the energy range 3 TeV to 10 PeV using the VENUS interaction program [1]. 10000 events were calculated for each combination of target and projectile and for each energy. Neutrons are expected to react in a very similar way as protons, and pions and nucleons are the predominant hadron species in EASs. The VENUS program was selected because we had used it in previous related studies [2] although it is now outdated (even in its author's opinion [3]) and although measurements by the KASCADE group have shown [4] that the hadron component of EASs is best described by the QGS model [5]. We feel that the results presented here are nevertheless adequate to reveal the important features of the problem.

The results of these simulations were then evaluated in the following way: Each event was first

PII S0920-5632(01)01222-1

searched for gamma rays above 1 TeV which was assumed to be the (sharp) threshold of the detector. Events with no gammas above this energy were assumed to escape detection. This yields immediately the (in)efficiency of an infinitely thick detector made of the respective material. Fig.1 displays the results. The efficiencies rise

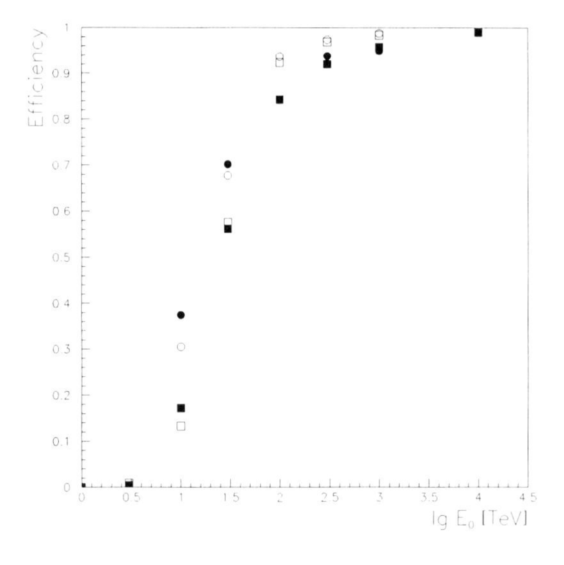

Figure 1. *Simulated efficiencies for an infinitely thick detector. Different symbols refer to different combinations of target and projectile, with squares representing protons and filled symbols carbon.*

from 0 at the assumed gamma ray threshold to values near 1 high above. Even at 10 times the threshold energy the probability of detection is clearly lower than 0.5. Also, the chance of detecting a pion is considerably larger than for a nucleon in the region above the gamma threshold. This appears reasonable in view of results reported previously that the fraction of energy going into electromagnetic particles is larger in pion induced reactions on light nuclei than in proton induced ones [2].

The other relevant quantity for the analysis of emulsion data is the fraction of the primary energy registered in the detector ('visible energy' or E_γ^h). This is displayed in Fig.2 for the 4 combinations of target and projectile. At all energies, pions convert a higher fraction of their energy into gammas, again in line with our previous results [2]. The fraction approaches c. 0.25 for high energy pions whereas the corresponding value for protons is smaller by about a third. The energy dependence of these quantities may appear surprising, especially the rise when approaching the gamma threshold. But this can be easily understood: An event will only be detected if at least one gamma is above 1 TeV. This then represents a large fraction of the primary energy. The shallow minimum visible at about an order of magnitude above threshold may be understood if the fraction of events detected increases faster than their visible energy.

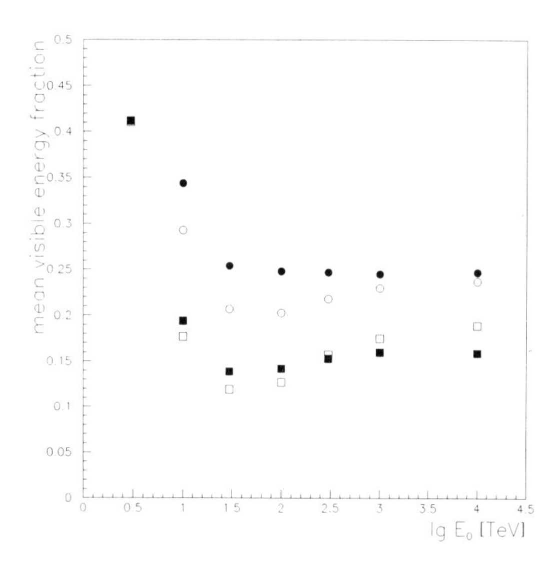

Figure 2. *Mean visible energy fraction versus primary energy. Different symbols refer to different combinations of target and projectile, with squares representing protons and filled symbols carbon.*

The dependence on the target material is apparently not dramatic.
A further relevant quantity is the fluctuation of the visible energy. Two examples of our respective results are shown in Fig. 3 and 4 for 4 energies from 10 TeV to 10 PeV. Both figures refer to 10000 simulated events. All distributions exhibit an exponential tail towards higher energy fractions which drops much steeper for the heavier target and projectile. For protons, even a broad maximum develops at higher primary energies. Hence the fluctuation is larger for incident pions.

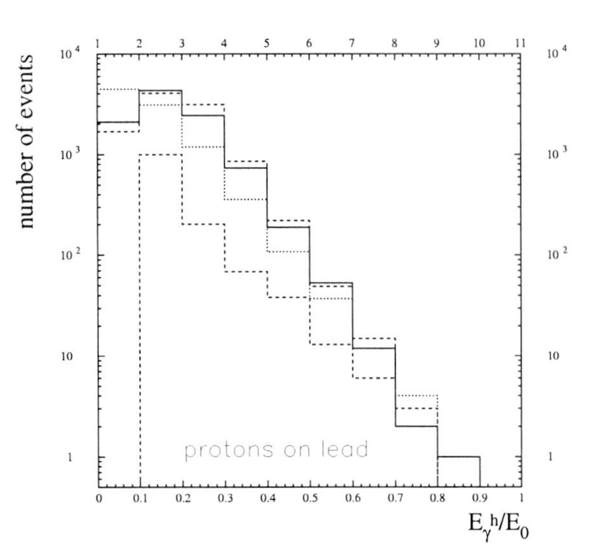

Figure 4. *Distribution of the visible energy fraction for protons on lead. Different curves represent different primary energies of 10, 100, 1000 and 10000 TeV, increasing from below. The event numbers refer to 10000 simulated events.*

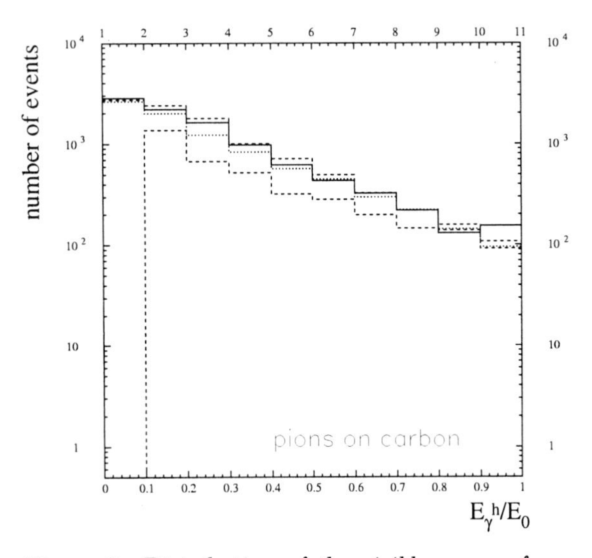

Figure 3. *Distribution of the visible energy fraction for pions on carbon. Different curves represent different primary energies of 10, 100, 1000 and 10000 TeV, increasing from below. The event numbers refer to 10000 simulated events.*

3. DISCUSSION

We feel that the effects discussed here should be taken into account when evaluating emulsion chamber data. The slowly rising efficiency will tend to suppress the number of hadron in the energy range one or two orders of magnitude above threshold. This implies, e.g., that the most probable number of hadrons in the famous first and most conspicuous Centauro event may be even larger than estimated [6]. The differences in visible energy between pion and nucleon projectiles add a further complication to the quantitative determination of the energy of a detected hadron but are in line with previous simulation results [2].
The results presented here are based on simulations with one model of high energy strong interactions. Therefore it is probably premature to draw quantitative conclusions from the data shown in the figures. As mentioned above the KASCADE collaboration have concluded that the QGS model developed by the Moscow group [5] describes tehir data best (at least in their lower energy range which covers most of the emulsion data).
It is also worth pointing out that our results are based on a very simple-minded model of the detector response. For really quantitative analyses,we believe that a more refined procedure, as

attempted presently by the Lodz group [7], is required.

REFERENCES

1. K. Werner, Phys. Rep. 232 (1993) 87
2. G. Schatz and J. Oehlschläger, Proc. VIIIth Int. Symp. High Energy Cosmic Ray Interact., Tokyo 1994
3. K. Werner, private communication, 2000
4. T. Antoni et al., KASCADE collaboration, J. Phys. G: Nucl. Part. Phys. 25 (1999) 2161
5. N. N. Kalmykov and S. S. Ostapchenko, Yad. Fiz. 56 (1993) 105.
6. C. M. G. Lattes, Y. Fujimoto and S. Hasegawa, Phys. Rep. 65 (1980) 151
7. A. Haungs, J, Kempa and J. Malinowski, these proceedings

ELSEVIER

Nuclear Physics B (Proc. Suppl.) 97 (2001) 142–145

www.elsevier.nl/locate/npe

A new detector for the measurement of the energy spectrum of cosmic ray nuclei in the TeV region*

J.R. Hörandel†, F.H. Gahbauer, G. Hermann, D. Müller, S.P. Swordy, S.P. Wakely
The Enrico Fermi Institute, The University of Chicago, 933 East 56th Street, Chicago, IL 60637, USA

A new instrument for direct measurements of the individual energy spectra of heavy cosmic-ray nuclei in the energy range from 10^{11} to several 10^{14} eV shall be described. The instrument performance is discussed, using first results from a 28-hour balloon flight and from an accelerator calibration.

1. INTRODUCTION

The individual energy spectra of heavy cosmic rays are poorly known above a TeV/nucleon. The goal of TRACER (Transition Radiation Array for Cosmic Energetic Rays) is to measure these spectra in the energy range from 10^{11} to several 10^{14} eV.

It is planned to achieve this goal in three stages: a standard balloon flight and two long-duration balloon flights.

The first balloon flight was conducted in September 1999 from Ft. Sumner/New Mexico. The payload has been 28 hours at operating altitude with less the 6.5 g/cm^2 residual atmosphere above the detector. All detector components were operating successfully. First results of this flight will be discussed in the following.

To increase the statistical quality of the data, a circum terrestrial flight of 14 days is planned for June 2001 from Fairbanks/Alaska. The currently used electronics, which was taken over from previous experiments, has limited dynamic range, and therefore restricts the charge range to nuclei from oxygen to iron. A new read-out system, allowing to cover the charge range from lithium to nickel, will be developed for another flight, planned for 2003.

2. DETECTOR SET-UP

TRACER measures the nuclear charge, the energy, and the trajectory through the instrument for each single cosmic ray nucleus. The particle energy is determined from measurements of the ionization loss in gases, and with a transition radiation detector (TRD). This approach permits the construction of large-area detectors without requiring an exorbitant detector mass. A schematic view of the detector system with a geometric factor of about 5 m^2 sr is shown in figure 1.

TRACER consists of eight double layers of 98 proportional tubes each which are oriented alternately in two orthogonal directions in order to determine the particle trajectory. The proportional tubes are 2 m long and 2 cm in diameter, consisting of aluminized Mylar with a wall thickness of 125 μm, and are filled with a xenon-methane mixture. The ionization losses of the particles are measured with the four upper double layers forming a proportional tube array. Four radiators of plastic fiber material, previously used in the CRN experiment [1], each followed by a double layer of proportional tubes, form a TRD to measure the particle energy.

Two scintillators ($200 \times 200 \times 0.5$ cm^3), placed on top and bottom of the detector stack, act as instrument trigger. In addition, the charge is determined through measurement of the specific ionization. A Čerenkov counter ($200 \times 200 \times 1$ cm^3) of acrylic plastic at the bottom of the detector is used to reject non-relativistic particles. The scintillation and Čerenkov counters are read out with photomultiplier tubes via wavelength shifter bars.

*Supported by NASA grant # NAG 5-5072.
†http://hoerandel.com

PII S0920-5632(01)01226-9

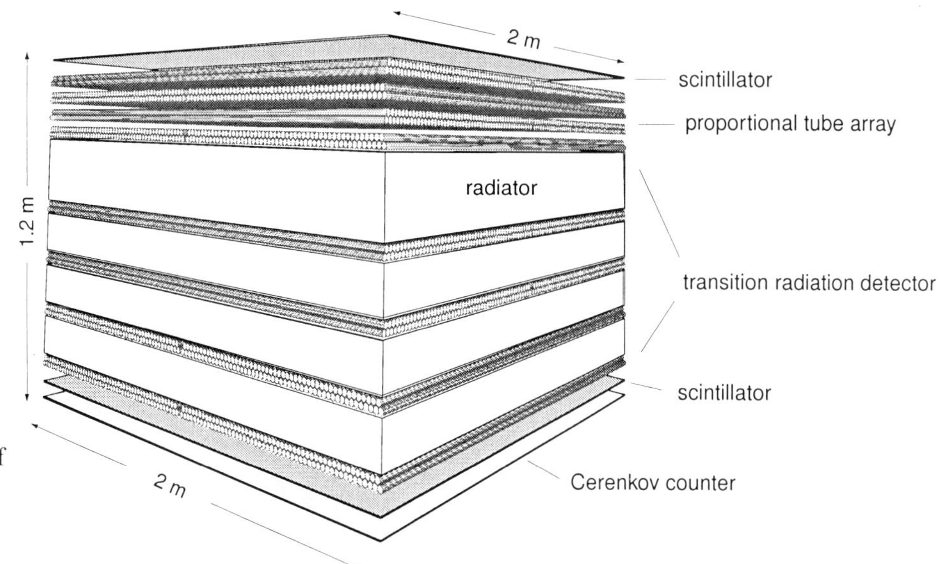

Figure 1. Schematic view of the TRACER detector system.

3. ACCELERATOR CALIBRATION

One of the main advantages of this detector system is the possibility of an absolute calibration at a test beam of singly charged particles. The response of the proportional tube array as well as the TRD system depends on the Lorentz factor $\gamma = E/mc^2$ of the particle. Scaling with Z^2 then permits one to predict the response to heavy nuclei.

The calibration was conducted from November 1999 to January 2000 in the fixed target area at Fermilab, using a secondary beam with energies from 3 up to 227 GeV.

As an example, the detector response for 35 GeV electrons is shown in figure 2. The energy deposit in the proportional tube array exhibits the typical Landau distribution. The signals in the TRD can be described as the superposition of a Landau function to describe the ionization losses and a second function to describe the energy deposit due to transition radiation photons. The parametrisation

$$f(\lambda) = \frac{C}{\sqrt{2\pi}} e^{-\frac{1}{2}(s\lambda + e^{-\lambda})} \text{ with } \lambda = \frac{\Delta E - E^*}{\xi}$$

with the parameters C, E^*, ξ, and s can be used to describe both components of the signal, ionization and transition radiation. The function gives the probability for an energy deposit ΔE, and the parameters s and E^* depend on the Lorentz factor. $f(\lambda)$ represents a Landau distribution for $s \equiv 1$ [2].

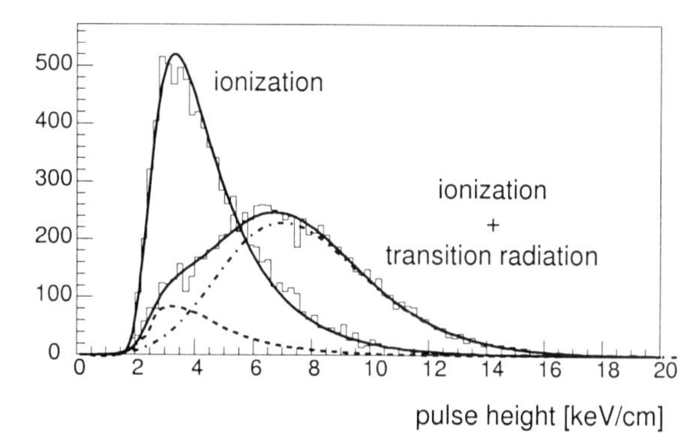

Figure 2. Measured energy deposit of 35 GeV electrons in the TRACER proportional tubes filled with a (50%,50%) xenon–methane–mixture.

4. INSTRUMENT PERFORMANCE

The analysis of the first balloon flight is presently (June 2000) in progress; first results are presented in the following.

Detailed Monte–Carlo calculations for the detector response of the complete TRACER system have been performed, using the detector simulation tool GEANT4 [3]. δ–ray production is an important contributor to the fluctuations of the energy deposit for heavy nuclei. As the present version of GEANT does not include δ–ray production for heavy ions, these, with charge Z, were simulated by superimposing $N = Z^2/\zeta^2$ lighter particles with charge ζ.

4.1. Signal fluctuations

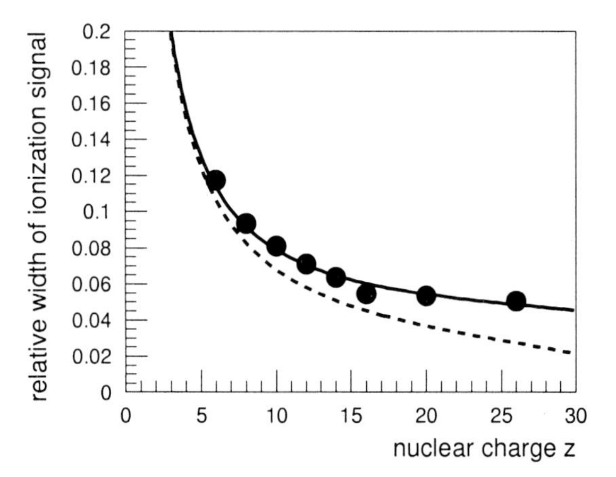

Figure 3. Measured relative width of ionization signal as function of the nuclear charge.

Once the track of a particle through the detector is known, the specific ionization of each particle can be measured. The relative width of the ionization signal in the proportional tubes is shown in figure 3 as function of the nuclear charge Z. The relative width of the signal decreases with Z and can be described by the function $\Gamma_\sigma^2(Z) = a^2/Z^2 + b^2$, with the parameters $a = 0.69$ and $b = 0.04$. The dashed line is the result of simulation calculations, including the intrinsic fluctuations of the energy deposit and the uncertainties introduced by the track reconstruction. Additional signal fluctuations of 4% due to the finite resolution of the proportional tubes and due to amplifier gain variations lead to the solid curve.

4.2. Track reconstruction

The perpendicular arrangement of the proportional tubes allows to reconstruct the trajectories of particles traversing the detector. Since the energy deposit in a proportional tube is proportional to the pathlength of the particle in this tube, the position of the trajectory relative to a tube can be obtained with a precision much better than a tube diameter. The spatial resolution of the track reconstruction is shown in figure 4 as function of the nuclear charge. The signal fluctuations of singly charged particles are relatively large (see figure 2). A maximum liklihood method taking the asymmetric Landau fluctuations into account is used to determine the particle trajectory in this case. It leads to a spacial resolution of 2.5 mm. The relative signal fluctuations decrease with the particle charge as demonstrated in figure 3. For heavy nuclei, the deposited energy exhibits a symmetric, Gaussian distribution, allowing to reconstruct the tracks with much better precision. Resolutions from 1 mm to 0.25 mm are achieved for oxygen and iron, respectively.

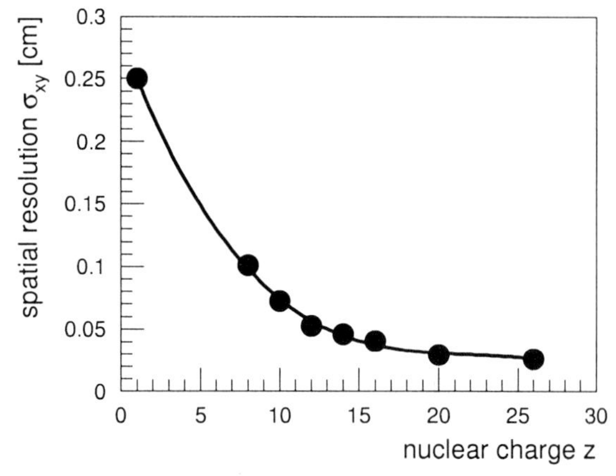

Figure 4. Spatial resolution of the track reconstruction as function of the nuclear charge.

4.3. Energy Resolution

The energy resolution of the detector system depends on the signal fluctuations as shown in figure 3, the relativistic rise of the ionization loss, and the magnitude of the transition radiation signal as a function of the Lorentz factor. Results from the balloon flight and the accelerator calibration lead to an energy resolution as shown in figure 5 as a function of the Lorentz factor γ for

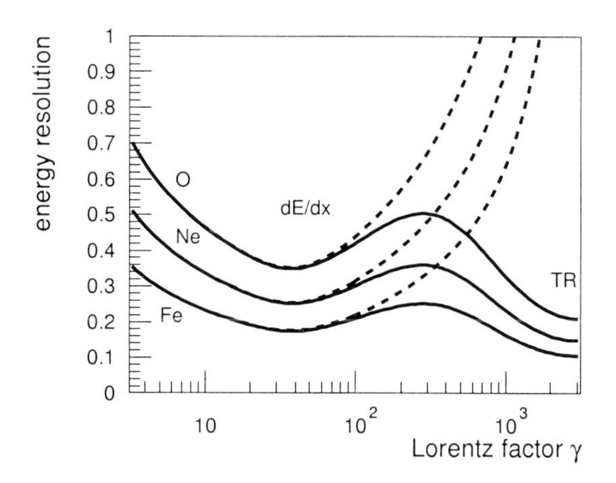

Figure 5. Energy resolution as function of the Lorentz factor γ of the cosmic-ray particles.

different species. The two detection techniques used in TRACER have their best energy resolution in two different energy regimes. The ionization measurements achieve best resolution around $\gamma = 32$ of about 18% for iron and 35% for oxygen. The transition radiation measurements allow more precise measurements beyond $\gamma = 1000$ with best resolutions of 11% for iron and 21% for oxygen.

4.4. Measured charge distribution

The two scintillators and the Čerenkov counter provide a signal essentially proportional to Z^2 and independent of the particle energy. Knowledge of the spatial response function of each counter as well as the non-linear response of the scintillators for higher charges allows to calculate the charge of the measured particles. The measured charge distribution for relativistic nuclei is shown in figure 6. The main cosmic-ray elements from boron to iron can be identified. The charge resolution is about 0.22 charge units for oxygen. It is remarkable that this resolution can be achieved with a 4 m^2 scintillator, just 5 mm thick, and read-out by just 12 photo multiplier tubes via wavelength shifter bars.

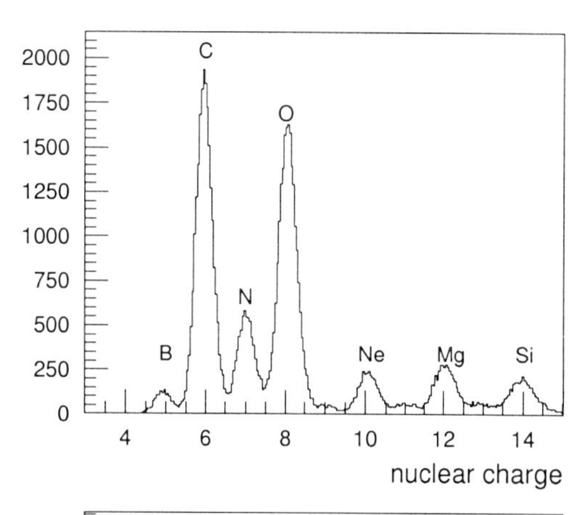

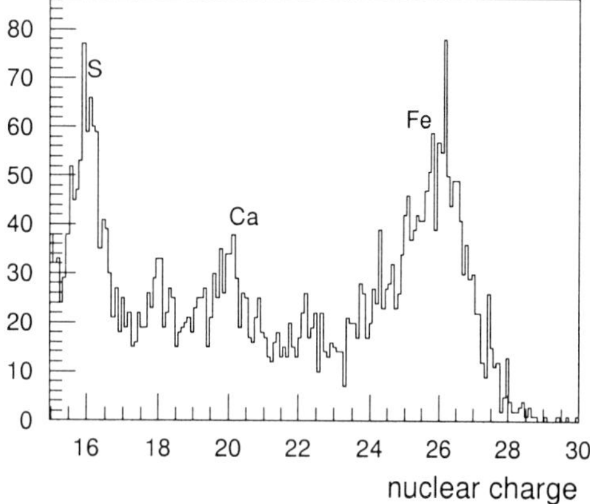

Figure 6. Measured charge distribution for relativistic cosmic-ray nuclei.

5. CONCLUSION

The successful balloon flight and the accelerator calibration of the detector show that the new instrument to measure the energy spectra of heavy cosmic-ray nuclei is working as expected. We expect to soon be able to present physics results from this investigations.

REFERENCES

1. J. L'Heureux et al., Nucl. Instr. and Meth. A 295 (1990) 246
2. F. Sauli, Principles of Operation of Multiwire Proportional and Drift Chambers, CERN 77-09 (1977)
3. http://wwwinfo.cern.ch/asd/geant4/

ELSEVIER

Nuclear Physics B (Proc. Suppl.) 97 (2001) 146–149

www.elsevier.nl/locate/npe

Characteristics of hadron-induced showers observed by the Pamir thick lead chambers

M.Tamada [a] and A.Ohsawa [b]

[a]Faculty of Science and Engineering, Kinki University, Osaka 577-8502, Japan

[b]Institute for Cosmic Ray Research, University of Tokyo, Chiba 277-8582, Japan

Structures of hadron-induced showers observed by the Pamir thick lead chambers are compared with simulations using VENUS, QGSJET, HDPM and the modified UA5 model for hadron-nucleus interactions. It is shown that none of the models of hadron-nucleus interactions used in the present analysis can describe all the characteristics of the observed two categories of hadron induced-showers, single-isolated and multi-hadrons, at the same time.

1. Introduction

A series of exposures of homogeneous-type thick lead emulsion chambers was carried out on the Pamir mountains by the Pamir collaboration experiment [1]. The homogeneous-type thick lead chamber has the advantage that we are able to observe the shower development of hadron-induced showers over a range of a few nuclear mean free paths, which enables us to study in detail characteristics of hadron-Pb interactions of hadron energy $E_h \geq 10^2$ TeV.

In Refs.[2] we discussed the inelasticity of hadron-Pb collisions at 10^{14} eV by analyzing the structure of observed hadron-induced showers. A parameter Z is defined as $Z \equiv E_1^{(\gamma)}/\sum E_i^{(\gamma)}$, where $E_1^{(\gamma)}$ is the released energy at the first interaction and $\Sigma E_i^{(\gamma)}$ the energy sum released at all the interactions during passage through the whole chamber. We compared the experimental distribution of Z with those obtained by the analytical calculations where various types of inelasticity distributions were assumed. We obtain $< k_{inel}^{h-Pb} >= 0.6^{+0.03}_{-0.05}$ for hadron-Pb collisions in the energy region of $E_h \simeq 10^{14}$ eV, appreciably smaller than usually considered. In the analysis the energies $E_i^{(\gamma)}$ were estimated by decomposing manually the observed shower transition into respective interactions. Therefore there still remains an ambiguity whether the experimental data can be compared directly with theoretical calculations or not. In order to make a direct comparison possible, we carry out simulations for hadron-induced showers and apply just the same procedure both to experimental and simulated data.

2. Pamir thick lead chambers

The homogeneous thick lead chamber is composed of 60 layers of lead plates of 1 cm thickness, equivalent to $\sim$ 3.5 nuclear mean free paths. Russian RT6-type X-ray films are inserted under every 1 cm lead plate except for the first 2 cm of lead. The darkness detection threshold, D_{th}, of the shower spot is $\sim$ 0.2, which corresponds to a shower energy of 2 $\sim$ 3 TeV approximately. Details of the chambers are given in Ref. [1].

A shower detected in the chamber is either a single-isolated shower, which is not accompanied by any other parallel shower of energy above the detection threshold, or a member of a bundle of parallel showers with the same arrival direction, which we call a cosmic-ray "family".

3. Simulations

We use the following four models for hadron-Pb interactions;
(1) VENUS [3], (2) QGSJET [4]: both of which are based on the Gribov-Regge theory of multiple Pomeron exchanges.
(3) HDPM [5] : a semi-empirical model extrapo-

PII S0920-5632(01)01230-0

lating experimental data based on the Dual Multichain Parton model and
(4) a phenomenological UA5 algorithm [6] modified for hadron-nucleus interaction using a geometrical approach.
We calculate the development of nuclear and electromagnetic cascades in lead initiated by hadron-Pb interactions under the following assumptions:
(1) the energy spectrum of hadrons arriving at the chamber is of power-law type, $I(\geq E) \propto E^{-\beta}$ with β =1.8 for single-isolated hadrons and β=1.2 for hadrons in a family,
(2) the zenith angle distribution of arriving hadrons is $I(\leq \cos\theta) \propto (\cos\theta)^{-8}$ and
(3) the total thickness of the chamber is 60 cmPb and the sensitive layers are inserted under every 1 cmPb.

The interaction mean free path of hadron-Pb interactions is assumed to be energy dependent, e.g., $\Lambda_{coll}^{p-Pb}(E)$ = 15.9 cmPb and $\Lambda_{coll}^{\pi-Pb}(E)$ = 17.5 cmPb at $E = 10^{14}$ eV.

Protons and pions of $E_h \geq$ 30 TeV, respectively, are sampled from the above energy spectra, and all hadrons, produced in the collisions during passage through the chamber, are followed until their energy falls below 80 GeV or they leave the chamber. For γ-rays of $E_\gamma \geq$ 1 GeV, which are mainly decay products of π^0's, we further calculate the three-dimensional electromagnetic cascade development using the Monte-Carlo code formulated by M. Okamoto and T. Shibata [7], in which the LPM effect is also taken into accounts. Electrons and photons are followed until their energies fall below 1 MeV. The electron number density, ρ_e, is converted to the local spot darkness, d, of X-ray film, by using the characteristic relation for the Russian RT6-type X-ray film, and finally we obtain the transition curve of the spot darkness D *vs.* depth T throughout the chamber.

4. Estimation of ΣE_γ and $E_1^{(\gamma)}$

4.1. Total released energy ΣE_γ

The total observed energy of the hadron, ΣE_γ, released in form of γ-rays during the passage through the chamber is estimated by the sum, ΣD_i, of the shower-spot darknesses $D_i \geq D_{min}$ in its longitudinal development. Here we set D_{min} = 0.3. We observe a fairly good correlation between ΣE_γ and ΣD_i as shown in Fig.1 for the QGSJET model. The other three models give similar results.

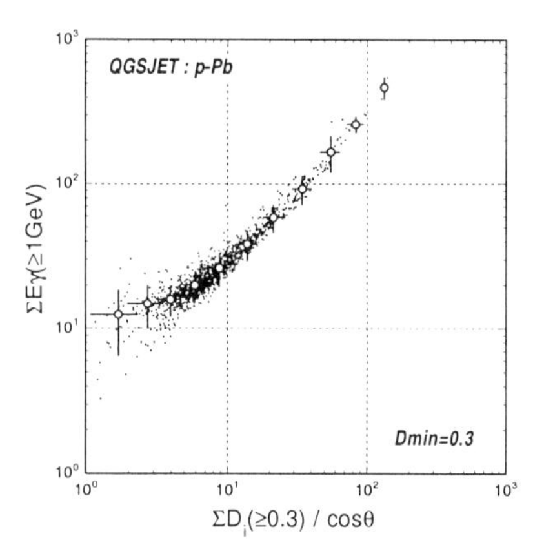

Figure 1. Diagram on ΣE_γ and $\Sigma D_i(D_i \geq 0.3)/\cos\theta$ for $p - Pb$ interactions. θ is the zenith angle of the incident hadron. Open circles show average dependence.

4.2. Energy $E_1^{(\gamma)}$ released in the first interaction

In order to extract the energy released in the first interaction from the whole shower transition curve of the event, the following fitting procedure is applied for the first 6 layers in which the spot darkness exceeds $D \geq 0.3$. From a the set of standard transition curves D^{std} *vs.* T, calculated for showers of electron-positron pair origin taking into account the exact chamber structure, the best fit is selected by choosing the energy value, $E_1^{(\gamma)}$, and the first pair-creation depth, ΔT, by a computer algorithm employing the gradient descent method for a search of the chi-square minimum. We also calculate the deviation of the observed shower transition from the best-fit by $\delta \equiv \Sigma(D_i - D_i^{std})/\sigma_i$, where the summation is made for the layers beyond the fit range and σ_i is the dispersion of darkness of the standard transition curve at the i-th layer. A large value of the deviation δ indicates that there are contributions from successive interactions.

Fig.2 shows a distribution of $E_1^{(\gamma)}/E_1^{true}(\gamma)$ where $E_1^{true}(\gamma)$ is the true energy released into γ-rays at the "first interaction". The distribution has a clear peak at $E_1^{(\gamma)}/E_1^{true}(\gamma) = 1.2$, indicating that the present procedure works but overestimates the energy released in the "first interaction" by $\sim$ 20% on average. The same fit procedure is also applied to γ-ray-induced pure electromagnetic cascades. The estimated energy is very close to the true energy in this case (Fig.2).

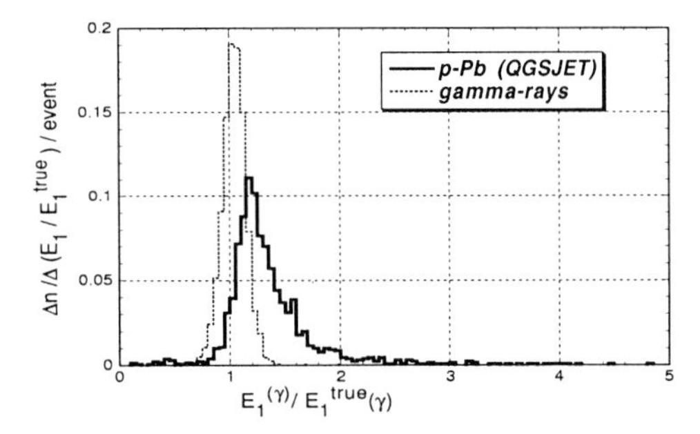

Figure 2. Distribution of $E_1^{(\gamma)}/E_1^{true}(\gamma)$, for $p-Pb$ interactions using the QGSJET model (solid histogram). The dotted histogram is for γ-ray induced pure electromagnetic cascades.

5. Selection of the events

Emulsion chambers detect both (e,γ)-induced and hadron-induced showers. Here we identify a shower as a hadron-induced one if $\Delta T \geq 4$ cmPb or $\delta \geq 50$. According to the simulation, almost all (e,γ)-induced showers are rejected by these criteria. We demand further that the darkness at shower maximum, D_{max}, in the shower transition is less than 3, because our standard procedure of photometric measurement is limited up to $D = 3 \sim 3.5$, and that the sum of spot darknesses, $\Sigma D_i(D_i \geq 0.3)/\cos\theta$, is larger than 8, which corresponds to a total released energy $\Sigma E_\gamma \geq 20-30$ TeV. In the Pamir thick lead chamber of 66 m^2·year exposure, we have 139 showers which satisfy the above criteria. Among them 54 are found as members of atmospheric families and the other 85 are isolated ones. The same selection criteria are also applied to simulated hadron-induced showers.

6. Distribution of $Z \equiv E_1^{(\gamma)}/\Sigma E_\gamma$

Applying the procedure described in the section 4, we obtain $E_1^{(\gamma)}$ and $\sum E_\gamma$ for the above selected events both in the experiment and in the simulations. We re-define the inelasticity-related parameter Z by $Z \equiv E_1^{(\gamma)}/\Sigma E_\gamma$. Fig.3 shows the experimental distributions of Z for single-isolated hadrons and Fig.4 for hadrons in families. On average hadrons in families have a smaller value of the parameter Z. The distributions are compared with those of simulated hadron-induced showers. In Fig.3 we assume that half of the incident hadrons are protons and the other half are pions both of which have the integral energy spectrum of the power index β=1.8. On the other hand, in Fig.4 we assume that all incident hadrons are pions with an energy spectrum of spectral index β=1.2. The dependence of the Z-distribution on the power index β and the nature of incident particles, however, is found to be small.

As is seen in Figure 3, the experimental distribution of single-isolated hadrons is described well by VENUS, QGSJET and the modified UA5 model, all of which give an average inelasticity of $< K >= 0.7 \sim 0.8$. [1] But the data deviate from the HDPM model which predicts a smaller inelasticity $< K >= 0.5 \sim 0.6$. For hadrons in families, on the contrary, more events populate the region of smaller Z, and the experimental distribution is close to that of the HDPM model and not compatible with those of the other three models. Thus none of the models can describe simultaneously the experimental Z-distribution of the two categories of hadrons. The experimental Z-distributions can be distorted by the poor accuracy in tracking of the shower spots. However, according to the examinations, this cannot be made responsible for the observed difference

[1]Here the inelasticity is given by $K \equiv 1 - x_l$ where x_l is defined by the energy fraction carried by the highest energy baryon ($\pi^\pm$ meson) in $p(\pi^\pm) - Pb$ interactions.

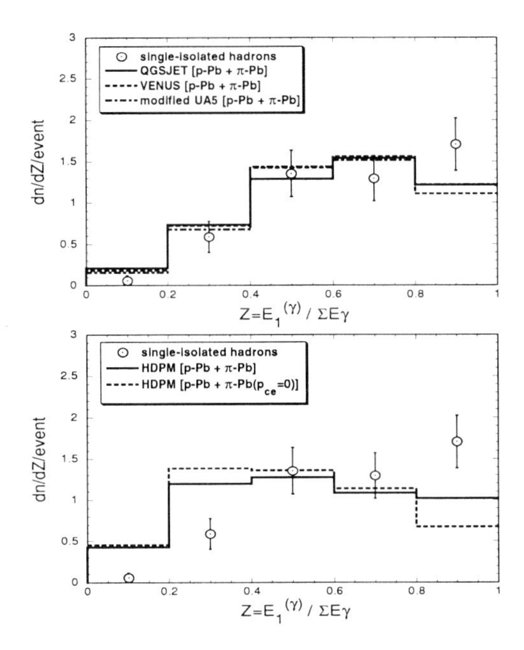

Figure 3. Distribution of $Z \equiv E_1^{(\gamma)}/\Sigma E_\gamma$ for single-isolated hadrons. The histograms are those for simulated hadron-induced showers. $p_{ce} = 0$ means no charge exchange processes in pion-Pb interactions.

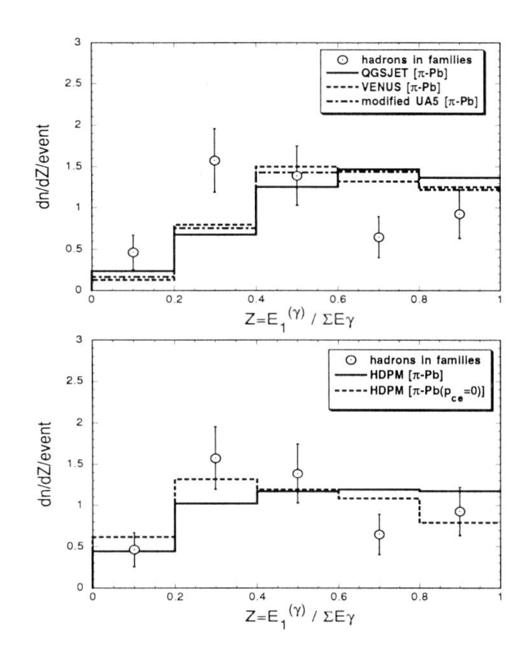

Figure 4. Distribution of Z for hadrons in families. The histograms are those for simulated hadron-induced showers.

of the Z distribution in the two categories of incident hadrons. One of the possible explanations of the observed difference between single-isolated hadrons and hadrons in families is to assume the existence of hadron bundles in high-energy cosmic-ray families. That is, if there exists a hadron bundle in which the mutual distance of the constituent hadrons is extremely small, *i.e.* less than $\sim$ 1 mm, we would possibly misidentify two independent hadron-induced showers as those produced by successive interactions. It would produce a smaller value of the parameter Z even if the average inelasticity is as large as that of single-isolated hadrons. The experimental data, however, require the number of those hadron bundles to be 3 $\sim$ 4 times more frequent than expected simply by chance. Extraordinary correlated hadron-bundles remind us of exotic shower-clusters, named "mini-clusters", in which hadrons and γ-rays are closely correlated, which are observed in high-energy cosmic-ray families detected in carbon-chambers of the Chacaltaya and Pamir experiments.

The authors would like to express their sincere thanks to all the members of the Pamir and Chacaltaya collaboration experiments.

REFERENCES

1. Chacaltaya and Pamir Collaboration (T.Arisawa et al.); *Nucl. Phys. B* **424** (1994) 241
2. C.R.A.Augusto, et al. : *Phys. Rev. D* **61**, (1999) 012003
3. K.Werner, *Phys. Rep* **232** (1993) 87
4. N.N.Kalmykov, S.S.Ostapchenko & A.I.Pavlov: *Bull. Russ. Acad. Sci. (physics)* **58** (1994) 1966,
5. J.N.Capdevielle: *J. Phys. G: Nucl. Part. Phys.* **15** (1989) 909
6. UA5 Collaboration (G.L.Alner et al.): *Nucl. Phys. B* **291** (1987) 445
7. M.Okamoto & T.Shibata: 1987, *Nucl. Instr. Meth.* **A257** (1987) 155

ELSEVIER

Nuclear Physics B (Proc. Suppl.) 97 (2001) 150–153

www.elsevier.nl/locate/npe

A study on the penetrating nature of gamma- and hadron-induced cascade showers in the two-storey emulsion chambers

M.Tamada [a]

[a]Faculty of Science and Engineering, Kinki University, Osaka 577-8502, Japan

Penetrating nature of the cascade showers observed in the Chacaltaya two-storey chamber is compared with that of simulated γ-ray- induced and hadron-induced cascade showers. It is shown that around 40 % of the penetrating showers are neither γ-ray-induced ones nor hadron-induced ones. A possible explanation is given in connection with "mini-clusters".

1. Introduction

The Chacaltaya and Pamir emulsion chamber experiments have shown that there exist unusual phenomena which are not yet observed in the present accelerator experiments from the analysis of high energy cosmic-ray families, a bundle of electromagnetic particles and hadrons produced in the nuclear and electromagnetic cascade process in the atmosphere [1]. Those are called "Centauro-species", the multiple hadron production without association of π^0-mesons. It is also reported that the nature of secondary particles is different from that of ordinary hadrons in those unusual phenomena. In Ref.[2] we discussed about unusual behavior of cascade development of the high energy showers in high-energy cosmic-ray families observed by Chacaltaya two-storey chamber no.19. We studied in detail how the shape of the cascade transition of the observed showers, which penetrate from the upper chamber down to the lower chamber, deviates from that of standard electromagnetic cascades expected in the uniform lead chamber. We found frequent existence of strong penetrating showers which became rejuvenated after passing through the target layer. The results were discussed in connection with "mini-clusters", clusters which consist of extraordinarily correlated γ-rays and hadrons [1,3]. Here we discuss the issue again comparing the shape of the cascade showers by those initiated by γ-rays and also by hadron interaction taking into account the exact structure of the Chacaltaya two-storey chamber.

2. Experimental data

Fig.1 shows the basic structure of Chacaltaya two-storey chamber no.19. The chamber consists of the upper chamber of 6 cmPb, the target layer of 23 cm carbon (petroleum pitch), wooden support of 5cm thick, the air gap of 158 cm height and the lower chamber of 8.4 cmPb. Four sensitive layers (X-ray film and nuclear emulsion plate) are inserted in the upper chamber and eight sensitive layers in the lower chamber. The chamber no.19, the area of which is 44.4 m^2 in the upper chamber and 32.4 m^2 in the lower chamber, was exposed 677 days at Mt. Chacaltaya. Showers detected in the upper chamber are mainly (e, γ)-induced ones with small admixture of hadron-induced ones. Showers detected in the lower chamber, on the other hand, are those initiated by nuclear interactions in the target layer ($C-jets$) and in the lead plates of the lower chamber itself ($Pb-jet-lower$). In the present analysis we pick up showers in the high energy cosmic-ray families with total visible energy greater than 100 TeV observed in the Chacaltaya chamber no.19.

3. Penetrating cascade showers

Some of the showers observed in the upper chamber can be followed down into the lower chamber. We define it as a penetrating shower when its spot darkness, D, is larger than 0.2 in at least two layers in the upper chamber and at least one layer in the lower chamber. A typical example of the shower transition of the penetrating

PII S0920-5632(01)01233-6

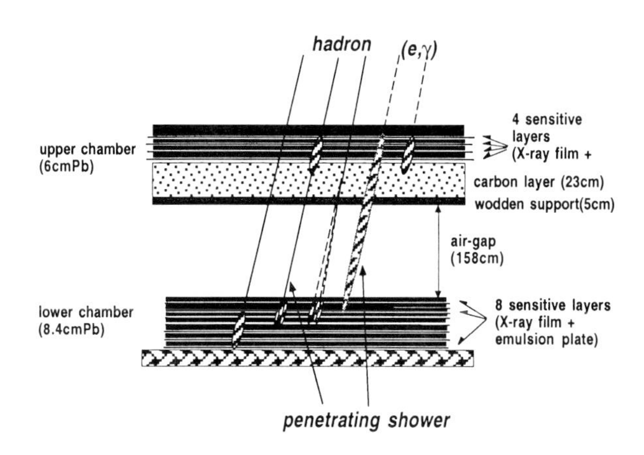

Figure 1. Illustration of Chacaltaya two-storey chamber no.19.

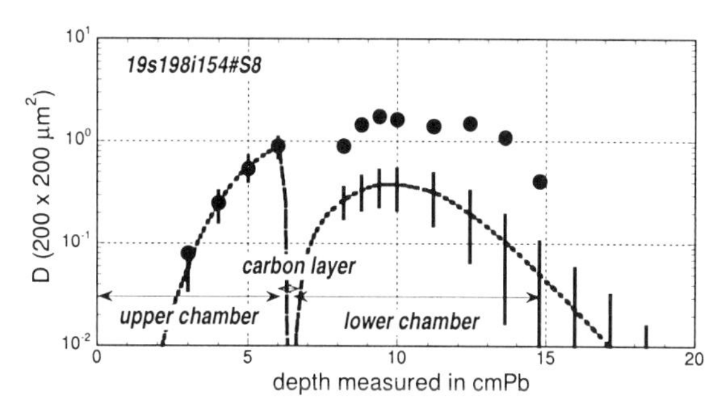

Figure 2. An example shower transition on spot darkness of the penetrating showers observed in Chacaltaya chamber no.19. The dotted curve is the best fit described in the text. Vertical bars are dispersion σ of the standard transition curve.

shower is shown in Fig.2. In Table 1 we summarize the number of showers with visible energy $E(\gamma) \geq 10$ TeV observed in the upper chamber, those observed only in the lower chamber $(C-jets, Pb-jets-lower)$ and those penetrating. Among 205 showers observed in the upper chamber, 85 penetrate into the lower chamber.

4. Simulations

We calculate a development of cascade showers in the two-storey chamber, of just the same structure as the Chacaltaya chamber no.19, initiated from a pure γ-ray and from a hadron-Pb (C) interaction.

4.1. Hadron-induced showers

We use two different models for hadron-nucleus interactions, one is QGSJET [4] based on the Gribov-Regge theory of multiple Pomeron exchanges and the other is the phenomenological UA5 algorithm [5] modified for hadron-nucleus interaction using a geometrical approach. In the nuclear cascade all hadrons, produced in the collisions during passage through the chamber, are followed until their energy falls below 80 GeV or they leave the chamber. The interaction mean free path of hadron-nucleus interactions is assumed to decrease with increasing interaction energy, e.g., $\Lambda^{p-Pb}(E) = 15.9$cmPb, $\Lambda^{\pi-Pb}(E) = 17.5$cmPb, $\Lambda^{p-C}(E) = 62.4$cmC and $\Lambda^{\pi-C}(E) = 79.4$cmC at $E = 10^{14}$eV.

For γ-rays of $E_\gamma \geq 1$ GeV, which are mainly decay products of π^0's produced in the collisions, we further calculate the three-dimensional electromagnetic cascade development in the chamber using the Monte-Carlo code formulated by M. Okamoto and T. Shibata [6], in which the LPM effect is also taken into account. Electrons and photons are followed until their energies fall below 1 MeV. The electron number density, ρ_e, is converted to the local spot darkness, d, of X-ray film, by using the characteristic relation for the N-type X-ray film, and finally we obtain the transition curve of the spot darkness D, measured by a $200 \times 200\mu\text{m}^2$ slit, *vs.* depth T throughout the chamber. The experimental error of the measurement of spot darkness D is also taken into account by adding noise ΔD in each spot darkness, where ΔD is sampled from Gaussian distribution with $\sigma_D = 0.1D$. Protons and pions of $E_h \geq 20$ TeV are sampled from the energy spectrum $I(\geq E_h) \propto E_h^{-1.2}$ and from zenith angular distribution $I(\leq \cos\theta) \propto (\cos\theta)^{-8}$.

4.2. γ-ray-induced showers

We also calculate electromagnetic cascade development in the chamber initiated by γ-rays us-

Table 1 Number of high energy showers, $E(\gamma) \geq$ 10 TeV, observed in the chamber no.19.

atmospheric families of $\Sigma E(\gamma) \geq 100$ TeV	41 events
(a) no. of showers observed in the upper chamber	205
(b) penetrating	85
(c) expected no. of penetrating γ-rays	47 ± 4
(d) non-γ-ray origin	38 ± 10
(e) no. of showers of observed only in the lower chamber ($C - jets, Pb - jets - lower$)	23
(f) expected no. of penetrating showers of hadronic origin	9 ± 2
(g) no. of anomalous penetrating showers	29 ± 10

ing the above mentioned Monte-Carlo code assuming the energy spectrum and zenith angle distribution of the γ-rays arriving at the chamber as $I(\geq E_\gamma) \propto E_\gamma^{-2}$ and $I(\leq \cos\theta) \propto (\cos\theta)^{-8}$ respectively.

5. Penetrating probability

According to the simulations, the penetrating probability of γ-ray- induced showers of $E_\gamma \geq 10$ TeV is found to be 0.23 ± 0.01. If we assume that all the 205 showers observed in the upper chamber are γ-ray-induced ones, the expected number of penetrating showers of γ-ray origin is then obtained as 47 ± 4. That is, around one half of the observed penetrating showers can be considered to be of γ-ray origin but the other half (38 ± 10) are of non-γ-ray origin. Possible origin of those which we can consider are hadron-induced showers. Now let's estimate the number of hadron-induced penetrating showers. All the showers detected only in the lower chamber can be considered to be hadron-induced ones ($C - jets$ and $Pb - jets - lower$). In the simulation calculations of hadron-induced showers we know the ratio between the number of penetrating hadron-induced showers and that of showers which are observed only in the lower chamber. The results are summarized in Table 2. As is seen in the table, the ratio is around 0.4 on the average. In the experiment, there are 23 showers which are observed only in the lower chamber. Then we can obtain the expected number of hadron-induced penetrating showers to be 9 ± 2. Among 38 penetrating showers of non-γ-ray origin, only 9 can be due to hadronic interaction in the upper chamber. Then we can conclude that the rest 29 ± 10 penetrating showers ($\sim$ 34 % of all the penetrating showers) are neither γ-ray-induced nor hadron-induced showers.

Table 2 Number of hadron-induced showers ($E_h(\gamma) \geq 10$ TeV).

	QGSJET		modified UA5	
incident	pion	proton	pion	proton
(1) penetrating	193	208	192	236
(2) visible only in the lower chamber	550	486	527	545
ratio=(1)/(2)	0.35	0.43	0.37	0.43

In each set of the calculations, 2,000 particles are sampled from the spectra described in Section 4.1.

6. Shape of the shower transition of the penetrating showers

Here we study how the shape of the shower transition of the penetrating showers differs from that of ordinary γ-ray induced cascades. [1] The following fitting procedure is applied for the layers of the upper chamber in which the spot darkness exceeds $D \geq 0.2$. From a set of standard transition curves D^{std} vs. T, calculated for show-

[1] In the Chacaltaya chamber no.19, full data of spot darkness of the penetrating showers are available at present for a half of the chamber which were measured in Japan. Hence the analysis of the shape of the shower transition is made using 56 penetrating showers in 22 cosmic-rays families.

ers of e^-e^+-pair origin, taking into account the exact structure of the two-storey chamber no.19, the best fit is selected out by choosing the energy value E_γ^{fit} and the first pair-creation depth ΔT by a computer algorithm employing the gradient descent method for a search of the chi-square minimum. We then calculate the deviation of the observed shower transition in the lower chamber from the best-fit by $\delta \equiv \Sigma(D_i - D_i^{std})/\sigma_i$, where the summation is made for the layer in the lower chamber. σ_i is the dispersion of the standard curve at the i-th layer. Fig.3 shows the integral distribution of δ for the events with $E_\gamma^{fit} \geq 5$ TeV. The same procedure is applied to the simulated penetrating showers of γ-ray origin and of hadronic origin. In the figure we show the expected distributions in the case that penetrating showers consist of a mixture of γ-ray-induced and pion-induced ones. As is seen in the figure, the experimental distribution is well described by assuming about 40% ~ 50% of penetrating showers are hadron-induced ones, i.e., non-γ-ray origin. The estimated fraction of hadron-induced penetrating showers does not depend much on the assumed model of hadron-nucleus interaction and on the nature of the incident particle.

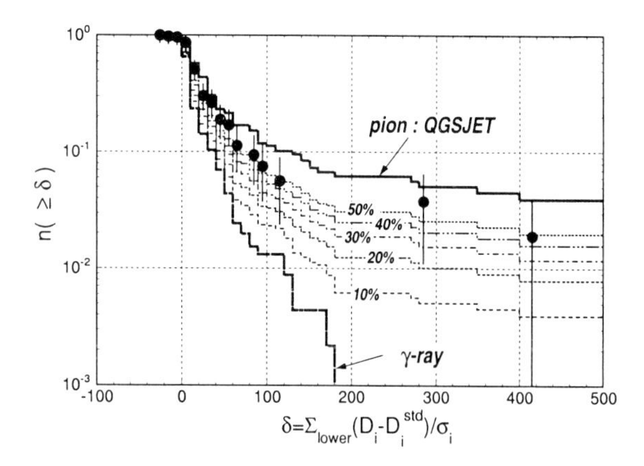

Figure 3. Integral distribution of $\delta \equiv \Sigma(D_i - D_i^{std})/\sigma_i$ for the penetrating showers. Thin histograms are for the case of mixture of pion-induced and γ-ray induced showers. Percentages attached to the histograms show the assumed fraction of the pion-induced penetrating showers.

7. Discussions

We have shown that around 34% of penetrating showers (~14% of the showers in the upper chamber) observed in the Chacaltaya two-storey chamber no.19 are neither γ-ray-induced nor hadron-induced showers. One of the possible explanations is to assume an existence of extremely collimated pair of a γ-ray and a hadron. That is, if the mutual distance between a γ-ray and a hadron is extremely small, e.g., less than ~ 1 mm, and the γ-ray-induced showers is observed in the upper chamber and the hadron-induced shower is observed in the lower chamber, we would possibly misidentify those two as a penetrating shower. If a γ-ray makes electromagnetic interactions in the atmosphere, we can observe several collimated (e, γ)-particles and a hadron as a "mini-cluster" which are often found in the exotic events. Possible existence of hadron-bundles in which the mutual distance of the constituent hadrons is extremely small is also discussed in the analysis of shower transition of high-energy hadronic showers observed in the Pamir thick lead chamber[7,8].

REFERENCES

1. Chacaltaya and Pamir Collaboration (L.T.Baradzei et al.); *Nucl. Phys. B* **370** (1992) 365
2. Y.Funayama & M.Tamada : *J. Phys. Soc. Japan* **55**, (1986) 2977
3. M.Tamada & Y.Funayama : *J. Phys. Soc. Japan* **55**, (1986) 2996
4. N.N.Kalmykov, S.S.Ostapchenko & A.I.Pavlov: *Bull. Russ. Acad. Sci. (physics)* **58** (1994) 1966,
5. UA5 Collaboration (G.L.Alner et al.): *Nucl. Phys. B* **291** (1987) 445
6. M.Okamoto & T.Shibata: *Nucl. Instr. Meth.* **A257** (1987) 155
7. M.Tamada & A.Ohsawa: *Nucl. Phys. B* **581** (2000) 73
8. M.Tamada & V.V.Kopenkin: *Nucl. Phys. B* **494** (1997) 3

ELSEVIER

Nuclear Physics B (Proc. Suppl.) 97 (2001) 154–157

www.elsevier.nl/locate/npe

Hadron/Gamma Identification of Showers Observed by EC's

Y. Fujimoto[a], V.V. Kopenkin[b], A. Ohsawa[c], M. Tamada[d]

[a]Advanced Research Center for Sci. and Eng., Waseda Univ., Shinjuku, Tokyo, 169 Japan

[b]Institute of Nuclear Physics, Moscow State University, 119899 Moscow, Russia

[c]Institute for Cosmic Ray Research, University of Tokyo, Kashiwa, Chiba, 277-8582 Japan

[d]Faculty of Science and Engineering, Kinki University, Higashi-Osaka, Osaka, 577-8582 Japan

Fluctuations of hadron- and gamma-induced showers, which are observed by emulsion chambers, are studied making use of simulated showers. An algorithm is formulated to analyze gamma- and hadron-induced showers. The algorithm is applied to the showers observed by Pamir thick lead chambers, and its results are presented.

1. Introduction

An emulsion chamber (EC) is sensitive to high energy (e, γ)'s and hadrons, incident upon the chamber, and can determine the positions and energies of these particles with high precision. It is used widely in the experiments to observe high energy cosmic rays, in particular in those at high mountains.

In this report we present an algorithm to analyze data by emulsion chambers and some results which are obtained by applying the algorithm to experimental data.

2. Showers observed by emulsion chamber

2.1. Emulsion chamber (EC)

EC, a multiple sandwich of lead plates (usually 1 cm thick each) and sensitive layers (X-ray films, etc.), detects cascade showers which are produced in the chamber by the particles incident upon the chamber.

A cascade shower traverses several centimeters of lead, and is recorded by several successive sensitive layers of EC. A cascade shower forms a small dark spot ($\sim$ 100 μm) on the X-ray film after development. Darkness of the spot, D, is measured by a microphotometer with a slit of 200×200 μm^2. EC detects spots with darkness exceeding ~ 0.1, which corresponds to the shower energy of $\sim$ 1 TeV. The transition of the spot darkness values along the depth of the chamber $D(i)$ (i : i-th sensitive layer), called "(shower) transition curve", enables us to determine energy and starting depth of the shower, by comparing it with those calculated on the basis of the cascade shower theory, taking into account the sensitivity curve of the X-ray film.

2.2. Showers observed by EC

(i) Gamma-showers and hadron-showers

Showers originated by (e, γ)'s are called "gamma-showers". Their shower transition curves are characterized by a single peak at shallow layers of EC. A hadron also originates a cascade shower through its nuclear collision with lead, which is called a "hadron-shower". Most of hadron-showers have starting points deep in the chamber, because of the long inelastic collision mean free path of hadrons in the chamber ($\sim$ 18.0 cm Pb). Sometimes the transition curve of a hadron-shower is not of a single peak but of multiple peaks, due to secondary collisions of the hadrons which are produced in the first collision.

(ii) Single-showers and family-showers

Observed showers are also classified into those of single arrival and those in a bundle. We call the former "single-showers" and the latter "family-showers". A family, a bundle of gamma- and hadron-showers with the same direction of incidence, is produced by nuclear interaction(s) in the atmosphere.

PII S0920-5632(01)01240-3

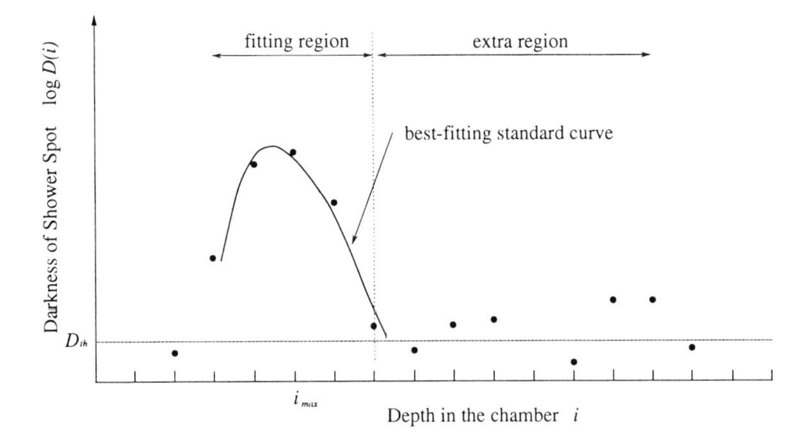

Figure 1. Illustration of fitting the standard transition curve to the data points. The best-fit is chosen by adjusting energy E and shower starting point Δt to minimize χ^2. (See Section 3.)

Among hadron-showers, single-showers consist of those originated by nucleons and pions with roughly equal abundance[1], while family-showers are mainly due to pions.

2.3. Gamma/hadron identification

Gamma- and hadron-showers differ mainly in the starting depth and in the peaks of the shower transition curve. However, fluctuation of shower development and experimental errors make shower identification less simple. So far we have made hadron-shower identification by the condition that the starting depth $\Delta t \geq 4(6)$ c.u. (cascade unit), and the number of hadron-showers with $\Delta t = 0 \sim 4(6)$ c.u. is estimated statistically[2]. This way is sufficient to discuss the intensities of (e, γ)'s and hadrons. It is the study of exotic events that urges us to identify respective showers.

3. Fitting procedure

The following is an algorithm to fit darkness data points $D(i)$ to the standard curves (Fig. 1):
(0) Standard curves by simulations.

[1]The intensities of nucleons and pions are approximately equal at mountain altitude.
[2]Hadron-showers with multiple peaks are rare because the chamber is not so thick so far.

Gamma-showers are produced for various primary energies E_0 and for various zenith angles of incidence $m \equiv \tan\theta$, by the simulation code of Ref.[1] which takes the chamber structure into account. The standard curves $D_{std}(i)$ and the dispersions σ_i are obtained for various E_0 and m by averaging them.
(1) "Fitting region" of $[n_1, n_2]$, where n_1 is the layer of the first data point and $n_2 = i_{max} + 2$ (i_{max} : the layer of the maximum darkness).
(2) Search for the minimum value of reduced chi square, defined as

$$\chi^2 \equiv 1/(n-2) \times \sum_i \{[D(i) - D_{std}(i)]/\sigma_i\}^2$$

by adjusting energy E_0 and the shower starting point Δt. The number n is the number of data points in the fitting region and consequently $n-2$ is the number of degrees of freedom. The energy of χ^2_{min} case is denoted as E_I.
(3) Calculation of the following quantities outside the fitting region, called "extra region" hereafter.

$\sum_i [D(i) - D_{std}(i)]$ (*Track length to estimate energy* E_X *in the extra region*)

$$\chi^2_{ex} \equiv \sum_i \{[D(i) - D_{std}(i)]/\sigma_i\}^2$$

(4) Estimation of the shower energy

$$E_{all} = E_I + E_X$$

Table 1
Statistics of showers in the experiment and in simulations.

(1) Experimental data		No. of showers for analysis
Single-showers		469
Family-showers		265
(2) Simulated data	Conditions	No. of showers
Gamma-showers	$E^{-2.8}dE$ ($E \geq 10$ TeV)	720
Hadron-showers	$E^{-2.8}dE$ ($E_{all} \geq 10$ TeV)	969

3.1. Statistics of showers

At first we examine the showers which are obtained by thick lead chambers exposed at Pamirs. These data are suitable for examination of showers, because the chamber is uniform (59 sensitive layers at every 1 cm Pb) and thick (60 cm Pb). The uniform chamber allows us to study shower development in detail and the thick chamber to detect high energy hadrons with almost 100 % efficiency. The showers in Table 1 are selected by the following criteria:

$D_{th} = 0.1$ (threshold darkness)
$E_{th} = 10.0$ TeV (threshold energy)
$m = \tan\theta \leq 1.5$ (angle of incidence)
data points ≥ 3
$D(i) \leq D_m$ ($D_m \sim 4.0$)

Gamma- and hadron-showers are produced by simulations (See Table 1.), where the code of Ref.[1] is used for electromagnetic cascade processes and the code QGSJET [2] is assumed for hadron-Pb collisions. The energies of the incident particles are sampled from the power-type energy distribution $E^{-\beta-1}dE$ ($\beta = 1.8$).

4. Fitting procedure to simulated gamma-showers

We examined the following points using gamma-showers from simulation.

(1) Optimum number of data points.

Varying the fitting range as $[i_{max} - n,\ i_{max} + n]$ ($n = 1, \cdots, 5$), we looked for the optimum number of data points for the fitting procedure. Our conclusion is that increase of data points does not necessarily mean good fitting or good energy estimation[3].

Table 2
Average value of reduced χ^2 minimum

Experiment	Single-showers	2.72
	Family-showers	3.03
Simulation	Gamma-showers	0.80
	Hadron-showers	1.08

(2) Effect of threshold darkness D_{th}

Decrease of D_{th} is equivalent to increase of data points. Hence we set $D_{th} = 0.1$.

(3) Validity of the fitting procedure

We applied the fitting procedure to the gamma-showers produced by simulations without restricting the fitting region, and obtained the average value of minimum reduced χ^2 as $< \chi^2_{min} >= 1.08$. This value means that the fitting procedure works satisfactorily, because $< \chi^2_{min} >= 1.0$ for an ideal case of minimum χ^2 search.

(4) E_{all} *vs.* E_0

The error in the energy estimation by our algorithm is ~ 10 %.

5. Fitting procedure in the fitting region

The fitting procedure in the fitting region is applied to both data of the experiment and of the simulations. The average values of χ^2_{min} in Table 2 show that the values by simulation are near 1.0 while those by experiment are far from 1.0. It indicates that there is another source of shower fluctuation in experimental data.

Probably it is the experimental error in darkness measurements, which is not taken into account in the simulations and which would be re-

[3]It is probably because neighboring data points, $D(i)$ and $D(i+1)$, are not completely independent. That is, a large electron number at the i-th layer means again a large number at the $(i+1)$-th.

duced by smoothing the data or by using the average values of the neighboring two data points. After smoothing, the χ^2_{min} distribution of the experiment comes close to that of the simulation, particularly in small χ^2_{min} region. That is, the value $< \chi^2_{min} >= 2.34$ becomes 1.42 after smoothing for single-showers[4].

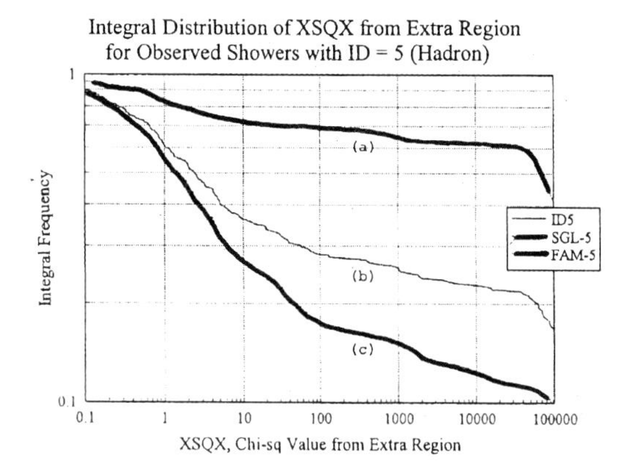

Figure 2. The distribution of χ^2_{ex} for the single-showers (c) and for family-showers (a), both with the starting depth $\Delta t \geq 6$ c.u. The curve (b) is the average of the both.

6. Fitting procedure in the extra region

Fitting procedure (the steps (3) and (4)) is applied to the extra region of the data from the experiment and simulations.

(i) χ^2_{ex} *vs.* E_X/E_0 by simulation data

The correlation between χ^2_{ex} and E_X/E_0 of gamma-showers and hadron-showers by simulations shows the following:

(1) Large χ^2_{ex} corresponds to $E_X/E_0 > 0$ and small χ^2_{ex} to $|E_X/E_0| \sim 0$, which signifies that a larger value of χ^2_{ex} represents real successive showers due to hadron interactions.

(2) A criterion $\chi^2_{ex} \geq 20$ is good for hadron-shower identification (See below).

(ii) Integral distributions of χ^2_{ex} of the simulation data

The χ^2_{ex} distributions of gamma-showers and hadron-showers by simulations are quite different, and show that the criterion of $\chi^2_{ex} \geq 20$ for hadron-shower identification works well. In the region of $\chi^2_{ex} \geq 20$ there exist 1.5 % of gamma-showers and 98 % of hadron-showers of the simulated events.

(iii) Integral distribution of χ^2_{ex} of the experimental data (hadron-showers)

Fig. 2 presents the χ^2_{ex} distributions of single-showers ((c) in Fig. 2) and of family-showers ((a) in Fig. 2), both with $\Delta t \geq 6$ c.u. The figure shows that both distributions are different in spite of the fact that all the showers are hadron-induced ones.

In summary we presented an algorithm to analyze showers, observed by emulsion chambers, based on the simulated showers, and applied it to those detected by Pamir thick lead chambers.

(1) The algorithm works satisfactorily to determine energies and starting points of the showers and to identify shower origin.

(2) Experimental data of hadron-induced showers, observed by Pamir thick lead chambers, show difference of shower transition between single-showers and family-showers. Discussion on the origin of the observed difference, methodological or physical, is under way, and one of possible descriptions is given by Tamada [3].

REFERENCES

1. M. Okamoto et al., Nucl. Instr. Meth. **A257** (1987) 155.
2. N.N. Kalmykov et al., Bull. Russ. Acad. Sci. **58** (1994) 1966.
3. M. Tamada et al., Talk in this symposium.

[4]The number of events decreased from 469 to 410, because some events do not fulfill the selection criteria after smoothing.

ELSEVIER

Nuclear Physics B (Proc. Suppl.) 97 (2001) 158–161

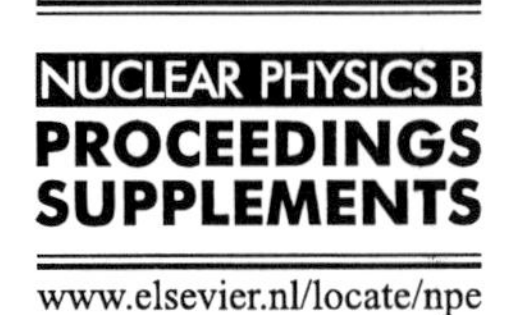

www.elsevier.nl/locate/npe

Study of Hadronic Component in Air Showers at Mt. Chacaltaya

C. Aguirre[a], H. Aoki[b], K. Hashimoto[c], K. Honda[d], N. Inoue[e], N. Kawasumi[c], Y. Maeda[f], N. Martinic[a], N. Ohmori[f], A. Ohsawa[g], K. Shinozaki[e], M. Tamada[h], R. Ticona[a] and I. Tsushima[c]

[a]Instituto de Investigaciones Fisicas, Universidad Mayor de San Andres, La Paz, Bolivia

[b]Faculty of Science, Soka University, Hachioji, Tokyo, 192-8577 Japan

[c]Faculty of Education, Yamanashi University, Kofu, 400-8510, Japan

[d]Faculty of Engineering, Yamanashi University, Kofu, 400-8511, Japan

[e]Faculty of Science, Saitama University, Urawa, 388-8570 Japan

[f]Faculty of Science, Kochi University, Kochi, 780-8520 Japan

[g]Institute for Cosmic Ray Research, University of Tokyo, Kashiwa, 277-8582 Japan

[h]Faculty of Science and Technology, Kinki University, Higashi-Osaka, 577-8502 Japan

An experiment of an air shower array, a hadron calorimeter (8 m^2) and an emulsion chamber (8 m^2, 15 cm Pb) is under way at Mt. Chacaltaya (5200 m above sea level, Bolivia), in order to study the hadron interactions and the primary cosmic rays in the energy region exceeding 10^{15} eV. The number of particles in the hadronic component in the air shower, which is detected by the hadron calorimeter, is not compatible with that obtained by simulations, indicating that violation of the Feynman scaling law is stronger at 10^{16} eV than the one assumed in the simulations. The average mass number of the primary cosmic rays, estimated from the distribution of the number of hadrons in the air shower, is $< \ln A >= 2.8 \pm 0.5$ at 10^{16} eV.

1. Introduction

An experiment of an air shower array, a hadron calorimeter and an emulsion chamber is being carried out at Mt. Chacaltaya (5200 m, Bolivia). The emulsion chamber detects high energy particles in the air shower and those are called "family". In this way the experiment simultaneously supplies data of the electron component in the air shower together with those of high energy particles in the air shower core. Air shower experiments and emulsion chamber experiments, which have been carried out independently so far, accumulate a large amount of data, respectively. Hence it is interesting and important to bridge the data by both experiments and a large scale of a new experimental set-up is not needed for this purpose. The present experiment enables us to study the structure of the air showers, the nuclear interactions and the primary cosmic rays in the energy region of $10^{15} \sim 10^{17}$ eV. Detailed knowledge of the air shower development is important in particular at present because there are experiments, running and in project, which intend to discriminate the air showers of gamma-origin from those of proton-origin by their inner structure. On the other hand the energy region of $10^{15} \sim 10^{17}$ eV is important both for particle physics and astrophysics. It is so in particle physics because this region is not covered by the existing accelerators and because there are several reports, experimental and theoretical, which point out the change of the nuclear interaction characteristics and/or existence of exotic phenomena [1,2]. It is so in astrophysics because there is a bend, called "knee", of the energy spectrum of the primary cosmic rays at $\sim 10^{15}$ eV.

PII S0920-5632(01)01243-9

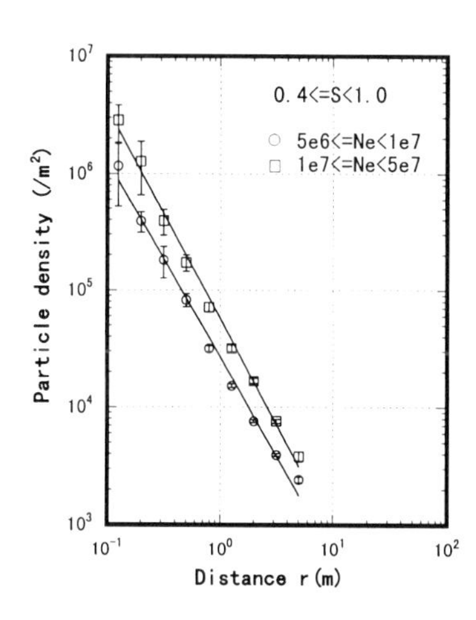

Figure 1. Lateral distribution of particle density for air showers of $N_e = 5 \times 10^6 \sim 10^7$ and $10^7 \sim 5 \times 10^7$. The air showers are grouped by their age parameters in respective size region.

2. Experiment

The experimental apparatus consists of the air shower array, the emulsion chamber and the hadron calorimeter [3]. The air shower array consists of 35 plastic scintillation detectors, which are distributed over a circular area of 50 m radius. The emulsion chamber and the hadron calorimeter are stored in AS EC room. The emulsion chamber consists of 32 units (50×50 cm^2 each), each of which is 15 cm Pb thick and contains 14 sensitive layers of X-ray films. The hadron calorimeter consists of 32 units of plastic scintillator ($50 \times 50 \times 5$ cm^3 each), which are located beneath the emulsion chamber.

The output from each unit of the hadron calorimeter is related to the energy deposit in the scintillator, and it is converted to charged particle number using the average energy loss of a single muon in the scintillator. The number of charged particles per area of 0.25 m^2, n_b, is called "particle density" hereafter.

The data produced by the air shower array and by the hadron calorimeter are recorded when at least one unit of the hadron calorimeter has the particle density $n_b \geq 10^3$ (particles/0.25 m^2). In this sense the mode of the run is called "hadron calorimeter triggering". The present data set is not biased by the triggering mode in the region of air shower size $Ne > 5 \times 10^6$.

The lateral distribution of the particle density is determined by applying the least square fitting to the curve,

$$n(r) = \frac{A}{r_0^2}\left(\frac{r}{r_0}\right)^{\alpha} \qquad (r_0 = 1 \text{ m}) \qquad (1)$$

where the parameters A and α are to be adjusted to the experimental data.

Figure 1 shows the average lateral distribution of the particle density for the air showers of the size $Ne = 5 \times 10^6 \sim 10^7$ and $10^7 \sim 5 \times 10^7$. The average density is obtained by calculating the particle densities at several distances, using parameter values A and α of eq.(1), which are determined by the least square fitting in respective events.

We estimate the energy spectrum of hadrons in the air shower from the lateral distribution of the particle density, detected by the hadron calorimeter, the detail of which is described in Ref.[4]. Figure 2 shows the differential energy spectrum of hadrons in the air showers with the size $Ne = 5 \times 10^6 \sim 10^7$, together with those obtained by the simulations. The simulations employ different models for multiple particle production: UA5 algorithm modified for hadron-nucleus collisions, VENUS[5], QGSJET[6] and HDPM[7]. A proton-dominant composition is assumed for the primary cosmic rays. The energy spectrum of hadrons obtained by the experiment is not consistent with those by simulations in the number, but consistent in the power index. If one assumes a heavy-dominant composition of the primary cosmic rays in the simulations, the number of hadrons increases and brings larger discrepancy between the experimental and simulated data.

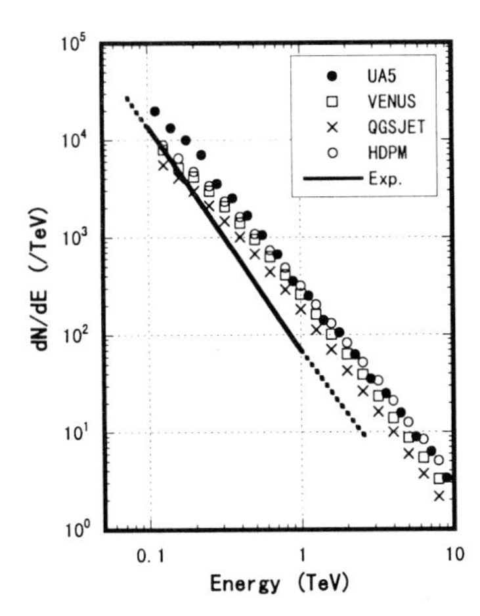

Figure 2. Differential energy spectrum of hadrons in air shower (the solid line) which is estimated from the lateral distribution of the particle density distribution. The size range of the air shower size is $N_e = 5 \times 10^6 \sim 10^7$. The predictions by various simulations are presented together. The assumptions for the primary cosmic rays are the same for all the cases. The sampling of the primary energy is made for $E_0 \geq= 5 \times 10^{15}$ eV, and the air showers with $N_e = 5 \times 10^6 \sim 10^7$ are collected.

3. Discussion and conclusion

3.1. Primary cosmic-ray composition at 10^{16} eV

Figure 3(a) shows the distribution of n_h (the differential number of hadrons at $E = 1$ TeV) for 62 air showers which have the size $Ne = 5\times10^6 \sim 10^7$. This air shower size corresponds to the primary energy of $E_0 = 10^{16}$ eV on average. Figure 3(b) shows the distribution of n_h obtained by a simulation using UA5 algorithm for nuclear interactions. In these figures we can see:

(1) The absolute value of n_h obtained by the experiment is different from that obtained in the simulation, which is pointed out in the energy spectrum of hadrons in Fig. 2.

(2) The n_h distribution obtained by the experiment is wider than that by simulations, probably due to the experimental errors.

(3) In the figure (b) by the simulations the proton-induced events occupy the left-hand side of the distribution and are distributed over 8 bins in the histogram.

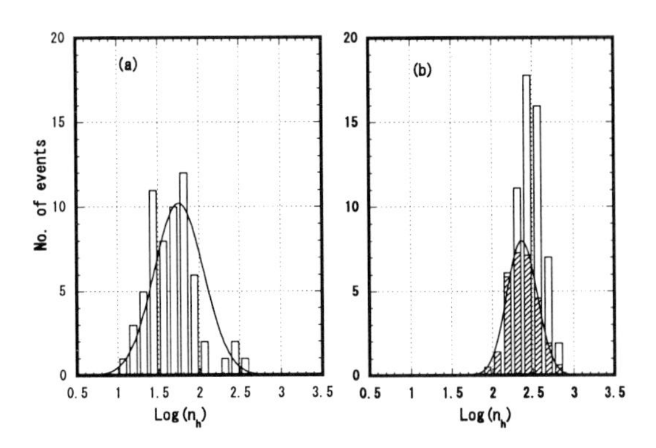

Figure 3. Distribution of n_h (a): The experimental data of 62 events. (b) : The UA5 simulation. The number of events is normalized to the experimental data. The hatched events are proton-induced ones.

Assuming the superposition model for the primary cosmic-ray particle of mass number A, we have a relation

$$< \log n_h^{(A)} > - < \log n_h^{(p)} >= (1-\sigma) < \log A >$$

where σ is the index of the energy spectrum of the hadrons in integral form. It means that, if we can know the average number of hadrons in the air shower $< \log n_h^{(p)} >$, due to the primary protons, we can obtain the average mass number of primary cosmic rays $< \log A >$ from the average number of hadrons in the air shower $< \log n_h^{(A)} >$.

By the point (3) seen in Fig. 3(b), we assume that the $n_h^{(p)}$ distribution of the experimental data has a peak at the fourth bin from the left-hand side of the distribution. Then we have $< \log n_h^{(p)} >= 1.44$ for the experimental data. Be-

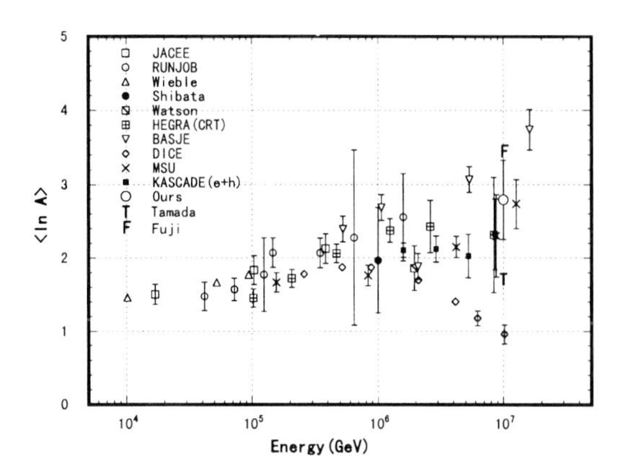

Figure 4. Average mass number of the primary cosmic rays, $< \ln A >$, which is estimated from the number of hadrons in the air shower. The data obtained by other experiments, compiled by KASCADE group, are shown together. The marks F and T, located at $E_0 = 10^7$ GeV, are those of typical heavy-dominant[8] and proton-dominant[4] compositions.

cause we have $< \log n_h^{(A)} >= 1.68 \pm 0.04$ from the experimental data and $\sigma = 0.8$ by the theoretical consideration, we have

$$< \log A >= 1.20 \pm 0.22 \quad or \quad < \ln A >= 2.8 \pm 0.5$$

One can see that the method of estimation is free from the characteristics of nuclear interaction. Figure 4 shows the average mass number of primary cosmic rays, which is estimated from the number of hadrons in air showers.

3.2. Nuclear interactions at 10^{16} eV

Figure 2 shows that the number of hadrons in the air shower is lower than that obtained from the simulations in the energy region of 10^{16} eV. This tendency is consistent with the relationship between the families and accompanied the air showers [1]. That is, the number of γ-rays in the family from the experiment is smaller than results from the simulation. It is important to point out that both data of γ-ray and hadrons are independent, because they are detected by different detectors.

Our argument is in the following way:
(1) Main assumptions in the simulations are on the hadron-air collisions and on the chemical composition of the primary cosmic rays. The experimental data indicate that at least one of the assumptions should be revised in the direction to make the energy subdivision more rapid.
(2) Heavy-dominant hypothesis of the primary cosmic rays, proposed currently, is not effective to remove the discrepancy.
(3) Hence we reach the conclusion that the nuclear interaction has different characteristics from those assumed in the simulations in the energy region of 10^{16} eV. That is, the multiple particle production in 10^{16} eV is of higher multiplicity, of softer energy spectrum of the produced particles, of larger inelasticity, etc., compared with those assumed in the simulation.

REFERENCES

1. N. Kawasumi et al., Phys. Rev. D53 (1970) 3634.
2. C. M. G. Lattes et al., Phys. Rep. No.65 (1980) 151.
3. C. Aguirre et al., Nucl. Phys. B(Proc. Suppl.) 75A (1999) 186.
4. C. Aguirre et al., ICRR-Report-460-2000-4, (2000).
5. K. Werner, Phys. Rep. 232 (1993) 87.
6. N. N. Kalmykov and S. S. Ostapchenko, Phys. At. Nucl. 56 (1993) 346.
7. J. N. Capdevielle, J. Phys. G 15 (1989) 909.
8. M. Amenomori, et al., Phys. Rev. D25,(1982) 2807.

ELSEVIER

Nuclear Physics B (Proc. Suppl.) 97 (2001) 162–164

www.elsevier.nl/locate/npe

Arrival Directions and Chemical Composition of Ultrahigh Energy Cosmic Rays

A.A. Mikhailov[a]

[a]Institute of Cosmophysical Research and Aeronomy, 31 Lenin Ave., 677891 Yakutsk, Russia

To estimate the chemical composition of ultrahigh energy cosmic rays we apply an approach using the well established magnitude and character of the galactic magnetic field and also both theoretical and experimental distributions of showers in galactic latitude. Arrival directions of cosmic rays in the energy region of (0.8-4)$\times 10^{19}$ eV in galactic latitude are consistent with theoretical calculations, if cosmic rays are mainly heavy nuclei. An excess flux of cosmic rays at $\sim 10^{19} eV$ from the galactic plane is found.

1. INTRODUCTION

The estimation of chemical composition of cosmic rays from the development of the depth of shower maximum using Yakutsk extensive air shower (EAS) array [1] data and model calculations of NN and pN interactions of ultrahigh energy particles shows that at energies E $\sim 10^{18}$ – $10^{19} eV$ in the primary radiation the protons are prevalent. The analogous estimation of chemical composition was made in [2] from Fly's Eye array data, where the authors assumed that at 10^{19} eV the protons were not less than 90%. Using a new model for particle interactions, the authors of [3] concluded from the development of the shower maximum that cosmic rays of E $< 3 \times 10^{18}$ were heavy nuclei and that above this energy the portion of heavy nuclei is $\sim$ 50%.

As seen from [1,3], there are contradictions in the estimations of the chemical composition. In the present paper we estimate the chemical composition of primary cosmic rays from the number of observed and expected showers in galactic latitude.

2. EXPERIMENTAL DATA

In [4] and other papers the authors found that cosmic rays of E $< 4 \times 10^{19}$ eV are mostly galactic. Here we study arrival directions of EAS detected by the Yakutsk array in the period 1974 - 1995. Data are arrival directions of 576 showers with energies (0.8-4)$\times 10^{19} eV$, zenith angles $< 60°$ and axes lying inside the array perimeter. The average energy of showers is 1.3$\times 10^{19}$ eV.

3. CALCULATIONS

We consider the model for the disc magnetic field which was suggested in [5] and constructed on the basis of measurements of Faraday's rotation in radio emissions from pulsars. The main component of the magnetic field is the azimuth component of magnitude $\sim$ 2 μG. The radial and z - components are one order of magnitude smaller. Except for the regular components, there exists the irregular component of 5 μG magnitude and 100 pc characteristic size. Sizes of the disc are: the radius is 15 kpc, half-height is 0.4 kpc. As in [6], we suggest the existence of a magnetic field outside of the galactic disc - in a halo of the Galaxy. This field has both regular and irregular components. The magnitude of the main azimuth regular component of the magnetic field in the galactic halo is $\sim 1 \times f(z) \mu$G, (where $f(z) = exp\left(\frac{|z|-0.4}{5kpc}\right)$), for the irregular component it is 1.5 μG, with a characteristic size of 500 pc and a half - height of the halo of 5 kpc.

We consider two assumptions for the distribution of cosmic ray sources: 1) sources are uniformly distributed over the galactic disc, 2) sources are pulsars (in [7] and other papers we found that cosmic ray sources were mostly pulsars). The distribution of pulsars in radius r and in height z across the Galaxy's disc is described

PII S0920-5632(01)01246-4

by a function:

$$f(r,z) = f_1(r) \times f_2(z),$$

where $f_1(r) = (1 - exp(r^2/8)) \times exp(-r^2/100)$, $f_2(z) = 1/0.46 \times exp(-|z|/0.23)$ and r and z are in kpc. The function $f_1(r)$ is found by an approximation of the observed distribution of pulsars [8] and the function $f_2(z)$ is taken from [8].

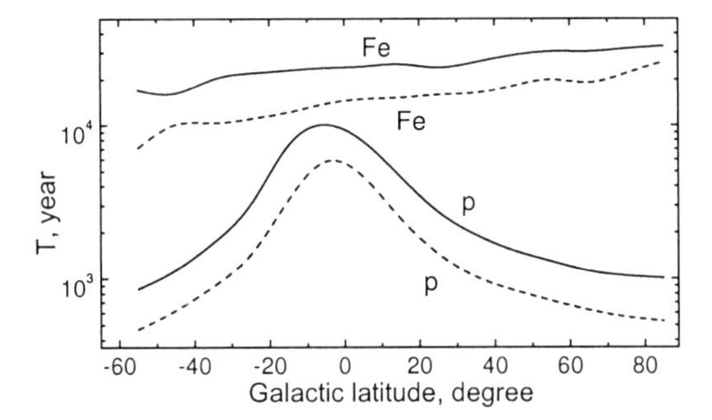

Figure 1. The path length (intensities) of protons (p) and iron nuclei (Fe) at the energy $1.3 \times 10^{19} eV$. The solid lines - the sources are distributed uniformly over the whole disc, the dashed lines - sources are the pulsars.

To determine the expected particle flux we calculated trajectories of antiparticles from the Earth to their exit beyond the halo. The antiparticle trajectories correspond to trajectories of positive charged particles moving from sources to the Earth. The expected particle flux in the given direction is assumed to be proportional to an antiparticle path length and a density of sources in the galactic disc. We consider the trajectories of antiprotons at energies of the 5×10^{17} eV and 1.3×10^{19} eV. The energy of antiprotons of 1.3×10^{19} eV corresponds to the average energy of considered showers, the energy 5×10^{17} eV corresponds to generation of showers by iron nuclei. The lifetimes of particles in the galactic disc, or expected intensities of particles depending on their arrival direction in galactic latitude, are shown in Fig.1.

If we determine the anisotropy of cosmic rays by the formula

$$\delta = (I(max) - I(min)/(I(max) + I(min)),$$

where I is the intensity (Fig.1), then independently of assumed sources, the anisotropy is expected to be equal to 100% and of a few percent for primary cosmic rays consisting of protons and iron nuclei, respectively. Note that the analogous conclusion has been obtained for other models of the magnetic field in the disc and halo [9–11].

4. DATA ANALYSIS

Fig.2 presents the distribution of the observed showers in galactic latitude b and the expected numbers of events for protons (p) and iron nuclei (Fe). The expected numbers of events were found from the expected intensity (Fig.1), taking into account an exposure of the Yakutsk EAS array at the celestial sphere and normalizing to the observed numbers of showers.

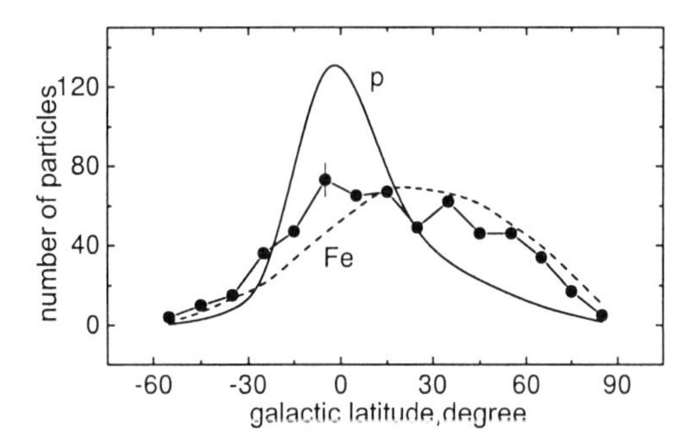

Figure 2. The observed (circles connected by a line) and expected numbers of EAS events in the case of protons (p) and iron nuclei (Fe) in galactic latitude b.

As a result of the normalization, the expected number of showers both from protons and iron nuclei for two different distributions of sources are coincident. As seen from Fig.2, the observed number of showers is not consistent with the expected one, if cosmic rays are only protons or only iron nuclei. The excess of events on the left side of the curve in the case of iron nuclei and the lack of events on the right side indicate that, except iron nuclei, there are nuclei lighter than iron. The estimation of the chemical composition in the case of protons and iron nuclei through the arrival direction of showers using the χ - square method testifies that cosmic rays consist predominantly of iron nuclei, the portion of which is $\geq 80\%$.

Note the number of observed showers with $(0.8\text{-}4)\times10^{19}$ eV within 3° of the galactic plane ($|\ b\ | < 3°$) exceeds the expected one: 58 showers instead of 34.4 events are observed, an excess is 4.0σ, where σ=(observed number - expected number)/(uncertainty of expected number). It confirms the conclusion in [4] that cosmic rays with energies $< 4 \times 10^{19}$eV are mostly galactic.

5. CONCLUSION

An estimation of the chemical composition of the primary radiation by the distribution of particles in galactic latitude shows that cosmic rays at E $\sim 10^{19}$ eV consist mainly of iron nuclei, their portion being $\geq 80\%$.

This work was supported by Russian Foundation for Fundamental Research (grant N 00-02-16325) and the Russian Ministry for Science (grant N 01-30).

REFERENCES

1. B. N. Afanasiev, M. N. Dyakonov, T.A. Egorov et al, Proc. Int. Symp. Extremely High Energy Cosmic Rays (Ed. M. Nagano), Inst. for Cosmic Ray Res., Tokyo (1996) 32.
2. D. Bird et al., Phys. Rev. Lett. 71 (1993) 4301.
3. A. W. Wolfendale and T. Wibig, Proc.26th ICRC, Salt Lake City, 3 (1999) 248.
4. N. N. Efimov, A. A. Mikhailov, A. D. Krasilnikov, Proc.21th ICRC, Adelaide, 2 (1990) 64.
5. R. J. Rand and S. R. Kulkarni, Astrophys. J., 343 (1989) 760.
6. V. S. Berezinsky, A. A. Mikhailov, S. I. Syrovatskii, Proc. 16th ICRC , Kyoto 2 (1979) 86.
7. A. A. Mikhailov, Nucl. Physics B. Proc. Suppl., 75A (1999) 359.
8. R. Manchester, J. Taylor, Pulsars. San Francisco.1977.
9. M. Giller, J. L. Osborne, J. Wdowczyk, M. J. Zelinska, Phys.G: Nucl. Part. Phys. 20 (1994) 1649.
10. V. S. Berezinsky, S. I. Grigoryeva, A. A. Mikhailov, et al. Proc. ICRR Int. Symp., Kofu (1991) 134.
11. R. Lampard, R. W. Clay, B. R. Dawson, A. G. Smith, Proc. 25th ICRC, Durban, 4 (1997) 193.

ELSEVIER

Nuclear Physics B (Proc. Suppl.) 97 (2001) 165–168

NUCLEAR PHYSICS B
PROCEEDINGS
SUPPLEMENTS
www.elsevier.nl/locate/npe

LAAS Network Observation of Air Showers

N. Ochi[a], T. Wada[a], Y. Yamashita[a], I. Yamamoto[b], T. Nakatsuka[c] and Large Area Air Shower (LAAS) group

[a]Department of Physics, Okayama University, Okayama 700-8530, Japan

[b]Okayama University of Science, Okayama 700-0005, Japan

[c]Okayama Shoka University, Okayama 700-8601, Japan

A network observation of air showers is in progress in Japan. Researchers from 10 institutions are taking part in the network, forming the Large Area Air Shower (LAAS) group. As of June 2000, eight air shower arrays (stations) are in operation, while five more stations are under construction. The stations are scattered over an enormous area of 130,000 km^2, however the Global Positioning System (GPS) provides comparable time stamps among them with an accuracy of one microsecond. The aim of the network is to find out correlations in primary cosmic rays just below the knee. The data from the planned Norikura array (2,770 m a.s.l.) is expected to reveal new aspects of correlated cosmic rays. The current status of the network and performance of arrays calculated by the air shower simulation code CORSIKA are described here.

1. INTRODUCTION

The presence of correlations in primary cosmic rays has not been considered well since the discovery of cosmic rays. One reason for this should be that conventional cosmic rays are thought to be completely randomized by the Galactic magnetic field by the time of their injection to the earth.

However, several groups reported correlations in ultra-high energy cosmic rays in recent years, by means of air shower arrays. Lateral structure (spatial correlation) in cosmic rays has been observed as the unusual increase of coincident triggers at four stations distant by more than 50 km [1], while longitudinal structure (time correlation) has been observed as the chaotic feature of arrival time intervals of air showers lasting more than 20 hours [2].

To measure such correlations more accurately and extensively, the Large Area Air Shower (LAAS) group was established in Japan. Thirteen air shower arrays form a network with an unprecedented enclosed area of 130,000 km^2. If correlations in primary cosmic rays are confirmed by our experiment with larger statistics, it must be important information for solving the mechanism of the formation and propagation of cosmic rays.

The network observations similar to ours are proposed recently by CERN [3] and by University of Washington [4]. However, the LAAS group covers much larger area than them and has been accumulating multi-stations data since 1996.

In this paper, we report the current status of the network and performance of arrays calculated by the air shower simulation code CORSIKA. Results of the coincident event analysis and the successive air shower analysis using four years' data of the LAAS group are also presented in the symposium [5,6].

2. NETWORK

The LAAS group consists of thirteen air shower arrays (stations). The sites are in the campuses of

Ashikaga Institute of Technology (AIT),
Hirosaki University (HU),
Kinki University (KU),
Kochi University (KCU),
Nara University of Industry (NUI),
Okayama University (OU),
Okayama University of Science (OUS),

0920-5632/01/$ – see front matter
PII S0920-5632(01)01248-8

Table 1
The coordinates and aspects of stations

Station	HU	AIT	PU	NO	NUI	KU1
Latitude (N)	40°35′	36°21′	35°35′	36°07′	34°35′	34°39′
Longitude (E)	140°29′	139°24′	139°20′	137°33′	135°41′	135°36′
# of Counters	5	7	*	*	7	5
Trig. Counters	5	*	*	*	7	5
Trig. Rate (/min)	0.42	*	*	*	0.20	0.27
Operating since	11/13/98	soon	*	soon	07/10/96	05/14/93

* not available yet

Table 1
The coordinates and aspects of stations (continued)

KU2	OCU2	OCU1	OUS1	OUS2	OU	KCU
34°39′	34°35′	34°48′	34°42′	34°42′	34°41′	33°33′
135°36′	135°31′	134°17′	133°56′	133°56′	133°55′	133°29′
7	10	22	8	4	8	8
7	3	4	4	4	5	8
0.22	2.8	1.2	0.53	*	0.41	*
07/10/98	06/97	1961	07/31/95	*	09/12/96	03/11/99†

* not available yet
† currently not operating

Osaka City University (OCU),
and Polytechnic University (PU).

Among these institutions, KU and OUS have two arrays each in their campuses, while OCU has an array at Mitsuishi as well as one in the campus. In addition to these arrays, an array at the Norikura Observatory (NO) (2,770 m a.s.l.) is under proposal. Mutual distances between them are more than 10 km except some combinations, so that plural stations can not be triggered by the same air shower. This is why we can measure spatial correlation in primary cosmic rays. As of June 2000, eight stations are in operation, while the reminder is under construction or repairing work.

Most stations use the same type of detector, which consists of a $50cm \times 50cm$ plastic scintillator and a photomultiplier (Hamamatsu, R329-02). 4 to 22 detectors are arranged at each station. The trigger conditions are different among stations; 3- to 8-fold coincidences are applied and yielding the trigger rates of 0.20 to 2.8 events/min. Most importantly for the network observation, each station has the Global Positioning System (GPS) as the common clock, so the arrival times of air showers can be recorded with an accuracy of $1\mu s$.

The geographical coordinates and some aspects of stations are summarized in Table 1. Table 2 shows mutual distances between them. For a more detailed description of each station and a graphical map of the LAAS group, see Ref. [7].

3. ARRAY PERFORMANCE CALCULATED BY CORSIKA

The angular accuracy, the trigger efficiency and the energy distribution for the OU-array and the NO-array are calculated by the air shower simulation code CORSIKA (Ver.5.624 with EGS, QGSJET and VENUS options).

The OU-array has five detectors for triggering, which are arranged at the corners and the center of a $10m \times 10m$ square. The arrival direction of a shower is determined by fitting a plane shower front to the observed time delays.

Artificial showers are produced by CORSIKA in the primary energy range of log E(TeV)=1.2

Table 2
Mutual distances between stations (km)

	HU	AIT	PU	NO	NUI	KU	OCU2	OCU1	OUS	OU	KCU
AIT	478.2										
PU	563.8	86.0									
NO	559.3	169.0	170.6								
NUI	787.9	390.3	350.2	238.3							
KU	786.8	394.2	356.1	239.7	10.6						
OCU2	796.4	404.0	365.3	249.9	15.9	10.0					
OCU1	842.3	495.0	467.6	329.7	130.2	120.9	114.6				
OUS	871.6	529.3	501.7	363.8	161.1	152.3	145.2	34.3			
OU	872.7	530.2	502.4	364.7	161.6	152.8	145.7	35.2	1.1		
KCU	997.2	623.6	581.6	467.2	233.4	229.4	219.5	157.1	133.9	132.8	

to 4.6 with the spectrum index of -2.7 below the knee and -3.2 above the knee. The zenith angles are randomly sampled from the range of $0°$ to $45°$. The composition of primary particles is assumed to be p(42%), He(17%), O(14%), Si(14%) and Fe(13%) [8]. We produce 5270 showers in total and scatter them randomly over an sufficiently large area.

First we estimate the energy threshold for a detector. This is done by adjusting the threshold to reproduce the observed trigger rate of the array. For OU-array, we get the threshold energy of 8.5 MeV. The observed distribution of χ^2-values associated with the plane fitting procedure is well reproduced by the simulation under this threshold value, so we have confidence in this value.

Figure 1 shows the simulated distribution of the angular distances between inputted and reconstructed shower angles. We define the angular accuracy as the angular distance that contains 68% of showers in this figure. For OU-array, this value is turned to be $7.0°$. Figure 2 shows the simulated trigger efficiency for showers whose cores are inside the OU-array. The efficiencies at 100, 1000 and 3980 TeV are calculated to be 1%, 41% and 87%, respectively. However, it should be noted that the OU-array can not determine accurately the core positions of real shower events owing to the small number of detectors. The energy distribution of showers which trigger the OU-array is shown in Figure 3. The acceptance of the array begins for energies of about 25 TeV, while the most probable energy of a detected shower is about 1 PeV, which is just below the knee of the primary cosmic ray spectrum. We can not estimate the primary energy on a shower-by-shower basis, again owing to the small number of detectors.

The configuration of NO-array has not been fixed yet, but here we assume that four detectors for triggering are arranged at the corners of a $8.0m \times 12.0m$ rectangle. The energy threshold of a detector is set to 10.0 MeV. For NO-array, 5397 showers in the primary energy range of log E(TeV)=1.0 to 4.0 are produced by CORSIKA. The angular accuracy calculated from Figure 1 is $8.0°$. The trigger efficiency of NO-array is not calculated yet. The energy distribution shown in Figure 3 indicates that the NO-array will collect 11 times as many showers as the OU-array per unit time and will be more sensitive to lower energies. This is due to the high altitude of the NO-array and must give new insight into correlated cosmic rays.

4. SUMMARY

The LAAS network observation consists of thirteen air shower arrays, which scatter over an enormous area of 130,000 km^2. The typical angular accuracy of an array calculated by CORSIKA is $7.0°$. The observed primary energy range is just below the knee of the primary cosmic ray spectrum. Eight stations are currently taking data and the large amount of data has already been accumulated since 1996. The network data is ex-

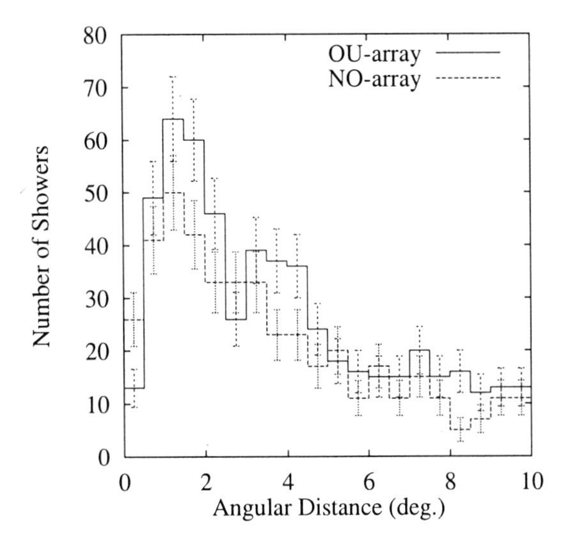

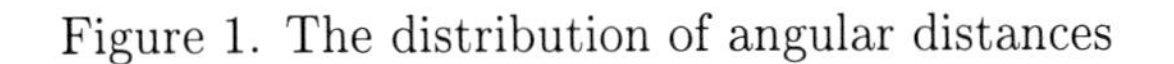

Figure 1. The distribution of angular distances

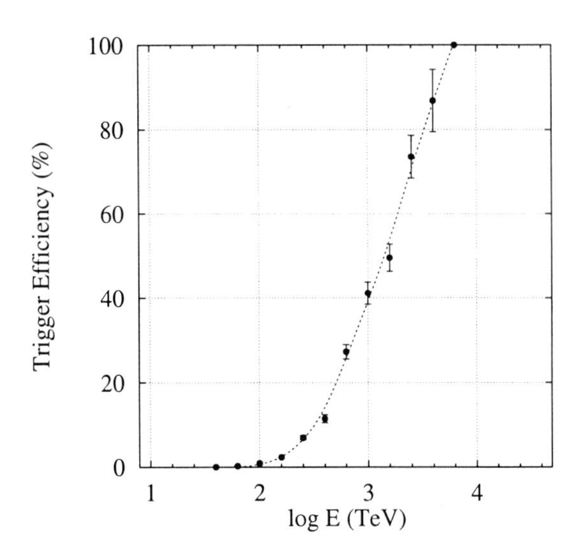

Figure 2. The trigger efficiency

pected to reveal new aspects of cosmic rays, that is, correlations in primary cosmic rays.

REFERENCES

1. O. Carrel and M. Martin, Phys. Lett. B325 (1994) 526.
2. T. Kitamura et al., Astrop. Phys., 6 (1997) 279.
3. C. Grupen et al., Proc. 25th Int. Cosmic Ray Conf. (Durban), (1997).
4. E. Zager et al., Proc. 26th Int. Cosmic Ray Conf. (Salt Lake City), 5 (1999) 341.
5. N. Ochi et al., "Search for Coincident Air Showers in the Network Observation", this symposium.
6. N. Ochi et al., "Anisotropy of Successive Air Showers", this symposium.
7. N. Ochi et al., Proc. 26th Int. Cosmic Ray Conf. (Salt Lake City), 2 (1999) 419.
8. N. Kawasumi et al., Phys. Rev. D53 (1996) 3534.

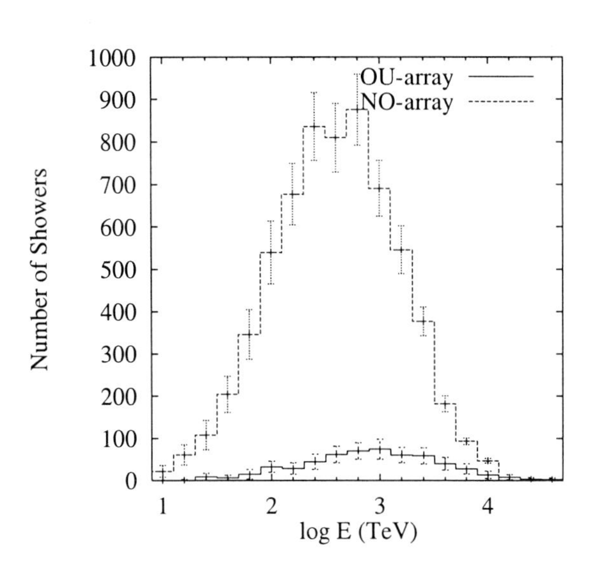

Figure 3. The energy distribution

ELSEVIER

Nuclear Physics B (Proc. Suppl.) 97 (2001) 169–172

www.elsevier.nl/locate/npe

Search for Coincident Air Showers in the Network Observation

N. Ochi[a], T. Wada[a], Y. Yamashita[a], I. Yamamoto[b], T. Nakatsuka[c] and Large Area Air Shower (LAAS) group

[a]Department of Physics, Okayama University, Okayama 700-8530, Japan

[b]Okayama University of Science, Okayama 700-0005, Japan

[c]Okayama Shoka University, Okayama 700-8601, Japan

Here we report results of the coincident event analysis, using air shower data taken at six stations of the Large Area Air Shower (LAAS) group in Japan. In this analysis we search for pairs of coincident air showers between stations, which are hypothetically induced by bursts of ultra-high-energy γ-ray sources or by secondary particles from interactions of extremely-high-energy cosmic rays with interstellar matter. From four years data (2.3×10^6 air showers) we find a pair of air showers with a very small time difference of 195 microseconds and an angular distance of 5.3 degrees, which is within the angular accuracy of our arrays, between two stations separated by 152 km. The chance probability of this event is 0.16. The Crab Nebula, a well-known ultra-high-energy γ-ray source, is within the angular accuracy from the arrival direction of this event. We expect that this event was induced by the ultra-high-energy γ-rays emitted in a burst of the Crab Nebula.

1. INTRODUCTION

When ultra-high-energy cosmic rays come to the earth, they interact in the atmosphere and produce extensive air showers. The lateral extent of an air shower at observation level becomes greater with energy of the primary cosmic ray, but it does not exceed several kilometers even for the highest energy cosmic ray ever observed. Thus, if air showers that are observed at different stations separated by more than 10 km have some correlations, they must be attributed to different but correlated primary particles.

Such correlations have been disregarded since the discovery of cosmic rays. However, the University of Geneva group performed an experiment which measured correlations in primary cosmic rays about ten years ago [1]. They observed secondary particles at four stations which were distant by more than 50 km each other (an enclosed area of 5,000 km^2) and found correlations with a typical time spread of about 0.3 ms.

This result motivated us to start the "coincident event analysis" described in this paper, using the air shower data taken by the Large Area Air Shower (LAAS) group in Japan. In this analysis, we search for pairs of coincident air showers between stations separated by up to 900 km. As compared to the Geneva group, we have the following advantages: (i) The LAAS group covers a much greater area of 130,000 km^2 and has thirteen stations (six stations' data are used in this paper). (ii) We already have four years data of simultaneous observation by multiple stations, while Geneva's analysis was based on about one year data. (iii) More important, each array of the LAAS group can determine arrival directions of air showers, being useful to select air showers induced by primary particles traveling from the same direction.

A network observation which aims at the same goal as ours has started recently at CERN [2]. It covers an intermediate scale of approximately 60 km^2. It would be expected that their results and ours will be complementary to each other.

The source of correlated cosmic rays is still an open question. It seems that charged particles can not be a propagator because they are deflected randomly by the Galactic magnetic field before coming to the earth. Ultra-high-energy γ-rays from bursts of their sources or secondary particles from interactions of extremely-high-energy

PII S0920-5632(01)01251-8

Table 1
The aspects of stations and data

Station Name	HU	NUI	KU1	KU2	OUS1_a	OUS1_b	OUS1_c	OU
Latitude (N)	40°35′	34°35′	34°39′	34°39′	34°42′	34°42′	34°42′	34°41′
Longitude (E)	140°29′	135°41′	135°36′	135°36′	133°56′	133°56′	133°56′	133°55′
Dist. from OU (km)	873	162	153	153	1	1	1	–
Dist. from KU1 (km)	787	11	–	0	152	152	152	153
Trigger Counters	5	7	5	7	4	8	4	5
Trigger Rate (/day)	594	309	461	317	753	314	737	589
Data Period	11/98 – 12/99	08/96 – 06/98	09/96 – 12/99	07/98 – 04/99	09/96 – 11/97	11/97 – 01/99	01/99 – 12/99	09/96 – 12/99
Number of Showers	231k	172k	457k	62k	280k	130k	266k	664k

cosmic rays with interstellar matter may have potential to induce correlated air showers. However, currently we have too little information about correlated cosmic rays and we need more extensive and detailed observation of them to construct concrete models.

2. NETWORK OBSERVATION

This paper describes the analysis of data collected at the six stations of the LAAS group. Table 1 shows the profiles of the stations. As for the LAAS group, a more detailed description can be found in Ref.[3,4]. In this section we describe some aspects of stations that are relevant for this analysis.

Each station has four to eight scintillation counters and the trigger conditions are different station-by-station: 4- to 8-fold coincidences, yielding the trigger rates of 309 to 753 showers/day. We note that the trigger condition of the OUS1-array was changed twice in four years, so we treat the three periods, OUS1_a, OUS1_b and OUS1_c independently. The estimated mean primary energy is about 1 PeV, which is just below the knee of the primary cosmic ray spectrum. The estimated angular accuracy is about 7.0 degrees.

Mutual distances between stations are, as shown in Table 1, more than 1 km except for the combination of KU1-KU2. Thus we treat the air shower data sets from different stations as independent except for KU1-KU2. Most importantly for this analysis, each station has the Global Positioning System (GPS) as the common clock, so the arrival times of air showers can be recorded with an accuracy of $1\mu s$.

This analysis is based on about 2.3×10^6 showers collected in the period from Aug. 1996 to Dec. 1999 with reconstructed zenith angles $\leq 45°$.

3. ANALYSIS AND RESULTS

First, we compare the arrival times of air showers taken at two stations and draw the distribution of the time differences. No directional restriction is imposed. This is performed separately for each combination of two stations out of eight. All of the resulting distributions are well described by an exponential, as expected from chance coincidences. No pair of air showers with significantly small time difference is found. In Figure 1, we show in the histogram the distribution of time differences summed for all combinations of stations. The broken line is an exponential function fitting the data. In this figure we excluded coincident events between KU1 and KU2, which have time differences of less than $2\mu s$ and must have been induced by identical air showers.

Though we do not find any sign of correlations in Figure 1, there is a possibility that correlated air showers are overwhelmed by chance coincidences. Thus, as the next step, we inspect pairs of air showers with the smallest time differences. Table 2 shows parameters of the pairs of air showers with time difference $\leq 200\mu s$. The smallest time difference of $44\mu s$ (event (k)) is observed between HU and NUI. However, even this time difference

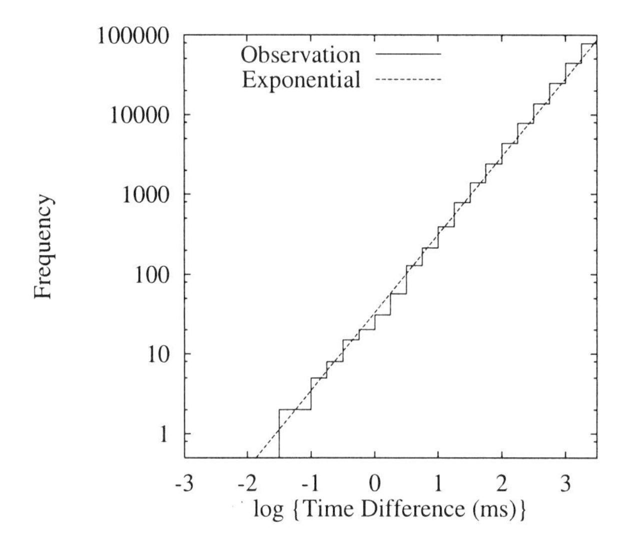

Figure 1. The distribution of time differences for all data

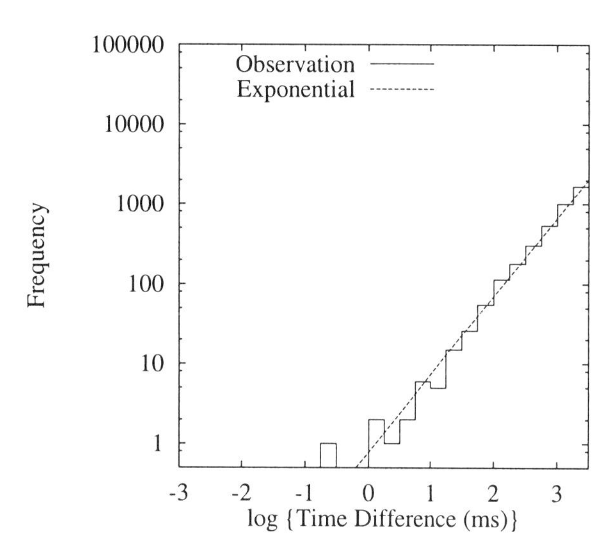

Figure 2. The distribution of time differences for pairs with angular distance $\leq 5.3°$

is not significantly small and is consistent with chance coincidence, as seen in Figure 1. The angular distance between the pair of air showers is a rather large value of 24.6 degrees, so we judge this event a chance coincidence and to have no physical implication.

The most notable event in Table 2 is (f), which is observed between OUS1_b and KU1. It has a very small time difference of $195\mu s$ and the smallest angular distance of 5.3 degrees, being the only event within the angular accuracy of our arrays. In Figure 2, we show the distribution of time differences for air shower pairs whose angular distances are equal or less than 5.3 degrees. The $195\mu s$ event is separated from chance coincidences with a small margin; the chance probability of this event is calculated as 0.16.

In Figure 3 the arrival directions of coincident air showers shown in Table 2 are plotted in equatorial coordinates. The broken line shows the Galactic plane and the cross indicates the position of the Crab Nebula, a well-known ultra-high-energy γ-ray source. Intriguingly the $195\mu s$ event is very adjacent to the Galactic plane and the Crab Nebula. Though we can not confirm that this event came exactly from the direction of the Crab Nebula due to poor angular accuracy, one possible interpretation is that this event was induced by the ultra-high-energy γ-rays from it.

Historically a lot of papers have been published about TeV and PeV emissions from the Crab. Most of them concluded that the Crab emits ultra-high-energy γ-rays only sporadically, not steadily. As an extreme example, there was an episode of simultaneous observation of a PeV energy burst by four different groups [5]. We expect that the $195\mu s$ event detected here was induced by ultra-high-energy γ-rays emitted in such a burst of the Crab.

4. SUMMARY

We search for coincident air showers between stations of the Large Area Air Shower group in Japan. A pair of air showers with a very small time difference of $195\mu s$ and the smallest angular distance of 5.3 degrees is found. The chance probability of this event is 0.16. Moreover, the arrival

Table 2
The parameters of coincident events (Time difference $\leq 200\mu s$)

ID	TD(μs)*	AD(deg)†	Stations	Date, Time‡	α (h)	δ (deg)	θ (deg)	ϕ (deg)
(a)	116	15.7	OUS1_a→	02/17/97	14.1	33.7	8.0	94.5
			OU	71985	14.0	49.3	16.8	26.1
(b)	86	28.5	OUS1_a→	08/22/97	21.4	30.4	8.0	239.4
			OU	49895	19.4	46.4	20.6	48.7
(c)	47	29.8	NUI→	10/30/97	6.4	17.2	25.6	232.3
			OU	62474	4.4	6.6	28.9	164.3
(d)	131	23.1	OUS1_b→	11/15/97	14.8	19.5	19.5	222.7
			OU	04728	13.7	37.7	3.7	33.4
(e)	106	38.0	OU→	12/20/97	7.0	28.9	53.8	282.8
			NUI	42754	4.6	14.3	31.0	235.5
(f)	195	5.3	OUS1_b→	09/24/98	4.9	17.8	18.9	151.1
			KU1	73332	5.4	18.6	16.4	166.8
(g)	67	11.9	KU1→	12/03/98	13.5	35.7	20.7	79.9
			OU	05019	13.1	24.7	28.2	102.5
(h)	113	50.0	OUS1_c→	01/13/99	22.8	21.5	16.0	217.3
			HU	20374	0.6	76.4	38.2	348.8
(i)	132	19.9	OUS1_c→	04/19/99	7.6	27.6	16.3	249.0
			NUI	27828	6.3	25.1	10.4	154.6
(j)	199	49.2	OU→	05/07/99	3.5	36.1	14.4	280.7
			OUS1_c	08671	1.2	-1.6	39.4	153.6
(k)	44	24.6	HU→	09/26/99	7.6	41.3	1.2	54.6
			NUI	79077	6.7	11.9	24.3	157.2

* Time difference, †Angular distance, ‡Date (mm/dd/yy) and time (in seconds)

direction of this event points to the Crab Nebula within angular accuracy. There is a possibility that this event was induced by the ultra-high-energy γ-rays from the Crab. If this hypothesis is true, this event provides important information on the mechanism of ultra-high-energy emission from the object.

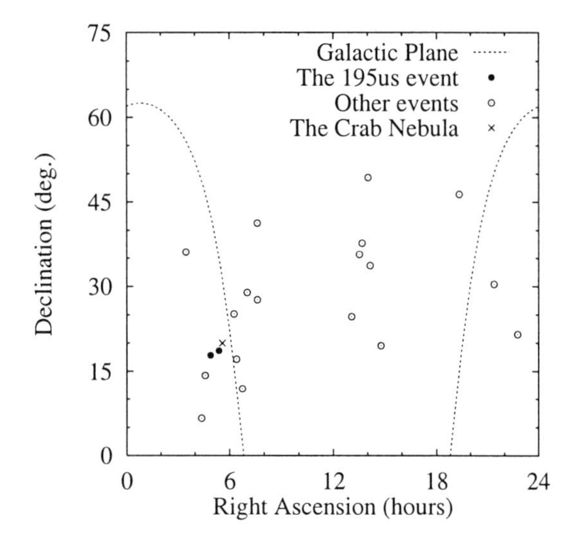

Figure 3. The direction of coincident events

REFERENCES

1. O. Carrel and M. Martin, Phys. Lett. B325 (1994) 526.
2. C. Grupen et al., Proc. 25th Int. Cosmic Ray Conf. (Durban), (1997).
3. N. Ochi et al., "LAAS Network Observation of Air Showers", this symposium.
4. N. Ochi et al., Proc. 26th Int. Cosmic Ray Conf. (Salt Lake City), 2 (1999) 419.
5. M. Rao and B. Sreekantan, Current Sci., 62 (1992) 617.

ELSEVIER

Nuclear Physics B (Proc. Suppl.) 97 (2001) 173–176

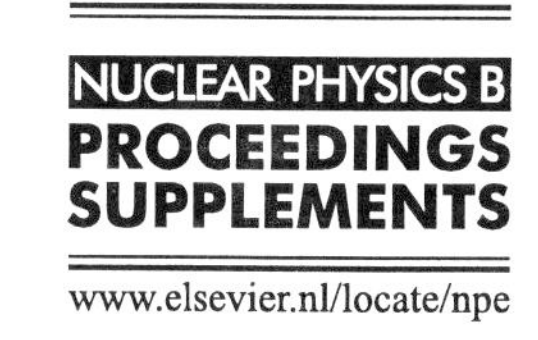

www.elsevier.nl/locate/npe

Anisotropy of Successive Air Showers

N. Ochi[a], T. Wada[a], Y. Yamashita[a], A. Ohashi[a], I. Yamamoto[b], T. Nakatsuka[c] and Large Area Air Shower (LAAS) group

[a]Department of Physics, Okayama University, Okayama 700-8530, Japan

[b]Okayama University of Science, Okayama 700-0005, Japan

[c]Okayama Shoka University, Okayama 700-8601, Japan

We have investigated the anisotropy of successive air shower (SAS) events, which we define as the detection of many air showers within a short time window, using data from six stations of the Large Area Air Shower (LAAS) group. On the criterion of 22 air showers within 20 minutes, five SAS events are found against 1.4 expected from the Poisson distribution in Okayama University station's data. From six stations' data, we find 24 SAS events in total. By plotting them in equatorial coordinates, it is revealed that SAS events are observed more frequently when the Galactic plane is around the zenith. This can be attributed to a hypothetical small flux of ultra-high-energy γ-rays from the direction of the Galactic plane superposed on conventional cosmic rays. If this hypothesis is true, the analytical procedure used here has potential to measure ultra-high-energy γ-ray sources by even small air shower arrays like ours.

1. INTRODUCTION

The ultra-high-energy cosmic rays, which induce extensive air showers in the atmosphere, come to the earth almost randomly after they lose the information of their sources by deflection in the Galactic magnetic field. However, there are several reports of the detection of non-random components in them. The most outstanding one is by Smith et al. [1]. They observed a "burst" of 32 air showers of the estimated mean energy 3 PeV within a time window of five minutes. The chance probability for this event is calculated as 10^{-35}. Fegan et al. [2] reported the detection of an unusual simultaneous increase of the air shower rate lasting 20 seconds at two stations separated by 250 km. The recent report by Katayose et al. [3] used a sophisticated algorithm to extract non-random components from the cosmic ray time series. Though they found no Smith-like events, they picked up five "clustered" events with moderate significances. The arrival directions of these events grouped about two regions on the Galactic plane, giving the chance probability of 0.016.

One possible particle responsible for these non-random components is an ultra-high-energy γ-ray, which is not affected by the Galactic magnetic field. Though the origin of the non-random cosmic rays is still an open question because of the lack of observation, it should have interesting and significant astrophysical implication. Currently we need more data and analyses focused on the non-randomness in cosmic rays.

In this paper, we report the results of the successive air shower (SAS) analysis, which is focused on the non-randomness in arrival times of air showers, using our own air shower data.

2. NETWORK OBSERVATION AND DATA

The air shower data used in this analysis have been collected by the six stations of the Large Area Air Shower (LAAS) group, which is described in Ref. [4]. Here we outline some aspects of stations, and see also Table 1. Each station has five to eight 0.25 m^2 scintillation counters arranged over a few hundreds of square meters. The trigger conditions are different among stations; 4- to 8-fold coincidences are applied and yielding the trigger rates of 306 to 725 showers/day. We note

PII S0920-5632(01)01255-5

Table 1
The aspects of stations and data

Station Name	HU	NUI	KU1	KU2	OUS1_a	OUS1_b	OUS1_c	OU
Latitude (N)	40°35′	34°35′	34°39′	34°39′	34°42′	34°42′	34°42′	34°41′
Longitude (E)	140°29′	135°41′	135°36′	135°36′	133°56′	133°56′	133°56′	133°55′
# of Counters	5	7	5	7	4	8	8	8
Trig. Counters	5	7	5	7	4	8	4	5
Trig. Rate (/day)†	567	306	451	317	725	309	704	566
Data Period	11/98 – 12/99	08/96 – 06/98	09/96 – 12/99	07/98 – 04/99	09/96 – 11/97	11/97 – 01/99	01/99 – 12/99	09/96 – 12/99
# of Showers†	221k	168k	447k	62k	270k	127k	254k	638k

† showers with zenith angle $\leq 45°$

that the trigger condition of the OUS1-array was changed twice in three years, so we treat the three periods, OUS1_a, OUS1_b and OUS1_c independently. The primary energy range observed by our arrays, estimated by CORSIKA, is from 50 TeV to 10 PeV with the mean energy at about 1 PeV, which is just below the knee of the primary cosmic ray spectrum. The angular accuracy is about 7.0 degrees. Each station is equipped with the Global Positioning System (GPS) as the common clock, which can record arrival times of air showers with an accuracy of $1\mu s$.

The data period used here is from August 1996 to December 1999. About 2.2×10^6 showers are analyzed in total. The restriction of the reconstructed zenith angle $\leq 45°$ is applied. It is confirmed for all stations' data that the arrival directions of all showers are distributed uniformly over the sky with fluctuation expected from statistics.

3. ANALYSIS AND RESULTS

In order to investigate whether our air shower data contain non-random components, we count the number of detected showers (N) within a time window. The width of a time window is fixed on 20 minutes, which is derived from the lasting time of one of Katayose's "clustered" events, without overlapping consecutive time windows. This is done for each station's data and the resulting N distributions for the OU-array and the OUS1_b-array are shown by points in Figure 1 and Figure 2, respectively. If the arrival times of air showers were completely random, these distributions should follow the Poisson distributions shown by histograms in the same figures. Because of the difference of trigger rates, the averages of (and, equivalently, the shape of) the distributions are different among stations. Here we are showing these two figures as representatives, but the same discussion is valid for the other stations.

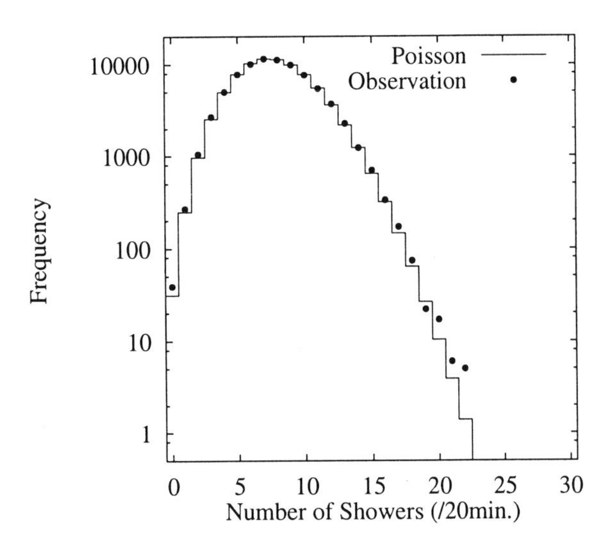

Figure 1. The N distribution for the OU-array

In Figure 1 and 2, good agreement between the observation and the Poisson distribution is

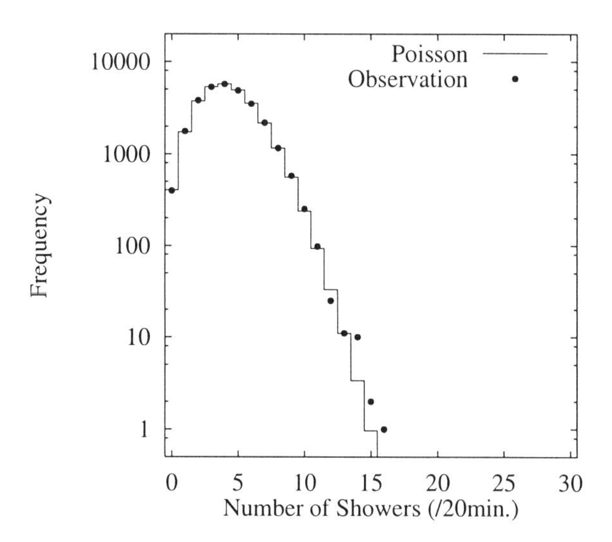

Figure 2. The N distribution for the OUS1_b-array

obtained, which means our air shower data are sufficiently stable. No "burst" events are found in any station's data. However, we can see small excess in the largest N bins. At the $N = 22$ bin of the OU-array, the largest deviation is gained; 5 events are observed against 1.4 expected. Hence we define the "successive air shower" (SAS) event as the detection of 22 or more air showers within a time window for the OU-array's data, expecting it to contain non-random components related to astrophysical origins. For other stations' data, we take the threshold value of N where the Poisson value diminishes below 1.4 ($N = 15$ for the OUS1_b-array). From six stations' data, we get 24 SAS events in total, which are tabulated in Table 2. The third column of Table 2 represents the date (mm/dd/yy) and time (in seconds) of the first shower of each SAS event. The fourth and fifth columns show the averages of right ascensions (RA) and declinations of all showers in each event, respectively.

To investigate whether these SAS events have a connection with the Galactic plane as "clustered" events by Katayose et al. do, we plot them

Table 2
The parameters of successive air shower events

Station	N	Date, Time(UT)	$\overline{\alpha}$	$\overline{\delta}$
OU	22	02/15/97, 26046	2.16	32.54
	22	06/26/97, 42828	15.41	32.63
	22	07/26/97, 54809	21.08	31.67
	22	08/24/97, 47654	20.21	35.63
	22	02/28/98, 72899	15.87	40.26
OUS1_a	25	12/05/96, 35670	0.03	30.19
	25	05/19/97, 51342	15.03	24.53
	27	09/12/97, 47681	21.38	33.01
	25	10/19/97, 30844	19.96	36.49
OUS1_b	15	11/14/97, 46351	2.21	34.50
	16	01/20/98, 46355	5.97	24.47
	15	08/15/98, 16450	11.30	40.55
OUS1_c	25	01/09/99, 42623	4.01	29.15
	24	03/03/99, 83463	18.99	35.36
	25	06/20/99, 85826	2.93	31.64
KU	20	10/14/99, 15213	14.80	28.78
KU2	15	07/19/98, 50554	19.46	34.16
	15	09/16/98, 80516	7.12	40.77
	15	01/10/99, 38668	2.86	34.48
	16	04/02/99, 13353	1.45	30.81
NUI	15	08/24/97, 78311	4.98	36.07
HU	21	12/16/98, 57433	7.37	35.91
	21	05/08/99, 79085	21.90	39.75
	21	09/20/99, 22638	16.13	37.08

in equatorial coordinates by the average values shown in Table 2, as Figure 3. The error bar attached to each point shows the standard deviation of RA and declinations of all showers in each event. The concentration of SAS events on the directions of the Galactic plane (the dashed line) is apparent, while another cluster exists about RA=15 hours.

4. DISCUSSION

The excess of the observed events in the largest N bins of Figure 1 and 2 (and for other stations, too) is small and is not so significant. We need more data to judge whether it is due to real non-random components in cosmic rays or due to fluctuation. Another pessimistic possibility for the excess is the unstable trigger rate in the years-

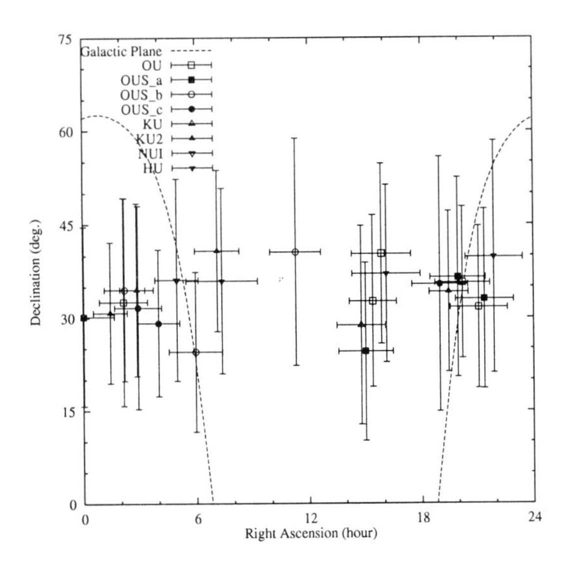

Figure 3. The arrival direction of SAS events

long data, and we will need a more sophisticated procedure to check this.

Assuming that the excess is due to real non-randomness, we direct our attention to Figure 3. The chance probability for the concentration on the SAS events to the directions of the Galactic plane (RA=5.4 and 20.2 hours at our latitude) is calculated as 0.038, including the events about RA=15 hours. This suggests that the SAS events are attributed to the emission from the Galactic plane. Since charged particles are deflected by the Galactic magnetic field, γ-rays seem the most likely particle responsible for the SAS events. However, it is not necessary that all air showers constituting the SAS events are induced by γ-rays from the Galactic plane. The superposition of only one or a few γ-rays on conventional cosmic rays is enough to make up the SAS events, because the difference of the number of showers belonging to the SAS events and the Poisson-distributed ones is small. This model is not in contradiction to the following unwanted facts:

- The attenuation length of photons in the energy range we are observing is less than 1 Mpc due to the absorption by the microwave background photons in space, so we can not expect a large flux of them.
- The points in Figure 3 are not exactly on the Galactic plane. These are calculated from all showers in each event, whereas we need only one or a few showers pointing to the Galactic plane. In fact, most of the SAS events contains a few showers each from the direction of the Galactic plane.

Unfortunately stations used here do not have muon detectors, so it is experimentally unknown whether the observed SAS events contain γ-ray induced showers.

If it is true that the SAS events are due to ultra-high-energy γ-rays, the analytical procedure used here, which is very simple, has potential to measure ultra-high-energy γ-ray sources by even small air shower arrays like ours. Looking at Figure 3 again, there is another clustering at about RA=15 hours. This may indicate a ultra-high-energy γ-ray source.

5. SUMMARY

We search for non-random components in the arrival time data of air showers. The distribution of the number of showers (N) within a time window of 20 minutes shows small excess in the largest N region compared to the Poisson distribution. By plotting the arrival directions of events in the excess region, the concentration on the directions of the Galactic plane is apparent. The chance probability for this is 0.077. We can interpret this as the consequence of the detection of ultra-high-energy γ-rays from the Galactic plane superposed on conventional cosmic rays.

REFERENCES

1. G. Smith et al., Phys. Rev. Lett., 50 (1983) 2110.
2. D. Fegan, B. McBreen and C.O'Sullivan, Phys. Rev. Lett., 51 (1983) 2341.
3. Y. Katayose et al., Nuovo Cimento C21 (1998) 299.
4. N. Ochi et al., "LAAS Network Observation of Air Showers", in this symposium.

ELSEVIER

Nuclear Physics B (Proc. Suppl.) 97 (2001) 177–180

www.elsevier.nl/locate/npe

Angular Distributions of Hadrons and Gammas in Pamir Emulsion Calorimeter

J.Kempa [a] *, J.Malinowski †, H.Bialobrzeska, M.Linke, A.Podgorzak

[a]Department of Experimental Physics, University of Lodz
90-236 Lodz, Poland, ul. Pomorska 149/153

Anomalies in angular distributions of hadrons and gamma quanta observed in the Pamir calorimeter are subject to a further more detailed analysis. In the present paper the experimental results which concern gamma quanta breaking through the upper part of the calorimeter into its lower part are presented. Hadrons are measured in the lower part of the chambers. The role of such cases in the observed anomalies will be discussed.

1. INTRODUCTION

The aims of this paper are the results of a further more detailed analysis and the presentation of zenith angle distributions of particles for different darkness threshold registered in carbon emulsion chambers of the Pamir experiment (4370 m a.s.l. - 600 g/cm^2). Special attention is paid to the comparison of data of the zenith angle distribution for different particle components (hadrons and gamma quanta) measured in slightly different chambers.

The traces of the hadron initiated cascades (in H-block) or gamma quanta and electrons (in Γ-block), which (the cascades) were registered in X-ray films, have been measured with the use of a photometer with diaphragms of a constant radius r, where r = 48, 84 or 140 μm. Optical densities measured in the experiment have been standarized and corrected following the procedures described in [1].

Apart from the optical densities, zenith and azimuth angles and the coordinates of the registered particles are measured.

2. EXPERIMENTAL DATA

The experimental data from chambers C141, C200 and C201, each of 24 m^2 area, have been measured totally or partly and used in the analysis [2].

The carbon emulsion chamber C141 has a typical construction, i.e. Γ-block: 6 cm Pb, H-block: 60 cm C (with density ρ=1.55 g/cm^3) and 5 cm of Pb below. The main registration layers are below 5 cm of Pb in Γ-block and below 4 cm of Pb in H-block.

The chambers C200 and C201 have an additional layer of carbon (thickness 5 cm) above the Γ-block.

Table 1 presents information about the analyzed chamber data.

In all presented distributions each bin of a histogram represents the multiplicity of a given component per area unit and per zenith angle unit measured in degrees. Differential zenith angle distributions of the hadrons registered in the experiment have been depicted in Figure 1. The comparison of the experimental with reconstructed data has been published in [3].

A simulation was made with the CORSIKA v.5.6 program [4] using Venus and QGSJET models. A similar form of the θ-distributions and a good agreement with experimental data has been noticed.

The differential zenith angle distributions of hadrons registered in chambers C200 and C201 have been shown for different thresholds of the darkness D in Figure 1. These chambers are different from the standard ones due to the presence of the additional layer of carbon above the

*Supported by the Rector of the University of Lodz (grant 2000/505/704)
†Supported by the Rector of the University of Lodz (grant 2000/505/709)

PII S0920-5632(01)01257-9

Table 1
Basic information about the experimental data.

Darkness threshold	C200+C201 (88/89) S=36m^2	C141 (85/86) S=24m^2	Energy threshold for hadrons [TeV]
for D>0.2	839	772	18.6
for D>0.3	378	371	26.4
for D>0.4	208	205	34.9
for D>0.5	125	121	44.0

Γ-block.

An additional peak can be seen for $\theta \in (5 - 15)^\circ$ in this distribution with a maximum at about $\theta \approx 8^o$. For the threshold of the darkness D > 0.2 the number of hadrons that contributed to the peak is about 15 % of the whole number of hadrons in the distribution. This effect is not observed in chamber C141.

The experimental data from chamber C200 have been measured by two persons independently to eliminate mistakes, while the measurements in chambers C201 and C141 have been made by a third person.

In our opinion, the observed effect cannot have been caused by zenith angle measurement errors. We estimate this error as $\Delta\theta \simeq 6^o$ for hadrons.

A different type of emulsion has been put in some chambers of standard construction since 1989. The zenith angle distributions of gamma quanta chambers of this type (data from chamber C207 have been analyzed) are similar in shape to those for chamber C141. That is why we can say that this effect does not depend on the type of emulsion.

The presence of the effect in the angular distribution does not depend on the procedure of the analysis. An additional peak in the distribution can be always observed, no matter from what part of the chamber the data were taken. In the H-block data the particles which entered the chamber from aside were omitted. The contribution of families is not important and omitting them does not change the distribution. The azimuth angle distributions of the analyzed data do not show any anomalies.

3. DISCUSSION OF THE RESULTS AND CONCLUSIONS

The angular distributions for two types of emulsion chambers and for different values of the threshold darknesses D were presented in the series of plots given in Figure 1. The greater the threshold value of the darknesses D, the stronger gets the observed effect of anomalies, which appear in the shape of the angular distributions of registered hadrons. In our opinion this fact is a strong argument which is in favour of the effect observed by us that there are anomalies in the angular distributions [1] for the calorimeters with an additional layer of carbon placed on top of the calorimeter.

No fully comprehensive interpretation of this effect can be found. We maintain a claim that there exists a difference observed in the paper [2] in the angular distributions of charged hadrons, kaons K^o_s and Λ. In the theoretical papers we were not able to reconstruct fully the effect observed in the experiment. In order to make it possible to understand the observed phenomenon better, we made the research on a part of the available experimental data searching for such events, in which the darkness in the Γ-block may be correlated with the darkness in the hadron block. We investigated the tracks observed in the area of 21 m^2 of emulsion films from calorimeters C200 and C201, which have an additional carbon layer on the top. In these films 1201 tracks in the Γ-block and 926 in the H-block which fulfilled a condition $\Theta \leq 12^o$ were found. The angles Θ and ϕ were correlated with the darknesses which were observed in both parts of the calorimeter. Only four such events were found, for which, in the er-

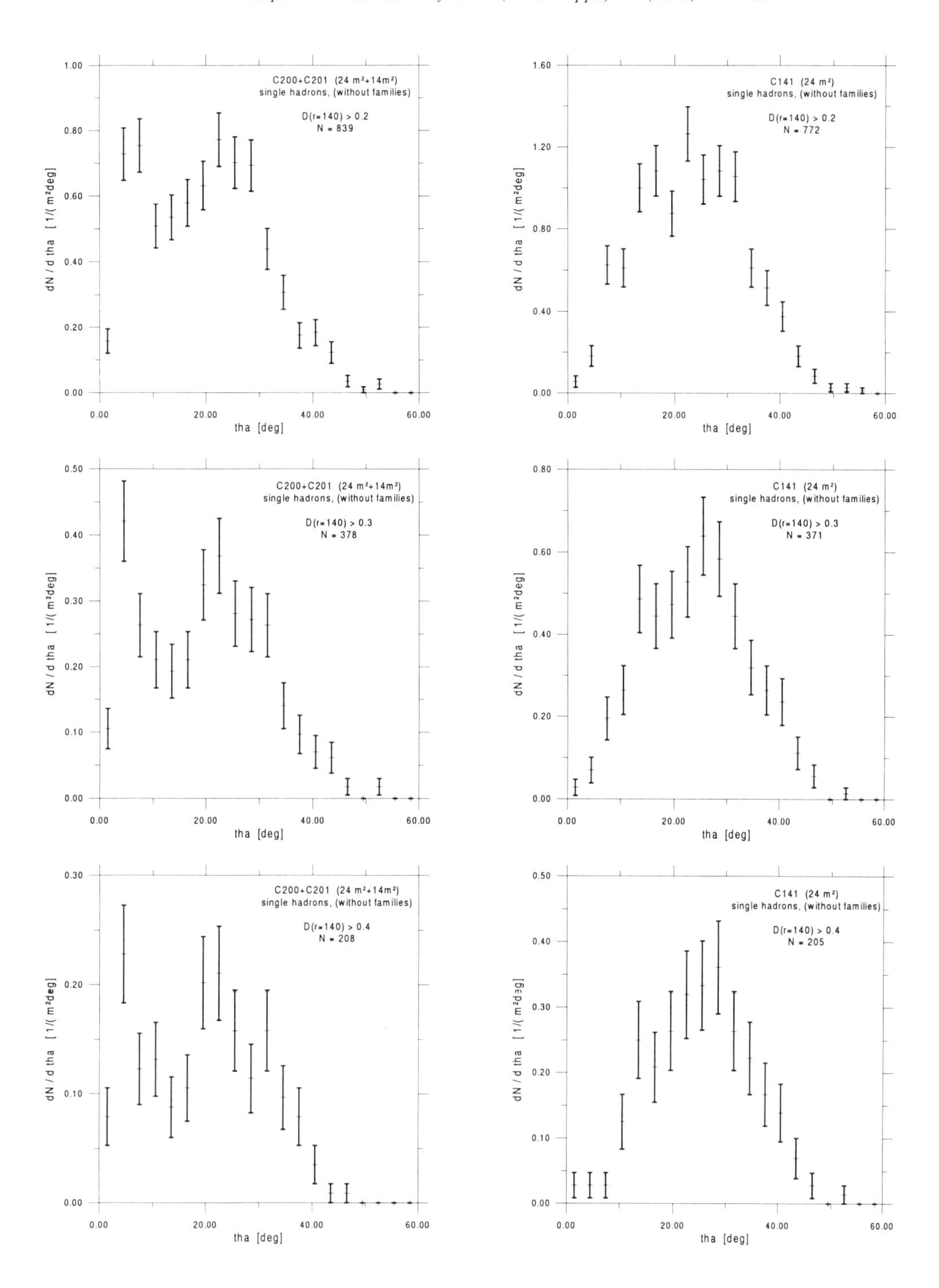
C200+C201 (24 m²+14m²)
single hadrons, (without families)
D(r=140) > 0.2
N = 839
C141 (24 m²)
single hadrons, (without families)
D(r=140) > 0.2
N = 772
D(r=140) > 0.3
N = 378
D(r=140) > 0.3
N = 371
D(r=140) > 0.4
N = 208
D(r=140) > 0.4
N = 205
dN / d tha [1/(m²deg]
tha [deg]

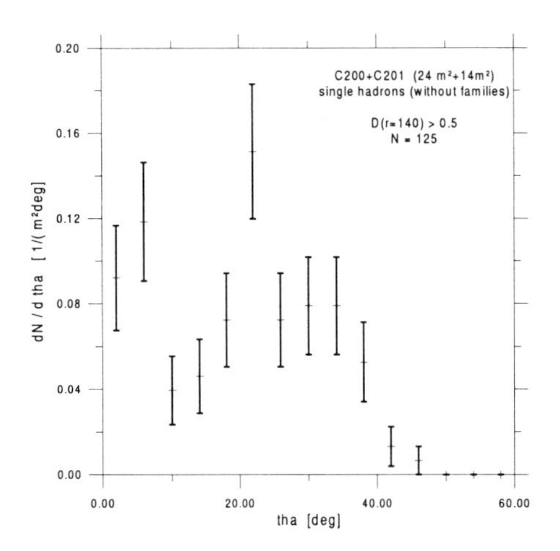

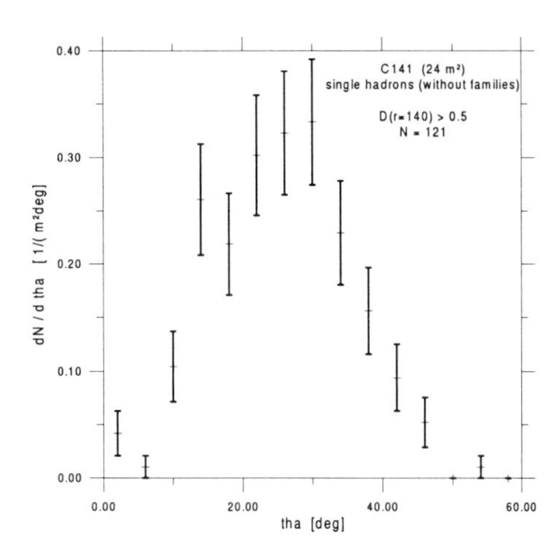

Figure 1. (Includes previous page.) Angular distributions for two types of emulsion calorimeters: at right - a calorimeter without an additional layer of carbon - C141; at left - a calorimeter with an additional carbon layer on top of the calorimeter (5 cm of C) - C200 and C201.

ror limits, it may be assumed that the cascades initiated and seen in the Γ-block continue in the hadron block. The number of such events is very small and we still are not sure whether or not these events are absolutely accidental. We may totally exclude the hypothesis that the observed anomalies in both blocks are responsible for the same cascades, passing from one block to another.

REFERENCES

1. Pamir Collaboration, Trudy FIAN, Moscow "Nauka", v.154 (1984)
2. J. Kempa et al., 26th ICRC Proc., Salt Lake City (USA) (1999) HE 1.2.10.
3. Bialobrzeska H. at al., Nucl. Phys. B (Proc. Suppl.) v.75A (1999).
4. Heck D. et al., FZKA 6019, Forschungszentrum Karlsruhe (1998).

ELSEVIER

Nuclear Physics B (Proc. Suppl.) 97 (2001) 181–184

Mass Composition of Primary Cosmic Rays for Energies of 10 - 1000 TeV on the Basis of the Pamir Experiment Results

J. Malinowski [a] *

[a]Department of Experimental Physics, University of Lodz
90-236 Lodz, Poland, ul. Pomorska 149/153
e-mail: malinow@krysia.uni.lodz.pl

Carbon emulsion chambers of the Pamir experiment at the altitude of 4370 m a.s.l. (600 g/cm^2) register secondary particles of cosmic ray. As an outcome of these data, the energy spectrum of hadrons has been estimated.

The spectra obtained have been compared with those obtained using another method of measurement (Pb-chamber) and energy estimation made in the Pamir experiment.

Mass composition and slope of the primary cosmic ray spectrum have been estimated from the comparison of experimental spectra with the results of simulations of Nuclear Electromagnetic Cascades (NEC) in the atmosphere.

1. INTRODUCTION

Distributions of the energy of hadrons registered in carbon chambers (C-chambers) in the Pamir experiment [1,2] at 600 g/cm^2 have been presented previously. The purpose of this paper is an attempt of making conclusions about the mass composition of primary cosmic rays in relation to the experimental results.

Our results have been compared with the spectrum obtained by means of another method in Pb-chamber of the Pamir experiment. Afterwards, the comparison of our data with other experiments has been shown. This is illustrated in Figure 1.

In order to compare experimental data from different altitudes in the atmosphere, the information about the attenuation mean free path of particles penetrating through the atmosphere, L_{att}, is needed. All data have been extrapolated to an altitude of 0 g/cm^2, corresponding to the top of the atmosphere. The estimation of L_{att} was obtained from simulations done for this purpose with the program CORSIKA v. 5.62 using QGSJet model [3]. Then data from the chosen experiments have been extrapolated to the top of the atmosphere as shown in Figure 2.

*Supported by University of Lodz (Rector's grant No 505/709).

2. EXPERIMENTAL DATA

The spectra of hadrons registered in C-chambers have been published in [1,2]. The results have a very good statistical background (basic information is shown in Table 1).

Table 1
Basic information about the experimental data

S [m^2]	T [year]	N_h(>E)
S=62m^2	0.9926	2275 (E_h >17.7TeV)
S=120m^2	0.9836	201 (E_h >70.6TeV)

The influence of methodical problems on the value of errors of parameter's estimators of analyzed distributions has been thoroughly estimated. All registered hadrons for chambers with area S=62 m^2 have been measured. Having increased the total area to 120 m^2, only registered high energy hadrons have been chosen. Therefore, two functions describing spectra for two thresholds 17.7TeV and 70TeV have been presented.

PII S0920-5632(01)01258-0

The results obtained (in $m^{-2}s^{-1}sr^{-1}$) were:

$$I(>E_h) = (2.79 \pm 0.06 \pm 0.85) \cdot 10^{-6} \cdot \left(\frac{E_h}{17.7TeV}\right)^{-2.01\pm0.04}$$

$$I(>E_h) = (1.28 \pm 0.09 \pm 0.38) \cdot 10^{-7} \cdot \left(\frac{E_h}{70TeV}\right)^{-2.07\pm0.18}$$

The first error in the intensity is a statistical one, the second is due to the estimation of errors of the experimental method. Experimental data have been normalized to unit area, time, solid angle and the efficiency of hadrons registration was taken into account, as described in [2]:

$$I(>E) = \frac{N(>E)}{S\ T\ \omega} \cdot \frac{m+2}{2\pi(1-\cos^{m+2}\Theta_o)}$$

where:

- S - total area of the chambers;
- T - exposure time;
- ω - efficiency of the chamber's registration ($\omega = 0.54$);
- m - exponent of the angular differential distribution of hadrons ($m = 6$);
- Θ_o - limiting zenithal angle ($\Theta_o = 60°$).

2.1. Comparison with Pb-chamber data

The first comparison has been done with data obtained by means of another method in Pb-chambers of the Pamir experiment [4]. From about 800 hadrons with energy E_h >20TeV measured in Pb-chambers the intensity of total spectrum at $I(>E_h) = (1.84 \pm 0.2) \cdot 10^{-6} m^{-2}s^{-1}sr^{-1}$ and its slope $\beta = -2.01 \pm 0.08$ have been obtained. It can be seen that the results from both these types of chambers correspond to the limits of statistical errors. This is illustrated in Figure 1.

2.2. Comparison with spectra from other experiments

In Figure 1 the same experimental data in different scales have been shown. The intensities presented are the values registered in experiments without any corrections or depth recalculations. For picture clarity statistical errors are marked only for the C-chamber spectrum (Pamir).

Data described as "KASCADE" (1000 g/cm^2) and "Mielke et al." have been taken for the energies in TeV's range. They are the final points of the spectra presented in the paper [5]. In the paper [1] these data have been compared with the spectra of hadrons of the Pamir experiment.

The three groups described as Tien Shan (681 g/cm^2) come from measurements using ionization calorimeter. Data for E<2TeV have been described in [6], the rest in [7]. As follows from the paper [7], these data should be treated as upper limit of registered hadron intensities because of methodical reasons.

Data of all particles registered in balloon and satellite experiments have been presented.

3. SIMULATIONS

For four different primary nuclei (H, He, CNO, Fe) the CORSIKA v. 5.62 program with QGSJet model has been used to calculations [3]. The simulations cover the energy range of 10^{13}eV - 10^{16}eV with slope $\gamma_H = 2.68$, $\gamma_{He} = 2.62$, $\gamma_N = 2.60$ and $\gamma_{Fe} = 2.60$. All secondary hadrons with energy larger than 5TeV at four observation levels (Chacaltaya, Pamir, Tien Shan and KASCADE experiments levels) are taken into account for further analysis.

The main contribution to the spectrum of observed hadrons comes from primary protons. Therefore, in this paper the distributions of hadrons deriving from primary nucleons have been analyzed. Obtained values for L_{att} have been presented in Table 2.

Table 2
L_{att} values estimated from simulations.

E [TeV]	10	25	50	100
L_{att} [g/cm^2]	99 ± 1	95 ± 1	92 ± 2	87 ± 3

The slope of the integral spectrum of simulated hadrons arriving at the Pamir level from primary

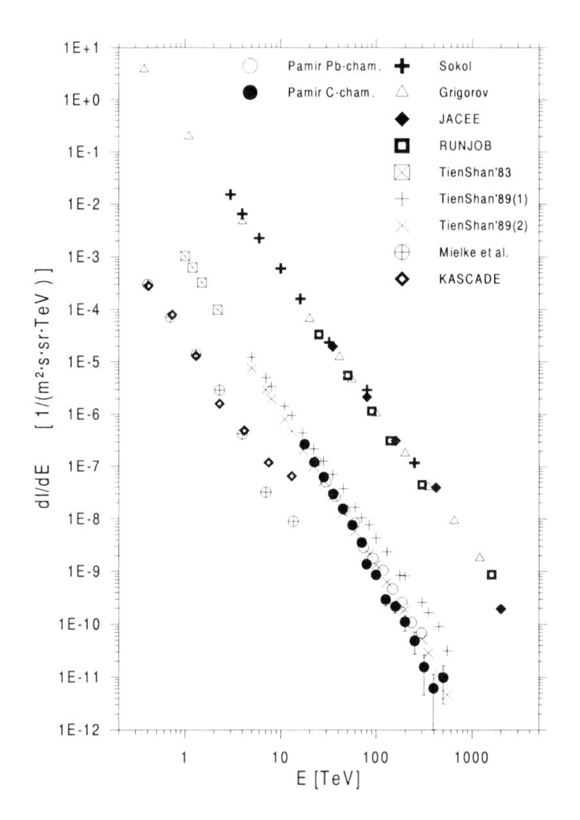

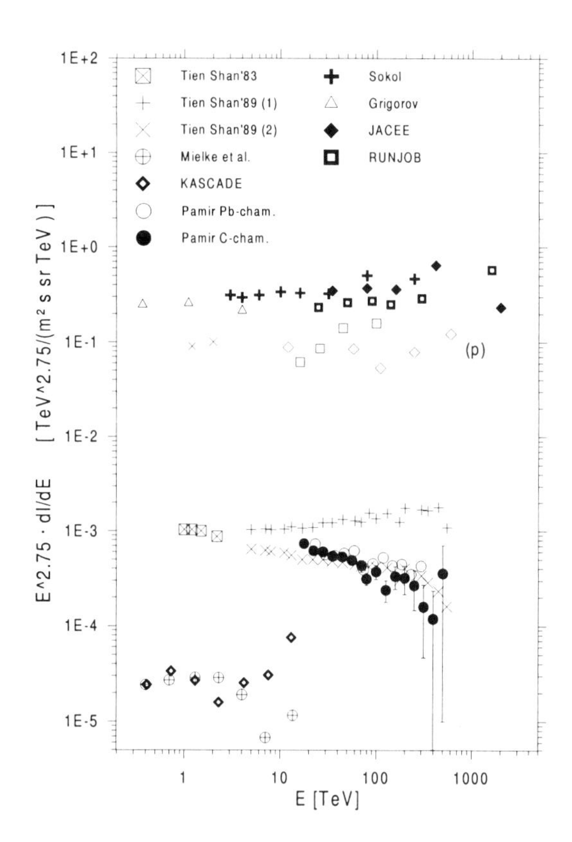

Figure 1. Experimental spectra of hadrons; on both figures - the same data are presented in a different way.

protons increases fluently with energy threshold from 1.45 (at E=20TeV) to 2.10 (at E=50TeV) to 2.93 (at E=70TeV).

From the calculations it results that primary nuclei with energies of the order of 10^3TeV are responsible for observed hadrons mainly with energies of tens of TeV. That is why spectra observed in our experiment enable drawing conclusions about primary spectrum with energies up to 10^3TeV.

4. DATA ANALYSIS

Data presented in Figure 1 have been extrapolated to the top of the atmosphere using data from Table 2 and have been shown in Figure 2. Comparison of recalculated spectra with spectra measured in direct measurements suggests that the estimation of attenuation length value was correct. L_{att} calculated from experimental data give very close values compared with those given in Table 2.

When normalizing the spectrum of simulated primary protons to the spectrum of protons observed by the JACEE experiment (with E=10TeV), we obtain simulated hadrons at the Pamir level with intensity around 2.5 times lower than the one observed in the experiment. It results also results from the simulations that for

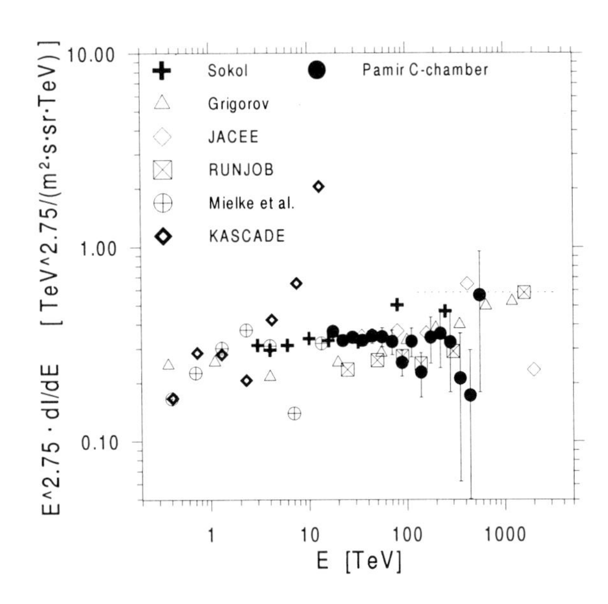

Figure 2. Experimental spectra of hadrons recalculated to 0 g/cm^2 level

each primary photon contributing to the spectrum of observed hadrons at this depth at the energy 10 TeV, there is also a contribution from every second He nucleus, every fourth CNO and every tenth Fe nucleus. This means that primary nuclei can increase the observed intensity of hadrons not more than 75% (it is also shown by data in the paper [8]).

To summarize, we can conclude that the primary spectrum up to 10^3TeV has substantially more protons than could be expected from the extrapolation of balloon data.

The slope of the energy spectrum of hadrons registered in the Pamir experiment is constant or slowly changing as function of the energy threshold. The slopes of simulated spectra behave differently. Thorough analysis of this parameter will probably enhance the conclusions about a great content of protons in primary spectrum at least up to the energy 10^3TeV.

5. CONCLUSIONS

The analysis presented here is rather more qualitative than quantitative. Nevertheless, it enables to state that primary spectrum for energy up to 10^3TeV includes substantially more protons than the extrapolation of data from direct measurements suggests.

The estimation of attenuation length resulting from both experimental data and simulations decreases with the energy from 99 g/cm^2 (for E=10TeV) to 87 g/cm^2 (for E=100TeV).

REFERENCES

1. J. Malinowski, Nucl. Phys. B (Proc. Supl.) v.75A (1999) 177-179.
2. J. Malinowski, Proc. of the 26th ICRC, Salt Lake City (1999) HE 1.2.11.
3. D. Heck et al., FZKA 6019, Forschungszentrum Karlsruhe (1998).
4. Pamir Coll., Bulletin de la Societe des Sciences et des Lettresde Lodz, vol.XII no 115 (1992) 71-92.
5. G. Schatz et al., Nucl. Phys. B (Proc. Supl.) v.60B (1998) 151-160.
6. D.S. Adamov et al., Proc. of the 18th ICRC, Bangalore (1983) HE 4-2 275-278.
7. A.P. Chubenko N.I.Nikolsky, Proc. of the 21th ICRC, Adelaide (1989) HE 2.1-15.
8. A.Haungs et al., "Test of High-Energy Interaction Model..", in this vol.

ELSEVIER

Nuclear Physics B (Proc. Suppl.) 97 (2001) 185–188

www.elsevier.nl/locate/npe

Hadronic structures of extensive air shower cores

J.Kempa[a] * and M.Samorski[b]

[a]Department of Experimental Physics, University of Lodz, Lodz, Poland

[b]Institut für Experimentelle und Angewandte Physik der Universität Kiel, D-24118 Kiel, Germany

The experimental results obtained from the Kiel EAS device have been made to study the properties of the core of the electromagnetic component as well as the hadron and muon components are measured by the underground neon hodoscope. Most energetic hadrons and hadron patterns at small distances from the core have been analysed for proton-like and iron-like showers.

1. INTRODUCTION

We continued to investigate the properties of shower cores for showers detected by the Kiel experiment [1]. Using the multifractal moment analysis for the structure of the density of charged particles near extensive air shower (EAS) cores we selected samples of showers significantly enriched in proton- and iron-initiated showers [2]. Then these two groups of selected showers have been used for the analysis of the core structure and muon density distribution near the core to find possible differences between the two groups. For the analysis described in this paper the 31 m^2 unshielded neon hodoscope has been used as the core detector of EAS charged particles and the 65 m^2 shielded neon hodoscope for hadrons and muons. Only EAS of $1.4 \cdot 10^5 \leq N_e \leq 5 \cdot 10^6$ and $0^o \leq \Theta \leq 30^o$ recorded with very good accuracy by the Kiel experiment have been used.

2. EXPERIMENTAL SETUP

Figure 1 gives a survey of the experimental setup of the Kiel EAS device. Details of this array have previously been published [1]. A total of 28 unshielded scintillation counters were available for the determination of shower sizes and core locations (we call this core position further as the scintillator core). 11 scintillation counters, connected to 22 fast-timing channels of about 1 ns time resolution, provide information on the arrival directions of the showers (the zenith angle error was about 1°).

The 31 m^2 neon hodoscope for the investigation of the electron core structure was located below a 2.5 g·cm^{-2} wooden roof. It incorporated 176400 neon flash tubes of 1 cm diameter each and enabled to localize the core position directly from the density distribution of the charged particles in the EAS core region as well as to investigate the fractal moments (we call that core position: the fractal core). The 65 m^2 underground neon hodoscope for the investigation of the hadron core structure and the lateral distribution of muons was located below 3.5 m of concrete and incorporated 367500 neon flash tubes.

3. DATA ANALYSIS

We performed the analysis of each individual EAS in the following way. For each EAS with the core falling into the unshielded neon hodoscope we estimated scintillator and fractal core positions. Then for such showers the position of the most energetic hadron as detected by the shielded neon hodoscope has been estimated. The relative distances between scintillator and fractal core positions for the two groupos of showers under discussion: for the iron-like showers and for the proton-like showers. The average distances are equal to 56 ± 7 cm and 54 ± 8 cm, respectively, and show no difference between the two groups of showers. The average distance between the position of the scintillator core and the most energetic

*Partly supported by the Rector of the Lodz University (grant 2000/505/704).

PII S0920-5632(01)01259-2

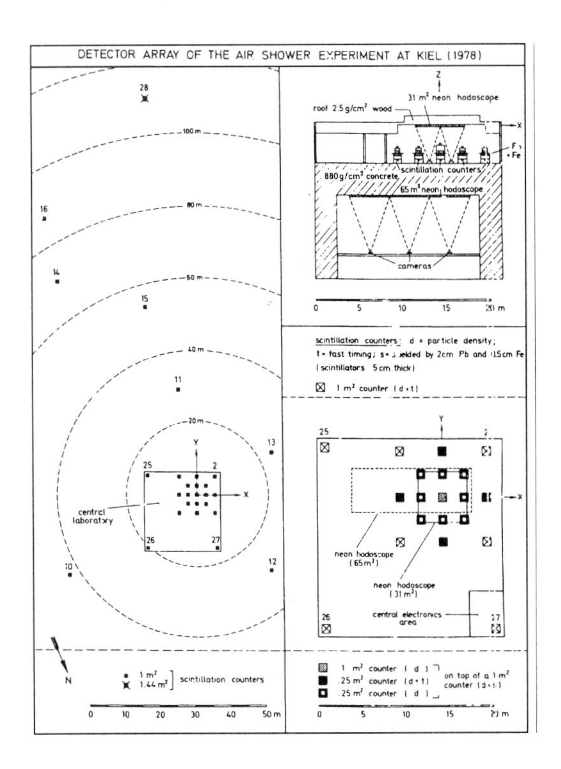

Figure 1. The Kiel EAS experiment.

hadrons are 135 ± 13 cm for the ironlike showers and 145 ± 13 cm for the proton-like showers. No dependence on the primary mass has been found.

The obtained values (135 ± 13 cm and 145 ± 13 cm) can be treated as an accuracy measure of the localization of the point, in which the symmetry axis of an extensive air showers falls: i.e., when deduced from the lateral density distributions of the particles measured by scintillators it is about 0.5 m, whereas it is about 1.5 m between this point and the localization of the hadron with the highest energy.

In proton-like and iron-like showers the total number of the flashed neon tubes in the cores was additionally investigated as well as the number of identified hadrons. Such a hadron identification was assumed whenever the number of the flashed neon tubes in the 10×10 cm^2 square was equal to at least 20% of the number registered for the highest energy hadron.

All the available shower cores were investigated in such a way. No statistically significant differences in proton-like and iron-like showers were observed.

Lateral muon density distributions for the two groups of showers of interest have been already published [3] for the range of distance between 3 m and 11 m from the core.

Taking into account that the muon densities obtained from our analysis are clearly the upper limit (due to hadron and delta-electron contamination), the observed agreements between our data and theoretical CORSIKA predictions are satisfactory. We would like to stress also again that the fluctuations of the muon densities observed experimentally are not the same: they are smaller for the iron-like showers than for the proton-like showers.

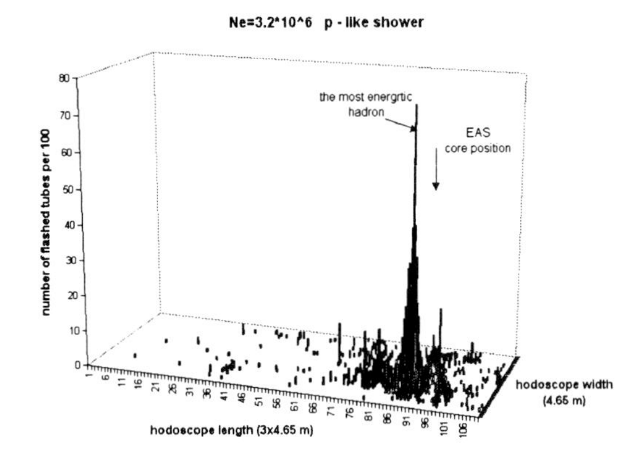

Figure 2. The lego-plot of the picture from the underground hodoscope.

4. ASYMETRIC SHOWERS

In most of the showers there are the cores with a clearly visible hadron of the highest energy (Figure 2). In this figure, (a lego-plot) the core and the hadron group around it with evidently lower energies, can be clearly noticed. Such a picture,

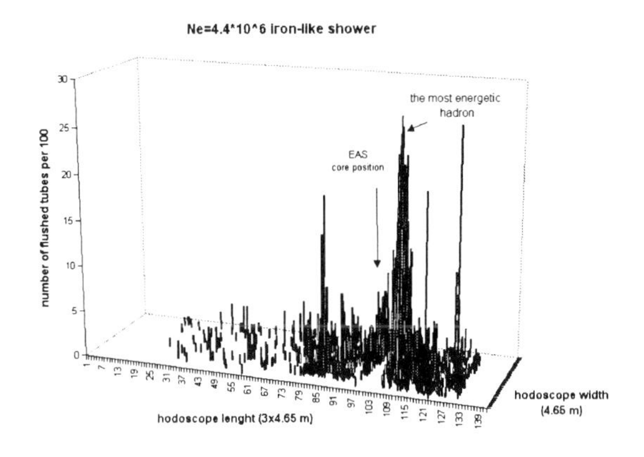

Figure 3. The lego-plot of the picture from the underground hodoscope.

with a characteristic gradient, is typical of the most of the showers. The observed asymmetry of the showers are shown in an exemplary iron-like shower in Figure 3. against the background of the main core there can be noticed the hadrons with relatively low energies. The following parameter was chosen for description of asymmetric showers:

$$\omega = \frac{n_l - n_p}{n_l + n_p} \ ,$$

where n_l and n_p denote the numbers of hadrons with the relatively high energies on the left (n_l) and on the right (n_p) of the main core. The events with $|\omega| \simeq 1$ are particularly interesting (Figure 4). There are $21.5 \pm 2.2\%$ of such events in the group of proton-like showers and $42.9 \pm 4.4\%$ in the iron-like group. This parameter is sensitive to the mass A of the nucleus, which is responsible for the initiation of a particular shower. It may happen in such an asymmetric shower that the energy of the hadron on the left or on the right of the shower axis is higher than the energy of the near axis hadron.

A somewhat "linear" lay-out of the highest energies hadrons, registered in the underground detector, may be noticed (Figure 5). However, this observation should be taken with a care as the shape of the detector itself is rectangular and such

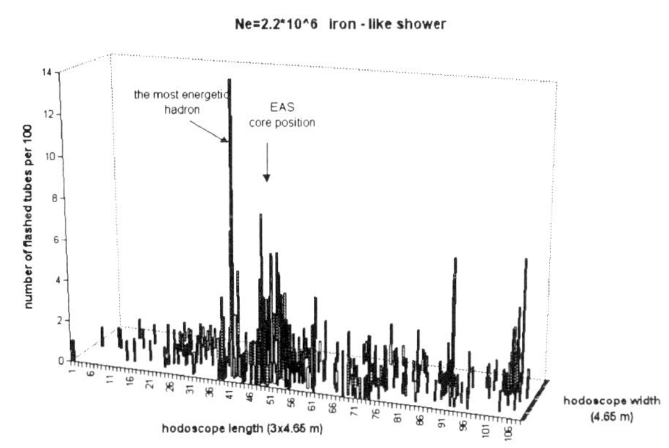

Figure 4. The lego-plot of the picture from the underground hodoscope.

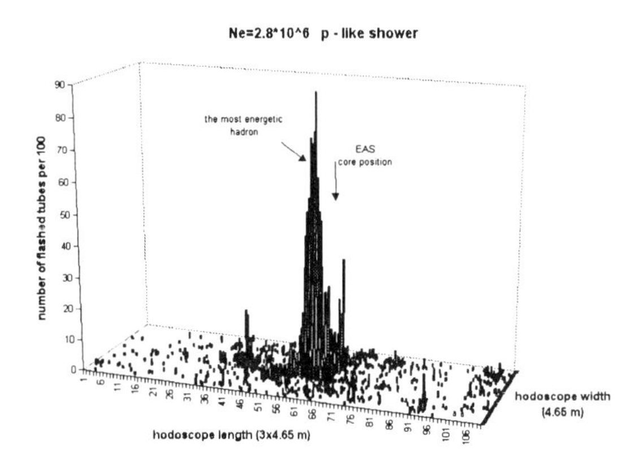

Figure 5. The lego-plot of the picture from the underground hodoscope.

"alignment" events may be artificially generated. Nevertheless, several such events were found although not all of them have hadrons lying exactly along one line. This fact confirms the observations made with the use of X-ray films in the Pamir calorimeter.

5. CONCLUSIONS

In the investigated group of several hundred showers the observed picture of the cores is slightly dependent on the primary nucleus mass. However, the asymmetric showers are more often noticed in iron-like shower group. It is intuitively obvious as there is a group of nucleons with energies E_p/A in the showers initiated by nuclei of the mass A. A weak trace is left by this bundle of nucleons in a form of the asymmetry as observed in the shower cores.

The groups of hadrons with the highest energies show tendency to align; the fact may be the confirmation for the observations made with the X-ray films.

REFERENCES

1. E. R. Bagge, M. Samorski, W. Stamm, Proc. 16th ICRC 13 (1979) 260
2. J. Kempa, M. Samorski, J. Phys. G: Nucl. Part. Phys. 24 (1998) 1039
3. J. Kempa, M. Samorski, Proc 26th ICRC 1 (1999) 282

ELSEVIER

Nuclear Physics B (Proc. Suppl.) 97 (2001) 189–192

www.elsevier.nl/locate/npe

New approach to separation of electromagnetic and hadron cascades and to energy measurement in detection of primary particles

R.A.Mukhamedshin on behalf of the INCA Collaboration [a*]

[a]Institute for Nuclear Research, Moscow 117312 Russia

A new method for separation of primary electrons and protons based on the difference in evaporated-neutron yield is proposed to investigate the primary electrons at $E_e > 1$ TeV with a total rejection factor of $\sim 10^{-5} - 10^{-6}$. A new approach to measure the energy of primary nuclei by analyzing the neutron yield in cascades is proposed to investigate primary cosmic rays in the "knee" range.

1. GOALS AND METHODS

One of the major goals of the INCA Project [1] is to study the spectrum of primary cosmic-ray (PCR) electrons at $E_e \gtrsim 1$ TeV. The available data are rather poor due to difficulties in separation of electron-initiated cascades against the proton-produced background. To reject electron-like proton-initiated cascades with a high efficiency, we propose to use the fact that the evaporated-neutron yield is much smaller in electromagnetic cascades than in hadron-induced cascades [2] .

The second goal is to study the PCR spectrum and composition in the "knee" range (1−10 PeV). Space measurements are usually carried out using calorimeters designed mainly of heavy substance. We propose, first, to use a light substance (polyethylene, e.g.) that permits to maximize the geometrical factor, and second, to measure energy of primary nuclei by analyzing the neutron yield in nuclear-electromagnetic cascades (NEC) in the ionization-neutron calorimeter (INCA).

To analyze these problems, we simulate processes in an INCA with a periodic structure, so that each layer contains lead and polyethylene (10 and 20 g/cm^2, respectively). The lead provides the intense neutron generation, while the light substance provides the NEC development. The modified SIMULNEC code was used [1,3]. In this work we demonstrate principal potentialities of the INCA.

*This work is supported by the Russian Foundation for Basic Research, projects no. 98-02-17157 and 00-15-96632.

PII S0920-5632(01)01260-9

2. SEPARATION OF PRIMARY ELECTRONS AND PROTONS

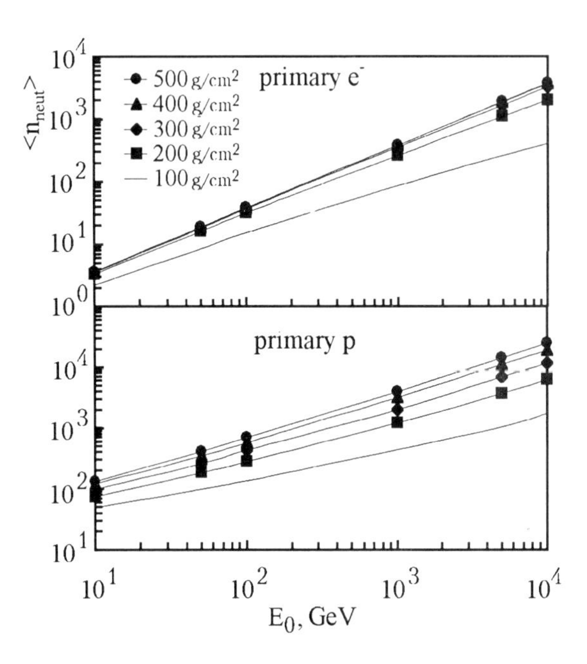

Figure 1. Energy dependence of the total neutron number $\langle n_{neut} \rangle$ varying INCA's thickness, in cascades initiated by electrons and protons.

Generally, procedures applied to separate electrons can be characterized by two rejection fac-

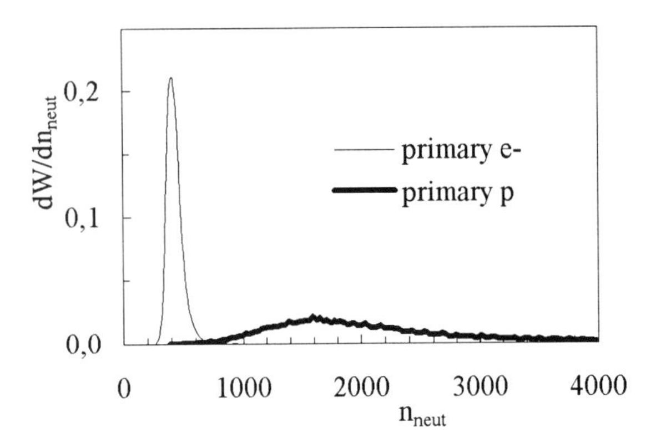

Figure 2. Distribution of neutron total multiplicity in cascades generated by electrons and protons in INCA with an effective thickness of 300 g/cm^2.

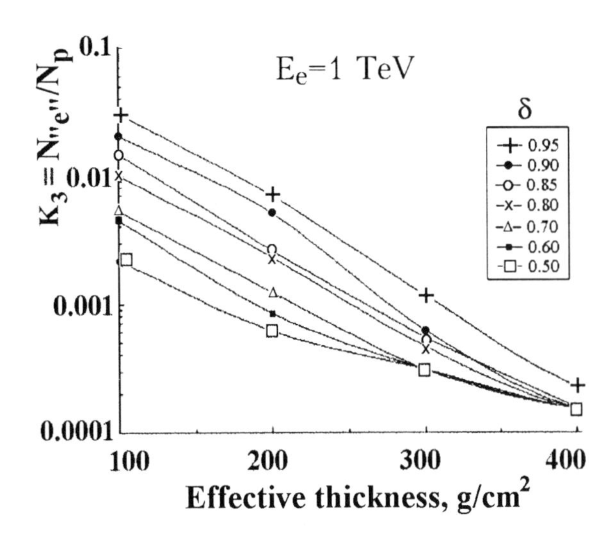

Figure 3. Rejection factor K_3 for electron-like proton-initiated cascades vs. thickness and electron detection efficiency δ.

tors $K_1 \simeq 1/20$ and $K_2 \simeq 1/15$ [4] related to:

(i) fixing the cascade starting point within the top lead layer and (ii) integral proton spectrum decreasing as $E_p^{-1.7}$.

We suggest a new powerful rejection factor associated with the neutron detection.

When accounting for the above factors, we can hereafter consider only proton-initiated cascades which start within the top 10 g/cm^2 lead layer and release the energy $E_\gamma = (0.8 - 1.2)E_e$.

Figure 1 shows the energy dependence of the total evaporated-neutron number integrated over various values of the INCA thickness in electron- and proton-initiated cascades. The difference for cascades of different origin is significant.

Figure 2 demonstrates typical distributions of neutron total multiplicity in cascades generated by protons and 1 TeV electrons at a thickness of 300 g/cm^2. Actually, the narrow distribution for electron-initiated cascades does not overlap the broad distribution for proton-initiated cascades. This feature is used to reject the later ones and can be described by the rejection factor K_3 which depends on both the absorber thickness and the efficiency δ of primary-electron detection (Fig. 3). Here δ is the fraction of electron-initiated events considered after cutting off their distribution's right wing, which could overlap the distribution of proton-initiated events. Even for a thin (100 g/cm^2) setup, $K_3 \approx 10^{-2}$ at $\delta = 0.8$ while at a thickness of more than 300 g/cm^2, $K_3 < 10^{-3}$. The efficiency of our criterion depends rather weakly on energy.

3. ENERGY MEASUREMENTS

In this work we attract attention to a possibility to apply the evaporated-neutron yield as a measure of cascade energy. As is seen in Fig. 4, the energy dependence of the total evaporated-neutron number $\langle n_{neut}^{tot} \rangle$ in proton-initiated cascades is linear over a wide energy range.

The difference in magnitudes of neutron yields in proton- and iron-initiated cascades does not exceed the factor of two beginning from ~ 200 g/cm^2 (Fig. 5). The multiple-peak behavior is related to INCA's periodic structure.

Figure 6 shows the cascade-length dependence of the $\sigma(n_{neut}^{tot})/\langle n_{neut}^{tot} \rangle$ ratio for the total neutron number in cascades initiated by 10 TeV and 1 PeV nuclei. As is seen, the standard deviation (i.e., the relative accuracy of energy measurements) at-

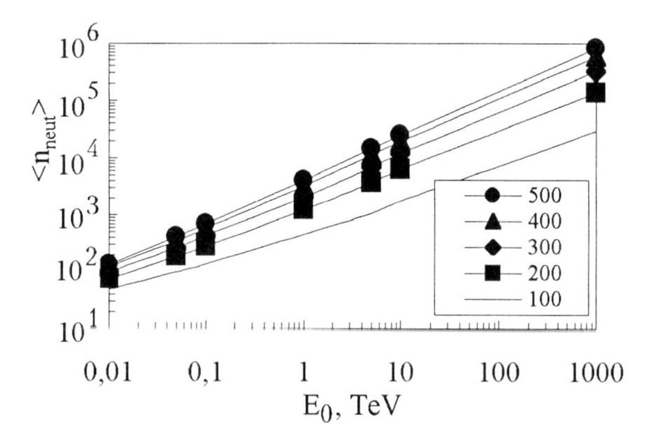

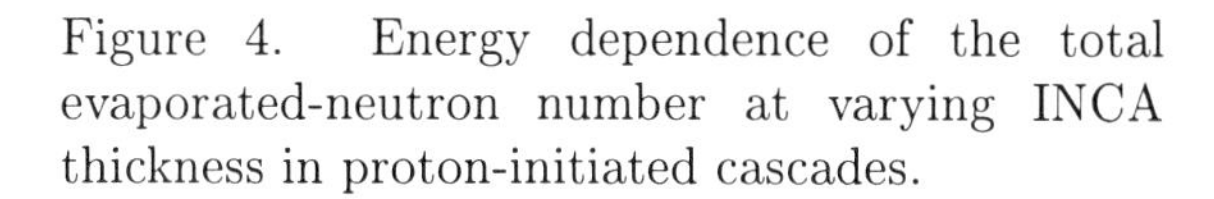

Figure 4. Energy dependence of the total evaporated-neutron number at varying INCA thickness in proton-initiated cascades.

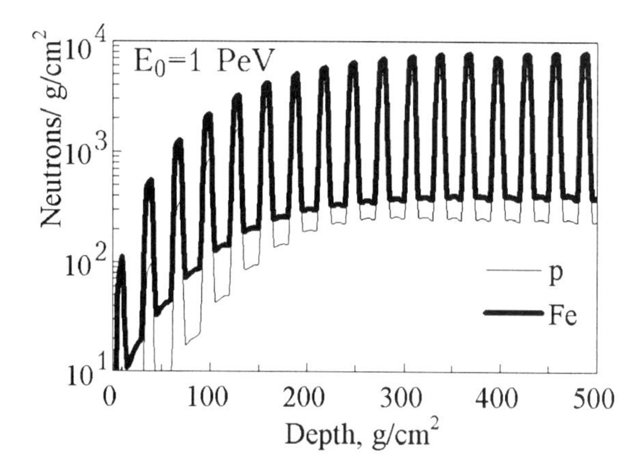

Figure 5. Depth dependence of the neutron yield in cascades initiated by protons and iron nuclei.

tains a reasonable value ($\sim$ 30% for protons and less for nuclei), if the cascade length is more than $\sim$ 300 g/cm^2.

4. STATUS

A module designed for balloon measurements of e^- spectrum (Fig. 7) was tested using 10 GeV e^- and π^- accelerator beams. Fig. 8 shows corresponding neutron-yield distributions. Note that these data cannot be directly compared with Figs. 1-6 because in this case: (1) there is no rejection of π^--induced cascades by criteria (a) K_1 (fixing the cascade starting point at the near-surface layer, i.e. pion interactions are distributed over all the module); (b) K_2 (energy realized into γ-rays is less than e^- energy and distributed over a wide range from 0 to 10 GeV); (2) efficiency of neutron counting is 7%, reducing the observed neutron number. So, the observed difference in neutron yield in e^-- and π^--produced cascades is much less than demonstrated in Figs. 1-6. However, the corresponding simulations, accounting for real experimental conditions, confirm our optimistic conclusions.

5. CONCLUSION

A new effective method for separation of primary electrons and protons with a total rejection factor of $\sim 10^{-5} - 10^{-6}$ is proposed to investigate the primary electron spectrum at $E_e > 1$ TeV.

A new method of energy measurement is proposed to study the energy spectrum of primary cosmic rays in the "knee" range.

A basic module of INCA's balloon version was tested. Results are in a good agreement with simulations.

REFERENCES

1. INCA Collaboration I,III & IV, Proc. 26th ICRC, Salt Lake City, (1999), 219,203 & 215.
2. Bezrukov L.B. *et al.*, Sov. J. Nucl. Phys. 1973, 17 98.
3. Mukhamedshin R.A. *et al.*, Bull. Soc. Sci. Lettr. Lodz, Ser. Rech. Def. 1994, XVI 137.
4. Aleksandrov K. V. *et al.*, Nucl. Phys. B (Proc. Suppl.) 75A (1999) 269

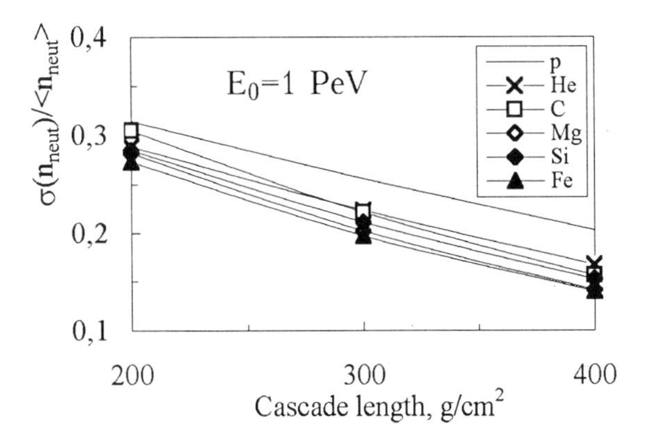

Figure 6. Cascade-length dependence of the $\sigma(n^{tot}_{neut})/\langle n^{tot}_{neut}\rangle$ ratio for the evaporated-neutron number integrated over the length of cascades.

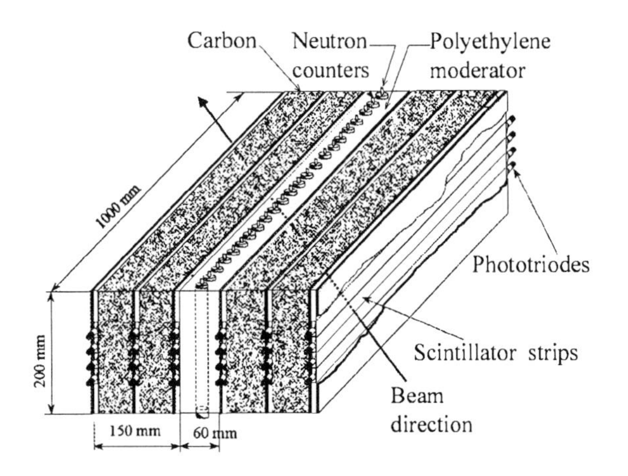

Figure 7. Basic module of INCA.

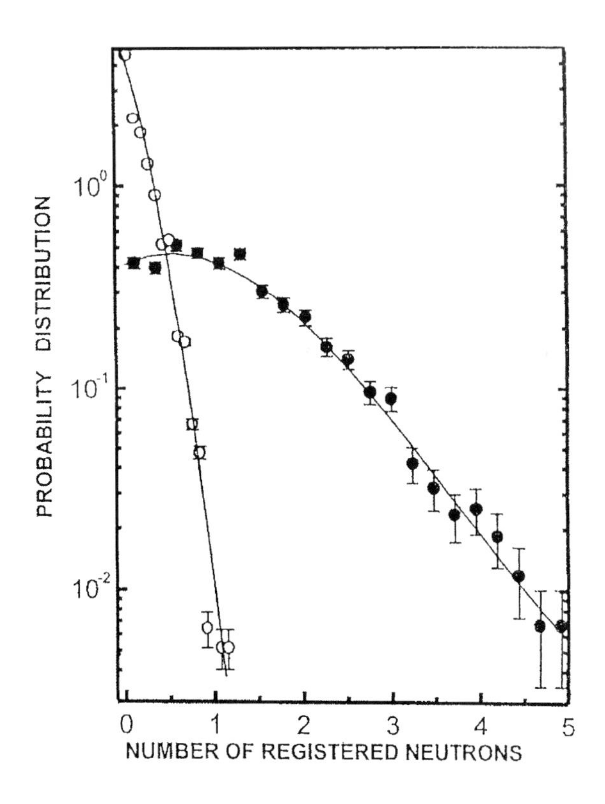

Figure 8. Neutron-yield distributions for electron- and pion-induced cascades (open and filled circles).

ELSEVIER

Nuclear Physics B (Proc. Suppl.) 97 (2001) 193–195

www.elsevier.nl/locate/npe

Using TOP-C for Commodity Parallel Computing in Cosmic Ray Physics Simulations

G. Alverson[a*], L. A. Anchordoqui[a†], G. Cooperman[b‡], V. Grinberg[b§], T. P. McCauley[a¶], T. Paul[a||], S. Reucroft[a**], and J. D. Swain[a††]

[a]Department of Physics, Northeastern University, Boston, MA 02115, USA

[b]College of Computer Science, Northeastern University, Boston, MA 02115, USA

TOP-C (Task Oriented Parallel C) is a freely available package for parallel computing. It is designed to be easy to learn and to have good tolerance for the high latencies that are common in commodity networks of computers. It has been successfully used in a wide range of examples, providing linear speedup with the number of computers. A brief overview of TOP-C is provided, along with recent experience with cosmic ray physics simulations.

1. Introduction

Ultra high energy cosmic rays are observed indirectly through detection of the extensive air showers that are produced when they travel through the atmosphere. To adequately interpret the measured observables and to be able to infer the properties of the incident primary particle, a full Monte Carlo treatment of the extensive air shower is neeeded. The CPU time required rises with the primary energy. For example, for primary energies around 10^{20} eV a shower contains about 10^{11} secondary particles. The amount of computing time required to follow all the particles seems to be prohibitive. Traditionally, sampling techniques are used to reduce the number of particles tracked [1].

In this article we describe an ongoing program to use commodity parallel computing for fast Monte Carlo simulations [2]. The aim is to go beyond the simple event-level parallelism which is commonly used today and actually run individual events faster than would be possible on a single workstation or PC.

*george.alverson@cern.ch
†doqui@hepmail.physics.neu.edu
‡gene@ccs.neu.edu
§victor@ccs.neu.edu
¶mccauley@hepmail.physics.neu.edu
||tom.paul@hepmail.physics.neu.edu
**stephen.reucroft@cern.ch
††john.swain@cern.ch

2. GEANT4

For a variety of reasons, in no small part driven by the wish to work with software which is likely to see use in the future, we decided to try to parallelize GEANT4 [3], the C++ rewrite of the older (FORTRAN77) GEANT3. GEANT4 is an object-oriented simulation package that provides general-purpose tools for defining and simulating detector geometry, material properties, particle transport and interactions, visualization, and all relevant physics processes. Its versatility allows it to be employed in applications beyond its traditional usage in High Energy Physics experiments, from the medical and biological sciences to Cosmic Ray Physics [4].

3. TOP-C

TOP-C (Task Oriented Parallel C) [5] was initially designed with two goals in mind:

1. to provide a framework for easily developing parallel applications;
2. to build in the ability to tolerate the high latency typically found on Beowulf clusters. [9]

The package is freely available [6]. The same application source code has been run under shared

[9]The term "Beowulf cluster" refers to a cluster of systems running Linux and connected by ethernet.

0920-5632/01/$ – see front matter
PII S0920-5632(01)01261-0

and distributed memory (SMP, IBM SP-2, NoW, Beowulf cluster). A sequential TOP-C library is also provided to ease debugging. The largest test to date was a computer construction of Janko's group over three months using approximately 100 nodes of an IBM SP-2 parallel computer at Cornell University [7].

The TOP-C programmer's model [5] is a master-slave architecture based on three key concepts:

1. *tasks* in the context of a master/slave architecture;
2. global *shared data* with lazy updates; and
3. *actions* to be taken after each task.

Task descriptions (task inputs) are generated on the master, and assigned to a slave. The slave executes the task and returns the result to the master. The master may update shared data on all processes. Such global updates take place on each slave after the slave completes its current task. The programmer's model for TOP-C is graphically described below.

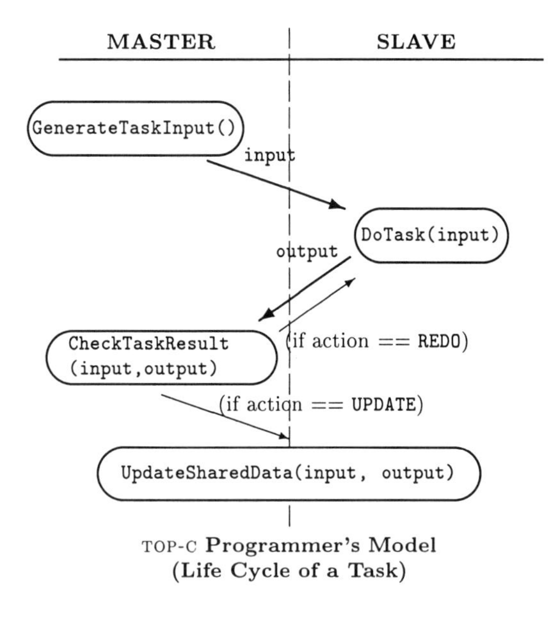

TOP-C Programmer's Model (Life Cycle of a Task)

4. Parallelization of GEANT4 Using TOP-C

The task-oriented approach of TOP-C is ideally suited to parallelizing legacy applications. The tactic in parallelizing GEANT4 was to perturb the existing software as little as possible and to modify just the section of the code which handles particle tracking and interaction (a frequent operation) to allow it to run on multiple CPU's. The largest difficulty was in *marshalling* and *unmarshalling* the C++ GEANT4 track objects that had to be passed to the slave processes. Marshalling is the process by which one produces a representation of an object in a contiguous buffer suitable for transfer over a network, and unmarshalling is the inverse process.

We developed a 6-step software methodology to incrementally parallelize GEANT4, allowing us to isolate individual issues. The six steps were:

1. the use of .icc (include) files to isolate the code from the original GEANT4 code;
2. collecting the code of the inner loop in a separate routine, `DoTask()`, whose input was a primary particle track, and whose output was the primary and its secondary particles;
3. marshalling and unmarshalling the C++ objects for particle tracks
4. integrating the marshalled versions of the particle tracks with the calls to `DoTask()`;
5. adding calls to TOP-C routines such as `TOPC_init()`, `TOPC_submit_task_input` and then testing as the marshalled particle tracks were sent across the network;
6. and finally adding `CheckTaskResult()`, which inspected the task output, and added the secondary tracks to the stack, for later processing by other slave processes.

Prior to the fifth step, all debugging was in a sequential setting. The maturity of the TOP-C library then allowed us to create fully functioning parallel code in less than a day.

5. Discussion

GEANT4 (approximately 100,000 lines of C++ code) was successfully parallelized using TOP-C. In the future we plan to perform timing tests on a long run using many processors. Initial results for the example described indicate that a single task in our application requires approximately 1 ms of CPU time. Hence, it will be essential to submit approximately 100 particles for a single slave process to compute in order to overcome network overhead. Optimization of the parallel implementaion is underway.

TOP-C seems to be well-suited to the problem of parallelizing GEANT4, and would likely be well-suited to other high energy physics and cosmic ray applications as well. Its flexibility and simplicity makes it possible to envision enormous speedups for GEANT4 within a single event, something not often considered in high energy experiments, but offering advantages over the usual event-by-event parallelism, especially during interactive data analysis and code or hardware design.

Of particular interest is the parallelization of existing cosmic ray simulation programs such as AIRES [8] and CORSIKA [9]. Although written in FORTRAN, such programs are in fact often converted to C for compilation using f2c, and can certainly be linked with other C programs, so we anticipate no major obstacles. We are always interested in collaboration with other groups who may have needs for the speedups that our methodology offers.

REFERENCES

1. A. M. Hillas, in *Proc. of the 16th International Cosmic Ray Conference*, Tokyo, Japan, 1979 (University of Tokyo, Tokyo, 1979), Vol.8,p.7.; updated in, *Proc. of the 17th International Cosmic Ray Conference*, Paris, France, 1981 (CEN, Saclay, 1981), Vol.8,p.183.
2. The first analyses were discussed in, G. Cooperman, L. Anchordoqui, V. Grinberg, T. McCauley, S. Reucroft and J. Swain, *Scalable Parallel Implementation of Geant4 Using Commodity Hardware and Task Oriented Parallel C*, in Proc. CHEP 2000 (only available in the CD version) [hep-ph/0001144].
3. http://wwwinfo.cern.ch/asd/geant/
4. L. A. Anchordoqui *et al.*, [astro-ph/0006141] and [astro-ph/0006142].
5. G. Cooperman, "TOP-C: A Task-Oriented Parallel C Interface", 5th *International Symposium on High Performance Distributed Computing* (HPDC-5), IEEE Press, 1996, pp. 141–150.
6. ftp://ftp.ccs.neu.edu/pub/people/gene/topc
7. G. Cooperman, W. Lempken, G. Michler and M. Weller, "A New Existence Proof of Janko's Simple Group J_4", *Progress In Mathematics* **173**, Birkhauser, 1999, pp. 161–175.
8. S. Sciutto, *Air Shower Simulations with the* AIRES *system*, in *Proc. XXVI International Cosmic Ray Conference*, (Eds. D. Kieda, M. Salamon, and B. Dingus, Salt Lake City, Utah, 1999) vol.1, p.411, [astro-ph/9905185] at http://xxx.lanl.gov.
9. D. Heck *et al.*, CORSIKA *(COsmic Ray Simulation for KASCADE)*, FZKA6019 (Forschungszentrum Karlsruhe) 1998; updated by D. Heck and J. Knapp, FZKA6097 (Forschungszentrum Karlsruhe) 1998.

ELSEVIER

Nuclear Physics B (Proc. Suppl.) 97 (2001) 196–198

www.elsevier.nl/locate/npe

Simulation of Water Čerenkov Detectors Using GEANT4

L. A. Anchordoqui[a*], T. P. McCauley[a†], T. Paul[a‡], S. Reucroft[a§], J. D. Swain[a¶], and L. Taylor[a||]

[a]Department of Physics, Northeastern University, Boston, MA 02115, USA

We present a detailed simulation of the performance of water Čerenkov detectors suitable for use in the Pierre Auger Observatory. Using GEANT4, a flexible object-oriented simulation program, including all known physics processes, has been developed. The program also allows interactive visualization, and can easily be modified for any experimental setup.

1. Introduction

Water Čerenkov detectors have proved to be superb devices for the study of cosmic air showers [1], and will constitute major components of future experiments such as the Pierre Auger Observatory [2]. Monte Carlo simulation of the response of such detectors to signal and background processes is crucial for proper interpretation of the data and to aid in development of reconstruction and analysis algorithms. To address these needs, experiments have prepared dedicated detector simulations [3]. There may be advantages, however, in exploiting some of the efforts which the High Energy Physics (HEP) community has directed at this problem. The GEANT3 package [4], for example, was developed in order to provide HEP experiments with generic tools for simulating the passage of particles through matter, but it also found substantial usage in the medical and biological sciences and in astronautics, and in fact has been studied for use in simulating water Čerenkov detectors for the Auger experiment [5]. In 1996 CERN initiated the GEANT4 project [6] with the goal of reproducing all the functionality of GEANT3 using an Object Oriented approach, as well as addressing some of the shortfalls of the older program. We have begun investigating the suitability of GEANT4 for the problem of simulating water Čerenkov detectors, and report here on the status of the work.

*doqui@hepmail.physics.neu.edu
†mccauley@hepmail.physics.neu.edu
‡tom.paul@hepmail.physics.neu.edu
§stephen.reucroft@cern.ch
¶john.swain@cern.ch
||lucas.taylor@cern.ch

2. GEANT4

Like its predecessor, GEANT4 provides a battery of tools to describe the geometry and material properties of an experimental setup, handle particle transport through materials and magnetic fields, and simulate particle decay and interactions with detector elements. All relevant physics processes have been included. For example, simulation of an electron traversing a material can include effects of ionisation, delta ray production, multiple Coulomb scattering, bremsstrahlung and Čerenkov radiation. Optical photons produced by processes such as Čerenkov radiation may then be subjected to Rayleigh scattering, absorption, and optical boundary interactions. We note that the calculation of reflection and transmission coefficients at material boundaries takes into account the polarization state of the photon, which is important in accounting for a photon's fate as it traverses multiple optical boundaries en route to a photocathode.

In addition to reproducing the functionality of GEANT3, GEANT4 aims to improve the procedures used for geometry definition, introduction of special physics processes, visualization, and optical processes. In addition, special effort has been made to ensure tracking precision over arbitrary scales, so the package may lend itself to simulation of extensive air showers as well as detailed detector response to ground particles. GEANT4 is written in C++ using an object ori-

PII S0920-5632(01)01262-2

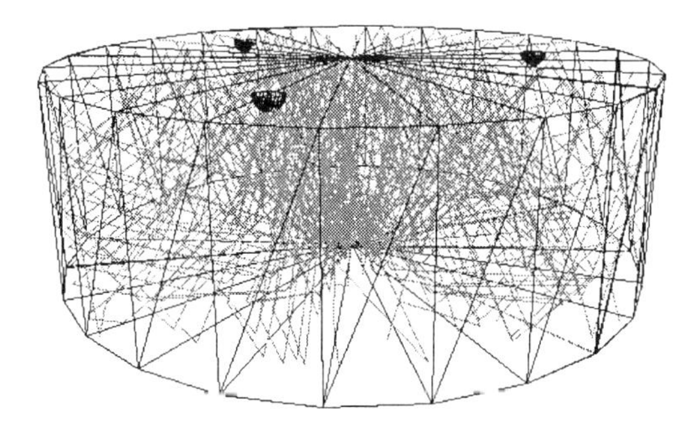

Figure 1. A muon entering a tank of water from above and radiating Čerenkov light. Housings for three photomultiplier tubes are visible at the top of the tank.

ented approach, and as such alleviates some of the shortcomings inherent in the procedural approach (FORTRAN77) of GEANT3.

3. Detector Geometry

GEANT4 strives to provide rather advanced tools for describing detector geometry, such as methods for interpreting files produced by Computer Aided Design (CAD) systems. However, for our application a first approximation to the geometry may be described in only a few lines of code. Figure 1 shows an example of an Auger-style Čerenkov detector which consists of a cylindrical tank filled with water. A reflective liner surrounds the water volume. Three hemispherical domes at the top of the tank house photomultiplier tubes, which contain sensitive photocathode volumes that register a hit when a photon strikes. Photons bounce about inside the tank until they are absorbed in the water, the tank walls, or until they enter a photomultiplier. The frequency-dependent optical properties of all these elements can be tuned to available data. In the figure, a 1 GeV muon is incident at the top of the tank, and some of the radiated Čerenkov photons are shown.

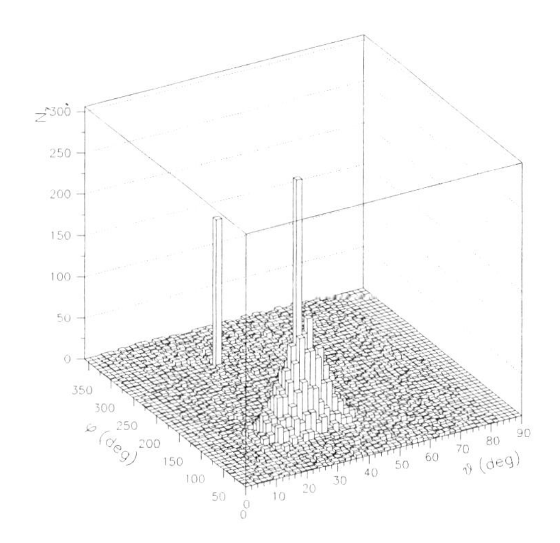

Figure 2. Distribution of reflected photons as a function of polar angles θ and ϕ for a beam incident at $\theta_{\text{incident}} = 30°$. This simulation uses the unified model of GEANT4 with specular and backscatter spikes, visible at $\phi = 90°$ and $270°$ respectively, a specular lobe centered at $\theta = 30°$, as well as uniform (Lambertian) reflection.

4. Optical Modeling

The detector geometry of Figure 1 includes a diffuse reflector to contain the Čerenkov light inside the volume of water. The use of Tyvek [7] as such a reflector has been explored by the Auger experiment, and it has been noted [7] that the treatment of photon interactions with a rough dielectric surface as available in GEANT3 does not appear to be sufficient to describe the experimentally observed reflection of light from this material. GEANT4 provides a more flexible optical model inspired by the work of Nayer *et al.* [8]. There it was observed that the principal features of both physical and geometrical optical models of surface reflection could be accommodated in a

[7]Tyvek is a trademark of DuPont.

so-called unified model which is applicable over a wide range of surface roughness and wavelengths.

The unified optical model in GEANT4 allows adjustment of parameters to control the relative contributions from: specular reflections about both the average surface normal and the normal of a microfacet at the surface; the diffuse or Lambertian reflection; a possible backscatter constant; overall surface reflectivity. Figure 2 shows the polar angular distribution of light bouncing off such a unified surface, with contributions from the various sources identified. Tuning of this model to experimental data [9] is under way.

5. Benchmarking

On a 450 MHz Pentium, our program requires roughly 9 seconds to simulate the tank response to a 1 GeV vertical muon[8]. We anticipate substantial improvement in this figure, possibly through application of TOP-C [10] or parameterization tools built into the GEANT4 framework.

6. Summary and Prospects

A prototype simulated water Čerenkov detector has been developed using GEANT4. We are optimistic about the utility of this package for a number of reasons.

- The general, extensive arsenal of tools provided will allow the detail inherent in the simulation to evolve as needs are identified. On the other hand, fast parameterizations can be employed in a straightforward manner wherever details might be found to be unnecessary.

- GEANT4 design is consistent with modern software trends. Though the product is relatively new and potholes in the road should be expected, the project enjoys a large development community, which bodes well for its longevity and evolution.

- Several shortcomings of GEANT3 have been addressed in the new version. For example the surface reflection model now available is more amenable to our problem than its predecessor.

[8]No optimization was used in the compilation

Acknowledgements

We would like to thank Peter Gumplinger for help implementing the unified surface model. This work was supported by the National Science Foundation and CONICET.

REFERENCES

1. M.A. Lawrence, R.J.O. Reid, and A.A. Watson, J.Phys.G.:Nucl.Phys., **17** (1991) 773.; M. Takeda *et al.*, Phys.Rev.Lett. **81** (1998) 1163.
2. http://www.auger.org
3. C. Pryke, “Instrumentation development and experimental design for a next generation detector of the highest energy cosmic rays,” Ph.D. Thesis, University of Leeds (1996);
4. R. Brun *et al.* “GEANT 3”, CERN DD/EE/84-1 (1987).
5. FCEyN and Tandar Groups, “Simulations with GEANT,” Auger Technical Note GAP-96-011 (1996); P. Bauleo *et al.* “Tank depth and muon/electromagnetic separation in a water Čerenkov detector,” Auger Technical Note GAP-96-029 (1996); J.R.T. de Mello Neto, “WTANK: A GEANT surface array simulation program,” Auger Technical Note GAP-98-020 (1998).
6. Website *http://wwwinfo.cern.ch/asd/geant4* .
7. F. Hasenbalg and D. Ravignani, “Tyvek Diffuse Reflectivity,” Auger Technical Note GAP-97-035 (1997).
8. S.K. Nayer, K. Ikeuchi, and T. Kanade, IEEE Trans. on Pattern Analysis and Machine Intelligence, **13** (1991) 611; A.Levin and C. Moisan, TRI-PP-96-64 (1996).
9. A. Filevich, P. Bauleo, H. Bianchi, J. Rodriguez Martino and G. Torlasco, Nucl. Instr. Meth., **A** 423 (1999) 108.
10. G. Cooperman, “TOP-C: A Task-Oriented Parallel C Interface”, 5th *International Symposium on High Performance Distributed Computing* (HPDC-5), IEEE Press, 1996, pp. 141–150.

ELSEVIER

Nuclear Physics B (Proc. Suppl.) 97 (2001) 199–202

NUCLEAR PHYSICS B
PROCEEDINGS
SUPPLEMENTS

www.elsevier.nl/locate/npe

Fractal Electromagnetic Showers

L. A. Anchordoqui[a*], M. Kirasirova[a†], T. P. McCauley[a‡], T. Paul[a§], S. Reucroft[a¶], and J. D. Swain[a||]

[a]Department of Physics, Northeastern University, Boston, MA 02115, USA

We study the self-similar structure of electromagnetic showers and introduce the notion of the fractal dimension of a shower. Studies underway of showers in various materials and at various energies are presented, and the range over which the fractal scaling behaviour is observed is discussed. Applications to fast shower simulations and identification, particularly in the context of extensive air showers, are also discussed.

1. Introduction

One of the most serious problems in the analysis of cosmic ray data is the complex and time-consuming nature of the codes used for shower simulation. In order to try to capture the detailed physics of the processes involved, it is customary to directly simulate [1,2] the multiplicative branching process whereby an initial particle gives rise to two or more secondary particles, each of which, in turn, initiates what is essentially its own shower, albeit now at lower energy.

Such a process can give rise to large fluctuations, and the final distributions of ground particles and their energies (as well as the longitudinal distribution of the shower as a whole) are difficult to model with simple parametrizations unless one is happy to settle for a description of the *mean* behaviour of the shower and forego knowledge of the fluctuations. Indeed, this is the leading reason that so much Monte Carlo time must be used for shower simulations: there are no simple analytical forms for the relevant distributions which can describe the fluctuations. The issue is a pressing one for experiments collecting large amounts of data which may be difficult to compare against theory in any form other than a large number of simulated events.

Here we report on the observation that electromagnetic showers display self-similar behaviour which can be described by a multifractal geometry and describe first steps towards formalizing this concept. Our eventual goal is to describe showers in terms of what we argue here is the relevant geometry: not one of smooth functions, but one which allows for irregular geometries which are better described in terms of fractals. We consider here only electromagnetic showers, but plan to study hadronic showers in future work.

2. Self-Similarity in Electromagnetic Showers

The idea that an electromagnetic shower should, in some sense, be a fractal is almost obvious. It is generated recursively from the two processes"

1. pair creation: $\gamma \to e^+e^-$ in the electric field of a nucleus and;

2. Bremsstrahlung: $e^\pm \to e^\pm\gamma$ as an electron or positron is deflected by the electric field of a nucleus

This is illustrated in figure 1 which shows the particles making up a shower produced by a 100 GeV electron entering a block of aluminum 150 cm long (radiation length 8.9 cm) as simulated using the GEANT4 program [3].

Each final state particle from an interaction effectively initiates its own electromagnetic shower, and each process has a similar cross section to occur in matter. As long as the energies involved are large compared with the energy required to create an electron-positron pair (and thus also large

*doqui@hepmail.physics.neu.edu
†kirasirova@hepmail.physics.neu.edu
‡mccauley@hepmail.physics.neu.edu
§tom.paul@hepmail.physics.neu.edu
¶stephen.reucroft@cern.ch
||john.swain@cern.ch

PII S0920-5632(01)01263-4

compared to atomic processes such as ionization), each step is much the same as the one before it, but at a reduced energy.

Figure 2 shows a slice through the block right at the far end with the point of intersection of each particle with the slice shown as a black dot whose radius is independent of energy. Here one clearly sees the shower core, with a diminishing density of particles with distance from the centre.

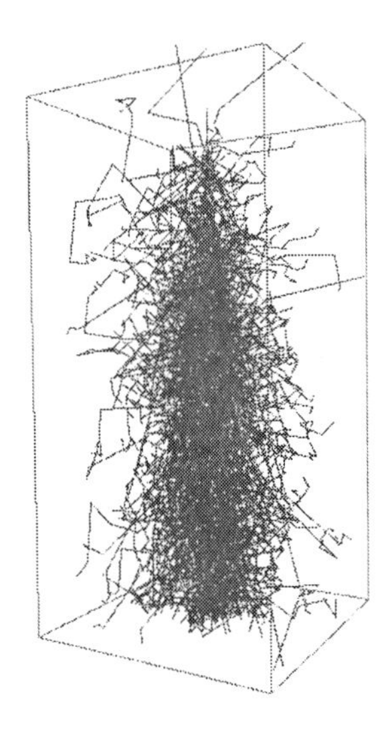

Figure 1. Three-dimensional view of the tracks making up an electromagnetic shower due to a 100 GeV electron entering an aluminum block 150 cm in length.

3. Fractals and Multifractals

There are many ways to characterize self-similar objects, but the most common and well-known way is in terms of fractal dimensions. There are many different concepts of fractal dimension which are useful, and perhaps the most obvious is that of mass dimension, D_M. The idea here is to see how the total energy $E_{TOT}(R)$ (considered now as a sort of weight) within a disk of radius R varies as R is changed. If the distribution were one-dimensional (a line of uniform energy deposited in the plane), one would find

$$E_{TOT}(R) \propto R^1 \qquad (1)$$

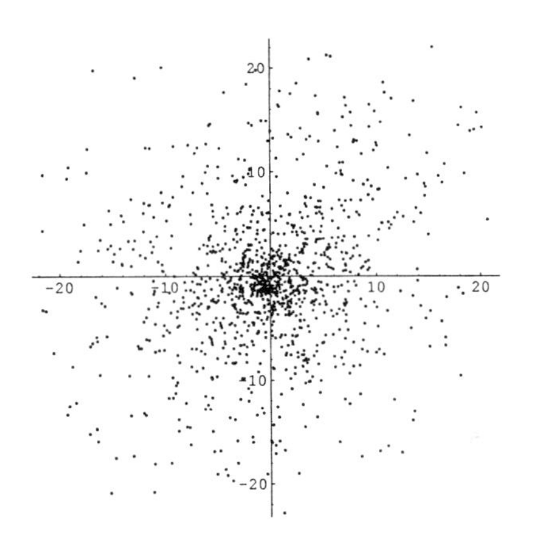

Figure 2. Points of intersection of particles with a slice 150 cm from the entry point of a 100 GeV electron into an aluminum block. Axes are perpendicular to the shower axis and are marked in centimetres.

and one would take the exponent in the foregoing equation to be the dimension of the distribution.

If the energy were uniformly distributed over the whole plane, one would find

$$E_{TOT}(R) \propto R^2 \qquad (2)$$

and conclude again that the exponent in the scaling law for the energy should be interpreted as the dimension of the distribution.

In the event that a scaling law of the form $E_{TOT}(R) \propto R^{D_M}$ holds for a non-integer D_M, we call D_M the "fractal mass dimension". A plot of $\log(E)$ as a function of $\log(R)$ will then have a slope in the limit of small R which is D_M.

Two points are important to keep in mind here: first that there are no true fractals in nature as there are always some smallest and largest value for variables in the problem beyond which scaling behaviour does not hold, and second that one must be careful to watch for systematic effects which can bias estimates of the dimension. Systematic effects which we have had to be wary of include the fact that early in the shower development the central core can contain particles which carry a large fraction of the initial energy and give the radial energy distribution a spike at small R which does not correspond to scaling behaviour.

In the case of the electromagnetic shower with the slice taken at the end of the shower at 150 cm, we look at the summed energy (scaled so that the total energy is 1) as a function of the fraction of the radius out (scaled so that the maximum radius is 1). This quantity we denote as $I(R|1)$ for reasons which will become clear later in the text. Plotting logarithms against logarithms (base 10), we find the distribution shown in figure 3. The

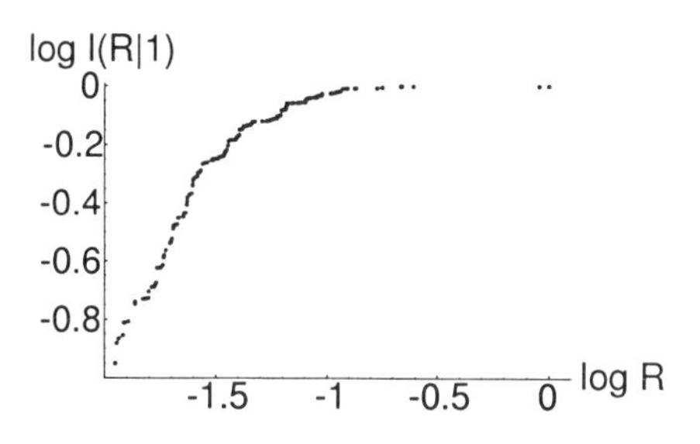

Figure 3. Summed energy as a function of radius from the shower centre. Logarithms are base 10.

first thing to notice is that the curve is reasonably approximated by a straight line at small radii. The second thing to notice is that the whole curve is *not* a straight line. At large radii we start to reach the physical boundaries of the shower and cannot expect scaling to hold.

In fact, even at very small radii, there is some anomalous structure which can be traced to the effects of very energetic particles very close to the core, which give an additional spike of energy to the distribution which cannot be expected to be a part of any overall scaling behaviour. This effect is more pronounced earlier in the shower.

The scaling properties of the shower are thus different in different parts of the plane, and in order to quantify this further, we study the scaling behaviour of cumulative moments of the energy distribution defined for $q > 0$ by

$$I(R|q) = \frac{\sum_{r<R} E_i^q}{\sum_{\text{all i}} E_i^q} \tag{3}$$

where E_i are the energies contained in a disk going out to radius R and the sum is taken over all particles within a distance $r < R$. What units are used is not important as we are only interested in the average scaling behaviour of the curves at small $R \to 0$. (As discussed earlier in the text, the region of very small R should be avoided for physical reasons, and we will avoid the subtleties of precise numerical analyses in this short communication.) For graphical purposes here, R is normalized so that the particle with the largest radial distance out is at $R = 1$ and the moments are defined so that their value at maximum radius is unity. We can then introduce an infinite family[4] of fractal dimensions D_q defined for $q > 0$ by

$$D_q = \lim_{R\to 0} \left\langle \frac{1}{q} \frac{\partial \log I(R|q)}{\partial \log R} \right\rangle \tag{4}$$

with the understanding that the limit must still lie in the scaling region in physical examples.

Figure 4 shows the scaling behaviour of moments of the electromagnetic shower corresponding to how the sums of the squares and cubes of the energy grow with distance. For a homogeneous and uniform fractal structure we expect the D_q to be equal. If not, then we describe the distribution as multifractal in that it requires more than one fractal dimension in order to characterize it. The associated D_q for small q estimated from finite differences in the scaling region are all approximately equal within the errors in the data here and approximately unity, suggesting a good degree of homogeneity. It is important to keep in mind that the results in this paper are presented for a full, realistic GEANT simulation, and include ionization, delta-ray, and other soft processes, so some care is needed in interpreting the results as if they corresponded to a pure electromagnetic shower generated only by pair creation and Bremsstrahlung (which is, of course, not realizable in nature).

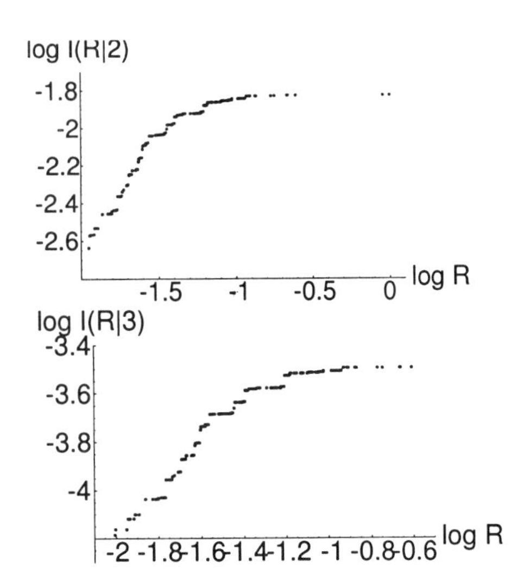

Figure 4. Total energy-squared (above) and total energy-cubed as functions of radius from the shower centre (see text for normalizations). Logarithms are base 10.

The definition of fractal dimensions can also be continued to $q \leq 0$, but this has some subtleties involved with the fact that as $q \to \infty$ the highest energy particles contribute most, while as $q \to -\infty$ the lower energy ones dominate. In particular, some care must be used with the D_q for $q < 0$ as they give high weights to softer particles which are not part of the hard shower process. These matters, as well as more precise results on dimensions including energy and material dependence will be presented elsewhere[5].

4. Further Work

Clearly space limitations make it impossible to cover the material as completely as one would like, but several points concerning work not discussed here are worth making. First of all, we expect fractal behaviour in all three dimensions, and in this discussion we have neglected the longitudinal scaling behaviour, where the full shower is made of many scaled and translated showers superimposed along the shower axis. In addition, there are clearly angular correlations and fluctuations, and studies can be made at a given fixed radius of the scaling behaviour of the shower as a function of the angular coordinate which we have integrated out in this discussion. The relation of these ideas to the concept of intermittency, especially as studied in hadronic jets has not escaped our notice and is currently under investigation.

One of the main goals of this work is to better understand the geometry of electromagnetic (and other) showers in order to try to parametrize them by the appropriate non-smooth basis functions, such as wavelets. Such a parametrization should allow the fast generation of showers without the attendant loss of information concerning large fluctuations[5] which has so far been handled only by the use of enormous computational resources.

5. Acknowledgements

We would like to thank the US National Science Foundation and CONICET, Argentina for support. We would also like to thank our collaborators on the Pierre Auger Project, as well as on L3 and CMS for useful discussions on electromagnetic calorimetry.

REFERENCES

1. S. Sciutto, *Air Shower Simulations with the* AIRES *system*, in *Proc. XXVI International Cosmic Ray Conference*, (Eds. D. Kieda, M. Salamon, and B. Dingus, Salt Lake City, Utah, 1999) vol.1, p.411, `astro-ph/9905185` at `http://xxx.lanl.gov`.
2. D. Heck *et al.*, CORSIKA *(COsmic Ray Simulation for KASCADE)*, FZKA6019 (Forschungszentrum Karlsruhe) 1998; updated by D. Heck and J. Knapp, FZKA6097 (Forschungszentrum Karlsruhe) 1998.
3. `http://wwwinfo.cern.ch/asd/geant4`
4. See, for example, "The Science of Fractal Images", eds. H. Peitgen and D. Saupe, Spinger-Verlag, 1988.
5. The authors, papers in preparation.

ELSEVIER

Nuclear Physics B (Proc. Suppl.) 97 (2001) 203–206

NUCLEAR PHYSICS B
PROCEEDINGS
SUPPLEMENTS

www.elsevier.nl/locate/npe

A pot of gold at the end of the cosmic "raynbow"?

L. A. Anchordoqui[a*], M. T. Dova[b†], T. P. McCauley[a‡], T. Paul[a§], S. Reucroft[a¶], and J. D. Swain[a||]

[a]Department of Physics, Northeastern University, Boston, MA 02115, USA

[b]Departamento de Física, UNLP, CC 67 (1900) La Plata, Argentina

We critically review the common belief that ultrahigh energy cosmic rays are protons or atomic nuclei with masses not exceeding that of iron. We find that heavier nuclei are indeed possible, and discuss possible sources and acceleration mechanisms for such primaries. We also show detailed simulations of extensive air showers produced by "superheavy" nuclei, and discuss prospects for their detection in future experiments.

The unambiguous detection of cosmic rays (CRs) with energies above 10^{20} eV (see [1] for a survey and bibliography on the subject) is a fact of outstanding astrophysical interest. As shown in the pioneering works of Greisen, Zatsepin, and Kuzmin [2], the possible sources and the accelerating mechanisms are constrained by the observed particle spectra due to the interaction with the universal radiation and magnetic fields on the way to the observer. The low flux of particles at the end of the spectrum (the typical rate of CRs above 10^{20} eV is one event/km^2/century) puts strong demands on the collection power of the experiments, such as can only be achieved by extended air shower detection arrays at ground level. This indirect method of detection bears a number of serious difficulties in determining the energy, mass and/or arrival direction of the primary particles. Astrophysical mechanisms to accelerate particles to energies of up to 10^{21-22} eV have been identified, but they require exceptional sites [3]. Essentially, the mechanisms rely on a net transfer of macroscopic kinetic energy of moving magnetized plasma to individual charged particles due to repeated collisionless scatterings with randomly moving inhomogeneities of the turbulent magnetic field, or with shocks in the medium. Since the high-energy cutoff of shock acceleration increases with the charge number of the nucleus Z, heavy ions would be nice candidates for ultra high-energy CRs. In this talk we review the current status of superheavy-nuclei as primaries of the cosmic radiation.

It has been generally thought that ^{56}Fe is a significant end product of stellar evolution and higher mass nuclei are rare in the cosmic radiation. Strictly speaking, the atomic abundances of middle-weight ($60 \leq A < 100$) and heavy-weight ($A > 100$) elements are approximately 3 and 5 orders of magnitude lower, respectively, than that of the iron group [4]. The explosion of the H-rich envelopes of type II supernovae has long been held responsible for the synthesis of the stable superheavy nuclides. There is observational evidence that these nuclides are synthesized in a variety of chemically peculiar Red Giants [5] and in special objects like FG Sagittae [6] or SN1987A [7]. Consequently, starbursts (astrophysical environments which comprise a considerable population of O and Red Giant stars [8], with a supernovae rate as high as 0.2 - 0.3 yr^{-1} [9]) appear as the natural sources able to produce relativistic super-heavy nuclei. Very recently, we have presented a comprehensive study of a possible nearby superheavy-nucleus-Zevatron [10]. We have shown that it is likely that nuclei heavier than iron with energies above a few PeV can escape from the dense core of a nearby starburst galaxy (like M82, NGC 253), and eventually be re-accelerated to superhigh energies ($E \geq 10^{20}$ eV) at the terminal shocks of

*doqui@hepmail.physics.neu.edu
†dova@venus.fisica.unlp. edu.ar
‡mccauley@hepmail.physics.neu.edu
§tom.paul@hepmail.physics.neu.edu
¶stephen.reucroft@cern.ch
||john.swain@cern.ch

0920-5632/01/$ – see front matter
PII S0920-5632(01)01264-6

galactic superwinds generated by the starburst.[7] Specifically, ions are diffusively accelerated up to a few PeV at single supernova shock waves in the nuclear region of the galaxy without suffering significant spallation reactions [12]. The cosmic ray outflow is convection dominated, and the typical residence time of the nuclei in the starburst results in $t \sim 1 \times 10^{11}$ s. Thus, the total path traveled is substantially shorter than the mean free path (which scales as $A^{-2/3}$) of a super-heavy nucleus (for details see [13]). Those which are able to escape from the central region without suffering catastrophic interactions could be eventually re-accelerated to superhigh energies at the terminal shocks of galactic superwinds generated by the starburst. In terms of parameters that can be determined from observations the maximum energy of the nuclei reads (again the reader is referred to [13]),

$$E_{\rm max} \approx \frac{1}{2}\, Ze\, B\, \frac{\dot{E}_{\rm sw}}{\dot{M}}\, \tau. \qquad (1)$$

The predicted kinetic energy and mass fluxes of the starburst of NGC 253 derived from the measured IR luminosity are $\dot{E}_{rmsw} \approx 2 \times 10^{42}$ erg s^{-1} and $\dot{M} \approx 1.2$ M$_\odot$ yr^{-1}, respectively [14]. The starburst age is estimated from numerical models that use theoretical evolutionary tracks for individual stars and make sums over the entire stellar population at each time in order to produce the galaxy luminosity as a function of time [15]. Fitting the observational data these models provide a range of suitable ages for the starburst phase that, in the case of NGC 253, goes from 5×10^7 to 1.6×10^8 yr (also valid for M82) [15]. These models must assume a given initial mass function (IMF), which usually is taken to be a power-law with a variety of slopes. Recent studies have shown that the same IMF can account for the properties of both NGC 253 and M82 [16]. Here we shall assume a conservative age $\tau = 50$ Myr. Finally, the radio and γ-ray emission from NGC 253 are well matched by models with $B \sim 50\mu$G [17]. All in all, the balance between the energy gains and synchrotron/photodisintegration losses [18] of a super-heavy nucleus (like gold) leads to a maximum energy of a few hundred EeV.[8]

Let us turn now to the analysis of the energy loses during propagation. The energetic nucleus is seen to lose energy mainly as a result of its photodisintegration. In the universal rest frame (in which the microwave backgroud radiation is at $3K$), the disintegration rate R of an extremely high energy nucleus with Lorentz factor Γ, propagating through an isotropic soft photon background of density n is given by [19],

$$R = \frac{1}{2\Gamma^2} \int_0^\infty dE\, \frac{n(E)}{E^2} \int_0^{2\Gamma E} dE'\, E'\, \sigma(E'), \qquad (2)$$

where primed quantities refer to the rest frame of the nucleus, and σ stands for the total photon absortion cross section. Above 10^{20} eV, the energy losses are dominated by collisions with the relic photons. The fractional energy loss around this energy is $R \sim 10^{-15}$ s^{-1}. With this in mind, it is straightforward to check that superheavy nuclei may reach the Earth (for details see Fig. 2 of Ref. [10]). In the rest of this report we shall discuss the characteristics of the extensive air showers that these nuclei may produce after interaction with the atmosphere.

Golden Shower Simulations: In order to perform the simulations we shall adopt the superposition model. This model assumes that an average shower produced by a nucleus with energy E and mass number A is indistinguishable from a superposition of A proton showers, each with energy E/A. We have generated several sets of ^{197}Au air shower simulations by means of the AIRES Monte Carlo code [20]. The sample was distributed in the energy range of 10^{18} up to $10^{20.5}$ eV. SIBYLL was used to reproduce hadronic collisions above 200 GeV [22]. All shower particles with energies above the following thresholds were tracked: 750 keV for gammas, 900 keV for electrons and positrons, 10 MeV for muons, 60 MeV for mesons and 120 MeV for nucleons and nuclei. The particles were injected vertically at the top of the atmosphere (100 km.a.s.l), and the surface de-

[7] It is important to stress that M82 is positioned close to the arrival direction of the highest CR event detected on Earth. This was first pointed out in [11].

[8] It should be stressed that for $A > 140$ the bulk solar-system abundance distribution peaks at $A = 195$ [21]. To make some estimates, we then refer our calculations to a gold nucleus.

tector array was put at a depth of 1036 g/cm^2, i.e., at sea level. Secondary particles of different types and all charged particles in individual showers were sorted according to their distance R from the shower axis.

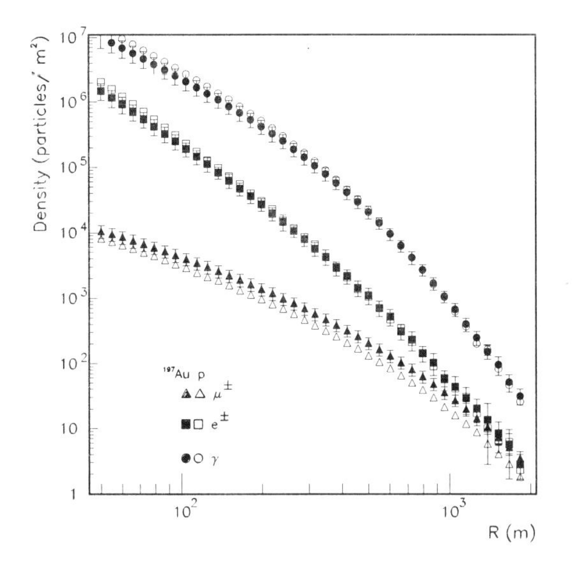

Figure 1. Ground lateral distributions of proton and ^{197}Au air-showers. The incident energy is $E = 3 \times 10^{20}$ eV.

In Fig. 1 we show the lateral distributions of different groups of secondary particles (we have considered separately γ, e^+e^-, and $\mu^+\mu^-$). One can see that the number of muons from the gold nucleus shower is greater than the number of muons from the proton shower.

As the cascade develops in the atmosphere, it grows until a maximum size (number of particles) is reached. The location in the atmosphere where the cascade has developed the maximum size is denoted by $X_{\max}$, with units of g cm^{-2}. For cascades of a given total energy, heavier nuclei have smaller $X_{\max}$ than nucleons because the shower is already subdivided into A nucleons when it enters the atmosphere. At 10^{20} eV, the $< X_{\max} >$ of a proton (gold) shower is ≈ 879 g/cm^2 (≈ 777 g/cm^2). A dust-grain has an even larger cross section, so it tends to interact sooner than protons and nuclei [24]. In Fig. 2, we compare the longitudinal profile of showers initiated by a proton, a gold-nucleus and a dust-grain. It is clearly seen how the $X_{\max}$ decreases when increasing the mass. The simulated gold shower is partially consistent with the Fly's Eye data. Furthermore, its longitudinal development better reproduces the data than protons or dust-grains. It should be remarked, however, that extensive air shower simulation depends on the hadronic interaction model [25]. We also point out that for the simulation, detector effects were not taken into account.

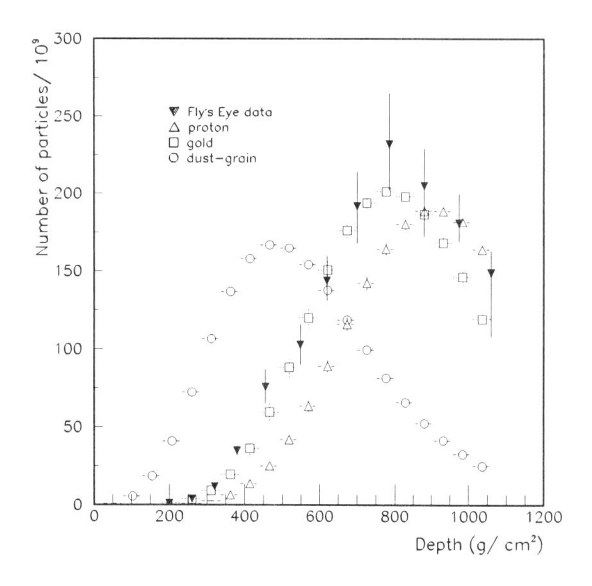

Figure 2. Longitudinal development of 3×10^{20} eV showers induced by a proton, a gold-nucleus and a dust-grain ($\log \Gamma = 4.5$), together with the data of the highest event recorded by Fly's Eye [23].

Even though the superheavy nucleus hypothesis is partially supported by data from the CASAMIA experiment [26], more data is certainly needed to verify this model. In order to significantly increase the statistics at the end of the spectrum, the Southern Auger Observatory is currently under construction [27]. It will consist of a surface array which will record the lateral and temporal distribution of shower particles, and an optical air fluorescence detector, which will observe the air shower development in the atmosphere. These two techniques provide complementary methods of extracting the required information from the shower to test the ideas discussed in this paper.

Note added in proof: After this talk was presented we have completed the air shower analysis by using a more inelastic hadronic model for the first generation of particles [28]. Definite conclusions on the nature of the highest energy Fly's Eye event cannot be reached, mainly because of large fluctuations from model to model. However, it should be stressed that even though the primary chemical composition remains hidden by the hadronic interaction model, it is evident that in both cases the shower profile is inconsistent with a proton, or a gamma-ray primary.

This work was partially supported by CONICET, Fundación Antorchas and the National Science Foundation.

REFERENCES

1. S. Yoshida, and H. Dai, J. Phys. G **24** (1998) 905.
2. K. Greisen, Phys. Rev. Lett. **16** (1966) 748; G. T. Zatsepin, and V. A. Kuzmin, Pis'ma Zh. Eksp. Teor. Fiz. **4** (1966) 114.
3. P. Bhattacharjee and G. Sigl, Phys. Rep. **327**, 109 (2000) 109, and references therein.
4. E. M. Burbidge, G. R. Burbidge, W. A. Fowler, and F. Hoyle, Rev. Mod. Phys. **29** (1957) 547.
5. V. V. Smith, in *Cosmic Abundances of matter*, AIP Conference Proceedings 183, ed. C. J. Waddington (New York, AIP 1989), p. 200.
6. G. E. Langer, R. P. Kraft, K. S. Anderson, Astrophys. J. **189** (1974) 509.
7. P. A. Mazzali, L. B. Lucy, K. Butler, Astron. Astrophys. **258** (1992) 399.
8. S. Sakai and B. Madore, *Detection of the Red Giant Branch Stars in M82 Using the Hubble Space Telescope*, [astro-ph/9906484].
9. See for instance, T. W. B. Muxlow et al., Mont. Not. Roc. Astron. Soc. **266** (1994) 455; J. S. Ulvestad and R. J. Antonucci, Astrophys. J. **448** (1997) 621; D. A. Forbes et al., Astrophys. J. **406** (1993) L11; R. de Grijs et al., *Supernova Remnants in the Fossil Starburt in M82*, [astro-ph/9909044].
10. L. A. Anchordoqui, M. T. Dova, T. P. McCauley, S. Reucroft and J. D. Swain, Phys. Lett. B **482** (2000) 343.
11. J. W. Elbert and P. Sommers, Astrophys. J. **441** (1995) 151.
12. P. O. Lagage and C. J. Cesarsky, Astron. Astrophys. **125** (1983) 249.
13. L. A. Anchordoqui, G. E. Romero, and J. A. Combi, Phys. Rev. D **60** (1999) 10300.
14. T. M. Heckman, L. Armus, and G. K. Miley, Ap. J. Suppl. **74** (1990) 833.
15. G. H. Rieke, et al., Ap. J. **238** (1980) 24.
16. C. W. Engelbracht, et al., Ap. J. **505** (1998) 639.
17. T. A. D. Paglione, A. P. Marscher, J. M. Jackson, and D. L. Bertsch, Ap. J. **460** (1996) 295.
18. R. J. Protheroe, [astro-ph/9812055].
19. F. W. Stecker, Phys. Rev. D **180** (1969) 1264.
20. S. J. Sciutto, in *Proc. XXVI International Cosmic Ray Conference*, Edts. D. Kieda, M. Salamon, and B. Dingus, Salt Lake City, Utah, 1999) vol.1, p.411.
21. E. Anders and N. Grevesse, Geochim. Cosmochim. Acta 53 (1989) 197.
22. R. S. Fletcher, T. K. Gaisser, P. Lipari and T. Stanev, Phys. Rev. D **50** (1994) 5710.
23. D. J. Bird et al., Astrophys. J. **441** (1995) 144.
24. L. A. Anchordoqui, Phys. Rev. D **61** (2000) 087302.
25. L. A. Anchordoqui, M. T. Dova, L. N. Epele and S. J. Sciutto, Phys. Rev. D **59** (1999) 094003.
26. M. A. K. Glasmacher et al., Astropart. Phys. **12** (1999) 1.
27. http://www.auger.org
28. L. A. Anchordoqui, M. Kirasirova, T. P. McCauley, S. Reucroft and J. Swain, Phys. Lett. B **492** (2000) 237.

ELSEVIER

Nuclear Physics B (Proc. Suppl.) 97 (2001) 207–210

www.elsevier.nl/locate/npe

Cosmic rays tell us on supernova explosion in the nearby interstellar space

Y. Stozhkov[a], V. Okhlopkov[b], P. Pokrevsky[a]

[a]Lebedev Physical Institute, Moscow, Russia

[b]Skobeltzyn Institute of Nuclear Physics, MSU, Russia

The data obtained in the long-term cosmic ray observations are analyzed. The analysis shows that a small negative trend about (-0.05 % /year) exists in the total flux of cosmic rays falling on the top of the atmosphere. Also, in the energy range of $E = (10^{11} - 10^{14})$ eV the sidereal anisotropy is observed. These experimental data could be explained if supernova explosion took place in the nearby interstellar space about $t = (5 \times 10^4 - 5 \times 10^5)$ years ago at the distance $r = (30 - 200)$ parsec from the solar system.

1. INTRODUCTION

The study of cosmic ray energy spectrum, sidereal anisotropy of their fluxes, and long-term modulation processes shows several peculiarities: there is a well-known "knee" in the high energy cosmic ray spectrum where the exponent of energy spectrum is changed from $\gamma = 2.75$ at $E < 3 \times 10^{15}$ eV to $\gamma = 3.1$ at $E > 3 \times 10^{15}$ eV; the sidereal anisotropy with the amplitude of $A = 5 \times 10^{-4}$ is observed in the energy range of $E = 10^{11} - 10^{14}$ eV [1]; the spectrum of protons in TeV-energy range is more steep in comparison with the spectra of other nuclei [2]; the long-term measurements of cosmic ray fluxes show the gradual decrease (negative trend) of their values [3].

These peculiarities could be understood if supernovae explosion is supposed to occur in the nearby interstellar space. The effect of the negative trend in cosmic rays was discovered recently and we shall discuss it in detail.

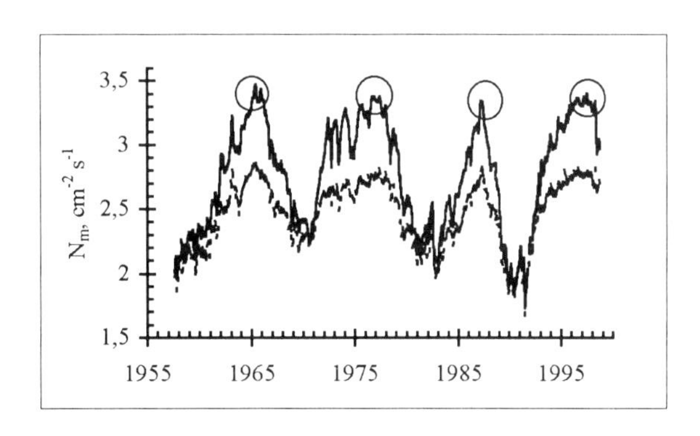

Figure 1. The time dependence of cosmic ray fluxes measured in the northern polar atmosphere (R_c=0.6 GV, solid curve) and at the mid-latitude (R_c=2.4 GV, dotted curve). The values of fluxes are given for Pfotzer maximum. The circles show the periods when maximum values of cosmic ray fluxes were observed.

2. EXPERIMENTAL DATA

The long-term cosmic ray measurements are performed on the ground level (over 60 years with ionization chambers and over 40 years with neutron monitors) and in the atmosphere (over 40 years with standard radiosondes). These observations span a wide energy range of primary particles from several hundreds of MeV (measurements in the stratosphere) to several tens of GeV (ionization chamber data). For analysis we select homogeneous sets of data. In Fig. 1 the monthly averages of maximum cosmic ray flux in the atmosphere (N_m) measured at the northern polar latitude (geomagnetic cutoff rigidity R_c=0.6 GV) and at the middle latitude (R_c=2.4 GV) from 1957 till the present time are shown. The periods of low solar activity are marked by circles. The Moscow neutron monitor data (R_c=2.4 GV) obtained during the same period are given in Fig.

PII S0920-5632(01)01265-8

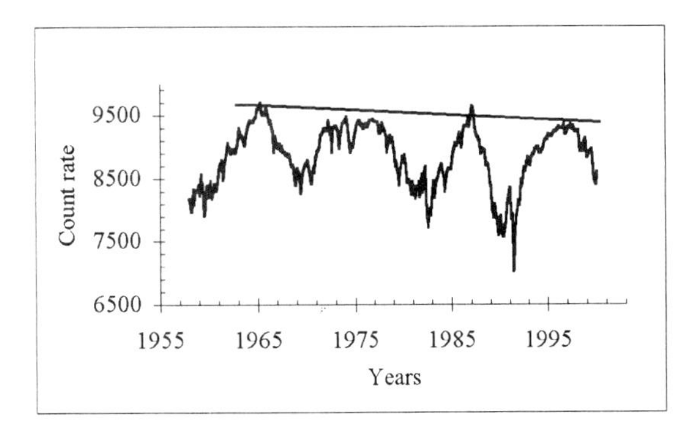

Figure 2. The cosmic ray fluxes measured by neutron monitor in Moscow (R_c=2.4 GV). The straight line calculated for maximal values of cosmic ray flux gives the negative trend $\delta = -(0.08-0.01)$ % /year.

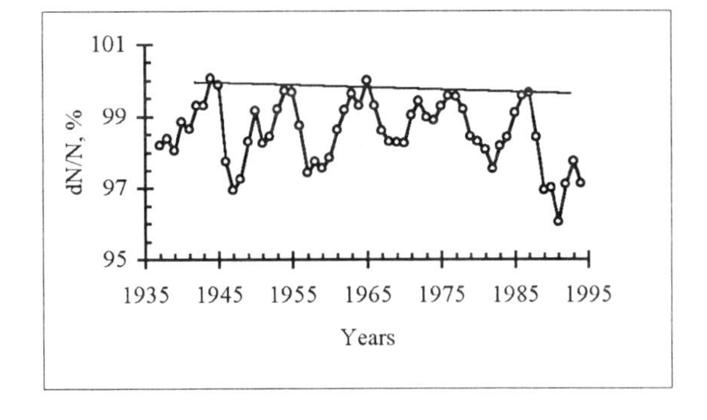

Figure 3. The changes in the ionization chamber count rate (yearly averages) relative to 1965 taken as 100 % versus time (adopted from Ahluwalia, 1997). The straight line calculated for periods of maximum values of cosmic ray fluxes (maximum values of dN/N) gives the negative trend $\delta = -(0.010 + 0.003)$ % /year.

2. The ionization chamber data obtained in the period of 1937-1993 are shown in Fig. 3 [4].

3. NEGATIVE TREND IN COSMIC RAY FLUX

Let us consider the values of cosmic ray flux in consecutive solar activity minima: 1964-1965, 1976-1977, 1986-1987, and 1996-1997. For ionization chamber data we have solar activity minimum periods in 1944-1945 and 1954-1955, additionally. As an example, in Fig. 4 the values of N_m obtained in the stratosphere in four consecutive solar activity minimum periods are shown in an enlarged scale. The maximum values of cosmic ray fluxes were observed in April-June 1965, September-October 1977, January-March 1987, and April-June 1997. The straight line passes through the maximum values of N_m. The cosmic ray flux decreases gradually from 1965 to 1997 with the rate δ=-(0.08-0.01) % /year. The data of neutron monitor and ionization chamber also show the negative trend with the rates δ=-(0.08-0.01) % /year and δ=-(0.010-0.003) % /year accordingly (see straight lines in Fig. 2 and 3). The analysis of the other neutron monitor data confirms the existence of the gradual decrease of cosmic ray flux from one solar activity minimum to another one (see Table 1).

The existence of the long-term negative trend in cosmic rays also follows from the analysis of cosmogenic isotope ^{10}Be concentration (a half-life of 1.5×10^6 years) in deep ice cores of Antarctica and Greenland and cosmogenic isotopes ^{14}C concentration (a half-life of 5730 years) in tree rings [5, 6]. These isotopes are produced in nuclear interactions of cosmic rays with nuclei of the atmosphere. The ^{10}Be and ^{14}C concentration data given in [5, 6] show the decrease of cosmic ray flux. From ^{10}Be data the value of the negative trend is δ=-(0.05-0.01) % /year in our century [5, 7] and δ=-0.02 % /year from ^{14}C data at the timescale about 25,000 years [6].

4. WHAT ARE THE REASONS OF COSMIC RAY NEGATIVE TREND?

The negative trend in cosmic ray flux observed in low solar activity periods (minimum solar activity) could arise from the respective changes in solar activity level during the periods under consideration or from the changes of cosmic ray flux

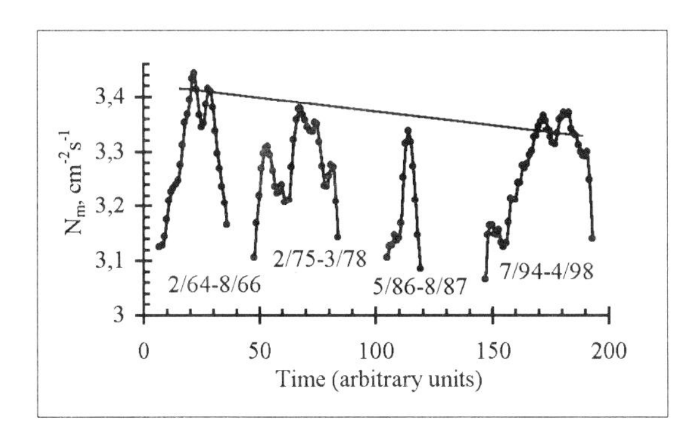

Figure 4. The 4 periods of maximum cosmic ray fluxes observed in the atmosphere are depicted on an enlarged scale (see circles in Fig. 1). The three month running values of N_m correspond to Pfotzer maximum. N_m were measured in the northern polar atmosphere (R_c= 0.6 GV). The beginning and the end of each period are given under the curves. The straight line calculated from the four maximum values of N_m by the least square method shows the decrease of N_m with the rate δ=-(0.08-0.01) % /year.

entering into our heliosphere from the nearby interstellar space.

We analyzed the long-term changes of some solar activity indices, parameters of interplanetary space, and some indices of the Earth's magnetic field in the solar activity minimum periods of 1964-1966, 1975-1978, 1986-1987, and 1995-1997 [8].

From these analyses it follows that the increases of solar activity level, interplanetary space parameters, or the Earth's magnetic field disturbances are not observed in solar activity minima periods from 1964 to 1997. Also there is no positive trend in the number of sunspots in the periods of solar activity minima from 1900 to 1997: $\delta R_z = -(0.001 - 0.002)\%/\mathrm{year}$. Thus, it is unlikely that our Sun or interplanetary space were responsible for the long-term negative trend observed in cosmic rays. Recently, the increase of aa-geomagnetic index in our century was found and the explanation of this effect was the following: our sun becomes more active and during the past 100 years the total solar magnetic flux leaving the sun was increased in 2.3 times [9]. But there is another explanation of the effect observed: the growth of aa-index is due to the gradual decrease of the Earth's dipole magnetic moment [10].

Table 1
The value of trend in cosmic ray flux δ according to neutron monitor data, ionization chamber data, and stratospheric measurements in Pfotzer maximum (str.)

Site	R_c, GV	δ % / year
Apatity	0.6	-0.08 ±0.01
Moscow	2.4	-0.08 ±0.01
Climax	3.0	-0.04 ±0.01
Ioniz. chamber	1.6-2.2	-0.01 ±0.003
Murmansk, str.	0.6	-0.08 ±0.01
Moscow, str.	2.4	-0.08 ±0.01

The another cause of the negative trend in cosmic rays could be the decrease of particle flux in the nearby interstellar space (the decrease of the particle flux on the modulation region boundary). Such decrease could take place if a explosion of a supernova had occurred not so far from the solar system (at the distance less than several hundreds of parsecs) [11-13].

The simple model with a point-like source of cosmic rays can explain the negative trend δ and siderial anisotropy A in cosmic rays. The solution of the spherical symmetric diffusion equation with convection of particles is

$$n(r,t) = \frac{N}{(2\pi Dt)^{3/2}} \times exp\left[-\frac{(r-ut)^2}{4Dt}\right] \quad (1)$$

where n is the charged particle concentration, N is the number of particle accelerated in the supernova explosion, t is the time elapsed after supernova explosion, r is the distance to the supernova, u is the average shock wave velocity, D is the diffusion coefficient [14]. From the expression (1) the values of the trend and anisotropy are defined as:

$$\delta = \frac{1}{n}\frac{dn}{dt} = -\frac{3}{2} \times \frac{1}{t} + \frac{r^2 - (ut)^2}{4Dt^2} \quad (2)$$

$$A = \frac{3D}{c} \times \frac{1}{n} \times \frac{dn}{dr} = \frac{3r}{2ct} \quad (3)$$

where c is the particle velocity [12, 15]. We take into account that there are two stages of shock wave propagation. During the first stage of free propagation the shock wave has a high velocity, $u = (1-2) \times 10^4$ km/s. The second stage corresponds to adiabatic propagation of a shock wave with subsonic velocity [11]. The results of calculations of time t and distance r of the possible supernova explosion and sidereal anisotropy A were made for two values of δ: $\delta = -0.01\%$/year and $\delta = -0.07\%$/year. For $\delta = -0.01\%$/year and $t = 3 \times 10^4 - 10^5$ years the typical values of distance and anisotropy are $r = 90 - 300$ pc and $A = 1.4 \times 10^{-2}$, for $\delta = -0.07\%$/yr and $t = 3 \times 10^4 - 10^5$ years these values are $r = 30-60$ pc and $A = 4.5 \times 10^{-3} - 2.1 \times 10^{-3}$. In the nearby interstellar space there are several stellar objects which could be sources of cosmic rays observed in the solar system: Geminga, Loop 1, Monogem, Vela, Spur.

Erlykin et al. [16] also came to the conclusion on the nearby supernova explosion on a base of the analysis of the "knee" peculiarities in the cosmic ray spectrum at $E = 3 \times 10^{15}$ eV. The possible softening of the proton spectrum in TeV-energy range (increase of the proton energy spectrum exponent from $\gamma = 2.65$ for protons with $E <1$ TeV to $\gamma = 3.1$ for protons with $E >1$ TeV) may also be attributed to the existence of a cosmic ray source in nearby interstellar space [3, 17].

5. CONCLUSION

To provide an explanation for the negative trend observed in cosmic rays we suggest that the nearby supernova explosion has occurred at the distance 30-150 parsec about $10^4 - 5 \times 10^5$ years ago. The distance and explosion time were evaluated in the scope of spherical symmetric model with a point-like source. In such a model the observed sidereal anisotropy of cosmic ray particles in the energy interval $E = 10^{11} - 10^{14}$ eV may be also explained. There are several celestial objects in the nearby interstellar space which could be responsible for such an explosion. This work was supported by the Russian Foundation for Basic Research (grant 98-02-16420).

REFERENCES

1. Munakata K. et.al., 1997, Proc. 26th ICRC, Durban, **2**, p. 153.
2. Grigorov N.L., 1990, J. Nuclear Phys., **51**, p. 157 (in Russian).
3. Stozhkov Y.I., 2000, J. Geophys. Res., **105**, p. 9
4. Ahluwalia H.S, 1997, J. Geoph. Res., **102**, p. 24,229.
5. Beer J. A. et al., 1990, Nature, **347**, p. 164.
6. Dergachev V.A., 1999, Izvestiya Rossiiskoy Akademii Nauk, ser. phys., **12** (in Russian).
7. Blinov A.V., 1997, Doctor dissertation, St.-Petersburg, Ioffe Physical-Technical Institute (in Russian).
8. Solar Geophysical Data, 1964 -1998.
9. Lockwood M. et al., 1999, Nature, **399**, p. 437.
10. Stozhkov Y. et al., 2000, J. Geophys. Res. (in press)
11. Amosov A.E. et al., 1991, Izvestiya Rossiiskoy Akademii Nauk, ser. phys., **55**, p. 2037 (in Russian).
12. Johnson P.A., 1993, Proc. 23rd ICRC, **2**, p. 362.
13. Sonett C.P. et al., 1997, Proc. 25th ICRC, **4**, p. 441.
14. Dorman L.I. and L.I. Miroshnichenko, 1964, Izvestiya Akademii Nauk SSSR, ser. phys., **28**,p. 678 (in Russian)
15. Hayakawa S., 1969, Cosmic ray physics, Monographs and Text in Physics and Astronomy edited by R.E. Mashak, New York, XXII.
16. Erlykin A.D. et al., 1998, Astropart. Phys., **8**, p. 283.
17. Watson A.A., 1997, Proc. 25th ICRC, **8**, p. 257.

ELSEVIER

Nuclear Physics B (Proc. Suppl.) 97 (2001) 211–214

www.elsevier.nl/locate/npe

Observation of γ-sources using a new reconstruction technique in the CLUE experiment

B. Bartoli[a], D. Bastieri[b], C. Bigongiari[d], R. Biral[c], M.A. Ciocci[d], M. Cresti[d], V. Dokoutchaeva[e], D. Kartashov[a], F. Liello[f], N. Malakhov[a], M. Mariotti[d], G. Marsella[g], A. Menzione[a], R. Paoletti[f], L. Peruzzo[d], A. Piccioli[f], R. Pegna[e], F. Rosso[e], A. Saggion[d], G. Sartori[d], C. Sbarra[d], A. Scribano[f], E. Smogailov[a], A. Stamerra[h], N. Turini[f]

[a]Dipartimento di Fisica Università di Napoli and INFN,sezione di Napoli, Italy

[b]Dipartimento di Fisica Università di Padova and INFN, sezione di Padova, Italy

[c]Dipartimento di Fisica Università di Pisa and INFN, sezione di Pisa, Italy

[d]Dipartimento di Fisica Università di Siena and INFN, sezione di Pisa, Italy

[e]INFN, sezione di Pisa, Italy

[f]Dipartimento di Fisica Università di Trieste and INFN, sezione di Trieste, Italy

[g]Dipartimento di Fisica Università di Lecce and INFN, sezione di Lecce, Italy

[h]Dipartimento di Fisica Università di Torino e INFN, sezione di Torino, Italy

The CLUE experiment, located in La Palma island at 2200 m a.s.l., is an array of 3×3 telescope, detecting the UV (190 – 230 nm) Čerenkov light produced by atmospheric showers. Since atmospheric absorption in the UV range is higher than in the visible range, CLUE cannot apply existing algorithms normally used in IACT experiments to determine primary cosmic ray direction. In this paper we present a new method developed by CLUE. The algorithm performances were evaluated using simulated showers. Using the new technique, preliminary results of last two years observational campaigns on the *Crab Nebula* and on *Markarian 421* are presented, showing a clear signal on both sources. The CLUE experiment collected also data with the telescopes aiming directly at the Moon: we expect improvements also on the *Moon Shadow* measurement adopting the new method.

1. Introduction

In the past, CLUE [1,2] used a maximum likelihood method approach [3] to determine primary cosmic ray direction comparing observed distributions of Čerenkov photons detected with simulated ones. The best angular resolution ($\Delta\alpha$ or $\Delta\beta$, see paragraph 2 for definitions) on shower direction, was 0.8° for energy $E > 2\ TeV$ and the reconstructed angle was affected by a large systematic error for off-axis showers (1° systematic error for showers with 2° off-axis angle). Recently CLUE developed a new reconstruction algorithm, called Thrust, adapting a method [4] widely used in accelerator physics experiments to determine particle jet direction and collimation.

2. Thrust method

The Thrust method relies on shower symmetry properties, assuming that the momentum of Čerenkov photons is distributed with axial symmetry around the primary direction. This method determines the shower direction estimating the unit vector which maximizes the overall longitudinal momentum of photons. Experimentally, an UV photon direction is associated, through a proper backtracking onto the parabolic mirror, to each charge cluster found in the CLUE chambers. We introduce a unit vector $\hat{n}_T$, called *Thrust axis*, and a scalar T, called *Thrust*, defined

PII S0920-5632(01)01266-X

by:

$$T = \frac{\sum_{k=1}^{N_c} Q_k \mid \hat{r}_k \cdot \hat{n}_T \mid}{\sum_{k=1}^{N_c} Q_k} \quad (1)$$

where N_c is the number of charge clusters, $\hat{r}_k$ is the unit vector associated to the kth cluster and Q_k its charge. The $\hat{n}_T$ which maximizes T is the Thrust estimate of the shower axis (primary cosmic ray direction). The unit vector components of each UV photon are given by:

$$\begin{cases} r_k^x = \alpha_k = \frac{X_k}{F} \\ r_k^y = \beta_k = \frac{Y_k}{F} \\ r_k^z = \gamma_k = \sqrt{1 - {\alpha_k}^2 - {\beta_k}^2} \end{cases}$$

where X_k and Y_k are the kth cluster centroid coordinates and F is the mirror focal length. The *Thrust axis* could be expressed as a function of zenith (θ) and azimuth (ϕ) angles:

$$\begin{aligned} \hat{n}_T &= (\alpha, \beta, \gamma) \\ &= (\sin(\theta)\cos(\phi), \sin(\theta)\sin(\phi), \cos(\theta)) \quad (2) \end{aligned}$$

Since our detector acceptance in α (or β) is $\pm$ 4°, it follows that $\theta \simeq \sqrt{\alpha^2 + \beta^2}$.

3. Test and performances

The method has been applied and tested on simulated showers samples after the full detector simulation. The MC showers were generated everywhere with an energy between 1 and 10 TeV, sampled according to primary cosmic ray spectrum (spectral index -2.7).

3.1. Fixed shower impact point

Firstly we evaluated the performances of this new method with respect to the old one, using proton and γ shower samples (11000 showers) where both samples were generated using *CORSIKA* [5] with a zenith angle between 0° and 4° (mirror axes are assumed vertical) and the shower impact point was fixed in the CLUE array center. The α and β angular resolution for triggered [1] vertical showers is 0.7° (old one was 0.8°)

[1] We required at least one cluster in three chambers.

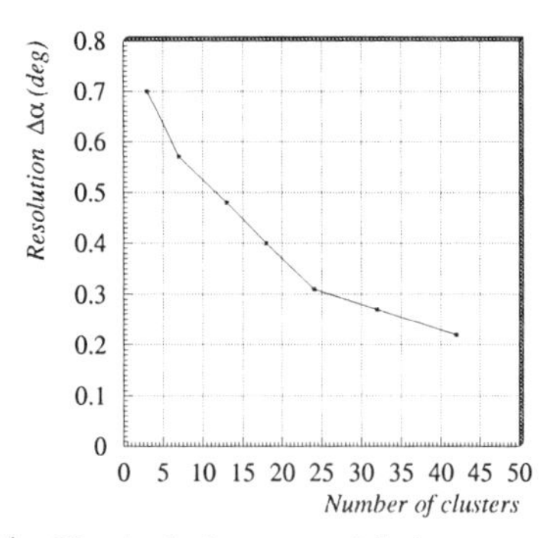

Figure 1. Vertical showers with impact point in the CLUE array center simulation: angular resolution on the reconstructed α angle versus the number of cluster in the event.

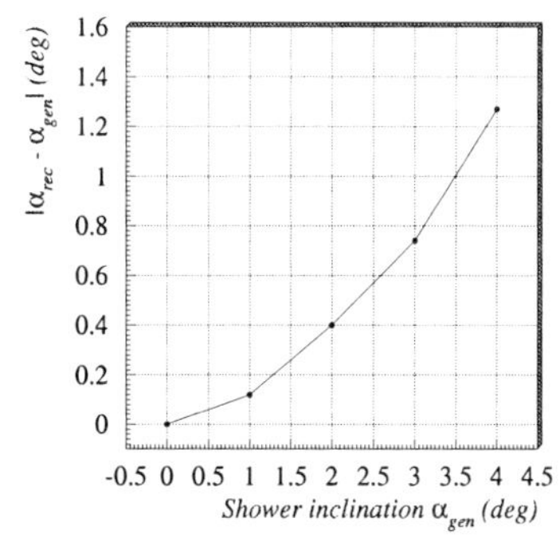

Figure 2. Showers with fixed impact point simulation: systematic on the reconstructed α angle (α_{gen}-α_{rec}) versus α angle (α_{gen}).

and it improves increasing the number of clusters, as expected (see Figure 1 as example). Furthermore, the Thrust angle α[2] is affected by a systematic error for off-axis showers with a much weaker dependence on off-axis angle (see Figure 2) than the old likelihood method: only 0.4° for showers with 2° off-axis angle. Those improved results on systematic and on resolution were found for showers originated both by protons and gammas.

3.2. Random shower impact point

To evaluate effects due to the impact parameter, we generated vertical showers with impact point randomized on a 300 m × 300 m square centered on the array center (250000 samples) using *HDS* [6]. From simulation, the trigger request selects only events impinging on a square of 120 m × 120 m. In that case, the resolution on

[2] A very similar result is obtained for the β angle.

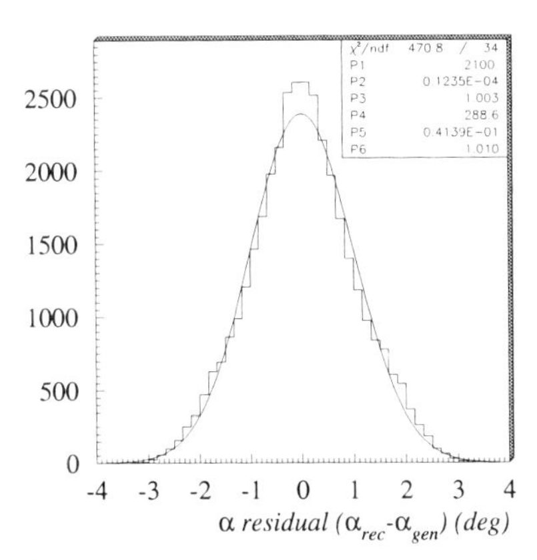

Figure 3. Showers with random impact point simulation: α residuals (α_{gen}-α_{rec}) and fitting curve (superimposed line).

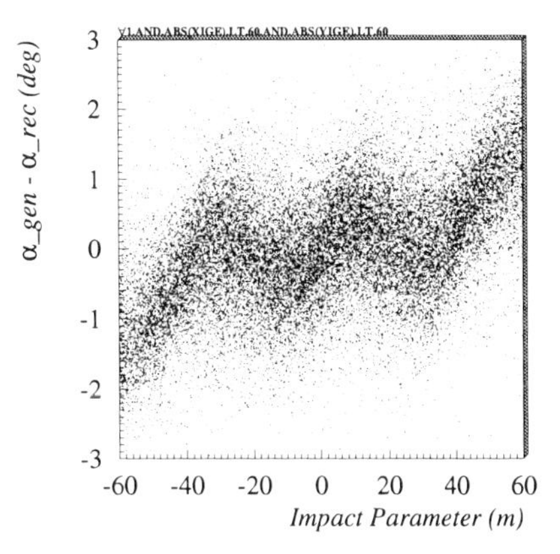

Figure 4. Showers with random impact point simulation: α residuals (α_{gen}-α_{rec}) versus the reconstructed impact parameter x coordinate.

α (β) is worse (Figure 3) then previous one (Figure 1). But we have found a clear correlation between α (β) residuals and the $x(y)$ coordinate of the reconstructed impact point (Figure 4). Using the correlation, we can apply a correction on the measured Thrust angles (α and β): a big improvement on the Thrust angular resolution on the non zero impact parameter showers is obtained (Figure 5).

The impact point coordinates of a shower were measured using a standard technique employed in high energy physics to measure particle lifetime. In Figure 6 are shown, as example, the residuals for the reconstructed impact parameter x coordinate[3].

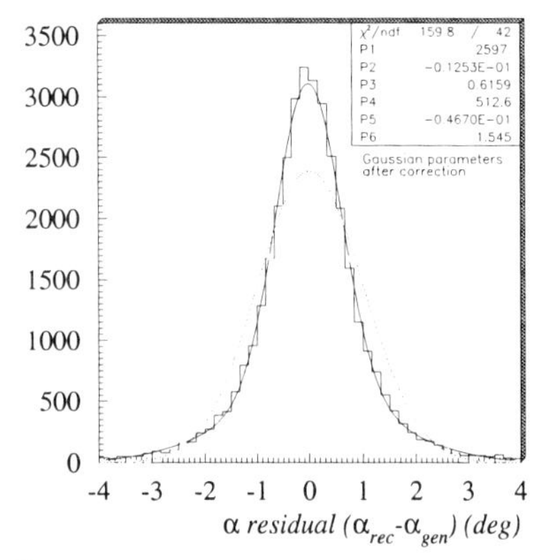

Figure 5. Showers with random impact point simulation: α angle residuals after corrections and corresponding fitting curve after (continuous line) and before corrections (dashed line).

4. Data analysis on sources

The data collection periods and their details are resumed in Table 1. Ghost runs are off-source runs taken tracing back, in right ascension, the source path. Before employing the reconstruction strategy outlined in section 2, the data underwent the standard clean-up [3] to remove odd behaviours caused by electronic noise and faults during data collection. The offline analysis is applied to all events that triggered with at least nine clusters.

A clear signal is visible from both sources (Figure 7 and Figure 8), even if no correction for non zero shower impact parameter was applied and

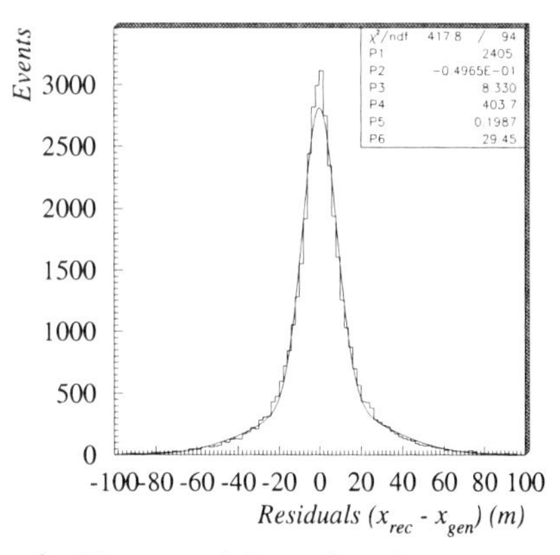

Figure 6. Showers with random impact point simulation: impact parameter x coordinate resolution.

[3]We obtain the same result for the y coordinate.

Table 1
Data acquisition information on used data. Beside real sources, data refer also to *ghost* sources and Moon.

	DAQ period	Entries	Time
Moon	Feb 98 - Apr 99	124580	138 *hr*
Markarian 421	Apr-May 98, Mar 2000	63000	86 *hr*
Crab Nebula	Feb-Mar 98, Feb 99, Nov 99-Mar 2000	68000	90 *hr*
*Mkn*421 - Ghost	Apr-May 98, Mar 2000	18000	30 *hr*
Crab - Ghost	Feb-Mar 98, Feb 99, Nov 99-Mar 2000	27000	42 *hr*

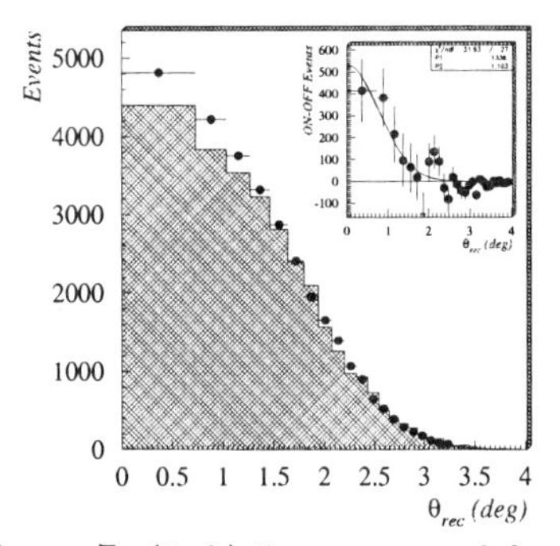

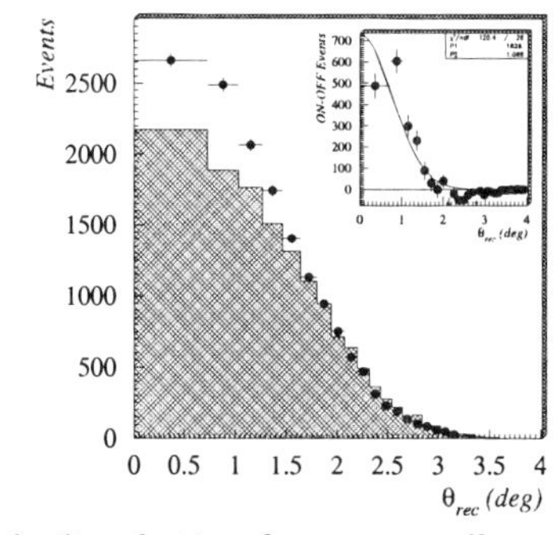

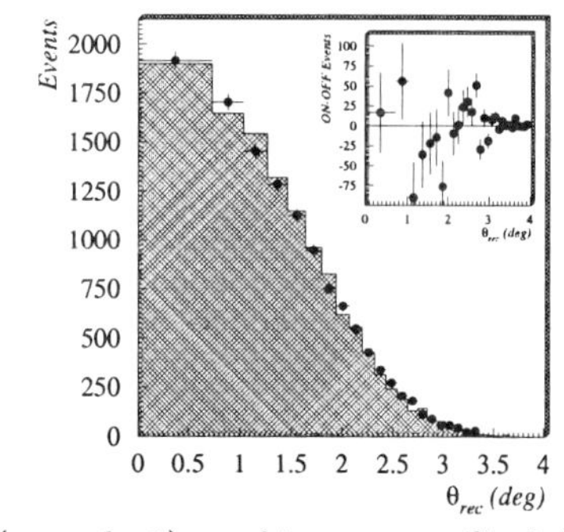

Figure 7. (Left) Reconstructed θ angle distribution for events collected (see tab. 1) tracking source *Crab* (dots) and for events collected tracking a region of the sky where no gamma ray source is known (filled area).
Figure 8. (Center) Reconstructed θ angle distribution for events collected (see Table 1) tracking source *Mrk*421 dots, and for events collected tracking a region of the sky where no gamma ray source is known, filled area.
Figure 9. (Right) Reconstructed θ angle distribution for events collected tracking source *Mrk*421 (dots) in a period where the source was inactive and for events collected tracking a region of the sky where no gamma ray source is known (filled area). In all figures the distributions are normalized in the region $\theta > 1.3°$. In all little squares the difference between on source and off source distributions is plotted.

only a small fraction of available data was used. Moreover, in a period where *Mrk*421 was not active, there is no signal from the source (Figure 9).

5. Conclusion

A new algorithm to estimate atmospheric shower direction is described and tested on simulated data. It does not depend on MC simulation and its performance is an improvement on the old likelihood method.
The Thrust method was applied succesfully on a sample of the data collected tracking the AGN *Mrk*421 and *Crab*. In a near future, we will apply the method and the correction for non zero shower impact parameter on the overall data collected and on the measurement of the *Moon shadow*. About the latter measurement, we are confident to improve the result obtained from the old method [3], by the new one.

REFERENCES

1. D. Alexandreas et al., *NIM* **A 409** (1998) 488-491
2. D. Alexandreas et al., *NIM* **A 409** (1998) 679-681
3. D. Bastieri et al.,The CLUE experiment running with 8 telescopes; observation of gamma sources and runs on Moon, in: B.L.Dingus et al., eds., *Proceedings of "GeV-TeV Gamma Ray Astrophysics Workshop"*. (2000) 436–440
4. S. Brandt et Al., *Phys. Lett.* **12** (1959) 57.
5. J.N. Knapp, D. Heck, Exstensive Air Shower Simulation with CORSIKA: A User's Guide (Version 5.61) (1998).
6. M.P. Kertzman and G.H. Sembronski, Computer Simulation Methods for investigating the Detection Characteristics of TeV air Čerenkov Telescope; *NIM* **A 343**(1994) 629–643.

ELSEVIER

Nuclear Physics B (Proc. Suppl.) 97 (2001) 215–218

TeV gamma-ray emission from point sources: galactic and extragalactic

V.G. Sinitsyna, T.P. Arsov, A.Y. Alaverdyan, I.A. Ivanov, S.I. Nikolsky, F.M. Musin, R.M. Mirzafatihov, G.F. Platonov, V.Y. Sinitsyna, A.N. Galitskov, I.V. Oblakov[a]

[a]P.N. Lebedev Physical Institute, Russian Academy of Science,
Leninsky pr., 53, Moscow 117924, Russia

Five years of TeV observations from two BL Lac objects Markarian 421, Markarian 501, Seyfert Galactic NGC and gamma-ray emission from Cygnus X-3 and supernova remnants Crab Nebula by SHALON-ALATOO observatory are presented. The observatory SHALON-ALATOO has just announced the TeV detection of NGC 1275 gamma-rays. Results obtained with the Mirror Cherenkov telescope SHALON-1 (mirror area more than 11.2 m^2) point out that, in view of the fact that at present a more intensive total flux of gamma-quanta with energies higher than 10^{12} eV is observed from Extragalactic sources in comparison with a near located source Crab Nebula (Galactic source), it is possible to assume that the major part of cosmic rays with E>10^{13} eV, observable in Earth space, also comes from Extragalaxy. The observable energy spectra of gamma - quanta in the energy interval 10^{12} – 10^{13} eV, from both close sources in our galaxy, and quasars, blazars and active galactic centers do not contradict too uniform proportionality $\sim E^{-2.4} \pm 0.1dE$. If one considers that this spectrum presents an energy spectrum of protons and nuclei of cosmic rays in local areas of their acceleration, then there is a problem in what processes a uniform spectrum of cosmic rays in the energy interval from $\sim 10^{11}$ eV to $> 3 \times 10^{19}$ eV is generated.

Research of local sources of gamma-quanta of super-high energies by any methods, including methods by mirror Cherenkov telescopes (Table 1) concerns rather a delicate problem of the nature of cosmic rays and accordingly the role of our Galaxy and Extragalaxy in their generation. The opinion, which may not have enough grounds, is widely distributed, that the cosmic rays before the break in their spectrum have a galactic origin (energy $<10^{15}$-10^{16}eV) and only at energies $> 10^{16}$eV the role of the Extragalaxy probably grows.

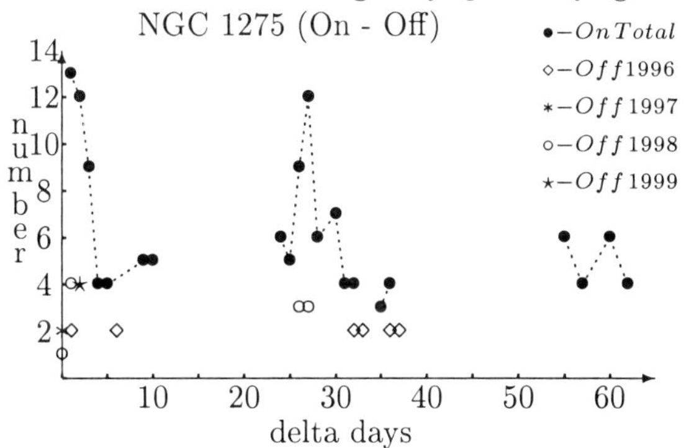

Few experimental data of gamma - astronomy of super high energies (gamma-astronomy is behind cosmic rays in the interval of relict cut off energies according to observed gamma-quanta) completely undermine mainly galactic origin of cosmic rays.

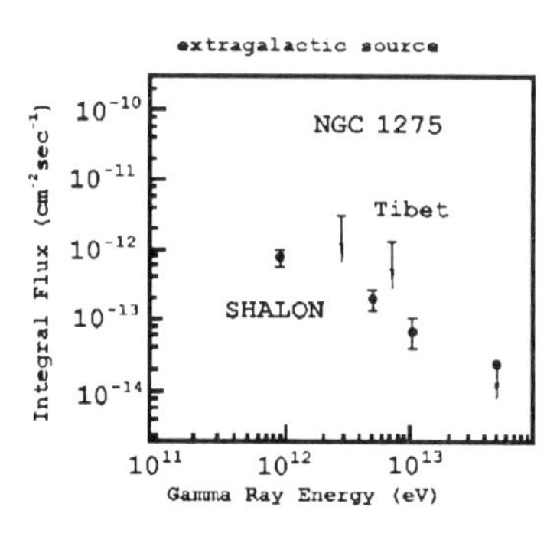

Figure 1. The spectrum of the gamma-radiation from AGN NGC 1275 and NGC 1275 On-Off diagram from 1996-1999.

PII S0920-5632(01)01267-1

Table 1
Experiments reported to have detected Markarian 421 and Markarian 501 (FIG 3,4)

Experiment, Altitude	Site, Country	Area m^2	Range of measurement	Full angle image	res.(^{o})• NPixel	Epoch beginning
Whipple,	Arizona,USA,UK-Irel	70	250GeV→ 12TeV		0.3^o•37	1984
2300m	$31^o41/N$ $110^o53/W$		5σ	3°	0.25^o•109	1993
				3°	0.25^o•151	1996
				4°	0.21^o•331	1997
SHALON,	ALATOO, Russia	11.2	800GeV→ 50TeV	8°	0.6^o•144	1992
3338 m	42^oN 75^oE					
HEGRA,	Canary Islands	8.5	700GeV→>10TeV	4.3^o	0.25^o•271	1994
2240 m	28.75^oN 17.89^oW	5	1.5TeV→>15TeV	3.25^o	0.25^o•127	
CAT,	French Pyrennees,	17,7	220GeV→ 10TeV	3.1^o	0.12^o•546	1996
1650 m	42^oN 2^oE					
TACTIC,	Mt.Abu, India,	9.5	700GeV→ 10TeV	2.6^o	0.31^o•81	1997
1300 m	24° 39/ N 72° 47/ E					
Telescope	Mt. Cedar,	6	600GeV→ 10TeV	4°	0.25^o•256	1998
Array,	Utah, U.S.A.,	7*6	3σ			
1600 m	40.33^o N 113.02^oW					
Tibet II EAS	Yangbajing,Tibet,	$3{,}7{*}10^4$	$>3\sigma$ above 10TeV	Angle	resolution:	
Array, 4300 m	China, 30.1^o N 90.53^o			1° at E_γ	=7TeV	
CASA-MIA,	Dugway, U.S.A.,	$23{*}10^4$	>45TeV	Angle	resolution:	
1450 m	40.2^o N 112.8^o W			0.15^o at $E\gamma$	=70TeV	

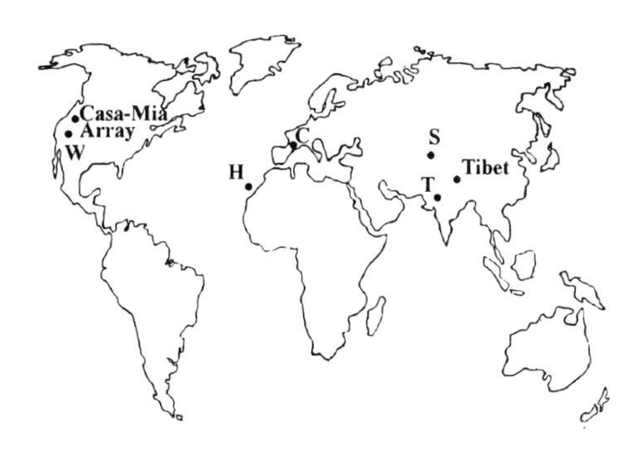

All extragalactic sources of gamma - quanta with the energies higher than 10^{12} eV detected at the present have the radiation intensity $10^6 - 10^8$ times larger than Crab Nebula. But we know that among all galactic sources the Crab Nebula is the brightest by an observable flux of gamma - quanta, and that is definitely connected with its nearness to the Solar system.

An out - of - atmosphere gamma - astronomy has detected a lot of sources of gamma - quanta with energies 10^8-10^9 eV, which are various by their nature, and in the researches of extensive air showers with energies of primary particles 10^{14}-10^{15}eV the flux of "no muons" or "no hadrons" showers (Table 2) was estimated. This leads us to suggest, that such showers are generated by primary gamma - quanta (the intensity of such gamma - quanta is 10^{-3}-10^{-4} of the flux of extensive showers generated by primary protons and nuclei of cosmic rays). Still it is unknown to what extend the observable gamma - quanta flux (10^{-3}-10^{-4} of a flux of protons and nuclei of cosmic rays) is a diffuse flux and a summary flux of quite definite sources, because of experimental accuracy of an estimation of coordinates of gamma - quanta of arrival direction that does not allow one to unambiguously connect such a shower with the coordinates of probable sources.

The extragalactic source NGC 1275 is observed for the first time by the telescope SHALON (E>

Table 2
The catalog of observations by SHALON telescope in TeV energies since 1994

Source	Type	Flux ($cm^{-2}s^{-1}$ at E>0.8TeV)	Distance	
	Galactic		Kpc	
Crab Nebula	Plerion	$(1.10\pm0.30)\bullet10^{-12}$	2.0	
Cygnus X-3	Binary	$(4.20\pm0.70)\bullet10^{-13}$	1.1	
Geminga	Supernova Remnant	$(0.48\pm0.17)\bullet10^{-12}$	0.25	
Tycho Brahe	Supernova Remnant	$(1.89\pm0.9)\bullet10^{-13}$	2.0 - 5.1	
	Extragalactic		mpc	z
Mkn 421	Blazar	$(0.63\pm0.14)\bullet10^{-12}$	124	0.031
Mkn 501	Blazar	$(0.86\pm0.13)\bullet10^{-12}$	135	0.034
NGC 1275	Seyfert Galaxy	$(0.78\pm0.13)\bullet10^{-12}$	71	0.013
3c454.3	Quasar	$(0.43\pm0.17)\bullet10^{-12}$	4685	0.859
1739+522	AGN	$(0.47\pm0.18)\bullet10^{-12}$	7500	1.375

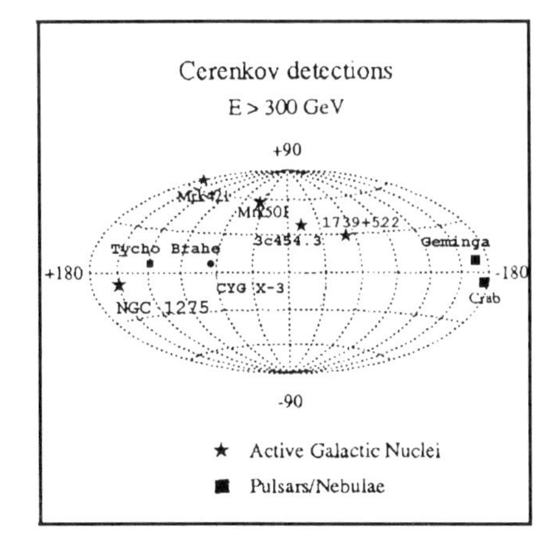

Figure 2. Sky map of gamma-quanta point sources in galactic coordinates that observed by gamma-telescope SHALON

0.8 TeV) with the flux $(0.78\pm0.13)10^{-12}cm^{-2}s^{-1}$, confirmed in observation on the Tibet [1-3]. The energy spectrum of the galactic and metagalactic sources (Fig. 1, 3, 4) at the energy range from 1TeV up to 50TeV is observed and the time analysis is carried out (Fig. 1, 4).

The data **On** and **Off** of extragalactic sources NGC 1275, Markarian 421, Markarian 501 and galactic sources Crab Nebula and Cygnus X-3 were the object of a time analysis. At data **On** in all sources the groups of peaks of the common average 5-10 days width were detected in 24-26 days (Fig. 1, 4). It is possible to interpret these peaks as periods connected with that the observations are carried out only in moonless nights.

As one can see from the presented data, the events **Off** are not more than 10%-15% of the events **On**. It means that the contribution of protons of cosmic rays in observable gamma - quanta with energy more than 0.8TeV does not exceed 10%-15%.

The sources and observation of gamma-quanta with energy $E \geq 1$ TeV of flux $F(E)dE \geq 10^{-12}dEcm^{-2}sec^{-1}$ are carried out by means of the telescope SHALON. The difference of distances to metagalactic objects and galactic objects at the above mentioned identical gamma-quanta flux with energy more than 1 TeV means that intensity of researched metagalactic sources Mkn501, Mkn421, and detected SHALON NGC1275, 3c454.3 and 1739+522 is larger than the intensity of observed fluxes of gamma-quanta with $E \geq 1$ TeV from Tycho Brahe and Geminga supernova remnants in our Galaxy.

The gamma-quanta energy spectrum in energy interval from 10^{12} to 10^{13} eV, observed with the help of mirror telescopes by Cherenkov radiation of electron-photon cascades in atmosphere from both supernova remnants in our galaxy and from various metagalactic objects (quasars, blazars, active galactic nuclei) can be presented in an

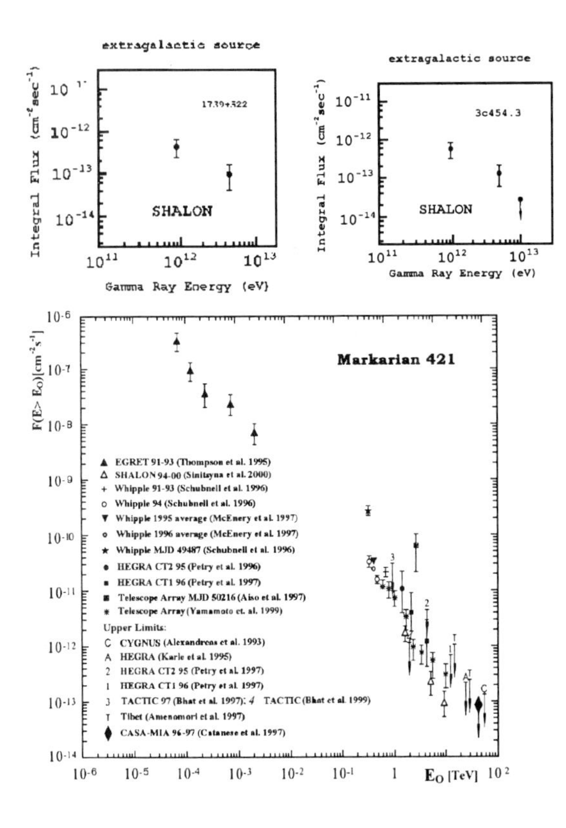

Figure 3. The spectrum of the gamma-radiation from AGN Markarian 421

uniform form: $F(E)dE \sim E^{-2.4\pm0.1}dE$. Thus, the difference of energy and time scales of the above listed gamma-quanta sources does not allow to choose some identical increase of proton and nuclei spectrum parameter with the purpose to obtain an uniform cosmic rays spectrum $F(E)dE \sim E^{-2.7\pm0.02}dE$ in the whole observable energy interval from iron nuclei relativistic energy $\sim 10^{11}$ eV up to resonant nuclei photosplitting at energy $> 3 \times 10^{19}$ in collisions with photons of relict radiation.

REFERENCES

1. Nikolsky S. I., Sinitsyna V. G., VANT, TFE(1331), (1987) 30. Proc. Workshop VHE Crimea, (1989) 11.
2. Sinitsyna V. G., Proc. Detector- VI, Snowbird, (1999) 205,293.
3. Amenomori M., et.al., 26ICRC, 3(1999)418.

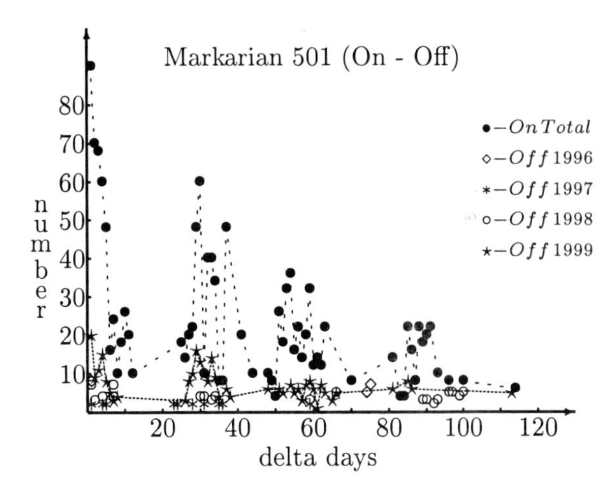

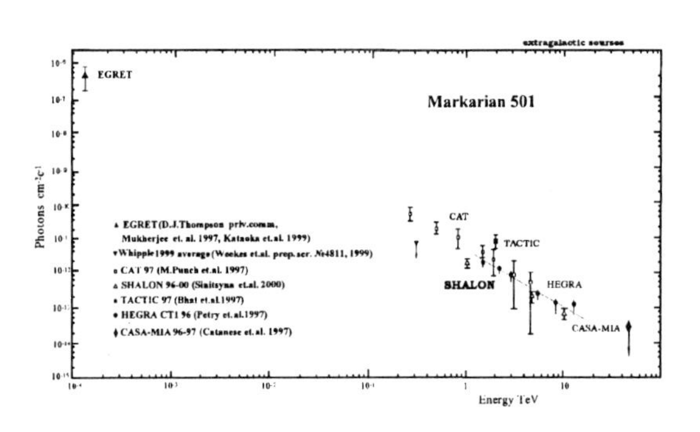

Figure 4. The spectrum of the gamma-radiation of AGN Markarian 501 and Markarian 501 **On-Off** diagram from 1996-1999.

ELSEVIER

Nuclear Physics B (Proc. Suppl.) 97 (2001) 219–222

www.elsevier.nl/locate/npe

Evidence of TeV gamma-ray radiation in supernova remnant Cygnus X-3

V.G. Sinitsyna, T.P. Arsov, A.Y. Alaverdyan, I.A. Ivanov, S.I. Nikolsky, F.M. Musin, R.M. Mirzafatihov, G.F. Platonov, V.Y. Sinitsyna, A.N. Galitskov, I.V. Oblakov [a] J.N. Capdevielle [b]

[a]P.N. Lebedev Physical Institute, Russian Academy of Science, Leninsky pr., 53, Moscow 117924, Russia

[b]Laboratoire de Physique Corpusculaire, Collège de France, F 75231 Paris Cedex 05, France

Since 1994 the telescope SHALON, at SHALON-ALATOO mountain observatory (height 3338m) detected very high energy gamma-rays from galactic objects Crab Nebula, Cygnus X-3, Geminga and Tycho Brahe. Time analysis shows that the contribution of protons of cosmic rays in observed gamma-quanta with energies higher than 0.8 TeV from the point sources of very high energy gamma-quanta does not exceed 10% – 15% (Fig. 6). The fluxes at energies above 0.8 TeV observed from the Crab Nebula are $(1.10 \pm 0.30) \times 10^{-12}$ $cm^{-2}s^{-1}$, from Cygnus X-3 are $(4.20 \pm 0.70) \times 10^{-13} cm^{-2}s^{-1}$, from Geminga are $(0.48 \pm 0.17) \times 10^{-12} cm^{-2}s^{-1}$ and from Tycho Brahe are $(1.89 \pm 0.90) \times 10^{-13} cm^{-2}s^{-1}$ (Fig. 5, 6, 7). Observed gamma-quanta fluxes from detected galactic and extragalactic sources are approximately equal (this is connected to the fact that the best gamma-telescopes of today are very flexible to the flux intensity). That means that for distances differing by a factor of 10^4, the radiating power of extragalactic objects is 10^8 higher. Taking into consideration a limited number of sources in our Galaxy in comparison to Methagalaxy, one can propose that cosmic rays of energies higher than 10^{12}eV are mainly of an extragalactic origin.

More than thirty years ago the project of the mirror Cherenkov telescope SHALON (Sinitsyna, 1987) was suggested and the first observations were started in 1991 at the ALATOO mountain observatory (Sinitsyna, 1992-1999) [1,2]. A distinctive property of the telescope is a large full angle due to a relatively large size of photomultipliers matrix (Fig. 1). This allows us to detect extensive air showers coming at distances up to 120 m from an optical axis of the telescope, increasing the statistics from the sources of gamma-quanta of very high energies. In addition, such a large full angle of an image matrix allows us to research an isotropic background of extensive air showers from charged particles of cosmic rays simultaneously with the observation of local sources of gamma-quanta at the same optical characteristics of atmosphere. This is particularly important because in our research of gamma-sources, the extensive air showers generated by gamma-quanta are selected not only according to the exceeding flux of showers in a small angle, but also according to the differences of the evaluation in the atmospheric depth of electron-photon cascades generated by protons and by nuclei of cosmic rays.

The Cherenkov imaging telescope SHALON - 1, equipped with a very high definition camera (144 pixels, full angle 8°), obtains data since 1992 at the height 3338m. We will discuss some results of observations of gamma-ray sources pointed out and methods of selections of gamma-rays and protons. The selection of showers produced by gamma-quanta from a background of showers produced by protons (Fig. 6) [1,2] is made according to the following criteria: 1) $\alpha < 20°$; 2) length/width>1.6 for γ; 3) relation of Cherenkov light intensity in pixels with maximum light to the light in eight pixels around it is $int_0 > 0.6$; 4) relation of Cherenkov light intensity in pixel with maximum light to light intensity in all pixels except nine in the center is $int_1 > 0.8$; 5) distance is < 3.5 pixels.

On figure 3 the experimental distribution of image parameters for proton and gamma showers data obtained at the telescope SHALON is shown. On the left, gamma-quanta (250) from

0920-5632/01/$ – see front matter
PII S0920-5632(01)01268-3

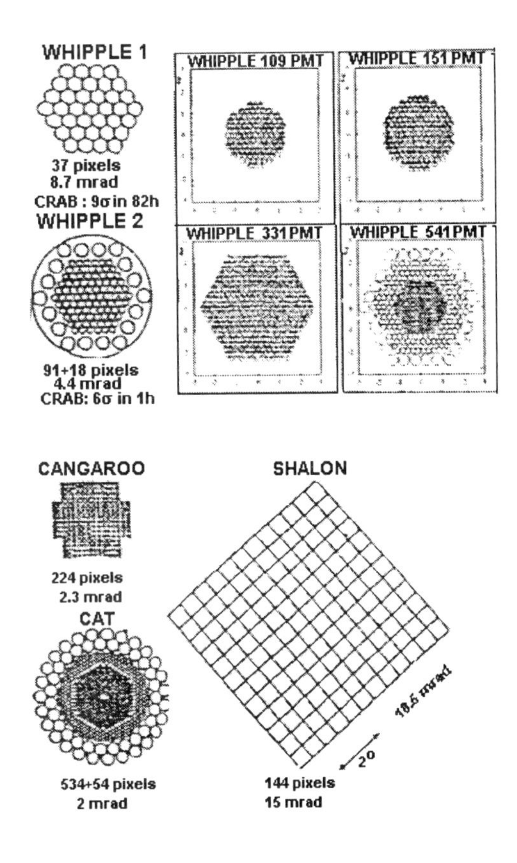

Figure 1. Pixel distribution at the focal plan of the 10m reflector: top left: 109 pixels (1993-1996); top right: 151 pixels (Dec.,1996); 331 pixels (Oct., 1997); 541 pixels (planned) [2,3]. Below: evolution of photomultiplier arrays to record images seen in Cherenkov light 144 pixels SHALON (1991) [1-4].

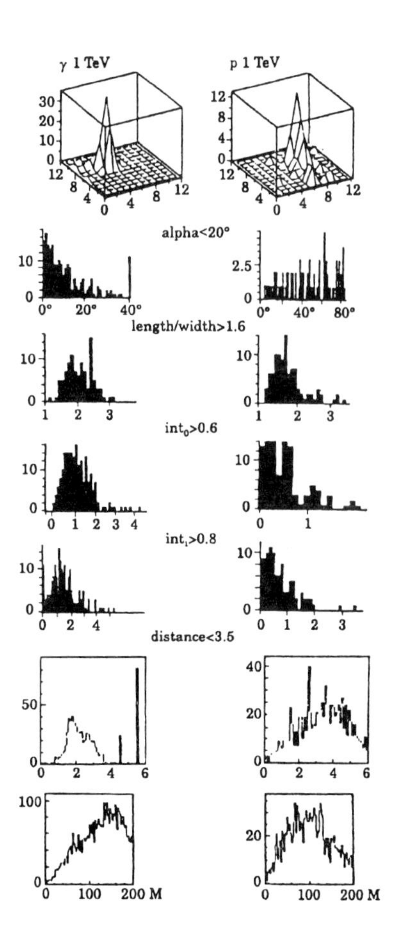

Figure 2. Monte Carlo distribution of image parameters of proton showers of 1 TeV.

point sources of SHALON observation are presented. On the right, cosmic ray protons (250) from zenith SHALON observation are shown. Figure 4 shows calculations for showers of protons carried out by means of CORSIKA, with a special program allowing to take into consideration the geometry of a telescope SHALON (height, Bx, By are equal 3338m, 27.5μT, 44.9μT, respectively). The minimal threshold energy, for which the history of cascade particles can be followed, is equal to 0.3GeV for hadrons and muons and 0.15GeV for electrons and photons. Primary particles fall vertically. In figure 4 the results of calculations by means of the program CORSIKA for protons with energies 2TeV and 5TeV of parameters int_0 and int_1 distributions are presented. Taking into consideration calculations carried out earlier (figure 2), the experimental distribution of gamma-quanta from point sources and from protons (figure 3), and the criteria $int_0 > 0.6$ and $int_1 > 0.8$ based on calculations carried out by means of program CORSIKA (figure 4), one can see, that the contribution of background proton events to gamma events is not more than 10%, i.e. 90% of

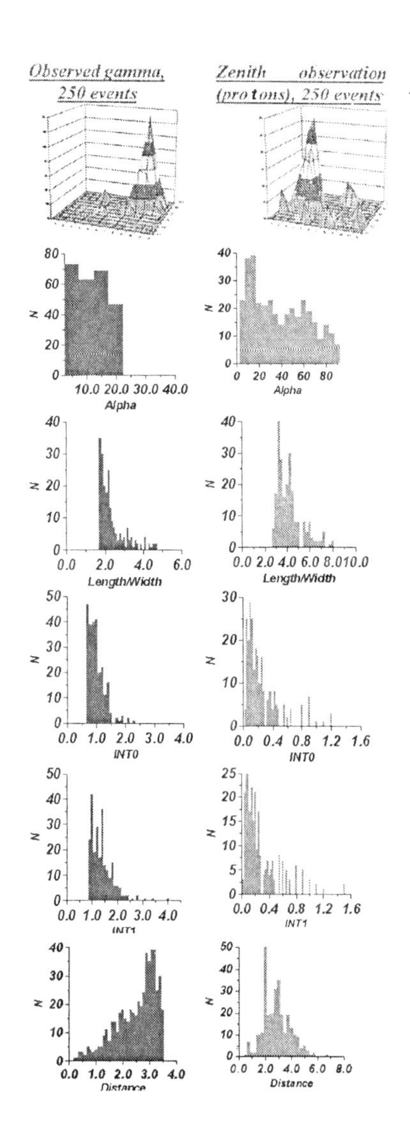

Figure 3. Experimental SHALON distribution of image parameters for gamma and proton showers with energy more than 0.8 TeV.

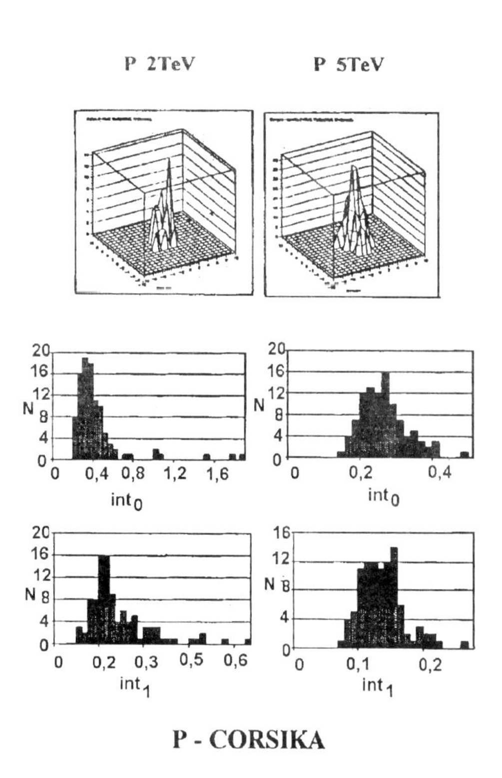

Figure 4. Distribution of image parameters for proton showers of 2 TeV and 5 TeV.

the background is cut, whereas the separation of gamma - quanta according to estimations carried out is not more than 6% (figure 3).

Maybe Crab Nebula is a typical source of Galactic cosmic rays and NGC 1275, Markarian 421 and Markarian 501 are typical sources of Extragalactic cosmic rays. Among different methods of search for local sources of gamma-quanta, the most important is the observation of Cherenkov radiation of electron - photon cascades generated in atmosphere by very high energy gamma - quanta, because a direction of Cherenkov radiation determines a direction of a source. The approximate equality of observable intensity of sources is connected with a limited time of observation of a proposed point source. The distances from Earth to galactic and extragalactic sources differ by a factor of 10^4, what means that extragalactic sources must have a luminosity of 10^8 higher in comparison to the galactic ones, in order to show similar fluxes.

REFERENCES

1. Nikolsky S.I., Sinitsyna V.G., VANT, Ser. TFE (1331), (1987) 30; Proc. VHE GRA, Crimea, (1989) 11
2. Hillas A.M., Nuovo Cimento, V.19C, N5, (1996) 701
3. Weekes T.C. at al., Harvard-Smithsonian Center for Astrophysics, Preprint Series N4450 (1996); N4811 (1999)
4. Sinitsyna V.G., Detector-V, ed. O.De Jager, Kruger Park, (1997) 136, 190; Detector- VI, ed. B.L.Dingus, M.N.Salamon, D.B.Kieda, Snowbird, (1999) 205, 293.

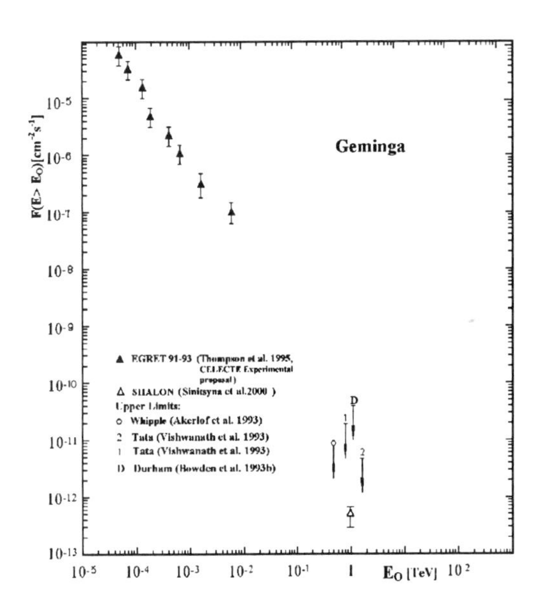

Figure 5. The spectrum of the gamma- radiation of extra- high energies from Geminga

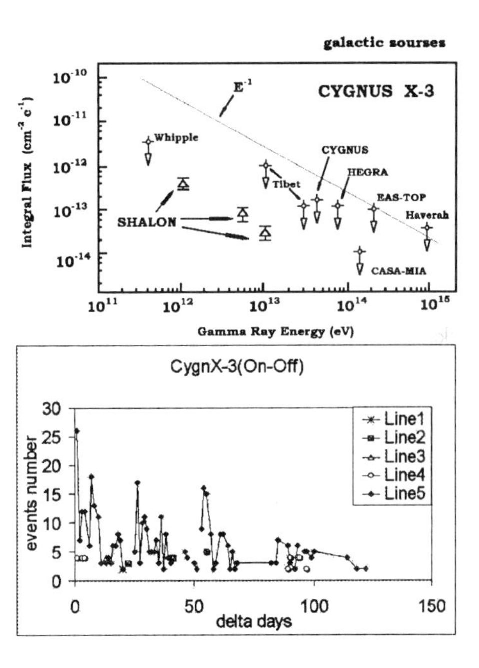

Figure 6. The spectrum of the gamma- radiation of extra-high energies from Cygnus X-3 and Cygnus X-3 **On–Off** diagram from 1996-1999: Line 5 - **On** events sum for 1996, 1997, 1998 and 1999; Lines 1,2,3,4 - **Off** events for 1996, 1997, 1998 and 1999 accordingly

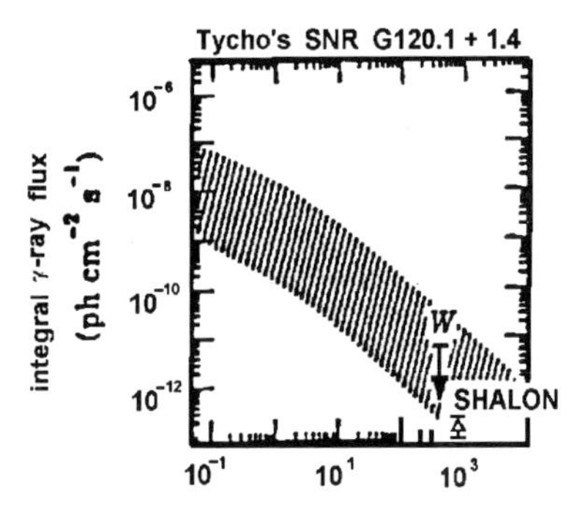

Figure 7. The spectrum of the gamma- radiation of extra-high energies from Tycho Brahe

ELSEVIER

Nuclear Physics B (Proc. Suppl.) 97 (2001) 223–226

www.elsevier.nl/locate/npe

Lateral Distributions, Localization Methods, $\rho(600)$, Size and Energy Determination in Giant EAS

J.N Capdevielle, C.Le Gall[a]*, J. Gawin, I. Kurp, M. Matraszek, B. Szabelska, J. Szabelski and T. Wibig[b]

[a]PCC, Collège de France, Paris

[b]The Andrzej Sołtan Institute for Nuclear Studies, 90-950 Łódź 1, Box447, Poland

The radial analytical structure functions used in different experiments for charged particles and muons are compared. A minimization method is used, according to each function, to optimize the axis localization, the size and $\rho(600)$ approach and finally the primary energy determination.
The procedure of evaluation has been employed for the giant EAS contained in Volcano Ranch and Yakutsk catalogues, as well as for the most energetic event published from AGASA.
The best results are obtained using a hypergeometric gaussian function (JNC function) for electrons (or charged).

1. Radial electron distribution from cascade theory

The structure functions *f(x)* in 3-dimensional cascade theory (where $x = r/r_M$, r being the distance to the core in meters), generally so normalized that $\int_0^\infty 2\pi x f(x)dx = 1$, are related to the electron density $\Delta_e(r)$ by $\Delta_e(r) = N_e f(x)/r_M^2$. The analytical parameterizations of numerical results from the solutions of diffusion equations or from Monte Carlo calculations are commonly classified following the earliest forms proposed:

$$\begin{aligned} f(x) &= 0.45(1/x+4)\exp(-4x^{2/3}) && (1)\\ &= c(s)x^{s-2}(x+1)^{s-4.5} && (2)\\ &= g(s)x^{s-a}(x+1)^{s-b}(1+dx)^{-c} && (3) \end{aligned}$$

The former approximation (Eq. 1), as quoted by Williams [1], was derived by Bethe from Molière's theory for small values of the argument x and for $s = 1$; this form was generalized by Nishimura and Kamata following the numerical values of their solutions of transport equations via Mellin's and Hankel's transformations in the complex plane and saddle point approximation to get the final real solutions [2]. The synthesis, so-called NKG formula [3], contained in Eq. 2 under a pair of power laws representing respectively the asymptotic tendencies (near and far from the shower core), with a simple normalization in terms of Euler's Beta function,

$$c(s) = \frac{\Gamma(4.5-s)}{2\pi\Gamma(s)\Gamma(4.5-2s)} \qquad (4)$$

became one of the most widely used radial distributions. The comparison to experimental results suggested however a more complex situation and some corrections such as the introduction of a new argument $x = r/mr_M$, m being a factor reducing Molière's radius, or a local age parameter $s(r)$, troublesome for the normalization of the structure function [4], [5]. In order to provide a better skewness than the transition between two power laws, we proposed later [6] the more complex relation of Eq. 3; such structure function, which is also a general form containing Eq. 2 for a particular set of parameters, has the advantage (for values of parameters respecting the conditions of convergence $s - a + 2 > 0$ and $c - 2s + b - 2 > 0$) to be exactly normalized in terms of Gaussian Hypergeometric function $F_{HG} = F(c, s-a+2, c+b-s; 1-d)$ by:

$$g(s) = \frac{\Gamma(c+b-s)}{2\pi\Gamma(s-a+2)} \times \frac{1}{\Gamma(c-2s+b+a-2)F_{HG}} \qquad (5)$$

*capdev@cdf.in2p3.fr

PII S0920-5632(01)01269-5

At large distances from axis, as emphasized by Particle Data Group, the description of the 3D-cascade transport by diffusion equations fails (small angle approximation in multiple Coulomb scattering, effect of single scattering, Landau approximation is not valid anymore) and the analytical descriptions can only be derived from Monte Carlo calculations.

2. Hypergeometric approach via simulation

The empirical distributions, such as AGASA function [7], as underlined by Vishwanath [8] enter in the category of Hypergeometric Gaussian functions [6]. We followed this approach and used the lateral distribution in the general form of structure function

$$f(x) = C_e \cdot x^{-\alpha} \cdot (1+x)^{-(\eta - \alpha)} \cdot (1+dx)^{-\beta} \quad (6)$$

with the conditions $2 - \alpha > 0$ and $\beta + \eta - 2 > 0$.

The exact value of the coefficient C_e equals

$$C_e = \frac{\Gamma(\beta + \eta - \alpha)}{2\pi \cdot \Gamma(2-\alpha) \cdot \Gamma(\beta + \eta - 2)} \cdot \frac{1}{F_{HG}}$$

where $F_{HG} = F_{HG}(\beta, 2-\alpha, \beta+\eta-\alpha; 1-d)$

The Hypergeometric Gaussian function can be easily calculated from the hypergeometric series:

$$F_{HG}(A,B,C;z) = \sum_{n=0}^{\infty} \frac{(A)_n (B)_n}{(C)_n n!} z^n,$$

$$C \neq 0, -1, -2, ...,$$

$$(A)_n = \Gamma(A+n)/\Gamma(A), \quad (A)_0 = 1$$

Equation 6 (JNC function) is equivalent to our version 3 containing the age parameter s with the following relations between respective coefficients: $x = \frac{r}{r_M}$, $d = \frac{r_M}{r_0}$, $s = 1.03$, $\alpha = a - s$, $\eta = b - s + \alpha$ $\beta = c$

We have adjusted with MINUIT the parameters of our hypergeometric function to the average lateral distributions of groups of 10 showers at $10^{20} eV$ induced by protons and by Fe nuclei, simulated with CORSIKA (QGSJET model) [10]. The fitting procedure uses N_e, r_M, r_0, a, b and β as free parameters to describe the lateral distribution of all charged particles (function JNC01) or only electrons (function JNC02). The value of s is taken from the longitudinal development simulated.

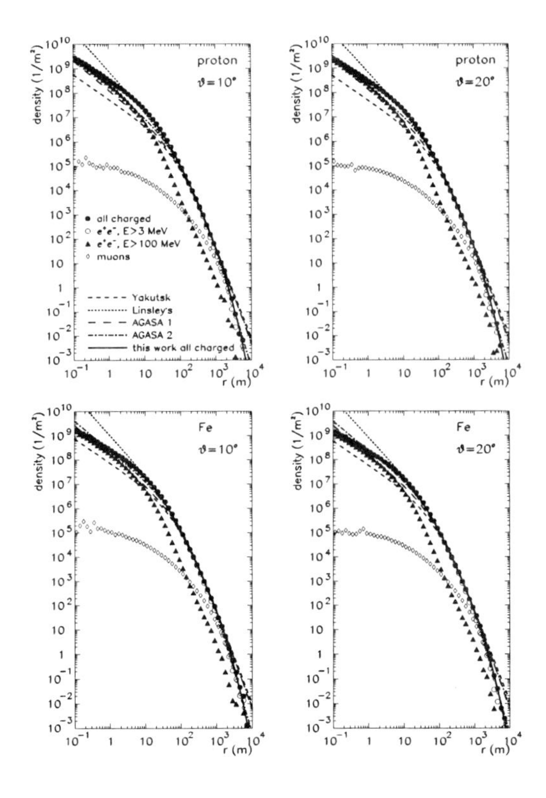

Figure 1. Fits to all charged particles lateral distribution from simulations (average from 10 EAS). Primary particle energy 10^{11} GeV. Lines are normalized to ϱ(600 m).

In each case, the adjustment has been performed with 50 points from the simulation distributed from 0.1 m up to 10 km from axis position. Results for charged particles (muons and electrons) are shown on Fig. 1 and in Table 1.

3. Analysis of experimental data

The advantages of JNC01 formula can be seen on fig. 1 and on Table 2.

Table 1
Best parameters to simulated e^+e^-+ muons (all charged) lateral distribution fit using JNC01 formula.

	proton 10°	proton 20°	Fe 10°	Fe 20°
$log_{10}N_e$	10.75	10.72	10.70	10.65
r_M	21.26	21.26	19.18	19.18
r_0	8785.	8785.	9536.	9536.
a	1.91	1.91	1.82	1.82
s	1.03	1.04	1.03	1.04
b	3.32	3.32	3.31	3.31
β	10.0	10.0	10.0	10.0

Table 2
Columns a present total number of charged particles N_e in 10^{10}, columns b the ratios E_0/N_e in GeV (E_0=10^{11} GeV) and columns c the ratios $\varrho(600)/N_e$ in 10^{-8} particles/m^2. $m(600) = \frac{d(log\ \varrho)}{d(log\ r)}$ at 600 m.

	proton 10°			proton 20°			iron 10°			iron 20°		
$\varrho(600)$	290 m^{-2}			318 m^{-2}			369 m^{-2}			356 m^{-2}		
$E_0/\varrho(600)$ (GeV m^2)	$3.4 \cdot 10^8$			$3.1 \cdot 10^8$			$2.7 \cdot 10^8$			$2.8 \cdot 10^8$		
$m(600)$	–3.9			–3.6			–3.6			–4.0		
fit	1a	1b	1c	2a	2b	2c	3a	3b	3c	4a	4b	4c
Yakutsk	1.8	5.6	1.6	1.7	5.9	1.9	2.3	4.3	1.6	1.9	5.3	1.8
Linsley's	8.2	1.2	0.3	8.0	1.3	0.3	10.5	0.9	0.3	8.9	1.1	0.4
AGASA #1	2.4	4.2	1.2	2.6	3.8	1.2	3.1	3.2	1.1	2.9	3.4	1.2
AGASA #2	3.3	3.0	0.8	3.6	2.8	0.8	4.2	2.4	0.8	4.0	2.5	0.8
this work	5.6	1.8	0.5	5.6	1.9	0.6	5.1	2.0	0.7	4.5	2.2	0.7

Table 3
Localization of the most energetic AGASA event [9]: χ^2, $\varrho(600)$, N_e and relative core distance from the original localization. Age parameter s fitted for A and B functions as 1.05 and 0.98, respectively.

	$\chi^2/23$	$\varrho(600)$ (1/m^2)	N_e (10^{10})	Δr (m)
original		892	7.84	0.00
A – JNC01	7.32	598	34.81	82.44
B – JNC02	8.24	599	17.57	98.83
C – Yakutsk	10.40	561	3.01	101.15
D – Linsley's	7.51	565	10.74	83.29
E – AGASA no. 1	6.90	580	6.43	77.10
F – AGASA no. 2	11.81	611	8.92	60.49

The major part of the particles is contained inside 200m from the axis and only the skewness of the hypergeometric function allows a reliable relation between size and density at 600m. This method has been applied to the showers contained in the catalogues of Volcano Ranch and Yakutsk. The core position has been obtained by minimization with Minuit program between different formulas available for lateral densities written versus the coordinates X, Y as

$$\varrho(r) = \varrho(\sqrt{(X - X_c)^2 + (Y - Y_c)^2}) \tag{7}$$

where the core coordinates X_c and Y_c are taken as two additive parameters in the minimization. The adjustments are generally improved when compared to the original treatments, turning to lower sizes (in the case of Yakutsk formula) and better approximation of the density at 600m. The situation of the most energetic event of AGASA [9] is given in table 3.

4. Conclusion

The hypergeometric approach gives a better accuracy in the interpolation of densities at 600-1000m, a more reliable estimation of the shower size (when the axis is in the array), a better axis localisation and finally a more correct constraint of the primary energy.

Acknowledgement

This work was partly supported (JNC and CLG) by INTAS contract 1339.

REFERENCES

1. R. W. Williams, *Phys.Rev.* (1947) **74** 1689
2. K. Kamata and J. Nishimura, *Suppl.Prog.Theor.Phys.* (1958) **6** 139
3. K. Greisen, *Ann.Rev. of Nuclear Science* (1960) **10** 63
4. J. N. Capdevielle and J. Gawin, *Jour. Phys. G* (1982) **8** 1317
5. M. Nagano et al., *Jour. of Phys.Soc.Japan* (1984) **53** 1667
6. J. N. Capdevielle and J. Procureur, *Proc. 18th ICRC (Bangalore)* (1983) **11** 307
7. M. Nagano et al., *J. Phys. G.: Nucl. Particle Phys.* (1992) **18** 423
8. P. R. Vishwanath, *Proc. 23rd ICRC, Rapporteur Papers (Calgary)* (1993) 384 World Scientific
9. S. Yoshida et al., *Astroparticle Physics* (1995) **3** 105
10. J. N. Capdevielle, C. Le Gall and K. Sanosyan, *Astroparticle Physics* (2000) **13** 259

ELSEVIER

Nuclear Physics B (Proc. Suppl.) 97 (2001) 227–230

www.elsevier.nl/locate/npe

CASTOR: Centauro And STrange Object Research in Nucleus–Nucleus Collisions at the LHC *

A.L.S. Angelis[a †], J. Bartke[b], M.Yu. Bogolyubsky[c], E. Gładysz-Dziaduś[b], Yu.V. Kharlov[c], A.B. Kurepin[d], A.I. Maevskaya[d], G. Mavromanolakis[a], A.D. Panagiotou[a], S.A. Sadovsky[c], P. Stefanski[b], Z. Włodarczyk[e]

[a]Nuclear and Particle Physics Division, University of Athens, Hellas.

[b]Institute of Nuclear Physics, Cracow, Poland.

[c]Institute for High Energy Physics, Protvino, Russia.

[d]Institute for Nuclear Research, Moscow, Russia.

[e]Institute of Physics, Pedagogical University, Kielce, Poland.

We present a phenomenological model which describes the formation of a Centauro fireball in the baryon-rich projectile fragmentation region in nucleus-nucleus interactions in the upper atmosphere and at the LHC, and its decay to non-strange baryons and Strangelets. Strangelets are assimilated to the "strongly penetrating component" frequently observed accompanying hadron-rich cosmic ray events. We describe the CASTOR subdetector for the ALICE experiment at the LHC. CASTOR will probe, in an event-by-event mode, the very forward, baryon-rich phase space $5.6 \leq \eta \leq 7.2$ in 5.5×A TeV central $Pb + Pb$ collisions. It will look for events with pronounced imbalance between hadronic and photonic content and for deeply penetrating objects. We present results of simulations for the response of the CASTOR calorimeter to the passage of Strangelets.

1. Introduction

The physics motivation to study the very forward phase space in nucleus–nucleus collisions stems from the potentially very rich field of new phenomena to be produced inside an environment with very high baryochemical potential. The study of this baryon-dense region, much denser than the highest baryon density attained at the AGS or SPS, will provide important information for the understanding of a quark gluon plasma (QGP) state at relatively low temperatures, with properties different from those expected in the higher temperature baryon-free region around mid-rapidity. It is thought that such a state could exist in the core of neutron stars.

The LHC with an energy equivalent to 10^{17} eV for a moving proton impinging upon one at rest will be the first accelerator to effectively probe the highest energy cosmic ray domain. Cosmic ray experiments have detected numerous most unusual events which have still not been understood. Such events, observed in the projectile fragmentation rapidity region, may be produced and studied at the LHC in controlled conditions. Here we mention the "Centauro" events and the "long-flying component".

Centauros [1] exhibit relatively small multiplicity, complete absence (or strong suppression) of the electromagnetic component and very high $\langle p_T \rangle$. Furthermore, a number of hadron-rich events are accompanied by a strongly penetrating component observed in the form of halo, strongly penetrating clusters [2,3] or long-living cascades, whose transition curves exhibit a characteristic form with many maxima [4,5]. These events have until now systematically defied all attempts at explanation in terms of conventional physics [6].

*Further information and full bibliography on the web at http://angelis.home.cern.ch/angelis/castor/Welcome.html
†Also at CERN, Geneva, Switzerland.

0920-5632/01/$ – see front matter
PII S0920-5632(01)01270-1

2. A model for Centauro and Strangelet formation

We have developed a model in which Centauros are considered to originate in the hadronization of a QGP fireball of very high baryon density ($\rho_b \gtrsim 2$ fm^{-3}) and baryochemical potential ($\mu_b >> m_n$), which is formed in ultrarelativistic nucleus–nucleus collisions in the upper atmosphere [7–9]. In this model the QGP fireball initially consists of u and d quarks and gluons. The very high baryochemical potential inhibits the creation of $u\bar{u}$ and $d\bar{d}$ quark pairs, resulting in the fragmentation of gluons predominantly into $s\bar{s}$ pairs. In the ensuing hadronization of the fireball this leads to the strong suppression of pions, and hence of photons, but allows kaons to be emitted carrying away strange antiquarks, positive charge, entropy, and lowering further the temperature. This process of strangeness distillation transforms the initial quark matter fireball into a slightly strange quark matter state. The finite excess of s quarks and their stabilizing effects, in addition to the large baryon density and binding energy and the very small volume, may prolong the lifetime of the Centauro fireball enabling it to reach mountain-top altitudes [10]. In the subsequent decay and hadronization of this state non-strange baryons and Strangelets will be formed. Simulations show that Strangelets could be identified with the strongly penetrating particles frequently seen accompanying hadron-rich cosmic ray events [11,12] In this manner, both the basic characteristics of the Centauro events (small multiplicities and extreme imbalance of electromagnetic to hadronic content) and the strongly penetrating component are naturally explained.

In a similar way, Centauro events may be produced in $Pb + Pb$ collisions at the LHC from the hadronization of a QGP state formed in the beam fragmentation region, where the baryon number is expected to be strongly concentrated [8,13]. In table 1 we compare characteristic quantities of Centauro and strongly penetrating components (Strangelets), either experimentally observed or calculated within the context of the above model, for cosmic ray interactions and for nucleus–nucleus interactions at the LHC.

Table 1
Average characteristic quantities of modeled Centauro events and Strangelets produced in Cosmic Rays and expected at the LHC.

Centauro	Cosmic Rays	LHC
Interaction	"$Fe + N$"	$Pb + Pb$
$\sqrt{s}$	$\gtrsim$ 6.76 TeV	5.5 TeV
Fireball mass	$\gtrsim$ 180 GeV	$\sim$ 500 GeV
y_{proj}	≥ 11	8.67
γ	$\geq 10^4$	$\simeq 300$
η_{cent}	9.9	$\simeq 5.6$
$\Delta\eta_{cent}$	1	$\simeq 0.8$
$< p_T >$	1.75 GeV	1.75 GeV (*)
Life-time	10^{-9} s	10^{-9} s (*)
Decay prob.	10 % (x $\geq$ 10 km)	1 % (x $\leq$ 1 m)
Strangeness	14	60 - 80
f_s (S/A)	$\simeq 0.2$	0.30 - 0.45
Z/A	$\simeq 0.4$	$\simeq 0.3$
Event rate	$\simeq$ 1 %	$\simeq$ 500/month
"Strangelet"	Cosmic Rays	LHC
Mass	$\simeq$ 7 - 15 GeV	10 - 80 GeV
Z	$\lesssim 0$	$\lesssim 0$
f_s	$\simeq 1$	$\simeq 1$

(*) assumed

3. The CASTOR detector

Based on the above considerations, we have designed the CASTOR (Centauro And STrange Object Research) subdetector [13,14] of the ALICE heavy ion experiment at the LHC, in order to study the very forward, baryon-dense phase space region. CASTOR will cover the pseudorapidity interval $5.6 \leq \eta \leq 7.2$ and will probe the maximum of the baryon number density. It will identify any effects connected with this condition and will complement the physics program pursued by the rest of the ALICE experiment in the essentially baryon-free mid-rapidity region. Figure 1 depicts the baryon number pseudorapidity distribution as predicted by the HIJING Monte-Carlo generator for an average central $Pb + Pb$ collision at the LHC, with the acceptance of CASTOR superimposed on the plot. Other Monte-Carlo codes give similar results. The CASTOR detector is designed to measure the hadronic and photonic contents of an interaction and to identify deeply penetrating objects, in an event-by-event mode. It will initially consist of a calorimeter.

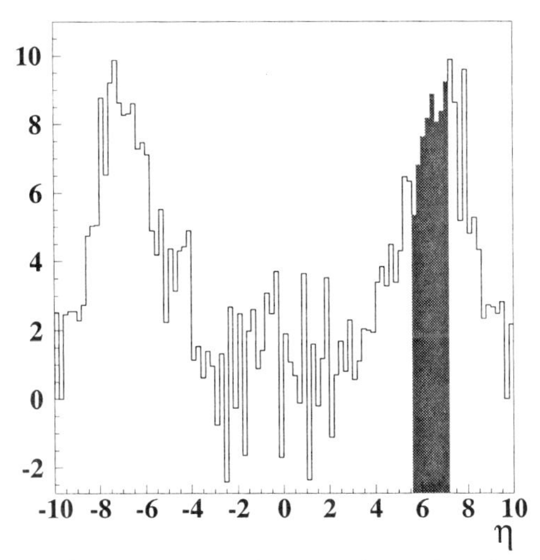

Figure 1. Baryon number pseudorapidity distribution predicted by HIJING for central $Pb+Pb$ collisions at the LHC. Superimposed is the acceptance of CASTOR.

A side view of the CASTOR calorimeter is depicted schematically in figure 2. It is azimuthally symmetric around the beam pipe, and is divided into 8 azimuthal sectors. It is longitudinally segmented so as to measure the profile of the formation and propagation of cascades and comprises electomagnetic and hadronic sections. The calorimeter is made of layers of active medium sandwiched between tungsten absorber plates. The active medium consists of planes of fused silica fibres and the signal is the Cherenkov light produced as they are traversed by the fast charged particles (for the most part e^+, e^-) in the cascades. The fibres are inclined at 45 degrees relative to the impinging particles in order to maximize the light output.

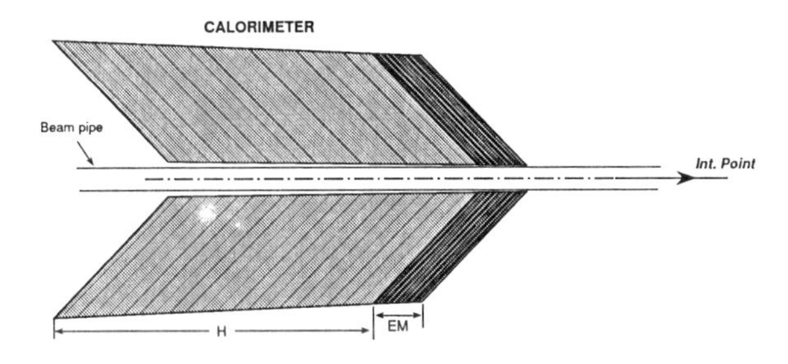

Figure 2. Schematic representation of the CASTOR calorimeter.

The calorimeter will be read out via air light guides made out of highly polished aluminium which is UV reflecting. The produced Cherenkov photons will propagate along the silica fibres to the lateral edges of the calorimeter where they will exit into the light guides. Inside the light guides they will be directed to photomultiplier tubes equipped with quartz photocathode entry windows to optimally match the wavelength of the Cherenkov light. It is envisaged to couple together the light output from groups of consecutive active layers into the same light guide for increased signal. Figure 3 depicts the total energy pseudorapidity distribution in an average central $Pb+Pb$ collision at the LHC as predicted by HIJING, with the CASTOR superimposed on the plot. The CASTOR calorimeter is expected to receive $\simeq$ 200 TeV per central event.

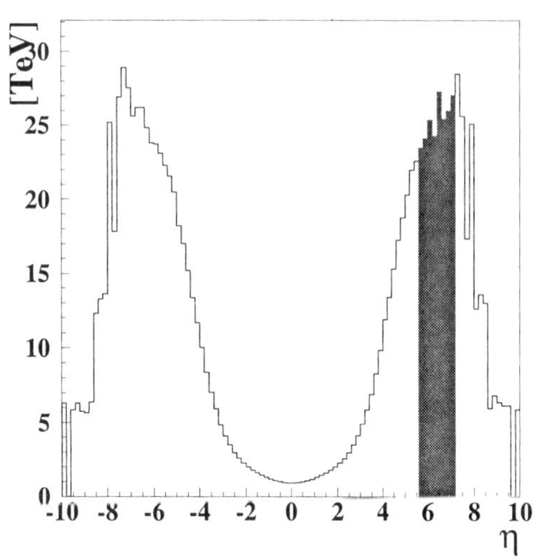

Figure 3. Energy pseudorapidity distribution predicted by HIJING for central $Pb+Pb$ collisions at the LHC. Superimposed is the acceptance of CASTOR.

4. Simulation of the CASTOR calorimeter performance

We have made detailed GEANT simulations of the performance of the CASTOR calorimeter for many different configurations. The results presented here refer to a design in which each octant is longitudinally segmented into 80 layers, the first 8 ($\simeq 14.7\ X_0$) made out of 5 mm thick W plates comprising the electromagnetic section and the remaining 72 ($\simeq 9.47\ \lambda_I$) made out of 10 mm thick W plates the hadronic section. The light output from groups of 4 consecutive sampling layers is coupled into the same light guide, giving a total of 20 readout channels along each octant. Figure 4 shows the total number of Cherenkov photons produced, retained and propagated inside the fibres, as a function of the incident particle energy for incident photons and hadrons from one central LHC $Pb + Pb$ HIJING event. About 210 Cherenkov photons per GeV are obtained for incident photons and 129 Cherenkov photons per GeV for incident hadrons.

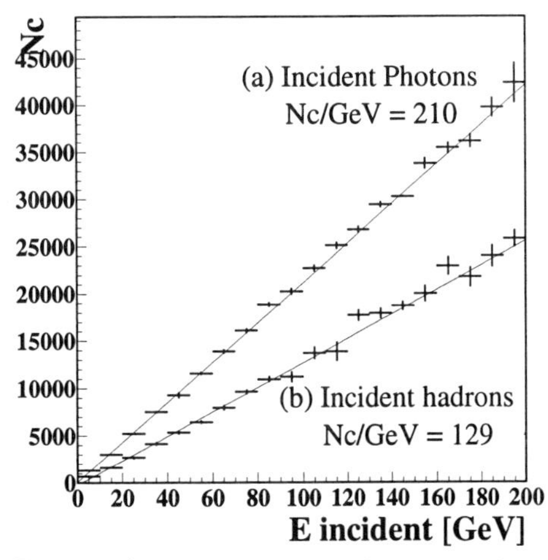

Figure 4. Simulation of the total number of Cherenkov photons vs. incident particle energy: (a) For incident photons, (b) For incident hadrons.

In addition we have simulated the propagation of Strangelets inside the CASTOR calorimeter using the method described in [15,11]. As an example figure 5 shows the response of the calorimeter to one central LHC $Pb + Pb$ HIJING event, to which has been added a Strangelet of $A_{str} = 20$, $E_{str} = 20$ TeV and $\mu_{str} = 600$ MeV (energy conservation has been applied). Figure 5a shows the energy deposition along the octant containing the Strangelet, while figure 5b shows the average of the energy deposition along the other octants. The study of such simulated events shows that the expected signal from a Strangelet is much larger than the background and that its transition curve displays long penetration and many maxima structure, as observed in cosmic ray events.

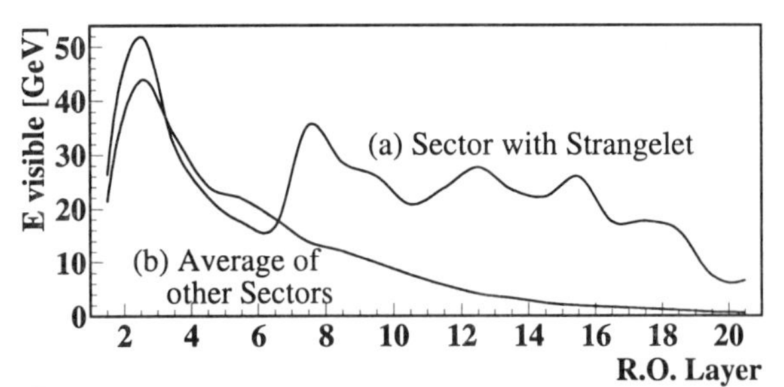

Figure 5. Simulation of the energy deposition in the readout layers of the CASTOR calorimeter: (a) In octant with Strangelet, (b) Average of other octants.

Acknowledgements

This work has been partly supported by the Hellenic General Secretariat for Research and Technology ΠΕΝΕΔ 1361.1674/31-1/95, the Polish State Committee for Scientific Research grants 2P03B 121 12 and SPUB P03/016/97, and the Russian Foundation for Fundamental Research grant 96-02-18306.

REFERENCES

1. C.M.G. Lates, Y. Fugimito and S. Hasegawa, Phys. Rep. **65** (1980) 151.
2. S. Hasegawa and M. Tamada, Nucl. Phys. **B474** (1996) 225.
3. L.T. Baradzei et al., Nucl. Phys. **B370** (1992) 365.
4. T. Arisawa et al., Nucl. Phys. **B424** (1994) 241.
5. Z. Buja et al., Proc. 17th ICRC, Paris 1981, Vol.11 p.104.
6. M. Tamada, Inst. Cosmic Ray Research, Univ. Tokyo ICRR-Report-454-99-12.
7. M.N. Asprouli, A.D. Panagiotou, E. Gładysz-Dziaduś, Astropart. Phys. **2** (1994) 167.
8. A.D. Panagiotou et al., Phys. Rev. **D45** (1992) 3134.
9. A.D. Panagiotou et al., Z. Phys. **A333** (1989) 355.
10. O.P. Theodoratou and A.D. Panagiotou Astropart. Phys. **13** (2000) 173.
11. E. Gładysz-Dziaduś and Z. Włodarczyk, J. Phys. G: Nucl. Part. Phys. **23** (1997) 2057.
12. E. Gładysz-Dziaduś and A.D. Panagiotou, Proc. Int. Symp. on Strangeness & Quark Matter, eds. G. Vassiliadis et al., World Scientific, 1995 p.265.
13. A.L.S. Angelis and A.D. Panagiotou, J. Phys. G: Nucl. Part. Phys. **23** (1997) 2069.
14. A.L.S. Angelis et al., CASTOR draft proposal, Internal note ALICE/CAS 97-07.
15. A.L.S. Angelis, J. Bartke, E. Gładysz-Dziaduś, Z. Włodarczyk, Eur. Phys. J. direct **C9** (1999) 1.

ELSEVIER

Nuclear Physics B (Proc. Suppl.) 97 (2001) 231–234

www.elsevier.nl/locate/npe

Diffuse reflectivity of Tyvek in air and water, and anisotropical effects

J.C. Arteaga Velázquez *, C. Vázquez López † and Arnulfo Zepeda ‡ [a]

[a]Departamento de Física, CINVESTAV, Apartado Postal 14-740,
07000, México, Distrito Federal, México

The reflectivity of Tyvek in air and water as a function of incidence and reflection angles is reported. The measurements were made for $\lambda = 488nm$ and $\lambda = 632.8nm$. Optical anisotropies of the material are revealed when a rotation is made about the normal vector of the sample.

1. Introduction

In the design of the water Cherenkov detectors of the Pierre Auger project it is important to consider the following aspects: The interaction of relativistic particles with water, the absorption length of light in water, the reflectivity of the material used to cover the inner surface of the detector and the efficiency of the photodetectors [1]. Tyvek will be used to cover the inner surface of the Cherenkov detectors, therefore, the information about the Tyvek optical properties is an important factor to take into account in simulations and reconstruction of ultra high energy cosmic rays detected through the Cherenkov light emitted by the air shower secondary particles penetrating the water detectors. Tyvek is made up of millions of polyethylene fibers randomly oriented that overlap, is chemically inert and resists water penetration [2]. We focused our attention on the diffuse reflectivity of Tyvek and the influence of the anisotropies of the material on these measurements.

2. Experimental set-up and procedure

The experimental design is shown in Fig. 1. Here the main measured parameters are the incidence angle θ_i, the reflection angle θ_r and the amount of reflected light in that direction I_t.

In order to get the data, we built an automatic system to measure the diffuse reflectivity as a function of θ_r with steps of 1.8^0 and for the following incident angles: $0^0, 15^0, 30^0, 45^0, 60^0$ and 75^0. As photodetector, an NPN silicon phototransistor was used, whose assembly was designed to allow to immerse it in water. We worked at two different wavelengths: one in the blue region ($\lambda = 488nm$, Argon laser) and another one in the red region ($\lambda = 632.8nm$, He-Ne laser). The distance between the laser and the sample was 35 cm while that between the Tyvek and the detector was 2.5 cm.

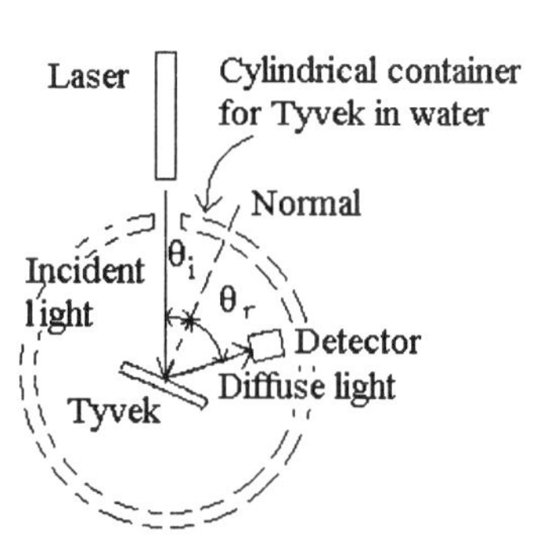

Figure 1. Experimental set-up.

For the measurements in pure water we used a cylindrical container whose inner surface was sprayed with black paint, in order to avoid undesirable reflections. This container was provided with a glass window to allow the incidence of the laser beam on the sample. For the measurements of the anisotropical effects we kept θ_i constant (75^0) and rotated the sample by an angle α about

*jarteaga@fis.cinvestav.mx
†cvazquez@fis.cinvestav.mx
‡zepeda@fis.cinvestav.mx

PII S0920-5632(01)01271-3

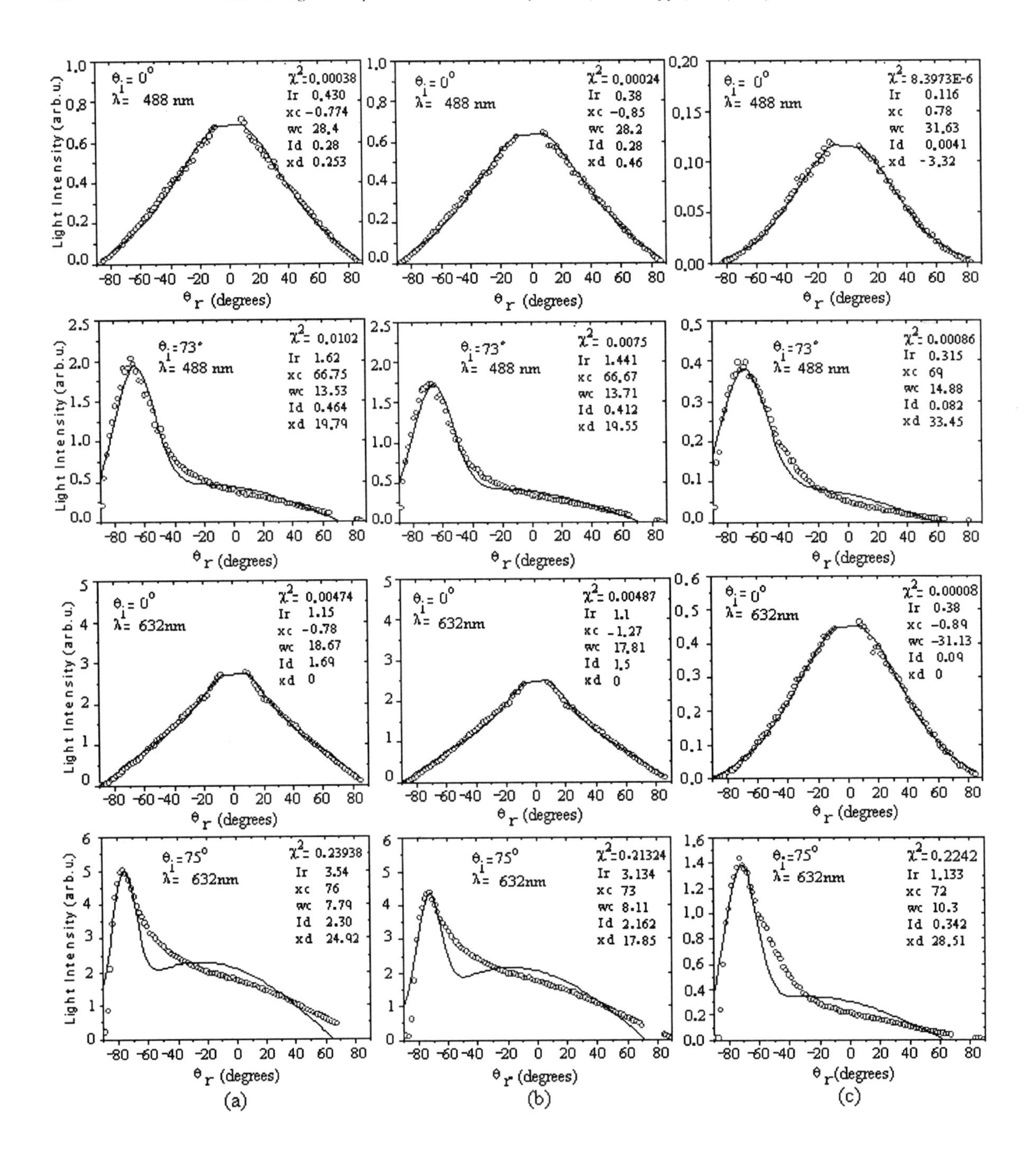

Figure 2. Diffuse reflectivity spectrum of Tyvek as a function of θ_r for $\lambda = 488nm, 632.8nm$, and $\theta_i = 0^o, 75^o$ for Tyvek: (a) in air, (b) inside the cylindrical container, and (c) in water. The circles represent the experimental data while the solid line the fit by using the relation (1).

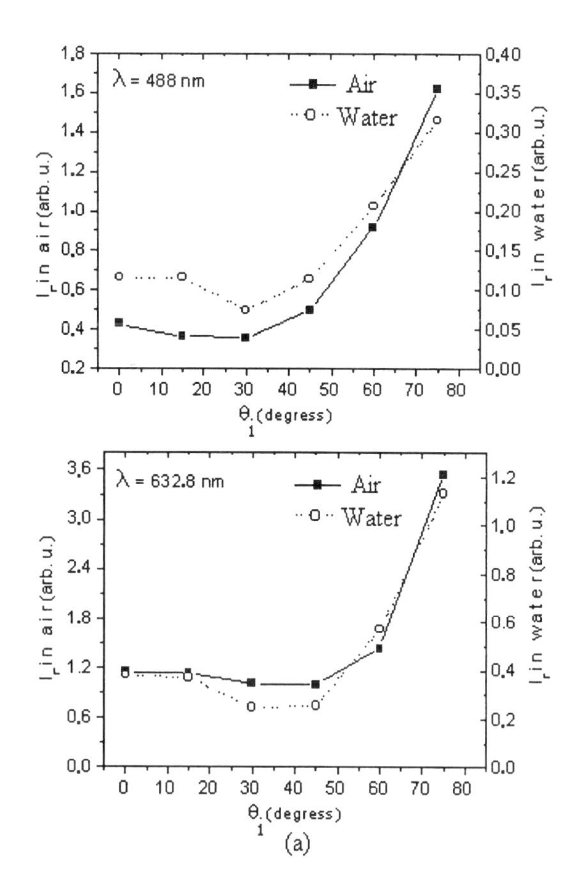

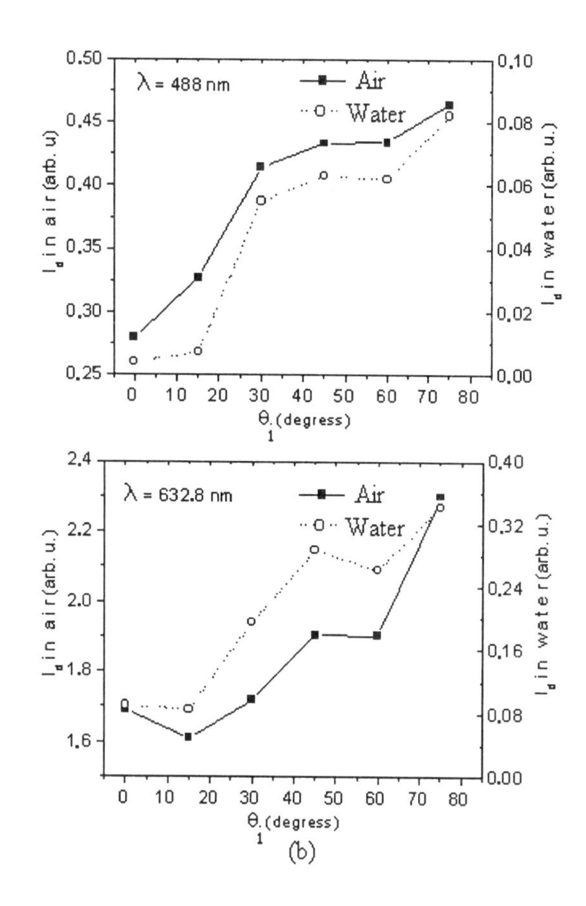

Figure 3. (a) I_r and (b) I_d as a function of θ_i

its normal vector.

3. Analysis and Results

The graphs for diffuse reflectivity of Tyvek in air and inside the cylindrical container with and without water are shown in Fig. 2 for $\theta_i = 0^0$ and 75^0. In Fig. 2, units are arbitrary but the same scale is used for all the cases. The difference in intensity when changing the laser beam arises from the spectral responsivity of the phototransistor and the nominal laser power. At $\lambda = 632.8nm$ the responsivity is $0.25 Amperes/Watt$ and the He-Ne laser nominal power is $2mW$, while at $\lambda = 488nm$ the responsivity is $0.18 Amperes/Watt$ and the Ar laser nominal power is $11mW$. In order to fit the data (see Fig. 2), we proposed the following relation:

$$I_t(\theta_i, \theta_r) = I_r e^{\frac{-(\theta_r + x_c)^2}{2w_c^2}} + I_d cos(\theta_r + x_d) \qquad (1)$$

by similarity with the relations proposed by Hasenbalg et al [3], and Filevich et al [4], where it is assumed a cosine function (suggested by Lambert's law) to fit the diffuse light, and a gaussian term to fit the peak (which has a specular contribution). Here, x_c gives us the position of the I_r component while x_d is a phase displacement of the cosine function.

The decrease of the diffuse reflectivity when the sample is immersed in water is mainly associated with the increase of the refractive index of the incidence medium, from 1 in air to 1.34 (for $\lambda = 488nm$) or 1.331 (for $\lambda = 632.8nm$) in water [5]. The I_r and I_d parameters are identified as the specular and diffuse components, respecti-

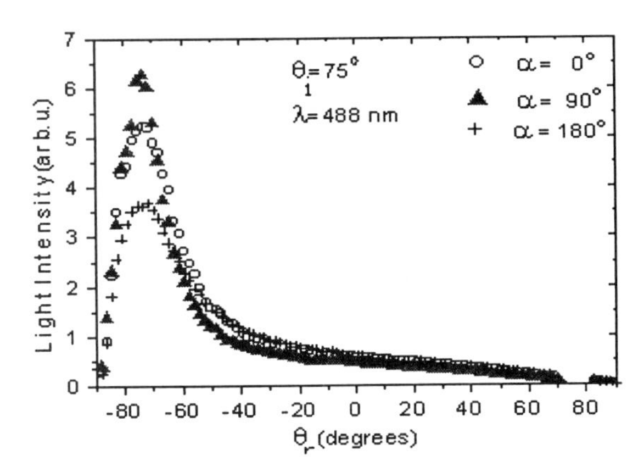

Figure 4. Optical anisotropic effects on the diffuse reflectivity.

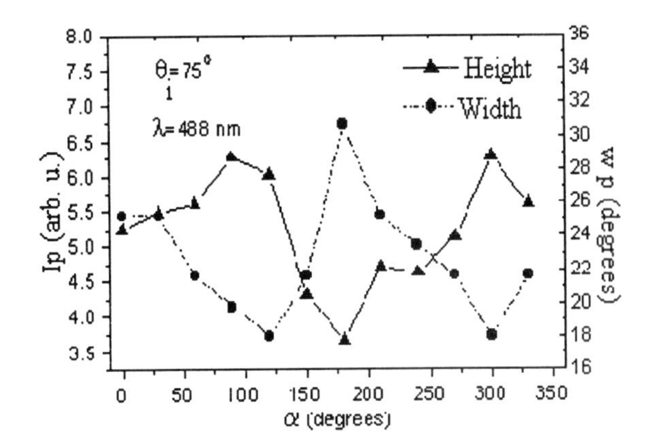

Figure 5. Changes of the diffuse reflectivity parameters under a rotation of the sample by an angle α.

vely, and are plotted in Fig. 3. From these curves, we can see that: 1)The specular component has a decreasing behavior from $\theta_i = 0^0$ to 30^0 or 45^0 (the difference can reach up to 10 %) and then, it increases when going from $\theta_i = 45^0$ to 75^0. This behavior is similar to that of the p component of the classical specular reflectivity between two media [6]. 2)The diffuse component has an opposite behavior. The optical anisotropies of Tyvek are shown in Fig. 4 and 5 for $\lambda = 488nm$. The shape of the graphs changes under a rotation α. The height (I_p) and half width (w_p) of the diffuse reflectivity as a function of α are shown in Fig. 5. Clearly, an anticorrelation between I_p and w_p is observed, which means that: The larger the specular component (I_d), the smaller the diffuse part (w_d). Note that I_p can decrease up to 43% of its maximum value.

4. Conclusions

By building a simple automatized system involving the control of a step motor and data acquisition, the diffuse reflectivity of Tyvek has been carefully obtained. The influence of optical anisotropies has been determined and should not be neglected in the simulation studies of the Cherenkov radiation involving Tyvek. This effect must be studied for different surface regions on Tyvek.

5. Acknowledgements

The authors thanks the technical assistance of B. Zendejas.

REFERENCES

1. C. Pryke, Performance simulations of a $10m^2$ water Cerenkov detector and comparison with experiment, GAP 97-004.
2. DuPont. http:// www.dupont.com, Tyvek is a DuPont registered trademark.
3. F. Hasenbalg and D. Ravignani, Tyvek Diffuse Reflectivity, GAP 97-035.
4. A. Filevich, P. Bauleo, H. Bianchi, J. Rodriguez Martino and G. Torlasco, Spectral-directional reflectivity of Tyvek immersed in water , GAP 97-065.
5. D.J. Segelstein, The Complex Refractive Index of Water, M.S. Thesis, Department of Physics, University of Missouri-Kansas City, 1981.
6. J. R. Reitz, F. J. Milford and R. W. Christy, Foundations of Electromagnetic Theory, Fourth Edition, Addison-Wesley.

ELSEVIER

Nuclear Physics B (Proc. Suppl.) 97 (2001) 235–238

www.elsevier.nl/locate/npe

Status of the solution to the solar neutrino problem based on non-standard neutrino interactions

S. Bergmann[a], M. M. Guzzo[b], P. C. de Holanda[b], P. I. Krastev[c] and H. Nunokawa[b]

[a]Department of Particle Physics
Weizmann Institute of Science, Rehovot 76100, Israel

[b] Instituto de Física Gleb Wataghin
Universidade Estadual de Campinas, UNICAMP
13083-970 Campinas SP, Brazil

[c]Department of Physics
University of Wisconsin, Madison, WI 53706, USA

We analyze the current status of the solution to the solar neutrino problem based both on: a) non-standard flavor changing neutrino interactions (FCNI) and b) non-universal flavor diagonal neutrino interactions (FDNI). We find that FCNI and FDNI with matter in the sun as well as in the earth provide a good fit not only to the total rate measured by all solar neutrino experiments but also to the day-night and seasonal variations of the event rate, as well as the recoil electron energy spectrum measured by the SuperKamiokande collaboration. This solution does not require massive neutrinos and neutrino mixing in vacuum. Stringent experimental constraints on FCNI from bounds on lepton flavor violating decays and on FDNI from limits on lepton universality violation rule out $\nu_e \to \nu_\mu$ transitions induced by New Physics as a solution to the solar neutrino problem. However, a solution involving $\nu_e \to \nu_\tau$ transitions is viable and could be tested independently by the upcoming B-factories if flavor violating tau decays would be observed at a rate close to the present upper bounds.

1. Introduction

In his seminal paper Wolfenstein [1] observed that non-standard neutrino interactions (NSNI) with matter can also generate neutrino oscillations. In particular this mechanism could be relevant to solar neutrinos interacting with the dense solar matter along their path from the core of the sun to its surface [2–8]. In this case the flavor changing neutrino interactions (FCNI) are responsible for the off-diagonal elements in the neutrino propagation matrix (similar to the $\Delta m^2 \sin^2 2\theta$ term induced by vacuum mixing). For massless neutrinos resonantly enhanced conversions can occur due to an interplay between the standard electroweak neutrino interactions and non-universal flavor diagonal neutrino interactions (FDNI) with matter [2,9].

In this paper we investigate the current status of the solution to the solar neutrino problem based on NSNI. Our analysis comprises both the measured total rates of Homestake [10], GALLEX [11], SAGE [12] and SuperKamiokande [13] and, for the first time in the context of NSNI, the full SuperKamiokande data set (corresponding to 825 effective days of operation) including the recoil electron spectrum and the day-night asymmetry.

For the solar input we take the solar neutrino fluxes and their uncertainties as predicted in the standard solar model by Bahcall and Pinsonneault (hereafter BP98 SSM) [14].

We find that non-standard neutrino interactions can provide a good fit to the solar neutrino data if there are rather large non-universal FDNI (of order $0.5\, G_F$) and small FCNI (of order a few times $10^{-3}\, G_F$).

0920-5632/01/\$ – see front matter
PII S0920-5632(01)01272-5

2. Neutrino flavor conversion induced by non-standard neutrino interactions

Any model beyond the standard electroweak theory that gives rise to the processes

$$\nu_e f \rightarrow \nu_\ell f\,, \quad (1)$$

$$\nu_\alpha f \rightarrow \nu_\alpha f\,, \quad (2)$$

where (here and below) $f = u, d, e$ and $\ell = \mu, \tau$ and $\alpha = e, \mu, \tau$, is potentially relevant for neutrino oscillations in the sun, since these processes modify the effective mass of neutrinos propagating in dense matter.

The matrix evolution for massless neutrinos that interact with matter via the standard weak interactions and the non-standard interactions in (1) and (2) is given by [2,3]:

$$U(\epsilon_{\nu_\ell}, \epsilon'_{\nu_\ell}) = \sqrt{2} G_F \begin{pmatrix} n_e(r) & \epsilon^f_{\nu_\ell} n_f(r) \\ \epsilon^f_{\nu_\ell} n_f(r) & \epsilon'^f_{\nu_\ell} n_f(r) \end{pmatrix} . \quad (3)$$

While W-exchange of ν_e with the background electrons gives rise to the well known forward scattering amplitude $\sqrt{2} G_F n_e(r)$, the FCNI in (1) induce a flavor changing forward scattering amplitude $\sqrt{2} G_F \epsilon^f_{\nu_\ell} n_f(r)$ and the non-universal FDNI are responsible for the flavor diagonal entry $\sqrt{2} G_F \epsilon'^f_{\nu_\ell} n_f(r)$ in eq. (3). Here

$$n_f(r) = \begin{cases} n_n(r) + 2n_p(r) & f = u \\ 2n_n(r) + n_p(r) & f = d \end{cases} \quad (4)$$

is the respective fermion number density at position r in terms of the proton [neutron] number density $n_p(r)$ [$n_n(r)$] and

$$\varepsilon = \epsilon^f_{\nu_\ell} \equiv \frac{G^f_{\nu_e \nu_\ell}}{G_F} \quad (5)$$

$$\varepsilon' = \epsilon'^f_{\nu_\ell} \equiv \frac{G^f_{\nu_\ell \nu_\ell} - G^f_{\nu_e \nu_e}}{G_F}\,, \quad (6)$$

describe, respectively, the relative strength of the FCNI in (1), and the new flavor diagonal, but non-universal interactions in (2). $G^f_{\nu_\alpha \nu_\beta}$ $(\alpha, \beta = e, \mu, \tau)$ denotes the effective coupling of the four-fermion operator

$$\mathcal{O}^f_\nu \equiv (\overline{\nu_\alpha} \nu_\beta)(\bar{f} f)\,. \quad (7)$$

that gives rise to such interactions.

3. Analysis of the solar neutrino data

In this section we present our analysis of the solution to the solar neutrino problem based on neutrino flavor conversions induced by NSNI in matter. Our main goal is to determine the values of ε and ε' that can explain the experimental observations without modifying the standard solar model predictions.

3.1. Rates

First we consider the data on the total event rate measured by the Chlorine (Cl) experiment [10], the Gallium (Ga) detectors GALLEX [11] and SAGE [12] and the water Cherenkov experiment SuperKamiokande (SK) [13]. We compute the allowed regions in parameter space according to the BP98 SSM [14], and we use the minimal χ^2 statistical treatment of the data following the analyses of Refs. [15,16].

In Fig. 1 the allowed regions in the parameter space of ϵ^d_ν and ϵ'^d_ν for neutrino scattering off d-quarks are shown at 90, 95 and 99 % confidence level (CL). The best fit point of this analysis is found at

$$\epsilon^d_\nu = 3.2 \times 10^{-3} \qquad \text{and} \qquad \epsilon'^d_\nu = 0.61\,, \quad (8)$$

with $\chi^2_{min} = 2.44$ for $4 - 2 = 2$ degrees of freedom (DOF).

Similar plots for scattering off u-quarks were obtained, and can be found on [17]. The best fit point of this analysis is found at

$$\epsilon^u_\nu = 1.32 \times 10^{-3} \qquad \text{and} \qquad \epsilon'^u_\nu = 0.43\,, \quad (9)$$

with $\chi^2_{min} = 2.64$ for two DOF.

3.2. Zenith Angle Data

Next, we consider the zenith angle dependence of the solar neutrino data of the SuperKamiokande experiment. NSNI with matter may affect the neutrino propagation through the earth resulting in a difference between the event rates during day and night time. The experimental results suggest an asymmetry between the total data collected during the day (1 bin) and the total data observed during the night (5 bins) [18].

In order to take into account the earth matter effect we also use a χ^2-function that characterizes

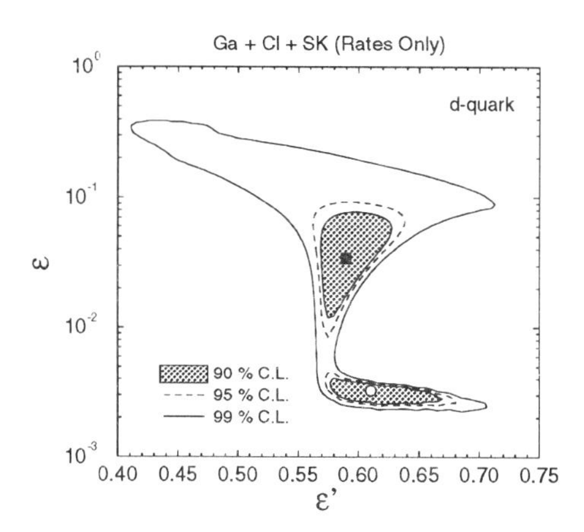

Figure 1. Region of $\varepsilon = \epsilon_\nu^d$ and $\varepsilon' = \epsilon'^d_\nu$ which can explain the total rates measured by the four solar neutrino experiments considered.

the deviations of the six measured bins from the predicted values of the rates.

The consequence of taking into account this zenith dependence can be seen in the next session, where the combined analysis is presented.

3.3. Combined Analysis

Our final result is the fit derived from the combined analysis of all presently available solar neutrino data. In Fig. 2 we show the allowed regions for $(\epsilon_\nu^d, \epsilon'^d_\nu)$ using both the results from the total rates from the Chlorine, GALLEX, SAGE and SuperKamiokande solar neutrino experiments together with the 6 bins from the SuperKamiokande zenith angle data discussed previously. Although adding the spectral information to our analysis does not change the shape of allowed regions nor the best fit points, it is included in order to determine the quality of the global fit.

For neutrino scattering off d-quarks the best fit for the combined data is obtained for

$$\epsilon_\nu^d = 0.028 \qquad \text{and} \qquad \epsilon'^d_\nu = 0.585 \tag{10}$$

with $\chi^2_{min} = 29.05$ for $28 - 4 = 24$ DOF, corresponding to a solution at the 22 % CL (see Fig. 2). For neutrino scattering off u-quarks the best fit for the combined data is obtained for

$$\epsilon_\nu^u = 0.0083 \qquad \text{and} \qquad \epsilon'^u_\nu = 0.425 \tag{11}$$

with $\chi^2_{min} = 28.45$ for $28 - 4 = 24$ DOF corresponding to a solution at the 24 % CL (see [17]).

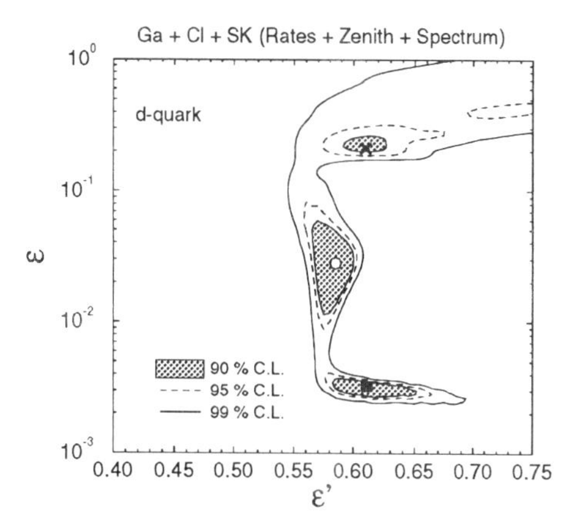

Figure 2. Region of $\varepsilon = \epsilon_\nu^d$ and $\varepsilon' = \epsilon'^d_\nu$ which can explain the total rates measured by the four solar neutrino experiments considered and the zenith angle dependence measured by SuperKamiokande.

We have performed the same calculations presented here for a free Boron-neutrino flux normalization, and for interactions with u-quarks. The results can be seen in [17]. We also present there an analysis of the phenomenological bounds on the parameters ε and ε', concluding that this solution is still viable if we consider the $\nu_e - \nu_\tau$ channel.

4. Conclusions

According to our χ^2 analysis non-standard neutrino interactions (NSNI) can provide a good fit to the solar neutrino data provided that there are rather large non-universal FDNI (of order $0.5\,G_F$) and small FCNI (of order $10^{-2} - 10^{-3}\,G_F$). The fit to the observed total rate, day-night asymmetry, seasonal variation and spectrum distortion of the recoil electron spectrum is comparable in quality to the one for standard neutrino oscillations.

Even though we consider the conventional oscillation mechanisms as the most plausible solutions to the solar neutrino problem, it is important to realize that in general New Physics in the neutrino sector include neutrino masses and mixing, as well as new neutrino interactions. While it is difficult to explain the atmospheric neutrino problem [19] and the LSND anomalies [20] by NSNI [21,22], we have shown in this paper that a solution of the solar neutrino problem in terms of NSNI is still viable. The ultimate goal is of course a direct experimental test of this solution. The upcoming solar neutrino experiments will provide a lot of new information which hopefully will reveal the true nature of the solar neutrino problem.

REFERENCES

1. L. Wolfenstein, Phys. Rev. **D17**, 2369 (1978).
2. M. M. Guzzo, A. Masiero and S. T. Petcov, Phys. Lett. **B 260**, 154 (1991).
3. V. Barger, R. J. N. Phillips and K. Whisnant, Phys. Rev. **D 44**, 1629 (1991).
4. E. Roulet, Phys. Rev. **D 44**, 935 (1991).
5. S. Degl'Innocenti and B. Ricci, Mod. Phys. Lett. **A 8**, 471 (1993).
6. G. L. Fogli and E. Lisi, Astroparticle Phys. **2**, 91 (1994).
7. P. I. Krastev and J. N. Bahcall, "FCNC solutions to the solar neutrino problem", hep-ph/9703267.
8. S. Bergmann, Nucl. Phys. **B 515**, 363 (1998) [hep-ph/9707398].
9. Resonant neutrino conversion induced by non-orthogonal massless neutrinos was first discussed by J.W.F. Valle in Phys. Lett. **B 199**, 432 (1987).
   However this mechanism can not induce a large effect on the solar neutrinos due to the stringent constraints on the model parameters. See also:
   P. Langacker and D. London, Phys. Rev. **D 38**, 886 (1988); *ibid* **38**, 907 (1988);
   H. Nunokawa, Y. -Z. Qian, A. Rossi and J. W. F. Valle, Phys. Rev. **D 54**, 4356 (1996) [hep-ph/9605301] and references therein;
   S. Bergmann and A. Kagan, Nucl. Phys. **B 538**, 368 (1999) [hep-ph/9803305].
10. T. B. Cleveland *et al.* (Homestake Collaboration), Astrophys. J. **496**, 505 (1998).
11. W. Hampel *et al.* (GALLEX Collaboration), Phys. Lett. **B 447**, 127 (1999).
12. J. N. Abdurashitov *et al.* (SAGE Collaboration), Phys. Rev. **C 60**, 055801 (1999).
13. Y. Fukuda et al. (SuperKamiokande Collaboration), Phys. Rev. Lett. **82**, 1810 (1999).
14. J. N. Bahcall, S. Basu and M. H. Pinsonneault, Phys. Lett. **B 433**, 1 (1998); see also J. N. Bahcall's home page, http://www.sns.ias.edu/~jnb.
15. G. L. Fogli, E. Lisi and D. Montanino, Phys. Rev. **D 49**, 3226 (1994).
16. G. L. Fogli and E. Lisi, Astropart. Phys. **3**, 185 (1995).
17. S. Bermann, M. M. Guzzo, P. C. de Holanda, P. I. Krastev and H. Nunokawa, Phys. Rev. **D 62**, 073001 (2000).
18. Y. Suzuki, "Solar Neutrinos", talk given at the Lepton Photon Conference, 1999.
19. For recent analysis, see *e.g.*:
    M. C. Gonzalez-Garcia, *et al.*, Phys. Rev. **D 58**, 033004 (1998);
    M. C. Gonzalez-Garcia, H. Nunokawa, O. L. G. Peres and J. W. F. Valle, Nucl. Phys. **B 543**, 3 (1999);
    N. Fornengo, M. C. Gonzalez-Garcia and J. W. F. Valle, [hep-ph/0002147].
20. C. Athanassopoulos *et al.* (LSND Collaboration), Phys. Rev. Lett. **77**, 3082 (1996); Phys. Rev. Lett. **81**, 1774 (1998).
21. S. Bergmann and Y. Grossman, Phys. Rev. **D 59**, 093005 (1999) [hep-ph/9809524].
22. S. Bergmann, Y. Grossman and D.M. Pierce, Phys. Rev. **D 61**, 53005 (2000) [hep-ph/9909390].

ELSEVIER

Nuclear Physics B (Proc. Suppl.) 97 (2001) 239–242

Muon Bremsstrahlung and Muonic Pair Production in Air Showers.

A. N. Cillis[a] and S. J. Sciutto[a*]

[a]Laboratorio de Física Teórica
Departamento de Física
Universidad Nacional de La Plata
C. C. 67 - 1900 La Plata
Argentina

The objective of this work is to report on the modifications in air shower development due to muon bremsstrahlung and muonic pair production. In order to do that we have implemented new muon bremsstrahlung and muonic pair production procedures in the AIRES air shower simulation system, and have used it to simulate ultra high energy showers in different conditions.

The influence of the mentioned processes in the global development of the air shower is important for primary particles of large zenith angles, while they do not introduce significant changes in the position of the shower maximum.

1. Introduction

The physics of ultra high energy cosmic rays plays an important role in our days. These primary particles can not be detected directly. When an ultra high energy astroparticle interacts with an atom of the Earth's atmosphere, it produces a number of secondary particles that continue interacting and generating more secondary particles. This process is generally known as the air shower.

We have developed a set of programs to simulate air showers and manage all the output data. Such simulating system is known as AIRES(AIRshower Extended Simulations)[1].

We started working on the topic of the electromagnetic processes in air showers some years ago. We analyzed the modification in the shower development due to the reduction of the electron bremsstrahlung and electron pair production by the LPM effect and the dielectric suppression [2]. We also studied the influence of the geomagnetic field in an air shower [3].

The main goal of this work is to analyze two processes that take place in an air shower: muon bremsstrahlung and muonic pair production (electron and positron). At energies high enough, these processes become more important than ionization, and therefore these mechanisms account for virtually all the energy losses for high energy muons.

In order to study the modifications that these effects introduce in an air shower, we developed new procedures for such mechanisms and incorporated them in the AIRES air shower simulation system [1]. The AIRES code has then been used as a realistic air shower simulator to generate the data used to make our analysis.

This work is organized as follows: we start in section 2 with a brief summary of the theory of the muon bremsstrahlung and muonic pair production; in section 3 we show the results of our simulations; finally we present our analysis and comments in the conclusion section.

2. Theory

2.1. Theory of the Muon Bremsstrahlung

The cross section for muon bremsstrahlung (MB) is calculated by the standard method of QED [4] similary as in the case of electron bremsstrahlung.

The first approach to the MB theory was due to Bethe and Heitler [5–7]. They considered in their calculation the screening of the atomic electrons.

*Fellow of CONICET (Argentina)

PII S0920-5632(01)01273-7

After this first formulation some corrections were introduced. Kelner, Kokoulin and Petrukhin [8] also included the bremsstrahlung with the atomic electrons. The nuclear form factor was investigated by Christy and Kusaka [9] for the first time and then by Erlykin [10]. Petrukhin and Shestakov [11] found that the influence of the nuclear form factor is more important than the ones predicted by the previous papers. This last results have been confirmed by Andreev et al. [12] who also considered the excitation of the nucleus.

2.2. Theory of Muonic Pair Production

In the lowest significant order of perturbation theory the muonic pair production(MPP) is a process of 4th order in QED.

Racah [13] was the first to calculate the MPP cross section in the relativistic region without taking into accuont the atomic and the nuclear form factor. Thereafter, Kelner [14] included the correction due to the screening of the atomic electrons. The analytical expression for any degree of screening was introduced by Kokoulin and Petrukhin [15]. Those authors also took into account [16] the correction due to the nuclear form factor. We wish to emphasize that the influence of the nuclear size is more important when the energy tranferred to the pair is large [16]. This last case is important for the air showers and therefore the nuclear size effect needs to be taken into account in the simulations.

3. Air Showers Simulations

We have calculated the total cross section, and equivalently, the mean free path (MFP), for both MB and MPP processes.

Figure 1 shows the MFP in g/cm^2 for MB, MPP and muon-nucleon interaction as functions of the kinetic energy of the initial muon. The information in this figure can be compared with the depth of the atmosphere. The vertical depth of the atmosphere is about 1000 g/cm^2 while in the case of very inclined showers (85° of zenith angle) is about 9000 g/cm^2. Therefore, the probabilities of MB and MPP will not be very large except in the case of large zenith angles. The muonic

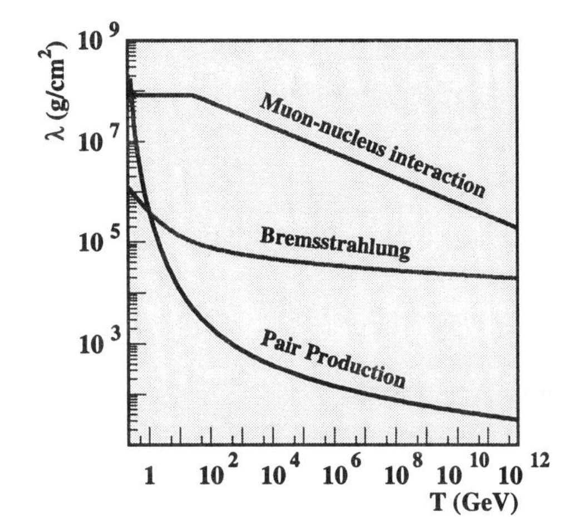

Figure 1. Mean free path: MB, MPP and muon nuclear interaction vs the kinetic energy of the initial muon.

component of the showers (number of muons divided by the number of electrons and gammas) at ground level becomes very important for zenith angles larger than 60° (see, for example, figure 2 in reference [3]); this is another reason to expect that the influence of both effects will be more appreciable under those conditions.

The MFP's for MB and MPP diminish when the kinetic energy of the muon increases and therefore, the influence of both process will be more important for large energies.

Due to the fact that the MFP of muon-nucleus interaction [2] is larger (more than one order of magnitude) than the corresponding for MB we do not take this effect into account in our calculations.

In order to analyze the influence of MB and MPP in the development of air showers initiated by ultra high astroparticles we have performed simulations using the AIRES program [1] with different initial conditions: primary particles (protons, irons, muons), primary energies (10^{18}, 10^{19} and 3×10^{20} eV) and zenith angles.

[2]We used the same parametrization of the total cross section used by GEANT[17]

To start with the analysis of the simulated data, let us consider the case of a single muon (eventually produced during the development of a given shower). This particle may generate a secondary shower if the processes of MB and MPP are taken into account. This effect is clearly illustrated in figure 2a where the longitudinal development of all charged particles is plotted versus the slant depth. In this case the initial particle is a muon of 10^{14} eV. When MB and MPP are not taken into account there is practically no production of new particles. On the other hand, when the effects are considered, it is possible to appreciate a secondary shower.

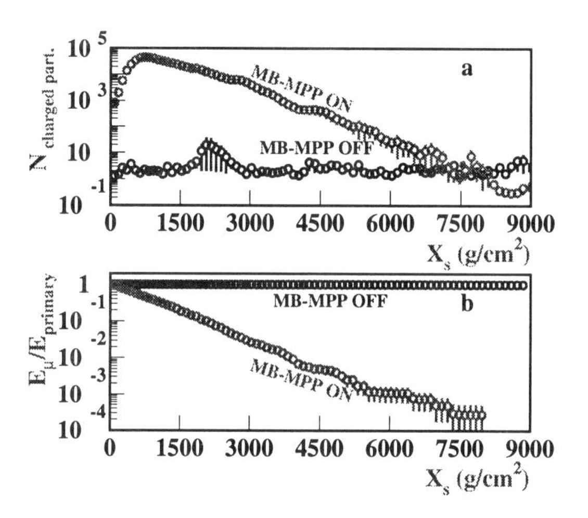

Figure 2. a)Longitudinal development of all charged particles vs the slant depth. Primary particle (Energy): muon (10^{14} eV). b)Longitudinal development of muon fractional energy vs the slant depth.

In figure 2b the longitudinal development of the muon fractional energy (muon energy/primary energy) is plotted. When the effects are not taken into account the muon practically does not loss energy during all its path, while if the effects are considered, the muon energy loss is significant.

We have also studied the modifications in the global observables of the shower, for example when the primary particle is a proton.

We have plotted in figures 3 and 4 different obsevables for 3×10^{20} eV proton shower with zenith angle 85°, taking and not taking into a account the effects of MB and MPP. Figure 3 represents the longitudinal development of electrons and positrons ($e^{+/-}$) versus the slant depth. The number of $e^{+/-}$ increase when the effects are taken into account. The differences between the two cases are clearly noticeable in the tail of the showers. The longitudinal development in energy

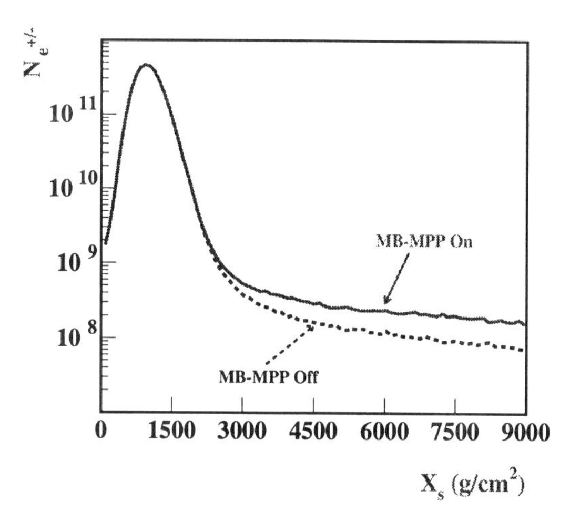

Figure 3. Longitudinal development of electrons and positrons vs the slant dept. Primary particle (Energy) proton (3×10^{20} eV).

of muons is plotted in figure 4. In this case the difference is about 15% at the tail. The energy of muons diminishes when the effects are considered.

For the longitudinal development of muons versus the slant depth the difference between the cases taking and not taking into account the effects is less significant, about 2 % (not plotted here).

The modification in the shower development due to the MB and MPP diminishes for small zenith angle of the primary particle. We did not find significant modification of the shower development for zenith angles smaller than 45°.

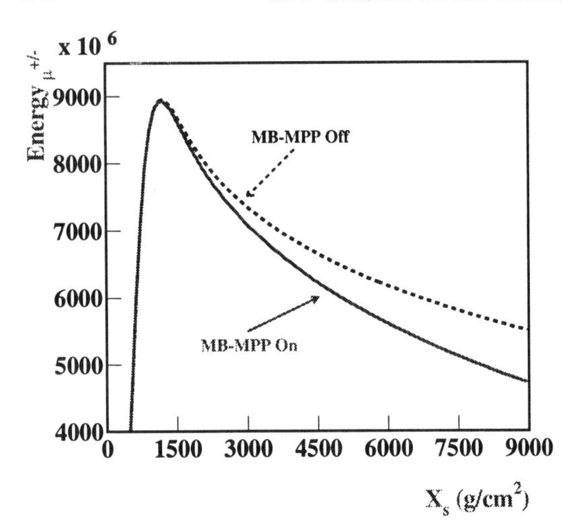

Figure 4. Longitudinal development in energy of muon vs the slant depth. Primary particle (Energy) proton (3×10^{20} eV).

4. Conclusions

Due to MB and MPP, a high energy muon in the shower can generate a secondary shower. This practically does not occur if the mentioned mechanisms are not taken into account.

The influence of MB and MPP in the total development of the air showers is more important for primary particle of large zenith angles. Below zenith angles of 45 degrees we do not observe any significant difference between the cases where these effects are or not taken into account.

The MB and MPP do not generate visible modifications in the position of the maximum development of the shower. The modification that we observe affect the shower development well past its maximum (tail of the shower).

We are performing more simulations in order to make a complete study of the influence of MB and MPP in an air shower [18].

5. Acknowledgments

This work was partially supported by CONICET, Agencia Nacional de Programación Científica and FOMEC program, Argentina.

REFERENCES

1. S. J. Sciutto, AIRES, a system for air shower simulations. User's manual and reference guide, version 2.2.0, preprint astro-ph 9911331 (1999). The AIRES software and documentation are also available electronically at the following Web address: `www.fisica.unlp.edu.ar/auger/aires`.
2. A. N. Cillis, H. Fanchiotti, C. A. Garcia Canal and S. J. Sciutto, Phys. Rev. D 59 (1999) 113012.
3. A. N. Cillis and S. J. Sciutto, Journal of Physics G: Nuclear and Particle Physics 26 Number 3 (2000) 309.
4. W. Greiner and J. Reinhardt, Quantum Electrodynamics, Springer-Verlag (1992).
5. H. A. Bethe, Proc. Cambridge Phil. Soc. 30 (1934) 524.
6. H. A. Bethe and W. Heitler, Proc. Roy. Soc. (London) 146 (1934) 83.
7. B. Rossi, High Energy Particles, Prentice Hall (1956).
8. S. R. Kelner, R. P. Kokoulin and A. A. Petrukhin, Phys. Atomic Nuclei 60 (1997) 576.
9. R. F. Christy and S. Kusaka, Phys. Rev. 59 (1941) 405.
10. A. D. Erlykin, Izv. Akad. Nauk SSSR, Ser. Fiz. (1965) 29 1757.
11. A. A. Petrukhin and V. V. Shestakov, Canadian Journal of Physics 46 (1967) s377.
12. Yu M. Andreyev, L. B. Bezrukov E. V. and Bugaev, Phys. Atomic Nuclei 57 (1994) 2066.
13. G. Racah, Nuovo Cimento 14 (1937) 93.
14. S. R. Kelner Yadernaya Fiz., 5 (1967) 1092.
15. R. P. Kokoulin and A. A. Petrukhin, Proc. 11th. Intern. Conf. on Cosmic Rays, Budapest, Acta Phys. Acad. Sci. Hung., 29 Suppl. 4 (1969) 277.
16. R. P. Kokoulin and A. A. Petrukhin, Proc. 12th. Intern. Conf. on Cosmic Rays, Hobart 6 (1971) 2436.
17. GEANT: CERN Program library Long Writeup W5013 (1994).
18. A. N. Cillis and S. J. Sciutto, in progress.

ELSEVIER

Nuclear Physics B (Proc. Suppl.) 97 (2001) 243–246

NUCLEAR PHYSICS B PROCEEDINGS SUPPLEMENTS
www.elsevier.nl/locate/npe

Multiplicity Spectra of Muon Bundles Deep Underground

A.E.Chudakov, A.S.Lidvansky, A.L.Tsyabuk, A.V.Voevodsky [a], and J.Szabelski[b]

[a]Institute for Nuclear Research
Russian Academy of Sciences
Moscow 117312, Russia

[b]The Andrej Soltan Institute of Nuclear Studies
Lodz, Poland

The multiplicity spectrum of muon bundles measured by the Baksan Underground Scintillation Telescope (BUST) is presented for the depths as large as 5000 hg/cm^2 and a threshold energy of 3.2 TeV. The experimental data are compared with predictions based on a few models of primary cosmic ray composition.

1. INTRODUCTION

Our muon bundle data cover an energy range from 10^{13} to $10^{17} eV$. The region of the knee of the primary cosmic ray spectrum at about $3 \cdot 10^{15} eV$ lies precisely near the center of this range. Thus, there is a hope of probing possible changes related to the knee (in particular, cosmic ray composition) by analyzing muon bundles. It should be emphasized that the muon bundle data could be especially useful for calibration and sewing together the data of direct measurements (balloons, satellites) and indirect observations of extensive air showers, since there is partial overlap in the ranges of both these methods. The general approach to the analysis was formulated in [1] and further developed in [2, 3]. In this paper, we present the data for large slant depths. Such data have already been obtained by some very deep underground facilities [4, 5, 6]. The Baksan telescope, located at a shallower depth, can reach comparable depths by analyzing strongly inclined bundles.

2. EXPERIMENTAL

The Baksan scintillation telescope [7] is located under a mountain slope in the underground cavity at a distance of 550 m from the entrance into a horizontal mine. Its effective depth is 850 hg/cm^2, and the effective threshold energy is 200 GeV. The depth of 850 hg/cm^2 is minimum, the slant depths can reach up to 6000 hg/cm^2 with corresponding energy threshold of about 6-10 TeV. The BUST is a four-storey building, nearly cubic in form (16.7 x 16.7 x 11.1 m^3). Four vertical sides and four horizontal planes are covered by liquid scintillation detectors of a standard type. The total number of detectors is 3150. A standard detector has dimensions of 0.7 x 0.7 x 0.3 m^3 and comprises an aluminum tank filled with liquid scintillator.

A specific trigger is used for recording muon bundle events. Preliminary processing of data is made by an on-line computer, which is also used for data acquisition for complex and rare events. Off-line computer program for muon bundle events includes reconstruction of trajectories, determination of multiplicity of a bundle, zenith and azimuth angles, and distances between parallel trajectories in the bundle. A rather strict criterion was used for reconstructing the direction of muon bundles. In order to include a particular muon bundle into the analysis, three planes of the telescope (at least one of them being an internal plane) should be crossed by each track.

3. RESULTS

The experimental multiplicity spectra were obtained at the Baksan telescope for the maximum depth of 5000 hg/cm^2. The running time of $2.42 \cdot 10^8$ s (7.7 years) is used in this work, which

PII S0920-5632(01)01274-9

Table 1
Experimental number of muon bundles N for each multiplicity m.

m	1	2	3	4	5	6	7	8	9
N	54595	1819	252	65	30	11	10	7	1

is three times more than in [3]. The arrival directions of muon bundles correspond to zenith angles of $50^o - 60^o$. The experimental multiplicity distribution is presented in the Table 1 for the above depth (threshold energy 3.2 TeV).

The calculations of expected muon multiplicity spectra have been made using the following assumptions:

- From the superposition model, the mean number of muons for a nucleus A is assumed to be $N_\mu = AF(E_p)$, where $F(E_p)$ is the mean number of muons for primary proton with energy E_p.
- The probability of obtaining exactly m muons obeys a Poisson law with a mean value N_μ. Obviously, it would be more appropriate to use a negative binomial distribution. However, as was shown in paper [8], multiplicity spectra calculated using these two types of distributions differ by no more than 15 %.
- The differential flux of each component is described in the form

$$P_i(A) = \rho_i A E^{-\gamma_i(A)} \quad (1)$$

where $\rho_i(A)$ is the weight, and $\gamma_i(A)$ is the spectral index of the i-th component.

- The mean number of muons of energy $E \geq E_{th}$ from a primary nucleon with energy E_p was taken from [9]:

$$F(E_p) \propto (\sec\theta) E_{th}^{-1} x^{-0.728} (1-x)^{7.73} \quad (2)$$

where E_{th} is the threshold energy and $x = E_{th}/E_p$.

- $\Delta(R)$ is the fraction of muons that hit the telescope when the core of a bundle is at a distance R from the telescope centre. $\Delta(R)$ is calculated using the lateral distribution function defined for this experiment in [3].

Then the probability that m muons hit the detecting area is equal to

$$I(m) = \sum_i \rho_i(A) E_{th}^{-\gamma_i+1} \int_0^\infty 2\pi R dR \times \int_0^1 \exp[-\overline{N}_\mu \Delta(R)] \cdot [\overline{N}_\mu \Delta(R)]^m x^{\gamma_i - 2} dx / m! \quad (3)$$

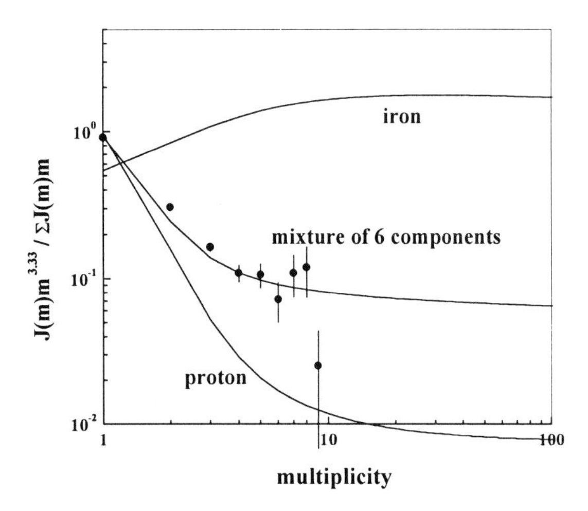

Figure 1. Multiplicity spectra from experiment (points) and calculation (curves).

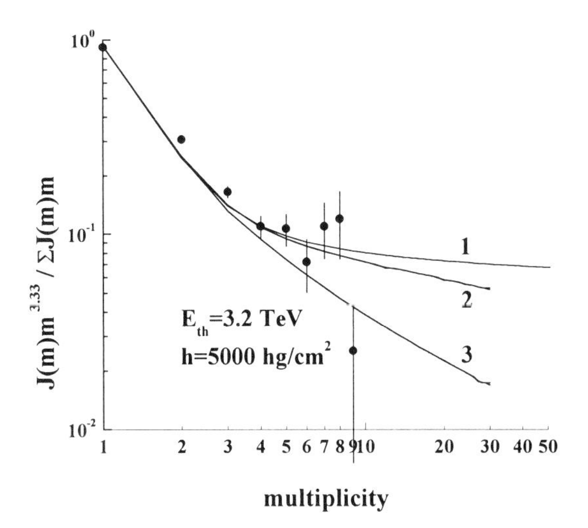

Figure 2. Multiplicity spectra from experiment (points) and calculation (curves).
1 - without a knee
2 - knee at fixed energy per nucleon
3 - knee at fixed energy per nucleus

The results of calculations and the comparison with experimental data measured by the BUST are shown in Fig. 1 for $E_{th} = 3.2\ TeV$ (maximum depth). The assumed weights of the components are as follows: $\rho_1 = 0.939$, $\rho_4 = 0.055$, $\rho_9 = 0.0009$, $\rho_{14} = 0.0035$, $\rho_{28} = 0.0011$, and $\rho_{56} = 0.0003$ at a fixed energy per nucleon, and $\rho_1 = 0.38$, $\rho_4 = 0.23$, $\rho_9 = 0.015$, $\rho_{14} = 0.13$, $\rho_{28} = 0.13$, and $\rho_{56} = 0.11$ at fixed energy per nucleus. The differential spectral index is $\gamma = 2.7$. The lines in this figure correspond to the calculations for pure protons and pure iron nuclei. The results for a mixture of all six components above of primary cosmic rays are also presented. It is seen that the experimental points slightly exceed the calculation results. As was shown in [2], we can evaluate the effective atomic mass number from Fig. 1. The experimental value is (see [2] for details)

$$(A_{eff})_{exp} = \left[\frac{I(m)\cdot m^{(\frac{\gamma}{\delta}+1)}}{\frac{(\gamma-\delta)}{\delta}\cdot\lambda(m)}\right]^{(\frac{\delta}{\gamma-\delta})}\cdot\frac{1}{N_0'} \quad (4)$$

where $\gamma = 1.7$ and $\delta = 0.728$. $N_0' = 0.021$ and $\lambda(m) = 1.44$ for $m = 6$ are calculated using equations (1-5) and the condition that $\lambda(m) \to \infty$. The value

$$(A_{eff})_{calc} = \left[\frac{\sum \rho_i A_i^{\frac{\gamma}{\delta}}}{\sum \rho_i A_i}\right]^{\frac{\delta}{\gamma-\delta}} \quad (5)$$

was calculated for the above mentioned composition and $(A_{eff})_{calc} = 4.9$, while $(A_{eff})_{exp} = 5.4$.

In Fig. 2, the experimental multiplicity spectrum and results of the calculations for the primary cosmic ray spectrum are shown. Three curves correspond to the cases without a knee (1), the knee at a fixed energy per nucleon, $E_{knee} = 3\cdot 10^{15}\ eV$ (2), and the knee at a fixed energy per nucleus (3). It is seen that spectrum (1) is fairly close to the experimental data, spectrum (2) does not contradict the data, and spectrum (3) is inconsistent with the experimental data.

4. CONCLUSION

The experimental data on multiple muon events measured by the BUST at large depths agree well with the suggestion of a constant mass composition and an integral exponent of primary power law spectrum $\gamma = 1.7$. The data do not contradict a knee in the primary spectrum at an energy of $3\cdot 10^{15} eV$ at the fixed energy per nucleon and are inconsistent with a knee at the fixed energy per nucleus. The experimental value of the effective atomic mass number is equal to 5.4, which is slightly larger than expected.

This work is supported by the Russian Foundation for Basic Research, project no. 99-02-16146.

REFERENCES

1. Chudakov A.E., Proc. 16th ICRC, Kyoto, 10 (1999), p. 192.
2. Budko E.V. et al., Proc. 19th ICRC, La Jolla, 8 (1985), p. 24.
3. Chudakov A.E. et al., Yadernaya Fizika (Russ. J. Nucl. Phys.) 56 (1993), p. 143.
4. Aglietta M. et al., Proc. 21st ICRC, Adelaide, 9 (1990), p.362.

5. Berger Ch. et al., Phys. Rev., D40 (1989), p. 2163.
6. Ambrosio M. et al. (MACRO Coll.), Proc. 24th ICRC, Roma, 2 (1995), p. 689.
7. Alekseyev E.N. et al., Proc. 16th ICRC, Kyoto, 10 (1979), p. 276.
8. Boziev S.N., Yadernaya Fizika (Russ. J. Nusl. Phys.) 52 (1990), p.500.
9. Bilokon H. et al., Proc. 21st ICRC, Adelaide, 9 (1990), p. 366.

ELSEVIER

Nuclear Physics B (Proc. Suppl.) 97 (2001) 247–250

www.elsevier.nl/locate/npe

The acceptance of fluorescence detectors for quasi horizontal showers induced by weak interacting particles

J. C. Díaz*[a], R. C. Shellard†[a] and M. G. Amaral‡[b]

[a] Centro Brasileiro de Pesquisas Físicas-CBPF
Rua Dr.Xavier Sigaud 150, Rio de Janeiro, RJ 22290-180 Brazil

[b] Instituto de Física, Universidade Federal Fluminense,
Av. Litorânea, s/n, Niterói, RJ 24210-340 Brazil

We have done an estimate of the acceptance for fluorescence detectors to atmospheric showers with zenith angles larger than 60 degrees and which penetrate deeply into the atmosphere. This class of showers have the signature of UHE neutrinos. We have used a fast simulation program for this type of detectors and give a lower bound to the acceptance.

1. Introduction:

The aim of this paper is to report on the evaluation of the acceptance of fluorescence detectors to quasi horizontal showers, that develop deep into the atmosphere. This class of showers have the typical signature of ultra high energy (UHE) neutrinos.

The Earth is opaque to UHE neutrinos [1], however, they will interact deeply in the atmosphere, in contrast to hadrons or electromagnetic particles. Neutrinos have a small cross section with matter in the atmosphere, so they have a homogeneous probability to interact at any point of it. The electromagnetic component of the normal showers, which are generated in the atmosphere, are suppressed sharply at slant depths which correspond to zenithal angles of about 60° [2].

A shower induced by a neutrino will behave like a hadronic one when deep inelastic charged or neutral current interactions are taken into account. Showers induced by an electron neutrino will have a superposition of an pure electromagnetic shower produced by the emerging electron superimposed on the hadronic shower. We have considered, as a first approximation, that the purely electromagnetic component of neutrino induced showers summed to the hadronic component can be described by the Gaisser-Hillas parametrization [3].

2. The atmosphere and detector simulation

We have simulated a fluorescence detector [4] with a single eye covering an azimuthal range of 360°, instrumented with mirrors and cameras which are identical to the preliminary design of the Pierre Auger Observatory [5]. The eye is made of 12 mirrors, each one covering roughly an angle of 30° in elevation from the horizon (actually the lower part of the mirror is raised by 2° from the horizon line) by 30° in azimuth width. The camera of each mirror is made up of 440 pixels, each covering a solid angle of about 1.5° squared. Each pixel has a hexagonal format and they are disposed in 22 rows with 20 pixels each. Each successive row is offset in relation to the previous one by half width of a pixel (an angle of 0.75°).

We use the Linsley parametrization for the atmosphere, which is based on the experimental data of the US standard atmosphere [6]. The density of the atmosphere is limited to the first 113km above the sea level and is considered zero above it in this parametrization. The vertical

*johana@lafex.cbpf.br
†shellard@lafex.cbpf.br
‡amaral@if.uff.br

PII S0920-5632(01)01275-0

depth, defined by

$$\chi_v(h) = \int_h^\infty \rho(z)\, dz, \tag{1}$$

is the quantity of matter (in g/cm^2) that a particle crosses as it travels vertically through the atmosphere. This relation can be inverted to obtain the value of the altitude (in km) corresponding to a vertical depth. The slant depth, defined by

$$\chi_s(z) = \int_z^\infty \rho(z_v)\, dl, \tag{2}$$

is the quantity of matter that a particle crosses during its actual trajectory. One may relate the slant depth to the vertical depth, for a given height, as

$$\chi_s(h) = \frac{\chi_v(h)}{\cos\theta}, \tag{3}$$

for values of θ, the zenith angle of the shower, smaller than 60°. Here one may ignore the curvature of the Earth. However, for larger values of the zenithal angle we must take it into account. The height of a particle that has traveled through the atmosphere towards a point O that lies at a distance l from the collision point is given by

$$H = R\left((1 + \frac{l^2}{R^2} + 2\cos\theta)^{0.5} - 1\right), \tag{4}$$

where θ is the zenithal angle of the shower line at the point O and R is the radius of the Earth. If $l << R$, then H can be approximated by

$$H = l\cos\theta + \frac{(l\sin\theta)^2}{2\,R}\,. \tag{5}$$

For $\theta < 60°$ the second term of the sum is negligible and we recover the usual expression for H. The effect of the curvature of the Earth is taken into account in the simulation program.

The simulation program generates the shower and consequently the intensity of the light with proper wavelength distribution. The intensity of the light emitted at each stage of the shower is proportional to the number of charged particles [7], with a factor which is reasonably independent of the height of the shower. It is given by

$$\frac{dN_\gamma}{dl} = N_f \times N_e, \tag{6}$$

where N_f is the fluorescence yield (number of photons/charged particle/meter) and N_e is the number of secondary particles produced in the shower. The phenomenon of fluorescence light production has very low efficiency, but this is compensated by the huge number of secondary particles produced in the shower. This number, which is a function of the slant depth of the point of emission, χ_s, is given, in the Gaisser-Hillas parametrization of the longitudinal shower development [3], by

$$\begin{aligned} N_e(\chi) &= N_{\max}\left(\frac{\chi - \chi_0}{\chi_{\max} - \chi_0}\right)^{(\chi_{\max}-\chi_0)/\lambda} \\ &\times \exp\left(\frac{(\chi_{\max} - \chi_0)}{\lambda}\right) \end{aligned} \tag{7}$$

where χ, $\chi_{\max}$ and χ_0 are given in units of g/cm^2, $\lambda = 70$ g/cm^2 and $N_{\max}$ is the shower size at its maximum development. The light, as it travels through the atmosphere, is attenuated, either by absorption or by scattering. Absorptive processes are important in the region of wavelengths below 290 nm and above 800 nm. Between these two limits the mechanism of attenuation is dominated by Rayleigh scattering [8] (scattering by molecules of the atmosphere) and Mie scattering [8][9] (scattering by natural and man-made aerosols). Rayleigh scattering is a stable phenomena, dependent on the atmosphere density profile, and which has an exponential profile that can be easily simulated. Mie scattering, on the other hand, is a highly variable phenomena, which depends on the particle composition, its distribution and on the vertical density profile. The aerosol scattering can be simulated, but its parametrization requires a continuous monitoring of the sky. The attenuation of light A_{TT} between two points at slant depth χ_1 and height h_1 and χ_2 and height h_2, is separated into the Rayleigh scattering, described by

$$T_{Ray} = \exp\left[-\frac{|\chi_1 - \chi_2|}{\chi_R}\left(\frac{400}{\lambda}\right)^4\right] \tag{8}$$

and the Mie scattering, by

$$T_{Mie} = \exp(-\frac{h_M}{l_M\cos\theta}\left(\exp\left(\frac{h_1}{h_M}\right) - \exp\left(\frac{h_2}{h_M}\right)\right))$$

with the cumulative effect

$$A_{TT} = T_{Ray} \times T_{Mie} \tag{9}$$

where χ_R is the mean free path for scattering at $\lambda = 400$ nm, with value $\chi_R = 2974$ g/cm^2, λ is the wavelength of the scattered light, l_M is the Mie scattering mean free path ($l_M \simeq 14$ km at $\lambda = 360$ nm), h_M is the aerosol scale high ($h_M \simeq 1.2$ km) and θ is the zenith angle.

We have calculated the acceptance of a single eye Fluorescent Detector to showers with varying energies and zenith angles larger than 60°. We have taken into account not only showers that hit the ground, but also those that cross the whole volume visible to the eye, without touching the ground or fall on the ground very far from the array area. Those are either horizontal showers or showers hitting the ground very far away.

In order to compute the neutrino acceptance of the fluorescence detector we have used the procedure in which we specify the triggering conditions, with the requirement that at least 5 neighbouring pixels are activated by the shower. We estimate the visible atmosphere volume to the fluorescence detector eye and generate quasi-horizontal showers in a larger volume. We generate the showers by requiring that the χ_{max} point is contained within the simulation volume, but disregard those showers which have the initial interaction point, χ_0, at less than 500 g/cm^2. We then compute the fraction of showers generated in this volume, which will trigger the detector, evaluating then the acceptance of the detector in units of water equivalent volume steradian (km^3.w.eq.sr). The fluorescence detectors have a limited duty cycle, being able to operate only on clear and moonless nights, about 10% of total time. We present our results, in this note, assuming a duty cycle of 100%, so that different weather conditions may be folded in the estimation. In order to compare the neutrino acceptance of fluorescence detectors to others, the duty cycle must be folded in. We have simulated 15 000 showers with energies ranging from 10^{18} to 10^{20}eV simulated in the effective volume shown in table 1. The position of the $\chi_{\max}$ for the showers were assigned to uniform random points within the generating volume, which has a cylindrical format. The position of $\chi_{\max}$ was distributed homogeneously in g/cm^2. This approach is justified for UHE neutrinos, since they can interact roughly with equal probability anywhere in the atmosphere, but the interaction depends on the amount of matter crossed, not linearly on the distance traveled. We have set a top value for the height of the simulation volume, set to 30 km. The zenith angles are generated randomly between 60° and 90° ($0 < \cos\theta < 0.5$) and the azimuth angles, between 0 and 2π.

Table 1
Effective volumes used to generate neutrino induced showers.

Energy	Radius	Height		Volume
(eV)	(km)	(km)	g/cm^2	(km^{3}w.eq.)
10^{18}	30	15	911.9	25.7
10^{19}	50	25	1009.8	79.3
10^{20}	80	25	1009.8	202.9

We will present, in this note, the results for the acceptance for showers with zenith angles larger than 60°, larger than 70° and larger than 80°, at different values of the energy, as $\chi_{\max}$ varies randomly over the simulation volume.

We present in figure 1 the acceptances for these showers at different energies, with the trigger requirement of 5 or more pixels activated at the detector, with the first interaction point χ_0 at values larger than 500 g/cm^2 and a 100% duty cycle.

3. Conclusions

We conclude in this note that the fluorescence detectors, as the one that will be installed at the Pierre Auger Observatory, have an acceptance to UHE neutrinos that will add significantly to that of the surface array detectors [10][11] when operating in hybrid mode.

From the analysis of our results we observe that even for showers with energies of 10^{18} eV, the acceptance of a single eye detector, with a 10% duty cycle is larger than a km^3.w.eq.sr. increasing to 6 km^3.w.eq.sr. at 10^{20} eV. This study refers to a single eye, with showers having a zenith angle of less than 90°. The Pierre Auger Observatory will have four eyes, one central as the one we simu-

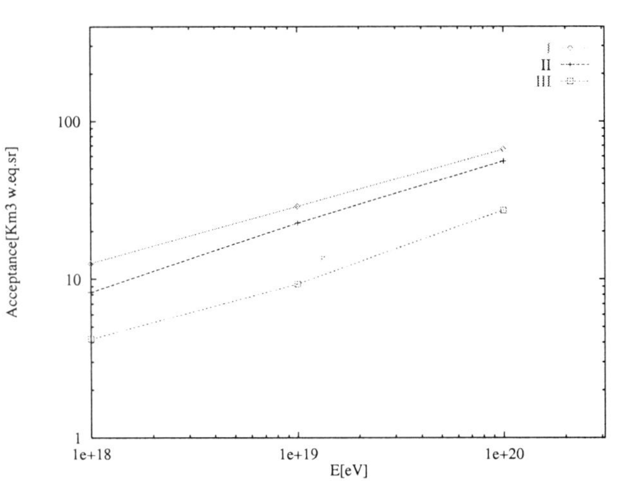

Figure 1. Acceptances (in units of km^3.w.eq.sr) for energies of $10^{18}, 10^{19}$ and 10^{20} eV for showers with zenith angles larger than 60° (diamonds), than 70° (cross) and than 80° (square).

lated and three others covering 180° in azimuth, on the periphery of the site, looking inwards to the array, adding the acceptance.

The result presented here is a lower value for the acceptance, once we have not included showers moving upwards, at low angles to the horizon. These class of showers where the neutrino interacts with the matter in the ground near the surface, will increase the values for the acceptances we have estimated here.

REFERENCES

1. R. Gandhi, C. Quigg, M. H. Reno, I. Sarcevic, *Phys. Rev.* **D58**, 093009 (1998).
2. A. Cillis, S. Sciutto, astro-ph/9908002 (1999).
3. T. K. Gaisser, A. M. Hillas, *Proc. 15th International Cosmic Ray Conference, Plovdiv*, (**8**), 353 (1977).
4. M.G. do Amaral, J. C. Diaz, R. C. Shellard, *The FDsim User's Guide*, GAP Note in preparation.
5. *The Pierre Auger Observatory Design Report* (Auger Collaboration, 2nd ed., Fermilab, Batavia, 1999).
6. S.J. Sciutto, *AIRES, a system for air shower simulation*, GAP-99-044 (1999).
7. B. R. Dawson, *Fluorescence detector techniques*, GAP-96-017 (1996).
8. R. M. Baltrusaitis, R. Cady, G. L. Cassiday, R. Cooper, J.W. Elbert, P. R. Gerhardy, S. Ko, E. C. Loh, M. Salamon, D. Steck, P. Sokolsky, *Nucl. Instrum. Meth.* **A240**, 410 (1985).
9. L. Eltermann, R. B. Toolin, *Handbook of Geophysics and Space Environments*, (1965).
10. P. Billoir, *Estimation of the acceptance of the Auger ground detector to quasi-horizontal showers induced deeply in the atmosphere*, GAP-97-049 (1997).
11. K. S. Capelle, J. W. Cronin, G. Parente, E. Zas, *Astropart. Phys.* **8**, 321 (1998)

ELSEVIER

Nuclear Physics B (Proc. Suppl.) 97 (2001) 251–254

www.elsevier.nl/locate/npe

Geometrical Reconstruction of UHE Cosmic Rays Detected by Fluorescent Detectors

J. C. Díaz*[a], R. C. Shellard†[a] and M. G. Amaral‡[b]

[a] Centro Brasileiro de Pesquisas Físicas-CBPF
Rua Dr.Xavier Sigaud 150, Rio de Janeiro, RJ 22290-180 Brazil

[b] Instituto de Física, Universidade Federal Fluminense,
Av. Litorânea, s/n, Niterói, RJ 24210-340 Brazil

We present the geometrical reconstruction of the events detected by fluorescence detectors (FD) similar to those of the Pierre Auger Observatory (PAO). The input of the geometrical reconstruction program is given by the output data of the FD simulation program, FD_sim. The efficiency of this program is computed by comparing its results with the data of the simulated shower. The implementation of the hybrid reconstruction is also discussed.

1. Introduction

The Pierre Auger Observatory (PAO) is a broad international effort to make a detailed study of cosmic rays at the highest energies. Two giant arrays of hybrid detectors have been proposed, one in the Province of Mendoza (Argentina), already in construction, and the other in the US state of Utah.

Each Observatory will consist of an array of 1600 surface detectors (SD) spread over 3000 km^2 and instrumented to measure the Cerenkov light produced by the passage of particles and four atmospheric fluorescence detectors (FD) viewing the volume above the surface array. These two different air shower detector techniques working together form a powerful instrument to search for the origin of ultra high energy cosmic rays.

2. The Reconstruction Program

The PAO will have two systems of detection, which are completely independent and complementary. The hybrid code must have two types of reconstruction, one for the data of the FD and the other for the data of the SD. The reconstruction of their data is synchronized by time. The programs use the data of the other detector, when available, operating in an hybrid mode reconstruction [1,2].

With the FD it is possible to map out the longitudinal development of the shower (shower size vs. atmospheric depth). This kind of mapping provides a calorimetric energy measurement and an estimate of the mass composition through the depth of shower maximum, χ_{max} [3].

The SD gives the arrival point of the shower and its timing. The energy is extracted from the data analysis of the radial distribution of the particle density. However, this analysis is model dependent and the information provided by the FD is crucial for a better reconstruction of the shower energy. Fluorescence detector data is available only for roughly 10% of the events. But this is sufficient for a cross check of the energy reconstruction when only SD data is used.

The main parameters to be reconstructed are the geometrical characteristics of the shower (arrival point and direction), the energy and number of particles, the chemical composition of the primary cosmic ray, and its origin.

In the case of the FD reconstruction code, we have a basic subroutine for the mono-ocular case, that is, when only one FD detects the shower. This subroutine with some variants can easily

*johana@lafex.cbpf.br
†shellard@lafex.cbpf.br
‡amaral@if.uff.br

PII S0920-5632(01)01276-2

be adapted for the case of stereo detection. Although the PAO will have 4 eyes, monocular reconstruction is more relevant now during the Engineering Array (EA) phase, since during this phase 40 SD tanks and a single eye will be installed in the southern site.

2.1. Input

In this preliminary stage, we only use simulated data obtained from the output file of the simulation program FDSIM [4]. This output file mimics as close as possible the output of the data acquisition system of the observatory. At this moment, this means that the data saved is only the counting of photoelectrons registered by the photomultipliers (PMT). From this data we find the shower detector plane (SDP) and the position of the line of the shower in the SDP. We use the following data as input parameters : a)the eyes that detect the shower; b)the identification of the each triggered pixel; c)the timing of the pixel, that is, the time at which a specific pixel starts taking data and the number of frames during which this pixel is still collecting data (where a frame is a time interval of 100 ns); d)the signal, that is, the total signal of a specific pixel and the partial signal of every frame in which data was collected.

In the hybrid reconstruction code we will also have to take into account data coming from the Surface Detector (SD). The use of the arrival time of the shower in the ground as well as its position will surely improve the reconstruction of the events.

2.2. The Shower Detector Plane (SDP)

When a shower crosses the field of view of a pixel, it triggers a series of pixels whose pointing directions form a great circle on the celestial sphere. For each viewing eye, we are going to use the FD data (triggered pixel directions) in order to find the best plane (SDP plane) that passes through the position of the eye and this great circle [5]. This procedure is performed by first moving our reference coordinates to that of the eye, and then by fitting the azimuthal and zenithal angles of the triggered pixels to a plane passing through the center of the coordinates (the position of the FD) with the equation $z = ax + by$ in 3D in order to obtain the vector perpendicular to the SDP.

We have two different ways of achieving this result: one that was developed by our group (method A1) and another one that was used by B. Dawson and collaborators in the reconstruction code for the Fly's Eye and that was adapted to our observatory by us (method A2).

In method A1 we use a linear fit for the plane. We do this fitting by using either the pixels direction, the pixels direction weighted by the number of frames (where a frame is a time window of 100 ns) or by its total signal. Even though we have not yet performed a detailed study of which method gives the best results, we are using the second alternative since it seems to give better results. After obtaining the correct equation we calculate the perpendicular vector to the SDP.

We can improve the results of both methods, A1 and A2, if we use a minimization program called *amoeba* [6]. Using the normal vector of the SDP calculated in the previous steps, for each model, and applying the minimization procedure to improve our fitting, we obtain two new sets of results (methods B1 and B2, respectively).

Since we first generate some showers using our simulation code FDSIM and then reconstruct them, we can estimate the efficiency between the methods A1 and A2 (without the minimization), B1 and B2 (with the addition of minimization) and the simulated normal vector to the SDP. We obtain small and similar errors for both methods (with and without minimization) when comparing with the simulated data of the FDSIM for the SDP.

We present in Table 1 the results for the mean values of the space angle error in degrees between the normal reconstructed and simulated vector of the SDP for all methods (A1 and A2, and B1 and B2) and for energies of 10^{19} eV and 10^{20} eV. We present also the results when 50% and 90% of the total data are taken into account and we show also the percentage of angular data that have deviations greater than 5°.

When we perform a monocular reconstruction (using information from only one eye) we use the timing information to determine the orientation of the shower within the SDP while for stereo

Type	50%	90%	$> 5°$
A1	0.51°	1.31°	5.66%
A2	0.59°	1.33°	5.78%
B1	0.55°	1.27°	5.41%
B2	0.55°	1.27°	5.28%

Table 1
Space angle difference (in degrees) between the normal reconstructed and the simulated vector of the SDP for the two methods studied, without and with minimization (A1, A2, B1 and B2, respectively).

reconstruction we use, as a first step, the intersection of the different SDPs to obtain the shower trajectory. From now on we will use only the reconstructed normal vector that was obtained using our method.

2.3. The Shower line in the SDP (1 Eye)

Once the SDP is known, we move to a 2D space. We change the directions of the pixels in 3D to directions in the SDP. Using the angle (Ω_i) that the pointing direction of the triggered pixel makes with the horizontal line (at the triggered time), the angle θ_i that this same pointing direction makes with the shower and the timing of the pixels we fit to Rp (distance from the shower to the FD) and the angle Ψ (angle between the shower and the ground in the SDP) using the geometry of the shower [2].

When we perform the fitting of Rp and Ψ we can make two different guesses for the trial values of these two quantities. In the first one (C1), we estimate the values of Rp and Ψ for the shower with the assumption that the axis of the shower is the basis of an isosceles triangle with the two sides being the pointing directions of the first and the last triggered PMT. In the second one (C2), that is used in the Fly's Eye code [7,8], the trial direction of the shower is calculated with a simple method that consists, more or less, in obtaining a direction perpendicular to the mean direction of the pixels. This is only a rough approximation, since they use a minimization to improve their results in the next stage.

In the two cases described in the last paragraph, we use a minimization program called amoeba in order to have a better shower direction in the SPD. We label the results obtained with the two different trial guesses (C1 and C2) and improved by the minimization procedure by D1 and D2, respectively.

We have estimated the efficiency of our reconstruction in the determination of the values of Rp and Ψ for the two cases C1 and C2. The results are not so good as those obtained for the efficiency of the SDP. Earlier studies [7,8] have shown that for showers detected with angles less than 35° the reconstruction of the shower line in the SDP is not so good. As the height of our detector is only 30° we may have some problems with monocular reconstruction.

2.4. The Shower line in the SDP (2 or more Eyes)

In the stereo reconstruction, as we have more data, we will be able to define in a best way the shower geometry. As a first approximation we only use the intersection of different SDPs in order to obtain the shower line included in the SDPs. We are going to follow the same steps we did in the case of monocular reconstruction.

We have estimated the efficiency of our calculations by calculating the distance between the reconstructed and the simulation ground point position and the difference (in degrees) between the reconstructed and simulated angles (both azimuthal and zenithal). Using the same philosophy we have applied in the monocular reconstruction, we have improved the geometry of the reconstructed events with the multiple reconstruction when compared to the mono-ocular reconstruction, in a similar way to what happened in other previous studies. Our results for the ground position points, the zenithal and azimuthal angles are presented in Table 2 .

Type	50%	90%	>750m	> 8°	> 4°
Shower Ground Position	97.8 m	245. m	13.9%	-	-
Azimuthal Angle	1.17°	2.87°	-	13.1%	-
Zenithal Angle	0.47°	1.31°	-	-	15.3%

Table 2
Distance between the simulated and reconstructed ground position of the shower. The third column shows the fraction of showers where the distance exceeds 750m. The others show the percentage of events where the azimuthal and zenithal angular deviations are greater than 8° and 4°.

3. Conclusions

We have made the first steps building the program of reconstruction of the events detected by the FD of the Pierre Auger Observatory. Our program reads the data and gives the geometrical reconstruction of the air shower. Anyway, there is still a lot of work to be done to improve our methods and to have a better reconstruction. Up to this moment, we have an ideal program and we need to improve it in order to have a realistic reconstruction for the real data, taking into account real noise and a realistic atmospheric attenuation treatment.

Our next step is to use the timing information for the multiple reconstruction and to use the information obtained from the SD to make an hybrid reconstruction and then be able to start the reconstruction of the energy of the primary high energy cosmic ray.

REFERENCES

1. Auger Collaboration, *Hybrid Detector Simulations*, in Pierre Auger Project Design Report, Fermilab, Batavia Ill. (1995).
2. B. R. Dawson, H. Y. Dai, P. Sommers, and S. Yoshida, Astropart. Phys. **5** 239 (1996).
3. B. R. Dawson, *Fluorescence Detector Techniques*, Auger Note GAP-96-017 (1996).
4. R. C. Shellard, J. C. Diaz, and M. G. Amaral, Auger Note (to appear).
5. R. M. Baltrusaits *et al.*, Nucl. Instrum. Methods **A240** 410 (1985).
6. W. A. Press, B. P. Flannery, S. A. Teukoslky, and W. T. Vetterling, *Numerical Recipes*, Cambridge, Cambridge University Press (1986).
7. J. W. Elbert, Proceedings of the Tokyo Workshop on Techniques for the Study of Extremely High Energy Cosmic Rays, 1993, ed. M. Nagano, University of Tokyo (1994), page 158.
8. C. R. Wilkinson, PHD Thesis, University of Adelaide (1998).

ELSEVIER

Nuclear Physics B (Proc. Suppl.) 97 (2001) 255–258

www.elsevier.nl/locate/npe

Artificial Neural Networks for the shower reconstruction of gamma-showers in the energy range [20-300] GeV

D. Dumora [a], G.M. Maneva [b], J. Procureur[a], D.A. Smith[a], P.P. Temnikov[b]

[a]Centre d'Etudes Nucleaires de Bordeaux-Gradignan, Université de Bordeaux 1
rue du Solarium, 33175 Gradignan-Cedex, France

[b]Institute for Nuclear Research and Nuclear Energy
72, Tzarigradsko chausee Blvd., 1784 Sofia, Bulgaria

A first approach to the shower reconstruction from the amplitudes of the Cerenkov photons is made using Artificial Neural Networks, (ANN). We show that ANN method gives good results for the photon-proton discrimination, the primary energy determination and encouraging perspectives for the determination of the location of the shower core.

1. Introduction

As is well known, the observation of photons in the primary cosmic radiation in the energy range [10-200] GeV has, up to now, not been made. The lower limit of this range corresponds to the highest energies of photons directly observed by satellites whereas some hundreds of GeV are the energy limits detected using Cerenkov imaging techniques on the ground. For the present time, two experiments try to resolve this deficiency: CELESTE [1], and STACEE [2]. In brief, the CELESTE experiment exploits a former solar plant. At present time 40 heliostats (each of them: 54 m^2) reflect the Cerenkov light in atmospheric showers towards the top of a tower at which some secondary optic and 40 photomultipliers are placed. Thus, the number and arrival time of photo-electrons detected by each heliostat can be measured. In such kind of experiments, the basic problem is, knowing the photo-electrons amplitudes and arrival times, to be able to determine the nature and energy of the primary particle inducing the shower and the location of the shower axis. The first idea for such work is to use the shower axial geometry as, for example, the determination of the core location using some geometric methods. However, it is important to notice that if these geometrical methods could lead to reasonable results for showers induced by primary photons with sufficient energies, (some hundred GeV), it would not be the same for lower energies because of the too large fluctuations. Taking into account these fluctuations of the measured parameters, the huge background noise induced by primary hadrons and the small number of photo-electrons seen by each heliostat (for primary photons with 30 GeV, three photo-electrons are seen, in average, by each heliostat), it is interesting to estimate the feasibility of employing Artificial Neural Networks (ANN) to solve the problem of the shower reconstruction. Such analysis of the Cerenkov light in atmospheric showers has already been made in the recent past for energies around TeV (for ASGAT [3] , for the Whipple observatory [4] [5] and for the indian TACTIC and MYSTIQUE experiments [6]). It has been concluded that for energies larger than some hundred GeV, results obtained using ANN were at least as accurate as χ^2 fitting techniques.

2. The ANN model.

Artificial neural networks represent a powerful substitution, non parametric and generally non linear, of the conventional methods of data analysis. They have been used successfully in many domains of technology and science and especially in High Energy Physics. That is why we chose the Lund program JETNET 3.1 [7] for the analy-

0920-5632/01/$ – see front matter
PII S0920-5632(01)01277-4

sis of the experimental data in gamma astronomy. The method used was supervised ANN, based on feed-forward Multilayer Perceptron with back-propagation updating rule. As a first approach our input layer consisted of 40 nodes, corresponding to 40 actually active heliostats. The input patterns are formed by the number of photoelectrons registered by any heliostat with the help of a corresponding photomultiplier. Some preliminary normalization of the input data was performed, transforming them to values close to 1. The exact architecture and internal parameters of ANN are described further, as they depend essentially on the problem to be solved (classification or function approximation). For the training, validation and testing procedures, we used simulated data produced by the CELESTE's team. These data have been obtained using the CORSIKA code [8], (version 5.20), adapted to the CELESTE experiment. Because of the limited values of the primary energies, the full Monte Carlo option has been used. 60769 gamma showers have been simulated between 20 and 300 GeV and 69350 proton showers, both following the primary energy spectrum with index 2.7 for protons and 2.0 for photons.

In the present work, we try to give answers to three basic questions:

a) what is the nature of the primary particle generating the shower, (photon or hadron) ?

b) in the photon case, what is its energy ?

c) what is the position of the shower axis ?

3. Nature of the primary particle

For this typical classification problem we have used ANN with 2 hidden layers, consisting of 30 and 15 hidden nodes respectively, and 1 single node in the output level. The activation function used in the training period was a "sigmoid" function (1 + tanh(x))/2, (where tanh(x)is the hyperbolic tangent). The target value of the output node was of a binary form (0-1) for the photons and protons respectively. For the selection criterium, primary has been claimed to be respectively photon or proton if the output parameter was smaller or larger than 0.5. Having used the normal updating for the backpropagation procedure, we have obtained a good discrimination between primary photons and protons. This means that the nature of 96% of showers are correctly selected using ANN. The fact is not surprising taking into account that the discrimination photon/proton for the primary particle is a pattern recognition problem, for which ANN are particularly efficient.

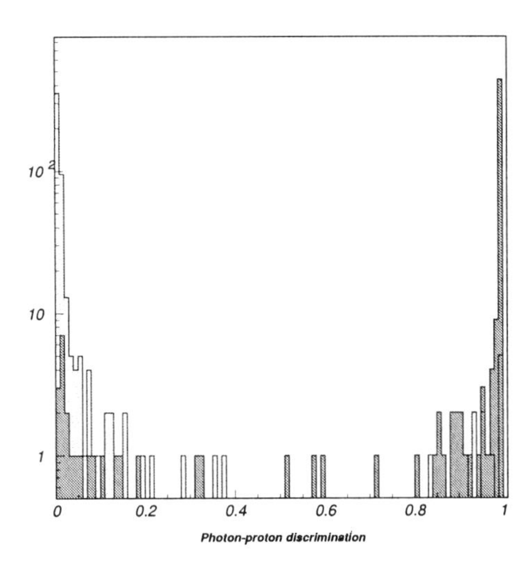

Figure 1. The photon-proton discrimination

Taking into account 500 photon and 500 proton showers following the primary energy spectra, the neural network output is shown in figure 1 on which, because of the log scale, we emphasized the 4% of the selection pollution. It can be seen that the split between the photon and proton peaks are quite well separated.

4. The primary energy determination

The first aim of the CELESTE experiment is to pick up among the cosmic projectiles generating EAS, primary photons with the lowest energy with a good energy determination event by event. The number of detected photoelectrons is roughly proportional to the primary energy. However, this theoretical proportionality is largely distorted by the number of unseen photo-electrons because of the very low density

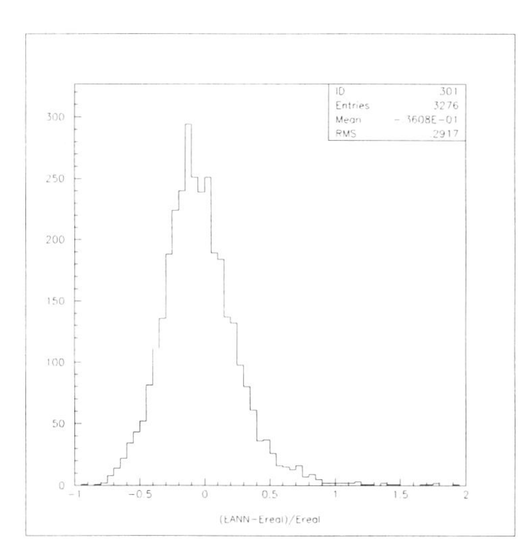

Figure 2. Relative energy difference

flux, (three photo-electrons seen, in average, by each heliostat in gamma showers of 30 GeV) and by undesirable effects of optics or electronics. This expected disorder could be reduced using ANN.

As training patterns for the supervised training stage we used a data set of 55000 simulated gamma-showers with energies between 20 and 300 GeV, following the primary energy spectrum. The test population consisted of 3276 showers of the same type. In this case only one hidden layer of 20 nodes and one node output layer were sufficient. The neuron transfer function applied was tanh(x). The problem in consideration needs good performance for the relative difference (EANN-Ereal)/Ereal (here Ereal is the real value and EANN the obtained value of the energy). As the error function minimized in the updating procedure was the summed square error, we noted that the value 1/Ereal as an output node contents lead to a better energy approximation. The relative energy resolution (EANN-Ereal)/Ereal is about 30% for primary gamma with energy greater than 20 GeV.

5. Location of the shower core

Independently of the nature and primary energy determination, the last parameter which has to be defined is the location of the shower axis. Taking into account the shape of the Cerenkov light front it is possible to locate its center using the time delay between the photons detected by different heliostats and, on the other side, the crossing of the shower axis with the ground level (foot). In the present work we investigate the possibility to apply the ANN-method for finding the shower impact position. For this problem we used two hidden layers, (with 15 and 5 nodes respectively), and two output nodes corresponding to the cartesian coordinates X, Y of the impact point. As a preliminary phase we have introduced coordinates of the impact point for any shower, relative to the weighted average of the photo-electrons, detected by the heliostats. The transfer function used was tanh(x), which naturally needed an additional normalization of the target output values in the interval (-0.5, 0.5). The distribution of the difference between the real (Xreal,Yreal) and obtained (XANN,YANN) coordinates of the impact position (in meters), for three different energy intervals, (over 30, over 50 and over 75 GeV), are shown in figures 3 and 4 with 6612 input events. It is seen that the distribution has the Gaussian form with zero mean values. It means that the method gives the unbiased estimation of the impact point coordinates. It is easy to notice that the quality of the method improves with the primary energy increase (the RMS of the deviation from the mean decreases to 25 m.) for initial photons with energy greater than 75 GeV.

6. Conclusion

The aim of the present work is to try to solve the inverse problem for showers recorded using the measurement of the Cerenkov front on the ground using artificial neural networks. Because of the specific low energy range of the CELESTE experiment including very low amplitudes and large fluctuations in the measurements, this project is not easy.

Using the JETNET-3 code and a very simple definition of the input parameters (number of photo-electrons seen by each heliostat), ANN are able to determine the nature of the primary

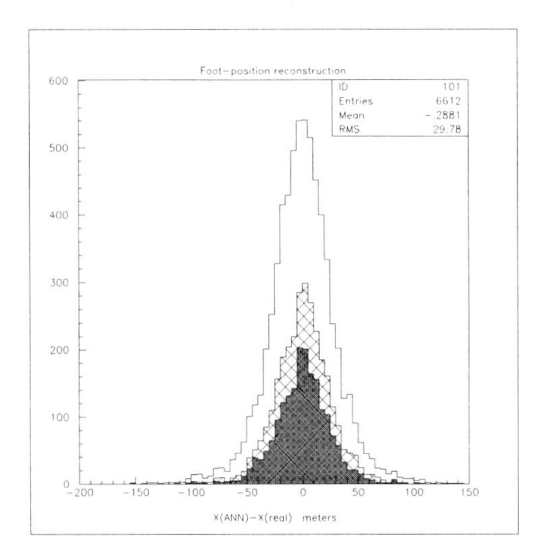

Figure 3. Difference of the estimated and exact abscissas of the core, Shaded: 50-300 GeV, RMS = 26m, Black: 75-300 GeV, RMS = 24m

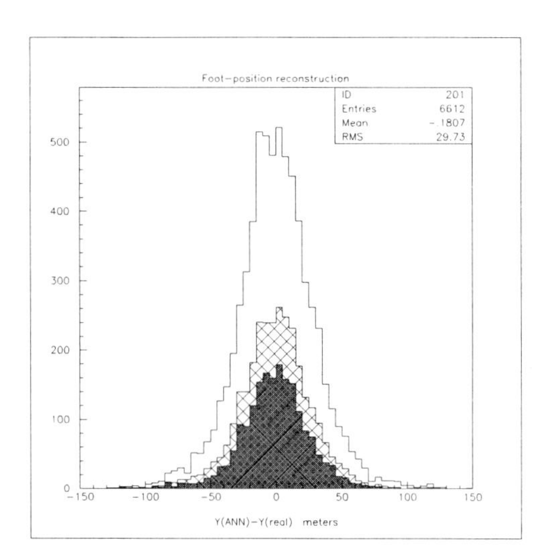

Figure 4. Difference of the estimated and exact ordinates of the core, Shaded: 50-300 GeV, RMS = 27m, Black: 75-300 GeV, RMS = 25m

particle at 96%, to estimate the primary energy inside the range [30-200] GeV within a relative error of 30% and to locate the shower axis position with an average error of about 30m for any of the cartesian coordinates. Of course, these results are relatively modest but it is important to notice that, using a so simple model for the definition of our input, these results are upper or equal to the corresponding obtained from the traditional χ^2 analysis. We would like to remark that these results were obtained using the whole available set of simulated gamma showers (in the energy region 20-300 Gev), without any preliminary rejection of the events - no limits on the number of heliostats "fired", no minimal number of the detected photons or any other "cuts" were performed. On the other side, we have the possibility to improve our logic including more information for our input, i.e. the location of each heliostat and the time delay of Cerenkov photons on mirrors. This will constitute the next step of the work.

REFERENCES

1. M. de Naurois, 1999, Proc. *26th International Cosmic Rays Conference*, Salt Lake City, USA, V. 5, 211
2. R. Ong and C.E. Covault, 1997, Proc. *Towards a Major Atmospheric Cerenkov Detector-V*, The Kruger National Park Workshop on TeV Gamma-Ray Astrophysics, Berg-en-Dal, South Africa , 247
3. G. Sembroski and M.P. Kertzman, 1991, Proc. *22nd International Cosmic Ray Conference*, V. 1, 500
4. F. Halzen, R.A. Vazquez, E. Zas, 1991, Proc. *22nd International Cosmic Ray Conference*, V. 1, 504
5. P.T. Reynolds, P. Moriarty, C.Masterson, D.J. Fegan, 1997, Proc. *Towards a Major Atmospheric Cerenkov Detector-V*, The Kruger National Park Workshop on TeV Gamma-Ray Astrophysics, Berg-en-Dal, South Africa, 362
6. C.L. Bhat, V.K. Dhar, A.K. Tickoo, R. Koul, 1995, Proc. *24th International Cosmic Ray Conference*, 1995, V. 3, 400
7. C. Peterson and T. Rögnvaldsson, 1993, LU TP 93-29, CERN-TH.7135/94
8. D. Heck, J. Knapp, J.N. Capdevielle, G. Schatz, T. Thouw, 1998, *Forschungszentrum Karlsruhe*, FZKA 6019

ELSEVIER

Nuclear Physics B (Proc. Suppl.) 97 (2001) 259–262

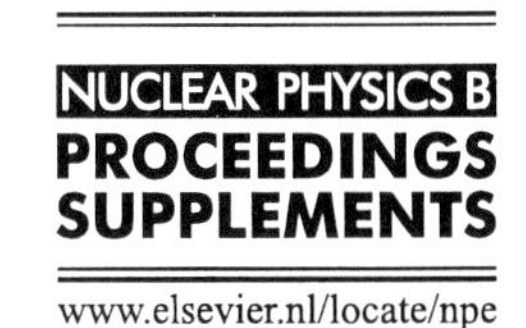

www.elsevier.nl/locate/npe

Behaviour of the EAS electron-photon and muon component characteristics in the knee region at mountain altitudes

V.S.Eganov[a], A.P. Garyaka[a], E.V. Korkotian[a], R.M. Martirosov[a], H.E.Sogoyan[a], M.Z.Zazyan[a], E.A.Mamidjanian[b], J. Procureur[c]

[a]Yerevan Physics Institute, Cosmic Ray Division, Alikhanian Brothers 2, 375036 Yerevan, Armenia

[b]P.N.Lebedev Institute, Leninsky pr. 56, Moscow 117924, Russia

[c]Centre d'Etudes Nucleaires de Bordeaux-Gradignan, Université de Bordeaux 1 Rue du Solarium, 33175 Gradignan-Cedex, France

The characteristics of the EAS electron-photon and muon components in showers with given sizes $10^5 \leq Ne \leq 10^7$ are obtained by the GAMMA installation at the Mt. Aragats (Armenia, 700 g/cm^2). The analysis of data was carried out having in mind the different explanations about the nature of the knee.

1. Introduction

At present, practically all experiments on research of the cosmic rays in the energy range 10^{14} - 10^{17} eV are pointed to investigate the nature of the *knee*. However, up to now there is no unequivocal explanation of the reason of this phenomena. The most popular interpretation of the *knee* is the existence of the *knee* in the primary cosmic ray energy spectrum at energy about $(3-5) \cdot 10^6$ GeV and, as a possible consequence, the change of the mass composition. There are different experimental analyses leading to opposite conclusions about primary composition after the *knee* (lighter, e.g.[1], normal, e.g.[2] or heavier, e.g.[3]).

At the same time there is an alternative explanation of the *knee*, also based on some inconsistencies in the EAS data, connected with a modification in the hadronic interaction properties [4] to the change of the number and energy spectrum of the secondaries.

We have to point out that only multiparameter procedures at analysis of the EAS characteristics allow some advance in understanding of the reasons of the existence of the *knee* in the spectrum on N_e.

In this paper we present the experimental results of the GAMMA installation also working in this field [5,6]. This will add some observations, which would clearly complement the data from KASCADE (near sea-level), CASA-MIA (1200 meters), and EAS-TOP (2000 meters).

2. Present status of the installation

The GAMMA installation is located on hillside of the Mt. Aragats in Armenia (3200 m a.s.l.) and is realized as a part of the project ANI [7]. In comparison with the other large scale experiments with the same goals, the high-altitude observation level of the GAMMA installation provides some advantages, in particular for minimizing the intrinsic fluctuations of the observables due to the stochastic character of the EAS development in atmosphere.

After some years spent to enlarge the effective area of the muon underground detector, which was necessary to elaborate methodical studies of the detector parameters, to investigate carefully the array response and to determine the precision of the shower parameter estimation, the GAMMA experiment is, now, effectively running.

GAMMA is a central type array and consists of two main parts:

(i) the surface part, for the registration of the

PII S0920-5632(01)01278-6

EAS electromagnetic component;

(ii) the large muon underground detector, to register the EAS muon component.

The surface scintillation array consists of 25 groups of 3 plastic scintillation detectors, each placed on concentric circles with radii of 17, 28, 50 and 70 m. Each detector has an effective area of $1\,m^2$ and thickness of 5 cm. They are distributed on the full area of $\approx 1.5 \cdot 10^4\,m^2$. An additional station with 20 detectors of the same type is placed at 135 m from the installation centre. Each of the 25 registration stations is equipped with the timing channel which allows to determine the angular coordinates of the shower axis.

The muon underground detector consists of 150 similar detectors with two muon energy thresholds: 60 detectors are placed in the hall ($E_\mu \geq$ 5GeV) and 90 in the tunnel ($E_\mu \geq 2.5$GeV). We have used the experimental data obtained during 3300 hours of operation time. The number of EAS with $N_e \geq 10^5$ and with zenith angle $\theta \leq 30°$ is $\sim$ 260.000. The effective area for the selection of the EAS axis is $\sim 5.000\,m^2$.

3. Results and discussion

The base of detectors used in the GAMMA installation are plastic scintillator with thickness 5 cm. It is well known that such a type of detector registers not only the EAS charged particles, but also electrons generated in the absorber above, as well as in the scintillator by EAS photons. Up to now there is not an exact estimation of the photon contribution $K_\gamma(r) = \rho_{sc}(r)/\rho_{ch}(r)$. In the energy range $10^5 - 10^7$ GeV these investigations show controversal results. Up to the distance of $100\,m$ from the shower core, some of them claim a smooth rise while others, a noticeable decrease in the photon contribution to the measured density. Calculations for the real conditions of the GAMMA installation have been performed taking into account the detector energy threshold and the photon conversion probability [8]. According to this work the photon contribution to the measured number of the charged particles is practically constant from 5 to 100 m and is $K_\gamma \approx 1.25$.

In our analysis we use the value $K_\gamma = 1.25$. The comparison between measured and simulated by CORSIKA code electron lateral distributions is shown in Figure 1 for the same values of $< N_e >$. For comparison with the experimen-

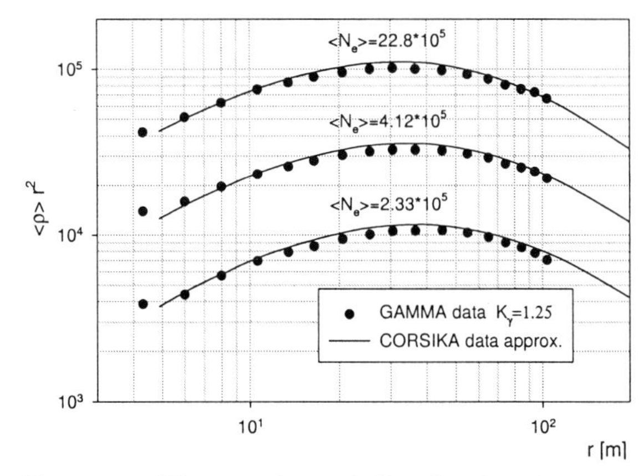

Figure 1. Electron lateral distributions in comparison with the CORSIKA simulation

tal data the normal primary mixed composition [9] for the simulation has been used. It can be seen a good agreement between them.

The differential size spectra is obtained (Figure 2) for the two zenith angle intervals at $< \theta >= 14°$ and $< \theta >= 31°$ to make the qualitative estimation of the behaviour of the spectra at different angles.

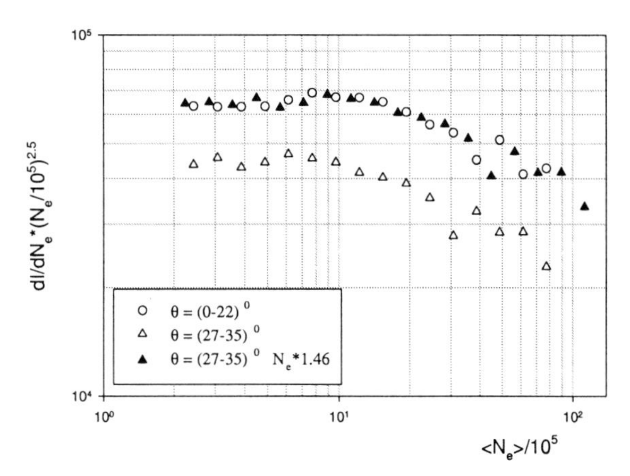

Figure 2. Differential size spectra at different zenith angles

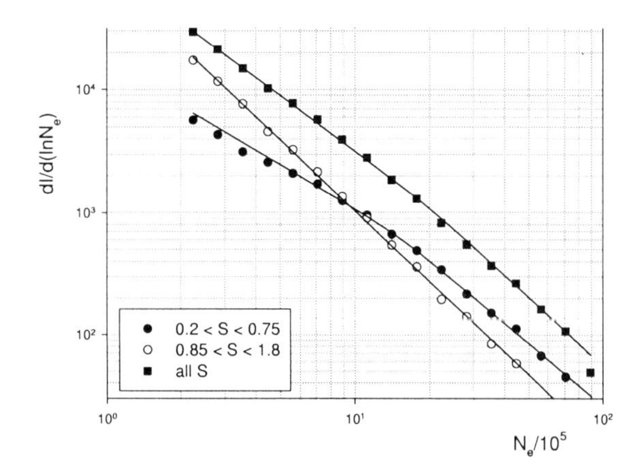

Figure 3. Differential size spectra at different intervals by age parameter

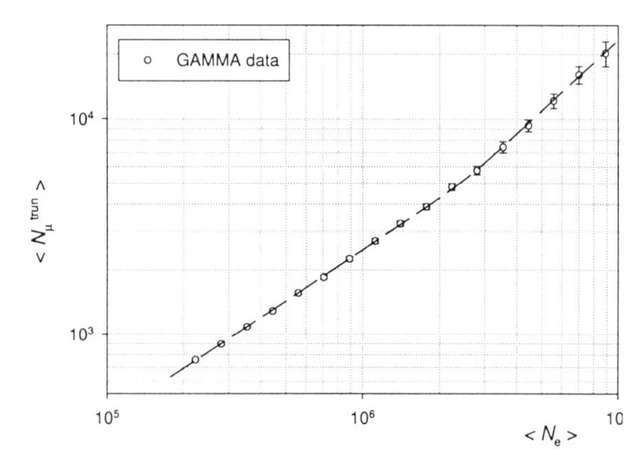

Figure 4. Average experimental truncated muon number $< N_\mu^{trun} >$ versus the EAS size $< N_e >$

They are practically parallel and coincide when data with $< \theta >= 31^0$ are shifted. Such behaviour corresponds to the shower absorption lenght of $\Lambda = (230 \pm 25) g \cdot cm^{-2}$ below the knee and $(230 \pm 40) g \cdot cm^{-2}$ above up to $N_e = 5 \cdot 10^6$. The permanency of the attenuation length is one evidence of the constant charged particle composition of the showers below and above the *knee* only and says nothing about the cause of the *knee*.

In Figure 3 the spectra are shown with various cuts of the age parameter S. Relatively *old* shower spectrum ($S > 0.85$) practically has no *knee* and is steeper than all shower spectrum. On the other hand, the *young* shower spectrum ($S < 0.75$) has the obvious *knee* and is flatter, than the all shower spectrum below the *knee* and is almost parallel to it above. These spectra may be explained with both hypothesises of the *knee* origin noted in the introduction, because of the region of $N_e \geq (1-2) \cdot 10^6$ is transitional. The *young* showers fraction does not arise above the *knee* because of change of the mass composition to heavier one in the model of primary *knee*. In the case of change of the hadron interaction of the primary protons the fraction of the heavy-like events will rise and the results will be similar.

The lateral distribution of muons in our experiment($E_\mu \geq 5$ GeV) is studied for distances $r < 52\,m$ from the shower core where about 33% of all muons are contained. In order to avoid an errors of extrapolation of the muon lateral distribution for $r > 52m$ we use the truncated muon number, which is the number of the muons in the ring $8m < r < 52m$. Figure 4 shows truncated muon number N_μ^{trun} dependence on N_e. Up to $N_e \approx 2 \cdot 10^6$ it can be fitted with the expression $N_\mu^{trun} = a \cdot N_e^{0.79}$. For larger N_e dependence becomes steeper. It should be noted that because of the shower size threshold $N_e > 10^5$ the registration of the shower with fixed muon size and 100% efficiency is possible at $N_\mu^{trun} > 10^3$. The change in $N_\mu^{trun} - N_e$ dependence takes place approximately at the N_e knee region.

4. Conclusions

The study of the EAS electrons and muons ($E_\mu > 5\,$GeV and $E_\mu > 2.5\,$GeV) by the GAMMA array gives reliable information about EAS characteristics in the shower size range of 10^5-10^7. Using the contribution of the EAS gamma-quanta in the form of $K_\gamma = 1.25 = const$ an agreement between CORSIKA simulation and experimental data is obtained.

Steepening of N_μ^{trun} depending on N_e at the same range by N_e confirm the more rapid shower development.With the object to extend a possibilities of the GAMMA installation it is expected to decrease the registration threshold by N_e and to enlarge an effective area of the shower selection with $N_e > 10^6$.

The present paper is based on the ANI collaboration data bank and express the point of view of the given group of co-authors.

5. Acknowledgments

The present investigations are embedded in a collaboration between the Moscow Lebedev Institute (Russia) and the Yerevan Physics Institute (Armenia).

We give thanks to the Staff of the Aragats Research Station for the assistance during the longterm maintenance of the GAMMA installation.

We are grateful to all colleagues of the Moscow Lebedev Institute and Yerevan Physics Institute who were taking part in the development and creation of the GAMMA installation. We would like to express our special gratitude to N.M. Nikolskaya for the creation of the sofware and to prof. S.I. Nikolsky for useful discussions.

We thank also prof. A. Chilingarian, prof. S. Ter-Antonyan for some useful remarks and comments and we do not forget P. Aguer from the CENBG, (CNRS-In2p3-France) whose help has been useful.

This work was supported by the grant 96-752 of the Armenian Ministry of Industry, by the Russian RFBR 96-02-18098 grant and by the Russian Atomic Energy Ministry.

REFERENCES

1. P.R. Blake and W.F. Nash, 1998, *J.Phys. G: Nucl. Part. Phys.*, **24**, 217
2. W Rhode *et al.*, 1996, *Nucl. Phys. B, Proc. Suppl.* , **48**, 491
3. M.A.K.Glasmacher, 1995, *Proc. 26th International Cosmic Rays Conference*, Salt Lake City, **3**, 129
4. S.I. Nikolsky, 1999, *Proc. 26th International Cosmic Rays Conference*, Salt Lake City, **1**, 159
5. V.S.Eganov *et al.*, 2000, *J. Phys. G: Nucl. Part. Phys.*, to be published
6. S.A. Arzumanian *et al.*, 1995, *Proc. 24th International Cosmic Rays Conference*, Roma, **1**,482
7. T.V. Danilova *et al.*, 1983, *Proc. 18th International Cosmic Rays Conference*, Bangalore, **8,** 104
8. S.V. Ter-Antonyan, 1998, *Private communication on Workshop ANI-98*
9. I.P. Ivanenko et al., 1988, *Astrophys. J., 268, 510*

ELSEVIER

Nuclear Physics B (Proc. Suppl.) 97 (2001) 263–266

www.elsevier.nl/locate/npe

Possibility to determine the mass composition from the GAMMA experimental data

A.P. Garyaka[a], R.M. Martirosov[a], J. Procureur[b], M.Z. Zazyan[a]

[a]Erevan Physics Institute
Br. Alikanian St.2, 375036 Erevan
Republic of Armenia

[b]Centre d'Etudes Nucleaires de Bordeaux-Gradignan, Université de Bordeaux 1
rue du Solarium, 33175 Gradignan-Cedex, France

Showers registered at the GAMMA installation (700 g.cm^{-2}, Armenia) are simulated using the CORSIKA code for different models of nuclear interactions. We show that we are able to select showers generated by primaries with different masses but with the same primary energy. After such selection and analysing the fluctuations of the muon numbers, it would be possible to obtain an unbiased estimation of the mass composition of primary radiation

1. Introduction

In spite of many works devoted to the determination of the primary mass composition, the uncertainties of the obtained results leave the question about the reason of the knee in the primary energy spectrum without any clear answer. This difficulty has different reasons. For example, a precise estimation of the mass and energy of primaries is not obvious. To enlarge the needed information, this implies the measurements of multiparameters in individual showers, what means to register not only the electromagnetic component, but hadrons and muons also. The shower size, N_e, is the easiest shower parameter to estimate from the measurements of the particle densities in EAS experiments. This is why, in all experiments, showers have been classified taking into account their size. It is well known that such a classification leads to undesirable consequences. One of them is that showers with given size are generated by primaries with different masses but with different energies too. Consequently, all information coming from showers of a given size are mixing primary energies, which is not the best way for the determination of the primary composition. In order to solve this problem a very simple method has been proposed [1] to select, at mountain altitude, showers generated by primaries with different masses but with the same energy. For example, the selection criterium for the depth of 700 g.cm^{-2} is to pick up showers with a constant value of the parameter $\alpha(70) = 70^2 \frac{\rho_e(70)}{f_{nkg}(10,S_{5-70})}$, were $\rho_e(70)$ is the density of charged particles measured at 70m from the shower axis, f_{nkg} is the well known Nishimura-Kamata-Greisen function, [2], and S_{5-70} is the local age measured at 5 m and 70 m from the shower axis. In fact this result has been modified comparing its stability using different simulation codes [3]. In this work it has been shown that the definition of the parameter α was only slightly depending on the models and choosing the CORSIKA code, [4], the selection parameter was redefined as $\alpha(135) = 135^2 \frac{\rho_e(135)}{f_{nkg}(3,S_{25-135})}$. However, it is important to notice that this definition has been obtained only taking into account simulated data. When the decison was taken to apply this shower selection to the GAMMA experiment, two difficulties appeared. The first one was the large fluctuations of the densities of charged particles measured at 135m from the shower axis. Indeed, in spite of a precise measurement of this density, (20m^2 of effective detectors), this large distance from the shower axis

PII S0920-5632(01)01279-8

involves large fluctuations and a measurement of $\alpha(135)$ with a bad accuracy. Another source of uncertainty has been in the local age measurement S_{25-135} which has been defined using densities measured at 25 m and 135 m from the shower axis. It has been verified that a small error in the shower axis location is inducing an additional error in the measurement of S_{25-135} and, consequently, in the evaluation of α. This is why it has been decided to include in the definition of α, an age parameter easily measurable, the well known S_{nkg}. Indeed, S_{nkg} is obtained fitting the charge particle densities for different distances from the shower axis and can be defined with a very good precision for all individual showers. In such a case, the following definition of the α parameter has been proposed: $\alpha\,(70) = 70^2 \frac{\rho_e(70)}{f_{nkg}(1,Snkg)}$.

2. Properties of the $\alpha(70)$ parameter

The general details of the α parameter have already been published and references are given in [1]. However, the primary energy E_0 versus the values of $\alpha\,(70)$ is shown in Figure 1 for primary protons and iron nuclei. These dependences were obtained from simulations using the CORSIKA code, [4], in its version 5.20 and with the QGS model for the hadron-hadron interactions. Furthermore, we have taken into account the uncertainties caused by the specific reception conditions of the GAMMA experiment as $\frac{\sigma_{rec}(E_0)}{\langle E_0 \rangle} = 25\%$. This figure shows clearly that showers picked up with constant values of $\alpha\,(70)$ are generated by primaries with the same energy and no mass dependence.

An other important test is to compare the measured and simulated values of $\alpha\,(70)$. In figure 2, the shower sizes Ne are drawn for showers selected with constant values of $\alpha\,(70)$. Experimental and simulated values are particularly in good agreement, which confirms the correct measurement of the α parameter. For the determination of the primary mass composition, it is important to verify that in showers picked up with same values of $\alpha\,(70)$, the experimental and simulated muon numbers are in agreement. Let us recall that one of the main specificities of the GAMMA exper-

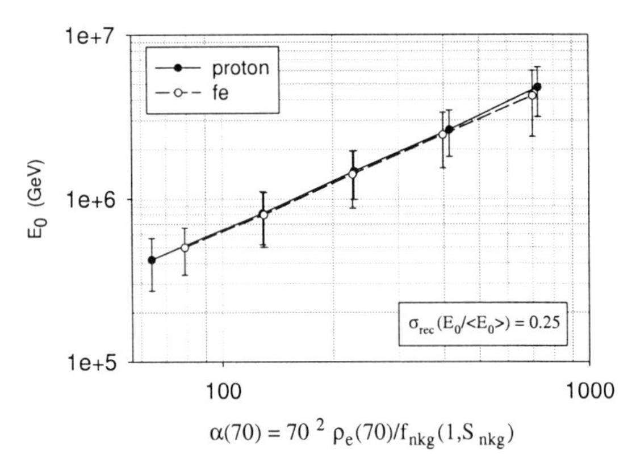

Figure 1. E_0 versus $\alpha(70)$

iment is the large effective area of underground muon detectors, (150m^2, [5]).

In order to estimated correctly the muon sizes from the measured densities, we define the truncated muon number, N_μ^{trunc}, representing the number of muons between 8 m and 52 m from the shower axis. The truncated muon numbers has two interests, the first one is that it is obtained without extrapolation of the muon densities far from the shower axis and it is well known that such extrapolation is not well defined. The second interest is that, because of the specific conditions of the muon measurement, [5], the muon energy threshold in this range of distance is well defined as 5 GeV [6]. The dependence of N_μ^{trunc} versus N_e is shown in Figure 3. Once again, the

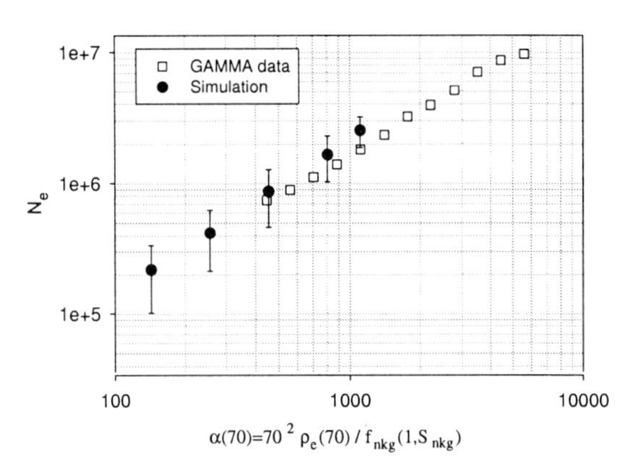

Figure 2. N_e versus $\alpha(70)$

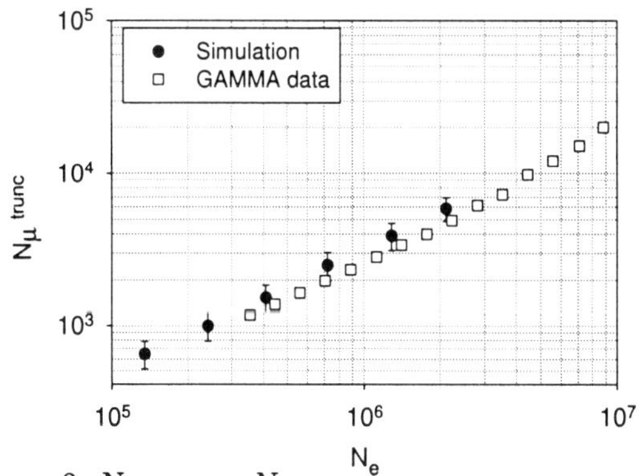

Figure 3. N_μ versus N_e

accordance between experiment and simulation is good. Finally, Figures 2 and 3 must confirm the correct agreement between experimental and simulated data for N_μ^{trunc} versus $\alpha\,(70)$.

3. Determination of the mass composition around the knee

The dependence of the EAS muon number with the primary mass is well known and has been already used for the elemental mass determination, for example in the Tien Shan experiment [7]. In the present work, we suggest to use this correlation but we applied it not for fixed size showers as before, but for EAS selected with given values of $\alpha\,(70)$, (i.e., fixed energy). For that, we classify the primary masses in five family groups, p, α, medium nuclei ($\langle A\rangle = 14$), heavy nuclei ($\langle A\rangle = 26$), and very heavy nuclei ($\langle A\rangle = 56$). Then, the basic idea of this method is, for $\alpha\,(70) = const.$, (E_0=const.), to determine by minimization the percentages α_i of the different mass families, ($i = 1,5$), in such a way that the experimental distribution of $K_\mu^{\exp} = N_\mu^{\exp}/\langle N_\mu^{\exp}\rangle$ will be fitted by $\sum \alpha_i W\left(N_\mu^i/\langle N_\mu^\Sigma\rangle\right)$ with $\langle N_\mu^\Sigma\rangle = \sum \alpha_i \langle N_\mu^i\rangle$.

The first approach is to check the sensitivity of the method using a pseudo-experimental distribution: light, [8], and heavy, [9] for showers selected with a given value of $\alpha\,(70)$. The distribution of $K_\mu = N_\mu/\langle N_\mu^\Sigma\rangle$ for these two compositions are shown in Figures 4 and 5.

Fitting these distributions with the five simulated distributions for each mass family and taking into account the reception condition with an experimental noise such that $\sigma_{rec}(N_\mu^{\exp}/ < N_\mu^{\exp} >)$ =0.35, we obtain the results given in Table 1.

4. Conclusion

The aim of the present work is to show that the GAMMA experiment gives the possibility to select showers generated by primaries with different masses but with the same primary energy. Furthermore, the large effective area of the muon detector, (150m^2), allows a good determination of the muon component. The primary mass determination based on the fluctuations of the truncated muon numbers gives good results using two pseudo-experimental data, one with a light mass composition, the second one with a heavy one. This method is being applied to the GAMMA experimental data which are, at present, under analysis.

One part of the present paper is based on the ANI collaboration data bank and express the point of view of the given group of co-authors.

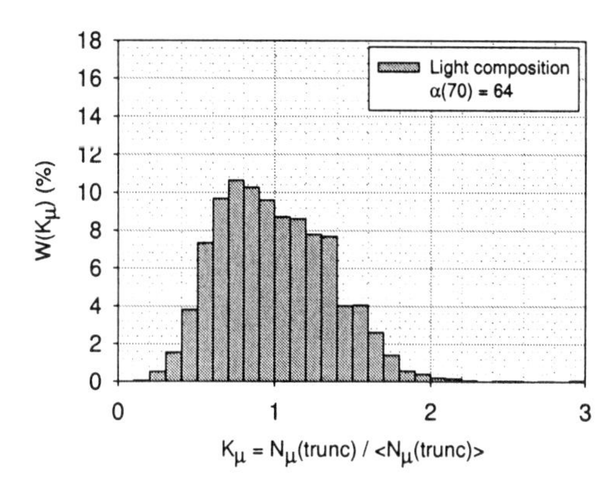

Figure 4. K_μ distribution for a light primary composition

Table 1
Test for the determination of the primary mass

A	p	α [4]	M [14]	H [24]	VH [56]
mass composition					
$\frac{lightcomposition[8]}{answer}$	$\frac{36\%}{(36\pm13)\%}$	$\frac{25\%}{(25\pm12)\%}$	$\frac{14\%}{(15\pm3)\%}$	$\frac{15\%}{(14\pm3)\%}$	$\frac{10\%}{(10\pm6)\%}$
$\frac{heavycomposition[9]}{answer}$	$\frac{16\%}{(17\pm2)\%}$	$\frac{8\%}{(9\pm2)\%}$	$\frac{10\%}{(7\pm2)\%}$	$\frac{27\%}{(30\pm3)\%}$	$\frac{39\%}{(37\pm3)\%}$

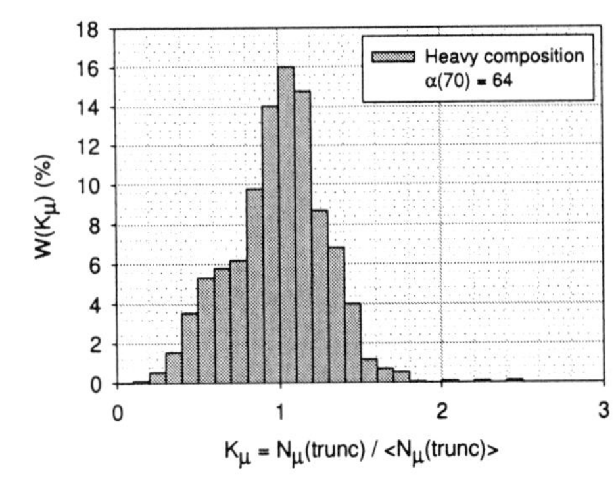

Figure 5. K_μ distribution for a heavy primary composition

REFERENCES

1. J. Procureur and J.N. Stamenov, 1995, *Nucl. Phys. B, Proc. Suppl.*, 39A, 242
2. Handbuch der Physik, Band XLVI/1, Kosmische Strahlung 1, 1961, 213
3. M. Brankova, R.M. Martirosov, N.M. Nikolskaya, V. Pavljouchenko, J. Procureur, J.N. Stamenov,1998, Proc. *25th European Cosmic Rays Symposium*, Madrid-Spain
4. D. Heck, J. Knapp, J.N. Capdevielle, G. Schatz, T. Thouw, 1998, *Forschungszentrum Karlsruhe*, FZKA 6019
5. V.S. Eganov, A.P. Garyaka, E.V. Korkotian, E.A. Mamidjanian, R.M. Martirosov, J. Procureur, H. Sogoyan, M.Z. Zazian, 2000, *J. Phys. G; Nucl. Part.*, (being printed)
6. A.P. Gariaka, R.M. Martirosov, J. Procureur, M.Z. Zazyan, 2000, This conference.
7. S.I. Nikolsky, J.N. Stamenov, S.Z. Ushev, 1984, *Zh. Eksp. Teor. Fiz.*, 87, 18
8. I.P. Ivanenko *et al.*, 1988, *Astrophys. J.*, 268, 510
9. Yodh G.B. *et al.*, 1984, *Phys. Rev. D29*, 5, 872

ELSEVIER

Nuclear Physics B (Proc. Suppl.) 97 (2001) 267–270

www.elsevier.nl/locate/npe

The "Knee" in the Primary Cosmic Ray Spectrum as Consequence of the Anomalous Diffusion of the Particles in the Fractal Interstellar Medium

A. A. Lagutin[a], Yu. A. Nikulin[a] and V. V. Uchaikin[b]

[a]Altai State University, 66, Dimitrov st., Barnaul, 656099, Russia

[b]Ulyanovsk State University, Institute for Theoretical Physics, 42, L. Tolstoy st., Ulyanovsk, 432700, Russia

The galactic cosmic ray spectrum is calculated on the assumption that interstellar medium is a fractal medium. Anomalous diffusion equation in terms of fractional derivatives is used for describing cosmic ray propagation. The calculations show that the "knee" in the primary cosmic rays spectrum appears due to fractal structure of the galactic magnetic fields. The source spectral exponent, found from experimental data in this approach, turns out to be equal to 2.9.

1. Introduction

The cosmic ray propagation is usually described in the frame of diffusion approach [1,2]. Without energy losses and nuclear interactions, the diffusion equation for concentration of the cosmic rays with energy E generated by sources distribution with density function $S(\mathbf{r},t,E)$ is of the form [1]

$$\frac{\partial N}{\partial t} = D(E)\Delta N(\mathbf{r},t,E) + S(\mathbf{r},t,E) \ , \qquad (1)$$

where diffusivity D is a parameter of the model. This equation is derived on the assumption that inhomogeneities of matter and field in the Galaxy have small-scale character.

However, during the last few decades many evidences of multi-scale inhomogeneities have been found [2–7]. Thus, the detailed analysis of supernova remnants shows that the existence of gas components with completely different physical parameters ($T_e \sim 5\,\mathrm{K}$–$10^6\,\mathrm{K}$, $n_e \sim 0.1\,\mathrm{m}^{-3}$–$10^3\,\mathrm{m}^{-3}$) in the same remnant can be understood only under condition that the unperturbed interstellar medium is non-homogeneous.

The irregularity of the matter density ρ, and correspondingly of the magnetic field strength $H \propto \rho^q$, $q \sim 1/3$–$1/2$ [5] with scale $l \lesssim 100\,\mathrm{pc}$–$150\,\mathrm{pc}$, and strictly speaking the inapplicability of the diffusion approach to media with sharp density variations [8,9] (see also discussion in [1]) stimulate the development of new models of cosmic ray propagation.

We proposed recently an anomalous diffusion (superdiffusion) model [10] for solution of the "knee" problem in the primary cosmic rays spectrum and the explanation of different values of the spectral exponent of protons and other nuclei at $E \sim 10^2\,\mathrm{GeV/nucleon}$–$10^5\,\mathrm{GeV/nucleon}$. The anomaly results from large free paths ("Lévy flights") of particles between magnetic domains. These paths are distributed according to an inverse power law, being an intrinsic property of fractal structures. In this paper we demonstrate the main results of the model.

2. Equations

The superdiffusion equation for cosmic rays concentration without energy losses and nuclear interactions has the form

$$\frac{\partial N}{\partial t} = -D(E,\alpha)(-\Delta)^{\frac{\alpha}{2}} N(\mathbf{r},t,E) + S(\mathbf{r},t,E) \ , \qquad (2)$$

where $D(E,\alpha)$ is the anomalous diffusivity, α is an exponent determined by the fractal structure of the medium, and $(-\Delta)^{\alpha/2}$ is the fractional Laplacian (called "Riss' operator" [11]).

The Green's function $G(\mathbf{r},t,E;\mathbf{r},t_0,E_0)$ satisfying the equation

0920-5632/01/$ – see front matter
PII S0920-5632(01)01280-4

$$\frac{\partial G}{\partial t} = -D(E,\alpha)(-\Delta)^{\alpha/2}G(\mathbf{r},t,E;\mathbf{r},t_0,E_0) + \delta(\mathbf{r}-\mathbf{r}_0)\delta(t-t_0)\delta(E-E_0) \quad (3)$$

and zero boundary conditions at infinity can be found by Fourier transformation:

$$G(\mathbf{r},E,t;\mathbf{r},t_0,E_0) = \delta(E-E_0)\big(D(E_0,\alpha)\tau\big)^{-3/\alpha} \times q_3^{(\alpha)}\left(|\mathbf{r}-\mathbf{r}_0|\big(D(E_0,\alpha)\tau\big)^{-1/\alpha}\right)\Theta(\tau) \ , \quad (4)$$

$$\tau = t - t_0 \ ,$$

$$\Theta(\tau) = \begin{cases} 1, & \tau > 0, \\ 0, & \tau < 0. \end{cases}$$

Here $q_3^{(\alpha)}(r)$ is the density of a three-dimensional spherically-symmetrical stable distribution with characteristic exponent $\alpha \le 2$ (see References [12, 13]).

Remember that $q_3^{(2)}(r)$ is the normal (Gaussian) distribution density, and $q_3^{(1)}(r)$ is the three-dimensional Cauchy density $[\pi(1+r^2)]^{-2}$. Other stable densities cannot be expressed by means of elementary functions, but there exist representations in terms of convergent and asymptotic series [13]:

$$q_3^{(\alpha)}(r) = \frac{1}{2\pi^2\alpha}\sum_{n=0}^{\infty}\frac{(-1)^n}{(2n+1)!}\Gamma\left(\frac{2n+3}{\alpha}\right)r^{2n}; \quad (5)$$

$$q_3^{(\alpha)}(r) = \frac{1}{2\pi^2 r}\sum_{n=1}^{\infty}\frac{(-1)^{n-1}}{n!}\Gamma(n\alpha+2) \times \sin\left(\frac{\pi\alpha n}{2}\right)r^{-n\alpha-2} \ . \quad (6)$$

Graphs of the stable densities are shown in Figure 1.

Let us note that not only the form of the superdiffusion packet differs from the normal case, but the rate of its widening $\propto \tau^{1/\alpha}$ is greater than in the case of normal diffusion ($\propto \tau^{1/2}$).

3. Spectra

Using Green's function (4) we can solve Equation (2) for sources, interesting for astrophysics.

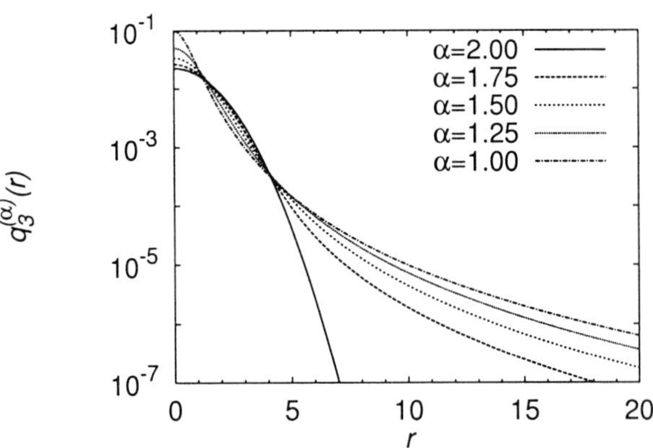

Figure 1. Three-dimensional stable densities $q_3^{(\alpha)}(r)$ for different values of α.

Thus, for a point pulse source with inverse power spectrum related to supernova bursts one has

$$S(\mathbf{r},t,E) = S_0 E^{-p}\delta(\mathbf{r})\Theta(T-t)\Theta(t) \ ,$$

the solution of Equation (2) being of the form

$$N(\mathbf{r},t,E) = S_0 E^{-p}\big(D(E,\alpha)\big)^{-3/\alpha} \times \int_{\max[0,t-T]}^{t} \tau^{-3/\alpha} q_3^{(\alpha)}\left(r\big(D(E,\alpha)\tau\big)^{-1/\alpha}\right)d\tau. \quad (7)$$

One must take into account that diffusivity for particles with charge Z and mass number A depends on rigidity as $D(R,\alpha) = D_0 R^{\delta}$. It follows from here that $D(E,\alpha) = D_0^{(\alpha)}E^{\delta}$, as well as $D_0^{(\alpha)} = D_0/Z^{\delta}$. Using the representation $N = N_0 E^{-\eta}$ and the property

$$\frac{dq_m^{(\alpha)}(r)}{dr} = -2\pi r q_{m+2}^{(\alpha)}(r)$$

of the stable law density [12], one obtains the spectral exponent for observed particles:

$$\eta = p + \frac{\delta}{\alpha}\Xi \ , \quad (8)$$

where

$$\Xi = 3 - \frac{2\pi r^2}{D(E,\alpha)^{2/\alpha}} \times$$

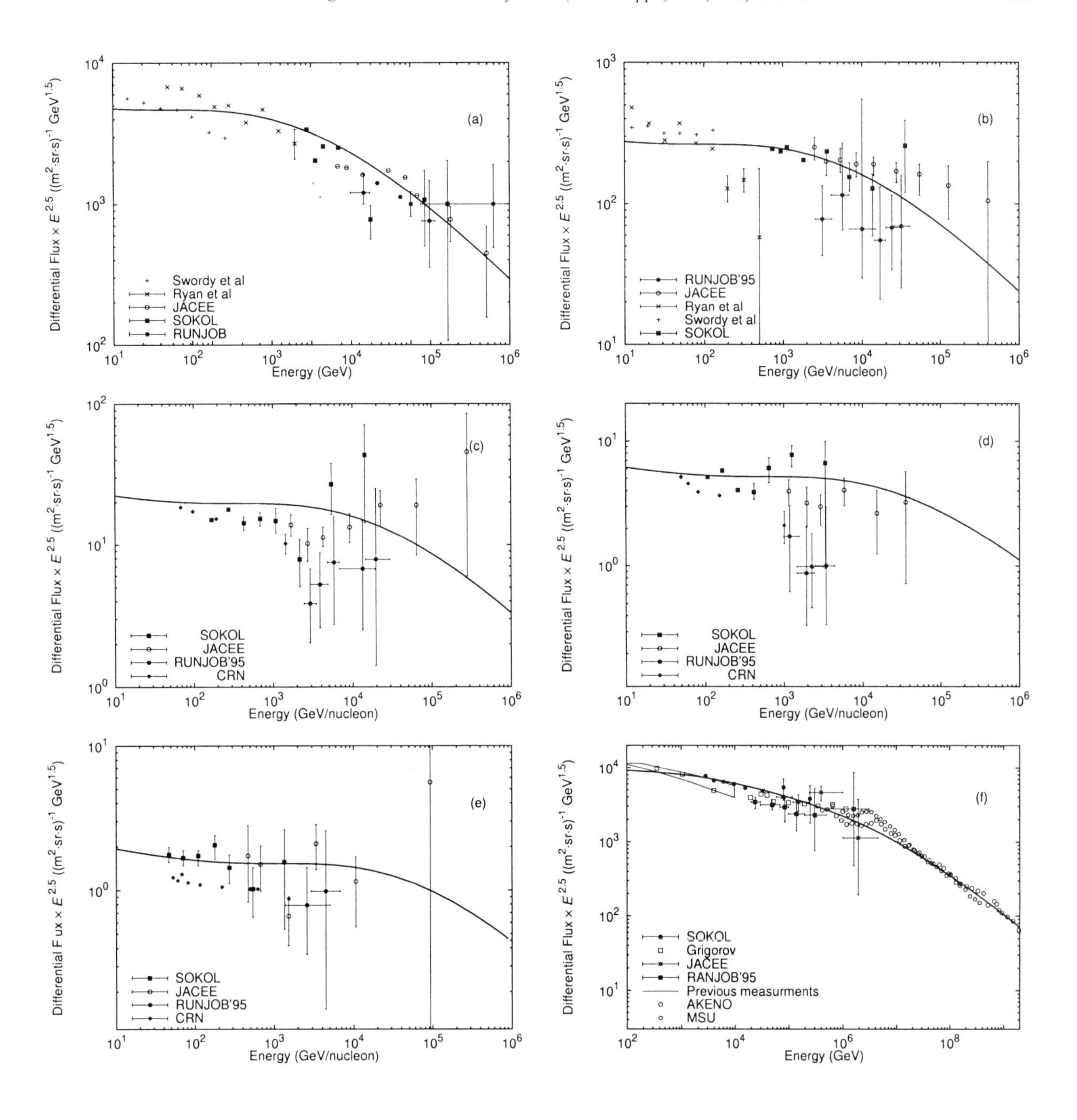

Figure 2. Comparison of our calculation of spectra (solid curve) with experimental data from direct measurements (review [19], and the MSU [20] and AKENO [21,14] arrays). We have: (a) proton spectrum, (b) helium nuclei spectrum, (c) CNO nuclei spectrum, (d) Ne-Si nuclei spectrum, (e) Fe group nuclei spectrum, and (f) all particle spectrum.

$$\frac{\int\limits_{\max[0,t-T]}^{t} \tau^{-5/\alpha} q_5^{(\alpha)}\left(|\mathbf{r}|\left(D(E,\alpha)\tau\right)^{-1/\alpha}\right) d\tau}{\int\limits_{\max[0,t-T]}^{t} \tau^{-3/\alpha} q_3^{(\alpha)}\left(|\mathbf{r}|\left(D(E,\alpha)\tau\right)^{-1/\alpha}\right) d\tau} . \quad (9)$$

Let E_0 be a solution of the equation $\Xi(E) = 0$. One can see from Equations (8) and (9) that at $E = E_0$ the spectral exponent for observed particles η is equal to the spectral exponent for produced particles p. At $E \ll E_0$ or $E \gg E_0$ we have correspondingly $\eta_{E \ll E_0} \approx p - \delta$, and $\eta_{E \gg E_0} \approx p + \delta$. In other words, the spectral exponent of observed particles increases with energy on 2δ. Let us also note that this property of the energy spectrum is missing in the normal diffusion case.

From experimental values at $E = E_0$, $\eta_{E \ll E_0}$ and $\eta_{E \gg E_0}$ one can find the main parameters of the model: D_0, δ and p. For example, taking $E_0 = 3 \cdot 10^4$ GeV, $\eta_{E \ll E_0} \approx 2.65$, and $\eta_{E \gg E_0} \approx 3.15$ [2,14], we obtain $p \approx 2.9$ and $\delta \approx 0.25$.

Our calculations show that the best fit of experimental data of direct measurements [15–19] of nuclei H, He, CNO, Ne-Si and Fe-group (found in review [19]) and all particles [14,19–21] may be got at $\alpha \approx 5/3$, $\delta = 0.25$ and $p \approx 2.9$ ($r \sim 200$ pc, $t \sim 10^5$ years, $T \sim t$) (see Figure 2).

4. Conclusions

1. The "knee" in primary cosmic ray spectrum at $E \sim 3 \cdot 10^6$ GeV and the observed distinction in spectral exponent of protons and other nuclei can be explained by superdiffusion propagation of cosmic rays in fractal interstellar medium.

2. The source spectral exponent found from experimental data in this approach is $p \approx 2.9$.

3. The energy dependence of particle spectra does not contradict the hypothesis that the sources of particles with $E \lesssim 10^{18}$ eV are the bursts of nearby supernovae in the last $\sim 10^5$ years.

This work was supported by RFBR grants 0001 00284 and 0002 17507 and the program "Integration" (project 2.1–252).

REFERENCES

1. Ginzburg V.L., Syrovatskii S.I. Origin of cosmic rays. Pergamon Press, 1964.
2. Berezinsky V.S., Bulanov S.V., Ginzburg V.L. et al. Astrophysics of cosmic rays. North Holland, Amsterdam, 1990.
3. Kaplan S.A., Pikelner S.B. Physics of the interstellar medium. M.:Nauka, 1979.
4. Lozinskaya T.A. Supernova and star wind: interactions with galactic gas. M.:Nauka, 1979.
5. Ruzmaikin A.A., Sokolov D.D., Shukurov A.M. Magnetic fields of Galaxies. Kluwer, Dordrecht, 1988.
6. Vainshtein S.I., Bykov A.M., Toptygin I.N. Turbulence, stream layers and shock wave in cosmic plasm. M.:Nauka, 1989 (in Russian).
7. Molchanov S.A., Ruzmaikin A.A., Sokolov D.D. In book Nonlinear waves: physics and astrophysics. M.:Nauka (1993) 47.
8. Davison B. Neutron transport theory. Oxford University Press, London, 1957.
9. Case K.M, Zweifel P.F. Linear transport theory. Addison-Wesley, Reading, MA, 1967.
10. Lagutin A.A., Nikulin Yu.A. Uchaikin V.V. Preprint ASU–2000/4, Barnaul, 2000.
11. Samko S.G., Kilbas A.A., Marichev O.I. Fractional integrals and derivations and some applications. Minsk: Nauka, 1987 (in Russian).
12. Zolotarev V.M., Uchaikin V.V., Saenko V.V., JETP, v.115 (1999) 1411.
13. Uchaikin V.V., Zolotarev V.M. Chance and Stability. VSP, Netherlands, Utrecht, 1999.
14. Yoshida S., Hayashida N., Honda K. et al, Astropar. Phys. v.3 (1995) 105.
15. Grigorov N.L., Nesterov V.E., Rapoport I.D. et al, Yadernaya Fizika v.11 (1970) 1058.
16. Ryan M.J., Ormes J.F., Balasubrahmanyan V.K., Phys. Rev. Lett. v.28 (1972) 985.
17. Ivanenko I.P., Rapoport I.D., Shestoperov V.Ya. at al, Pis'ma v JETP v.48 (1988) 468.
18. Asakimori K., Burnett T.H., Cherry M.L. et al, Proc. 23 ICRC. Calgary v.2 (1993) 21.
19. Shibata T. Nucl.Phys.B. v.75A (1999) 22.
20. Fomin Yu.A., Khristiansen G.B., Kulikov G. V. et al. Proc 22 ICRC. Dublin v.2 (1991) 85.
21. Nagano M., Hara T., Hatano Y. et al. J. Phys. G: Nucl. Phys. v.10 (1984) 1295.

ELSEVIER

Nuclear Physics B (Proc. Suppl.) 97 (2001) 271–273

www.elsevier.nl/locate/npe

Application of Markov model of random medium for the interpretation of experimental results on muon intensity measurements at underground installations

A.A. Lagutin, V.V. Uchaikin, A.V. Yushkov, V.V. Makarov [a]

[a]Theoretical Physics Department, Altai State University,
Dimitrova st. 66, 656099 Barnaul, Russia

Incomplete information on overburden rock composition is the source of systematic error arising in underground muon experiments. As was shown in our previous works this problem requires a probabilistical approach, namely the introduction of a statistical ensemble of media which includes all possible medium realizations. As an example of such an ensemble we proposed the Markov two-phase model of random medium and applied it to the KGF ground. In this paper we continue the discussion of this problem and consider further possibilities of application of this model for estimating the influence of ground properties uncertainties on experimental results.

1. Introduction

In analysis of data obtained in underground muon experiments, it is appropriate to use average characteristics of the muon component, remembering that fluctuations of these characteristics can be reduced to minimum, provided the statistics of the experiment is rather high, and that there are theoretical approaches to evaluate fluctuations numerically. Quite a different situation arises in account of overburden rock properties: one has to put up with the use of average density, $\langle Z \rangle$, $\langle A \rangle$, $\langle Z/A \rangle$, and $\langle Z^2/A \rangle$, although knowing that this set of values may be not the optimal one and varies depending on azimuthal and zenith angle of arrived muons, which, therefore, traverse media with differing properties. This case requires the operation with a set of characteristics, even while registering isodimensional muons (*e.g.* vertical ones), because of lack of geological information and, to less extent, due to Coulomb scattering of muons, which may, coming from different directions, consequently get into the detector from the same solid angle. The logical way out from this situation resides in the advent of a statistical ensemble of media, in which the whole set of possible medium realizations, with the probability given on it, is considered. In previous papers [1,2] we proposed, as an example of such an ensemble, the model of Markov random medium and applied it for the description of the KGF ground. In this paper we consider the possibilities of the further use of this model, which can serve as a tool for estimation of systematic errors, presenting in underground experimental data because of lack of geological information.

2. Markov model of random medium

In Markov model of random medium, physical properties of medium realizations are described by a finite number of step functions. Constancy of the function on some interval Δx means that it is filled with homogeneous substance of a particular type. The heterogeneous medium is formed by combination of these substances (which we call "phases"). One-phase medium is homogeneous, and a two-phase one, the simplest case of this model, consists of the alternating different-length sections of both phases. The random medium realizations differ from each other in borders positions and in a sequence of phases filling intervals between borders. The statistics of these realizations is described by a Markov process with continuous time, the role of which is played by the spatial variable x. The probability that at a depth x in a given realization the phase is in state

0920-5632/01/$ – see front matter
PII S0920-5632(01)01281-6

k is given by

$$p_k(x) = \sum_i p_i(0)p_{i\to k}(x), \quad (1)$$

where $p_{i\to k}(x)$ is the probability of the $i \to k$ transition in a distance x and satisfies the equations [3]

$$\frac{\partial p_{i\to k}}{\partial x} + \sigma_k p_{i\to k} = \sum_{j\neq k} p_{i\to j}\sigma_{j\to k}, \quad (2)$$

with $p_{i\to k}(0) = \delta_{ik}$. In this equation $\sigma_{j\to k}dx$ is the probability of a transition from phase j to phase k in a distance dx and $\sigma_k = \sum_{j\neq k}\sigma_{k\to j}$. In case of a statistically homogeneous (p_k = const) two-phase medium it is easy to find the following expression for p_k:

$$p_k(x) = p_k(0) = \frac{\lambda_k}{\lambda_1+\lambda_2}, \ \lambda_k = 1/\sigma_k.$$

The formal likeness of equation (2) to the transport equation along a direction x for particles of different types allowed us to give another useful interpretation of the Markov random medium. In this interpretation $p_k(0)$ is the probability of birth of a k type particle at point $x = 0$, $p_k(x)$ is the probability that the particle passing through point x is of type k, $\sigma_{j\to k}dx$ is the probability of the particle type changing from j to k in a distance dx. Besides, the free path of the ith type particle is associated with the layer of medium filled with ith phase and the points of interaction at the beginning and at the end of the path are associated with borders of ith phase. We named these particles "phasons"[3]. The phason interpretation of the Markov medium lets us use the concept of an importance in respect of phasons. Thus the random mass thickness of the (x,l) layer, beginning the in phase i

$$\tau_i(x,l) = \int_x^l \rho(x')dx'$$

can be considered as linear stochastic importance of phason i at position x. The equations for the probability density function of this importance have the following form [3]:

$$-\frac{\partial\Psi_i(\tau|x)}{\partial x} + \rho_i\frac{\partial\Psi_i(\tau|x)}{\partial\tau} + \sigma_i\Psi_i(\tau|x) =$$

$$= \sum_{j\neq i}\sigma_{i\to j}\Psi_j(\tau|x), \quad \Psi_i(\tau|l) = \delta(\tau).$$

For the statistically homogeneous two-phase medium these equations have the exact analytical solutions giving the formula for the cumulative distribution function in the range $\rho_2 s \le \tau \le \rho_1 s$ [4]

$$F(\tau|s) = p_1\left\{1 - e^{-u}[1 + +2\int_0^{\sqrt{uv}} dx I_1(2x)e^{-x^2/u}]\right\} + +p_2 e^{-v}[1 + 2\int_0^{\sqrt{vu}} dx I_1(2x)e^{-x^2/v}], \quad (3)$$

where I_1 is the modified Bessel function,

$$u = \sigma_1\frac{\tau-\rho_2 s}{\rho_1-\rho_2}, v = \sigma_2\frac{\rho_1 s-\tau}{\rho_1-\rho_2}.$$

3. Results and discussion

The KGF rock is a convenient case for application of the Markov model of a two-phase random medium, since it consists of alternating layers of hornblende schist and pegmatite (quartz) [5], and one can construct a statistical ensemble of this ground by Monte-Carlo simulation and obtain distribution of mass thickness for depth of 2760 meters, as was done in our previous papers [1,2]. The next step is to pick out suitable parameters for the model by the comparison of this distribution with calculations by formula (3). We chose the following set of parameters (optimal in our point of view): λ_1 = 500 m (hornblende schist, ρ_1 = 3.02 g/cm^3) and λ_2 = 56 m (pegmatite, ρ_2 = 2.65 g/cm^3). With the use of these parameters in formula (3) and our depth-intensity relations, calculated in the framework of adjoint approach [2], we obtained distributions of muon intensities for different depths of the KGF rock and compared them with the fit of experimental data [6]. In Table 1 we give, as an example, the confidence limits $I_{0.025}$–$I_{0.975}$ for two muon spectra: $I_{\pi,K+\nu}$ is the muon spectra from decays of pions and kaons [7] and $I_{\pi,K+pr+\nu}$ is the just mentioned spectrum plus prompt muons contribution from [8](recombination quark-parton model). In both spectra it is accounted the neutrino-induced muon flux $I_\nu^\mu = 2.98 \cdot 10^{-13}$ $(\mathrm{cm}^2 \cdot \mathrm{sec} \cdot \mathrm{sr})^{-1}$ [9]. From these data one can see that the spread of muon intensities reaches 30%, and, although the best agreement with experimental fit shows π, K-decays spectrum, this fact does not allow to make a definite conclusion on the contribution of prompt muons, since experimental data may contain such uncertainties.

The use of the approach developed here, for other underground installations, encounters many problems. Usually information about distribution and composition of heterogeneities at ground depth is not so detailed as in the KGF ground case. For example at Mont Blanc laboratory [10] and at Baksan neutrino observatory [11] thorough data are available merely for the location depth of the apparatus. It is apparent that in such events it remains to make only the most general assumptions about possible phases, their thicknesses and depths of occurrence, that can follow from geological analogies, for example. A reduction of the information leads to the expansion of the statistical ensemble of media and, hence, to that of muon intensities. Besides, only experimenters have enough information on all characteristics of registered muon events, and, thus, they have great advantage in performing of analysis, similar to the above described one.

Table 1
Experimental fit I_{exp} $(\mathrm{cm}^2 \cdot \mathrm{sec} \cdot \mathrm{sr})^{-1}$[6] and confidence limits $I_{0.025}$–$I_{0.975}$ for spectra $I_{\pi,K+\nu}$ and $I_{\pi,K+pr+\nu}$ $(\mathrm{cm}^2 \cdot \mathrm{sec} \cdot \mathrm{sr})^{-1}$ for different depths of the KGF rock

s, m	I_{exp}	$I_{0.025}$–$I_{0.975}$	
		$I_{\pi,K+\nu}$	$I_{\pi,K+pr+\nu}$
2760	$2.75 \cdot 10^{-11}$	$2.72 \cdot 10^{-11}$	$3.04 \cdot 10^{-11}$
		$3.50 \cdot 10^{-11}$	$3.88 \cdot 10^{-11}$
3000	$1.27 \cdot 10^{-11}$	$1.21 \cdot 10^{-11}$	$1.39 \cdot 10^{-11}$
		$1.57 \cdot 10^{-11}$	$1.79 \cdot 10^{-11}$
3400	$3.62 \cdot 10^{-12}$	$3.36 \cdot 10^{-12}$	$4.05 \cdot 10^{-12}$
		$4.35 \cdot 10^{-12}$	$5.19 \cdot 10^{-12}$
3600	$1.96 \cdot 10^{-12}$	$1.87 \cdot 10^{-12}$	$2.30 \cdot 10^{-12}$
		$2.39 \cdot 10^{-12}$	$2.91 \cdot 10^{-12}$

REFERENCES

1. A.A. Lagutin, V.V. Uchaikin, A.V. Yushkov, Proc. of 24th ICRC, 1 (1995) 605.
2. A.A. Lagutin, V.A. Litvinov, V.V. Uchaikin, The sensitivity theory in the cosmic ray physics, Barnaul, 1995 (in Russian).
3. V.V. Uchaikin, A.A. Lagutin, Stochastic importance, Moscow, Energoatomizdat, 1992 (in Russian).
4. Levermore C.D. et al., J.Mat.Phys. 27 (1986) 2526.
5. Miyake S., Narasimham V.S., P.V. Ramana Murthy, Nuovo Cim. 32 (1964) 1505.
6. Krishnaswamy M.R. et al., Proc. of 15th ICRC, 6 (1977) 85.
7. L. Byambajardal, V.A. Naumov, S.I. Sinegovsky, Scient. Trans. of the Inst. of Phys. and Tech., Mongolian Academy of Sciences, Ulan Bator, 27 (1989) 44.
8. E.V. Bugaev, V.A. Naumov, S.I. Sinegovsky et al., Nuovo Cim. 12 (1989) 41.
9. M. Aglietta, B. Alpat, E.D. Alyea et al. (LVD Collaboration), Astropart. Phys. 3 (1995) 311.
10. C. Castagnoli, O. Saavedra, Nuovo Sim. 9c (1986) 111.
11. E.N. Alekseyev, V.V. Alexeyenko, Yu.M. Andreyev et al., Proc. of 16th ICRC, 10 (1979) 276.

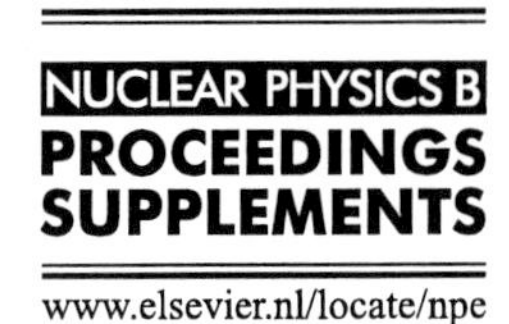

ELSEVIER

Nuclear Physics B (Proc. Suppl.) 97 (2001) 274–277

NUCLEAR PHYSICS B PROCEEDINGS SUPPLEMENTS
www.elsevier.nl/locate/npe

Lateral Distribution of Electrons in EAS at Superhigh Energies: Predictions and Experimental Data

A. A. Lagutin, R. I. Raikin[a]

[a]Department of Theoretical Physics, Altai State University,
66, Dimitrova str., Barnaul, 656099, Russia

In this paper we present a new approximation for the lateral distribution function of electrons in extensive air showers based on the scaling property established in our recent works. The detailed comparisons of theoretical predictions for the lateral distribution of charged particles in EAS at superhigh energies obtained with the use of new LDF of electrons with the experimental data of AGASA and Yakutsk array are carried out. The influence of the scintillation detector response and the atmospheric temperature effect on the shape of the lateral distribution of electrons is discussed in brief.

1. INTRODUCTION

Knowledge of the correct lateral distribution function (LDF) of extensive air shower (EAS) electrons and muons is of great importance for EAS research. The problem of fast calculations of LDF, proper and at the same time adequate to the experimental at very large distances from the shower axis ($r \geq 1$ km), is not solved yet. The reliable results for such big distances are necessary for the interpretation of experimental data on giant air showers and for new experiments designing. The prevalent approach to super-high energy EAS simulation provides the analytical description of electromagnetic subshowers based on the different well known modifications of NKG formula obtained for distances up to some hundreds of meters from core location. In this case the formal extrapolation to radial distances about 1 km and farther is a source of mistakes.

In our recent works [1,2] we established the new scaling property of the lateral distribution of electrons in both the electromagnetic cascade and extensive air shower. These results allow to get reliable data on the electron LDF up to $r \sim (25 \div 30)R_{\text{m.s.r.}}$, where $R_{\text{m.s.r.}}$ is the mean square radius of the electron component of the shower. It was shown that if we use the mean square radius as a radial scale parameter instead of the Molière unit R_M, the electron LDF is invariant with respect to the primary energy and the age of the cascade. Now we present the new formula for the lateral distribution function of extensive air shower electrons. We carried out detailed comparisons of our predictions for the charged particles LDF obtained with the use of the proposed formula for LDF of electrons with the latest experimental data of AGASA and Yakutsk array. The influence of the scintillation detector response and the temperature effect under the specific atmospheric conditions on the shape of the lateral distribution of electrons is discussed in brief.

2. NEW APPROXIMATION OF LATERAL DISTRIBUTION OF ELECTRONS

Our calculations [1–3] were carried out in the framework of quark-gluon string model for vertical showers initiated by primary protons with energies $E_0 = (10^5 \div 10^9)$ GeV for observation depths $t_{\text{obs}} = (614 \div 1030)$ g/cm^2 in the model of standard atmosphere. According to our calculational results, the electron local density at the distance r from the EAS core can be presented in a form:

$$\rho_e(r) = \frac{C_e N_e}{R^2_{\text{m.s.r.}}} \left(\frac{r}{R_{\text{m.s.r.}}}\right)^{-\alpha} \left(1 + \frac{r}{R_{\text{m.s.r.}}}\right)^{\alpha-\beta} \times$$

0920-5632/01/$ – see front matter
PII S0920-5632(01)01282-8

$$\times\left(1+\left(\frac{r}{10\,R_{\text{m.s.r.}}}\right)^2\right)^{-\delta}, \ \text{m}^{-2}, \quad (1)$$

where $C_e = 0.28$ is a normalization factor, N_e is the total number of electrons at the observation depth t_{obs}, $\alpha = 1.2$, $\beta = 4.53$, $\delta = 0.6$. The mean square radius can be found using the following approximations:

$$R_{\text{m.s.r.}}(E_0, t_{\text{obs}}) = \frac{\rho_0}{\rho_{\text{obs}}} \cdot A(t_{\text{obs}}) \times \\ \times \exp\left\{\frac{B(t_{\text{obs}})}{\ln(E_0/10^9\ \text{eV})}\right\}, \text{m}, \quad (2)$$

$$A = 6.69 \cdot 10^{-2}\, t_{\text{obs}} - 5.25, \\ B = 13.37\left[1 - 581.3 \times \right. \\ \left. \times \exp\{-1.44 \cdot 10^{-2}\, t_{\text{obs}}\}\right]. \quad (3)$$

Here ρ_{obs} is the density of the atmosphere at the observation level t_{obs}, $\rho_0 = 1.225 \cdot 10^{-3}$ g/cm^3.

Note that the dependence of the shape of the lateral distribution function on the energy, observation depth and primary particle (in superposition model for primary nuclei) is described by the variation of the single parameter $R_{\text{m.s.r.}}$. This is the essential difference between the LDF (1) and modified Linsley functions which are traditionally used for the approximation of experimental data of the biggest shower arrays AGASA and Yakutsk (see [4,5]).

3. COMPARISONS OF THEORETICAL RESULTS WITH EXPERIMENTAL DATA

The problem of correct comparisons of theoretical results on electron LDF with experimental data is very complicated because it is necessary to take into account the lateral distribution of muons (including particles with energies less than the threshold of muon detectors), detector response and some other features of the considered experiment, for example, the atmospheric temperature conditions of Yakutsk array, which are essentially different from the standard atmosphere.

Here we consider only near vertical showers with relatively small fraction of muons. Firstly, in order to estimate the charged particles density measured by the scintillation detector ρ_s we use the following relation:

$$\rho_s = k_{\text{sc}}\rho_e + k_\mu \rho_\mu, \quad (4)$$

where ρ_e is the electron density, calculated from eq. (1)-(3) with extrapolation up to extremely high energies, ρ_μ is the muon density, $k_{sc} = 1.1$ [6] is the conversion factor to the scintillation detector response and $k_\mu = 1.9$ [7] takes into account low energetic muons and electrons produced from the decay of muons. In order to clearly recognize the effects which are caused by the electron component, we use for ρ_μ the experimental muon densities measured by muon detectors at the considered array. Note that the muon lateral distributions measured by AGASA and Yakutsk array are in good agreement at $E_0 \leq (1 \div 3) \times 10^{18}$ eV [7].

3.1. AGASA

We obtained the following relation between $\rho_s(600)$ calculated from eq. (4) and primary energy for the atmospheric depth of AGASA

$$E_0 = 2.29 \cdot 10^{17} \rho_s(600)^{1.01}\ \text{eV}, \quad (5)$$

which agrees quite well with the AGASA relation [8]

$$E_0 = 2.03 \cdot 10^{17} \rho_s(600)^{1.02}\ \text{eV}. \quad (6)$$

In fig. 1 our results on ρ_s are presented in comparison with AGASA experimental data. It is seen that our results are in good agreement with the experiment but somewhat steeper than the approximation function used at AGASA.

3.2. Yakutsk Array

The energy conversion formula as calculated for Yakutsk atmospheric level (1020 g/cm^2) is

$$E_0 = 2.81 \cdot 10^{17} \rho_s(600)^{1.01}\ \text{eV}. \quad (7)$$

In order to directly compare with our results obtained under the standard atmospheric conditions, the Yakutsk conversion formula $E_0 = 4.80 \cdot 10^{17} \rho_s(600)^{1.0}$eV [5] should be recalculated. The method traditionally used for such correction is to take into account the ratio of the Molière units at the observation depth. The corresponding correction coefficient is $\rho_s(600; R_M^{\text{Y}})/\rho_s(600; R_M) =$

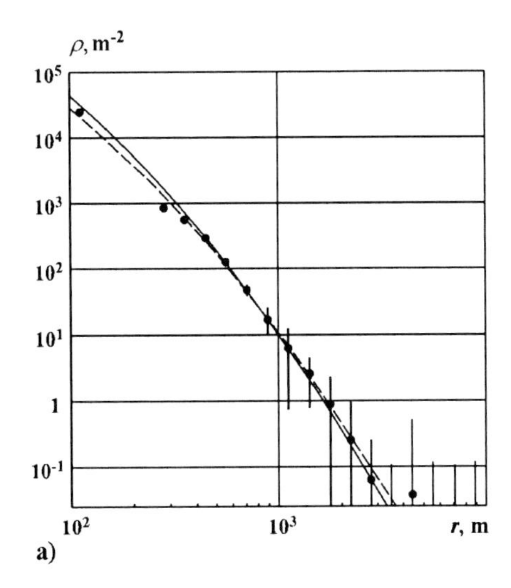

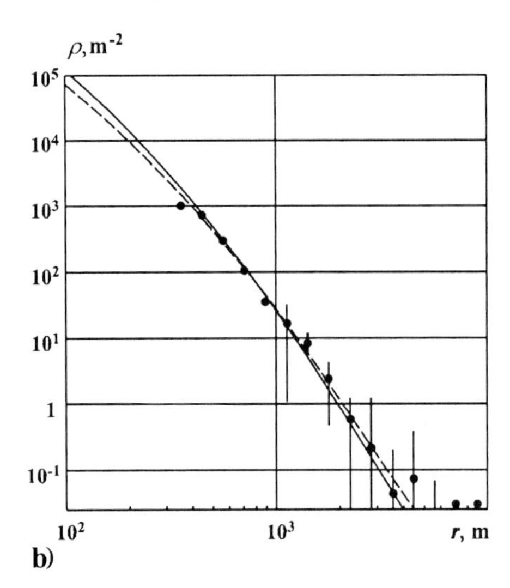

Figure 1. Lateral distributions of charged particles in EAS with $\langle \sec\theta \rangle \leq 1.1$ measured by AGASA [4]. a) $E_0 = 10^{19.3}$ eV, b) $E_0 = 10^{19.7}$ eV. Dashed lines – approximation function used at AGASA. Solid lines – our results

0.80, where $R_M = 81$ m for standard atmosphere, $R_M^Y = 68$ m for typical Yakutsk conditions (we use here the Yakutsk approximation formula for LDF of charged particles [9]). So we derive the new relation

$$E_0 = 3.85 \cdot 10^{17} \rho_s(600)^{1.0} \text{ eV}. \quad (8)$$

It is seen from the comparison of eq. (7) and (8) that the discrepancy is still essential. Note that approximately the same difference (by a factor ~ 1.4) occurs in comparisons of Yakutsk data with a number of theoretical results, for example, with QGSJET data [5]. There is evidence [10] that the correct simulations of energy deposition in the detectors (this result obtained using the SIBYLL-MOCCA program) leads to the additional correction of Yakutsk primary energy spectrum by a factor $\times 0.7$. After that, the relations turn out to be in satisfactory agreement to each other.

Fig. 2 shows Yakutsk experimental data on the LDF of charged particles in comparison with our calculational results obtained for the same $\rho_s(600)$. It is seen that our calculational function is steeper than the experimental data at $E_0 = 2 \cdot 10^{18}$ eV as well as at $E_0 = 2 \cdot 10^{19}$ eV. If we take into consideration the additional steepening of experimental data as a consequence of the specific atmospheric conditions for the Yakutsk array, the difference becomes essential.

4. DISCUSSION

Let us mention some additional factors that could partially compensate the differences between our calculational results and experimental data.

First of all it should be noted that all our calculations were carried out for primary protons. In the case of a heavier primary composition, we obtain a flatter distribution function.

In [2] we presented the results on the transition effect of EAS electrons in scintillation detectors of different thickness. It was shown that the correction factor k_{sc} depends on the radial distance and has a minimum around $r = 50$ m. So the correct description of the scintillators response leads to a slightly flatter distribution function at large distances than usual method does.

In our latest paper [11] we investigated the in-

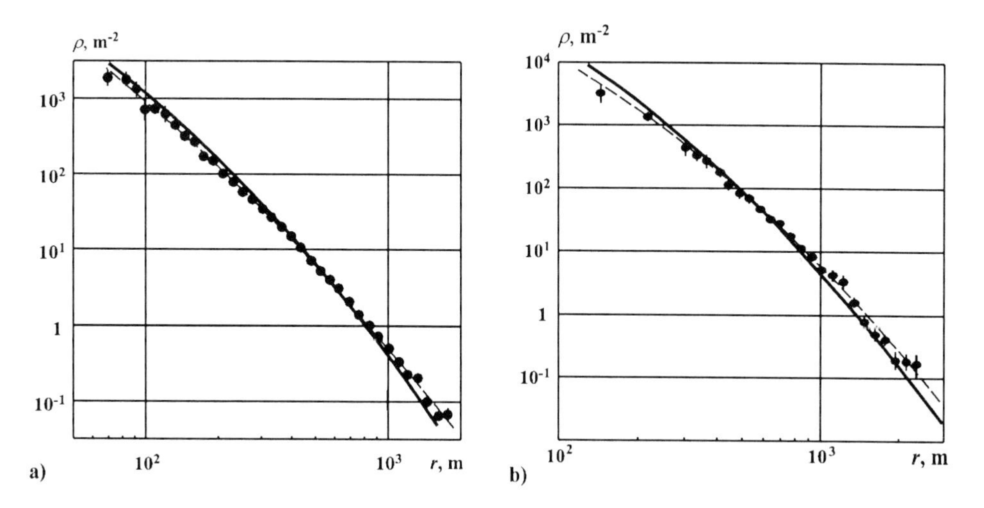

Figure 2. Lateral distributions of charged particles in EAS with $\langle\cos\theta\rangle \geq 0.98$ measured by Yakutsk array [5]. a) $E_0 = 2 \cdot 10^{18}$ eV, b) $E_0 = 2 \cdot 10^{19}$ eV. Dashed lines – approximation function used at Yakutsk. Solid line – our calculational results

fluence of the variation of the temperature profile of the atmosphere on the shape of the LDF of electrons (atmospheric temperature effect). It was shown that the temperature effect depends on primary energy due to the influence of inhomogeneity of the atmosphere on the mean square radius of electromagnetic subshowers. So in case of large variations of the temperature profile, the scale transformation of LDF using the Molière unit at the observation level is not adequate to describe the change of the LDF slope.

5. CONCLUSIONS

In this paper we propose a new scaling lateral distribution function of EAS electrons. The calculational predictions for LDF of charged particles are in a good agreement with the experimental data of AGASA. The comparisons of our results with the data of Yakutsk array demonstrate some inconsistency, which requires additional investigations.

REFERENCES

1. Lagutin A.A. et al. Proc. 25th ICRC, V.6, (1997) 285.
2. Lagutin A.A. et al. Nucl. Phys. B (Proc. Suppl.) 75 A, (1999) 290.
3. Lagutin A.A. et al. Preprint/Saitama University, Urawa, Japan; No 1 (1997).
4. Hayashida N. et al. Proc. 26th ICRC, V.1, (1999) 353.
5. Glushkov A.V. et al. Proc. 26th ICRC, V.1, (1999) 399.
6. Teshima M. et al. J. Phys. G: Nucl. Phys., 12 (1986) 1097.
7. Glushkov A.V. et al. Yad. Fiz., 58, (1995) 1265.
8. Dai H.Y. et al. J. Phys. G: Nucl. Phys., 14 (1988) 793.
9. Diakonov M.N. et al. Cosmic radiation of extremely high energy. Novosibirsk, Nauka, (1991) (in Russian)
10. Hillas A.M. Nucl. Phys. B (Proc. Suppl.), 75 A, (1999) 109.
11. Lagutin A.A. et al. Preprint/ASU; No 2000/2, (2000).

ELSEVIER

Nuclear Physics B (Proc. Suppl.) 97 (2001) 278–281

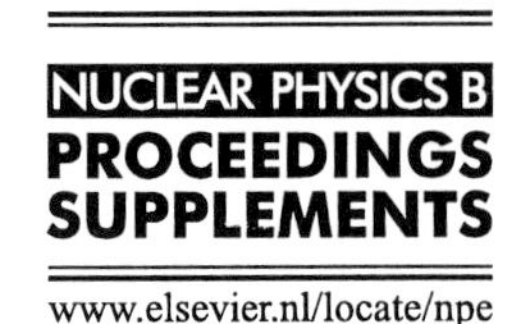

www.elsevier.nl/locate/npe

Connection Between UHECR's and Supermassive Relic Particles

L. L. Lengruber* and R. Rosenfeld†

Instituto de Física Teórica - UNESP
rua Pamplona, 145 - 01405-900 São Paulo (SP) Brazil

One of the models proposed for the origin of ultra high energy cosmic rays (UHECR's) suggests that these events are the decay products of relic superheavy metastable particles, which we call X particles. These particles can be produced in the reheating period following the inflationary epoch of the early Universe. We study this possibility and obtain constraints on some parameters such as the lifetime and direct couplings of the X-particle to the inflaton field from the requirement that they are responsible for the observed UHECR flux.

1. Introduction

According to conventional theories, cosmic rays are produced by the acceleration of charged particles in intense electromagnetic fields of astrophysical objects. This conventional wisdom, however, has some difficulties in explaining the observed flux of ultra high energy cosmic rays (UHECR) with energies above 10^{20} eV [1–4]. Protons at these high energies should quickly degrade their energy due to interactions with the cosmic microwave background (CMB), the so-called GZK limit [5–7]. Simulations [8] show that in order that protons with energy larger than 10^{20} eV arrive at Earth they must have travelled less than $\approx$100 Mpc, the attenuation length. Therefore, if their origin is astrophysical, the sources must be located within this distance. However, there are no candidate sources so far. Furthermore it is very difficult to find mechanisms to generate and extract such high energy particles from astrophysical sources.

A new scenario, the so-called top-down scenario, was then proposed to explain these events. According to this scenario, UHECR's would originate from the decay of superheavy ($m_X \gg 10^{20}$ eV) metastable particles. These particles would decay in a volume of 100Mpc around the Earth and thus the GZK limit would not apply. There are different mechanisms for the production of these particles [9–11], but here we will study their production in the reheating period following the inflationary epoch of the early Universe. Therefore, the lifetime of these particles must be greater than the age of the universe ($\tau_X \geq 10^{10}$ years).

2. Observed UHECR flux and X particle abundance

We will now derive a relation among the abundance Ω_X, the lifetime τ_X and the mass M_X of the X-particle in order to provide the observed flux of UHECR's on Earth. Following Sigl and Bhattacharjee [10], we will assume for definiteness that the UHECR's are photons, in which case the attenuation length from scattering off the cosmic microwave background is $l(E_\gamma) \approx 10$ Mpc.

The photon flux generated by the decay of an uniform distribution of X-particles is given by:

$$J_\gamma = \frac{1}{4\pi}\, l(E_\gamma)\, \dot{n}_X\, \frac{dN_\gamma}{dE_\gamma} \; , \tag{1}$$

where dN_γ/dE_γ is the photon injection spectrum from X-decay and $\dot{n}_X (= n_X/\tau_X)$ is the X decay rate.

Imposing that Equation (1) generates the observed UHECR flux (with $E \simeq 10^{20}$ eV),

$$J \approx 10^{-27}\, (\mathrm{m^2\, sr\, s\, GeV})^{-1} \; ,$$

and using dN_γ/dE_γ from a QCD model [10] one obtains:

$$\dot{n}_X = 3.7 \times 10^{-46} \left(\frac{M_X}{10^{12}\,\mathrm{GeV}}\right)^{-\frac{1}{2}} \mathrm{cm^{-3} s^{-1}} \; . \tag{2}$$

*leticia@ift.unesp.br
†rosenfel@ift.unesp.br

PII S0920-5632(01)01283-X

The relation between $\Omega_X h^2$ and τ_X is given by:

$$\tau_X = \frac{\Omega_X \rho_c}{\dot{n}_X M_X} = 10^{22}(\Omega_X h^2)\left(\frac{10^{12}\mathrm{GeV}}{M_X}\right)^{\frac{1}{2}} \mathrm{years}, \quad (3)$$

where $\Omega_X = \rho_x/\rho_c$, ρ_X is the X-particle mass density, $\rho_c \simeq 10^{-5}h^2\,\mathrm{GeV\,cm^{-3}}$ is the critical density, and h is the present value of Hubble's constant in units of $100\,\mathrm{km\,s^{-1}Mpc^{-1}}$.

If X-particles are a fraction of the cold dark matter (CDM), they must concentrate in galactic haloes and should not be uniformly distributed in the universe, as assumed in Equation (1). In this case the correct equation to describe the flux of UHECR's is

$$J^h_\gamma = \frac{1}{4\pi} R^h \dot{n}^h_X \frac{dN_\gamma}{dE_\gamma}\,, \quad (4)$$

where R^h is the radius of the galactic halo and $\dot{n}^h_X$ is the decay rate of the X-particles clustered in the halo.

This clustered distribution of X-particles also implies a modification of Equation (3):

$$\tau_X = 10^{22} f(\Omega_X h^2)\left(\frac{10^{12}\,\mathrm{GeV}}{M_X}\right)^{\frac{1}{2}} \mathrm{years}\,, \quad (5)$$

where

$$\begin{aligned} f &= \frac{J^h_\gamma}{J_\gamma} = \frac{n^h_X R^h}{n_X l(E_\gamma)} = \\ &= 1.5\times 10^3 \left(\frac{0.2}{\Omega_{CDM}h^2}\right)\left(\frac{R^h}{100\,\mathrm{Kpc}}\right)\times \\ &\quad \left(\frac{10\,\mathrm{Mpc}}{l(E_\gamma)}\right)\left(\frac{\rho^h_{CDM}}{0.3\,\mathrm{GeV\,cm^{-3}}}\right) \end{aligned}$$

and ρ^h_{CDM} is the cold dark matter energy density in the halo.

Considering $M_X = 10^{12}\,\mathrm{GeV}$, for a minimum τ_X of 10^{10} years and a maximum $\Omega_X h^2$ of the order of 1, we obtain the following limits (for $f \approx 10^3$):

- for $(\Omega_X h^2) \simeq 1 \rightarrow \tau_X \simeq 10^{25}$ years
- for $\tau_X \simeq 10^{10}$ years $\rightarrow (\Omega_X h^2) \simeq 10^{-15}$

3. Production of X-particles in the reheating phase

In the inflationary model, after the exponential expansion of the universe due to the vaccum energy of the inflaton field ϕ, particles are created by the oscillation of the inflaton field around its minimum, and the universe is reheated. In the case of relativistic particles (radiation), they are produced due to inflaton decay. For the case of the superheavy particle X, it could be created by either high energy photons or direct inflaton decay $\phi \rightarrow X\bar{X}$, if such a coupling is allowed. The study of the latter mechanism is the main result of this work.

In order to study quantitatively X-particle production during reheating, we have to solve the coupled differential Boltzmann equations for the energy densities of X-particles (ρ_X), radiation (ρ_R) and inflaton (ρ_ϕ):

$$\begin{aligned} &\dot{\rho_\phi} + 3H\rho_\phi + \Gamma_\phi = 0 \\ &\dot{\rho_R} + 4H\rho_R - (1 - B_X)\Gamma_\phi\rho_\phi \\ &\qquad - \frac{<\sigma|v|>}{M_X}[(\rho_X)^2 - (\rho_X^{EQ})^2] = 0 \qquad (6) \\ &\dot{\rho_X} + 3H\rho_X - B_X\Gamma_\phi\rho_\phi \\ &\qquad + \frac{<\sigma|v|>}{M_X}[(\rho_X)^2 - (\rho_X^{EQ})^2] = 0 \end{aligned}$$

In the above equation H is the expansion rate of the universe, Γ_ϕ is the inflaton total width, B_X is the branching ratio for the $\phi \rightarrow X\bar{X}$ decay, $<\sigma|v|>$ is the thermal average of the X annihilation cross secton times the Moller flux factor and ρ_X^{EQ} is the equilibrium value for the X-particle energy density.

It is convenient to work with co-moving and dimensionless variables and therefore we defined appropriate variables [11] and solved the equations numerically for the scaled variable $X \equiv \rho_X {M_X}^{-1} a^3$, where $a(t)$ is the scale factor. In this case the new independent variable is $x = aM_\phi$ (M_ϕ is the ϕ-field quantum mass).

For illustrative purposes we show in Figure 1 the behaviour of $X(x)$ for a given set of parameters. For that figure we used $M_\phi = 10^{13}\,\mathrm{GeV}$, $M_X = 10^{12}\,\mathrm{GeV}$ and $<\sigma|v|> = 10^{-26}\,\mathrm{GeV^{-2}}$.

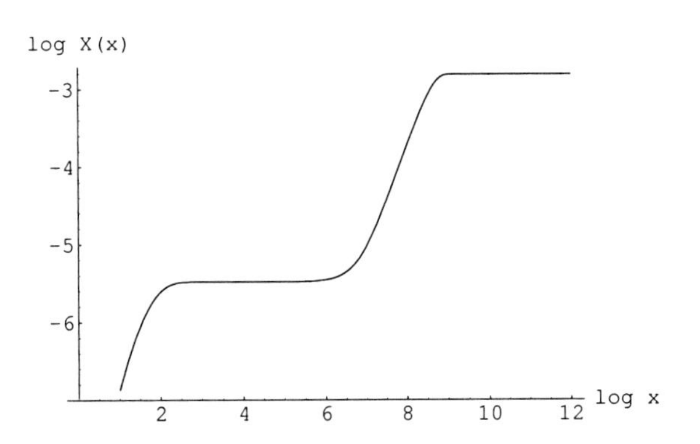

Figure 1. Evolution of X ($\equiv \rho_X M_X^{-1} a^3$) as a function of the modified scale factor, x. The two plateaus illustrate the two mechanisms of X-particle production: the first one ($X \approx 5 \times 10^{-6}$) is due to production of X-particle from high energy photons and the second one ($X \approx 10^{-3}$) corresponds to the production due to inflaton decay $\phi \to X\bar{X}$.

Γ_ϕ is obtained as a function of the reheating temperature T_{RH} assuming instantaneous conversion of the inflaton energy density into radiation:

$$\Gamma_\phi = 1.93 \times \left(\frac{T_{RH}}{10^9\,\mathrm{GeV}}\right)^2 \mathrm{GeV}\,.$$

We used $T_{RH} = 10^9$ GeV, which is high enough to generate the first plateau but at the same time does not lead to overproduction of gravitinos [12]. For this value of T_{RH}, the branching ratio B_X has to be very small (we used $B_X = 10^{-15}$) in order not to have a value of Ω_X much greater than one today.

4. Finding Ω_X

We can find the X-particle mass density Ω_X today using the present value ($x = x_0$) of the radiation abundance Ω_R ($\Omega_R h^2 = 4.3 \times 10^{-5}$):

$$\frac{\Omega_X h^2}{\Omega_R h^2} = \frac{\rho_X(x_0)}{\rho_R(x_0)} \tag{7}$$

In thermal equilibrium, $T \propto {g_*}^{-1/3} a^{-1}$ [13], and hence ρ_R can be written as:

$$\rho_R(x) = \rho_R(x_{RH}) \left(\frac{g_*(x_{RH})}{g_*(x)}\right)^{1/3} \left(\frac{x_{RH}}{x}\right)^4 \tag{8}$$

where g_* is the effective number of degrees of freedom (we used $g_*(x_{RH}) = 200$).

Substituting the above equation into Equation (7) and recalling that $\rho_X \propto a^{-3}$, we get:

$$\begin{aligned}\Omega_X h^2 &= \Omega_R h^2 \frac{\rho_X(x_{RH})}{\rho_R(x_{RH})} \frac{T_{RH}}{T_0} \\ &= 1.5 \times 10^{18} \left(\frac{T_{RH}}{10^9\,\mathrm{GeV}}\right) \frac{X_f M_X M_\phi^3}{x_i^3 H_i^2 M_{pl}^2}\,,\end{aligned} \tag{9}$$

where M_{pl} is the Planck mass and $T_0 = 2.37 \times 10^{-13}$ GeV is the present-epoch temperature. We also used the definition of H in the reheating epoch (H_{RH}) and assumed that the X-particles are already decoupled from the thermal bath at reheating and therefore

$$\rho_X(x_{RH}) = X_f {x_{RH}}^{-3} M_\phi M_X.$$

The subscript i refers to the epoch before inflaton decay. For chaotic inflation models we have $H_i = M_\phi$. As initial condition for the differential equations we choose $x_i = 1$.

X_f is the final value of $X(x)$, and in the example of Figure 1 is the value of the second plateau, $X_f = 1.5 \times 10^{-3}$, which results in $\Omega_X h^2 = 180$.

From Equation (9) we note that one could decrease Ω_X by decreasing the reheat temperature T_{RH}. Therefore, larger branching ratios are allowed in this case. Models with low reheat temperatures (down to 10 MeV) were recently analyzed in [14]. Even if this possibility eliminates the X-particle production by $\gamma\gamma \to X\bar{X}$ (the first plateau), the direct inflaton decay proceeds as usual and is the only operative mechanism in this case.

In order to understand the behaviour of Ω_X with B_X and T_{RH}, we rewrite Equation (9) using the relation:

$$\begin{aligned}\rho_X(T_{RH}) &= \frac{\rho_\phi(x_{RH})}{M_\phi} B_X M_X \\ &= \frac{3}{8\pi} \frac{M_{pl}^2}{M_\phi} H_{RH}^2 B_X M_X\,.\end{aligned} \tag{10}$$

Then, using the instantaneous reheating approximation ($\rho_R(T_{RH}) = (\pi^2/30)g_*(T_{RH})^4$), we find:

$$\Omega_X h^2 = 2.06 \times 10^7 B_X \left(\frac{T_{RH}}{1\,\mathrm{GeV}}\right) . \quad (11)$$

We can equate this expression to Equation (5) in order to find the allowed values for B_X as a function of T_{RH} and τ_X (for $M_X = 10^{12}$ GeV):

$$B_X = 4.9 \times 10^{-30} \frac{\tau_X}{T_{RH}} \frac{1}{f} , \quad (12)$$

for T_{RH} expressed in GeV's and τ_X expressed in years.

We show in Figure 2 the limits for B_X in the case $f = 10^3$.

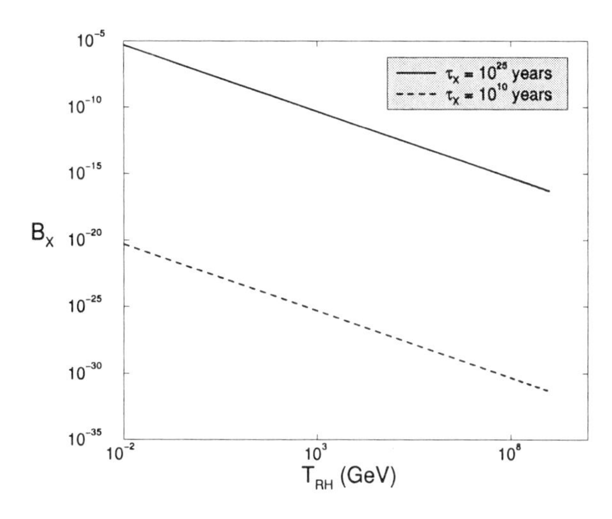

Figure 2. B_X as a function of T_{RH} for two limiting values of τ_X.

Unfortunately, the estimate of Equation (11) seems to lead to abundances one order of magnitude smaller than the ones obtained by solving the relevant Boltzmann equations numerically. This issue is still being investigated.

5. Conclusion

We have investigated the constraints on the properties of supermassive relic metastable particles demanding that their decay products are responsible for the observed UHECR flux. In particular, we found the relevant range for the branching ratio of inflaton decay into the X-particle as a function of the reheat temperature using a simplified analytical estimate. Detailed numerical results will be presented elsewhere.

Acknowledgments

We thank our collaborators A. H. Campos, H. Reis and R. Sato for many important discussions. This work was sponsored by FAPESP and CNPq.

REFERENCES

1. J. Linsley, *Phys. Rev. Lett. 10 (1963) 146.*
2. M. A. Lawrence et al., *J. Phys. G Nucl. Part. Phys. 17 (1991) 733.*
3. N. Hayashida et al., *Phys. Rev. Lett. 73 (1994) 3491.*
4. D. J. Bird et al., *Phys. Rev. Lett. 71 (1993) 3401.*
5. K.Greisen, *Phys. Rev. Lett. 16 (1966) 748.*
6. G. T. Zatsepin and V. A. Kuzmin, *JETP Lett. 4 (1966) 78.*
7. P. H. Frampton, *e-print astro-ph/9804219.*
8. F. A. Aharonian and J. W. Cronin, *Phys. Rev. D 50 (1994) 1892.*
9. V. Kuzmin and I. Tkachev, *JETP Lett. 68 (1998) 271; Phys. Rept. 320 (1999) 199.*
10. P. Bhattacharjee and G. Sigl, *Phys. Rept. 327 (2000) 109.*
11. D. J. H. Chung, E. W. Kolb, A. Riotto, *Phys. Rev. D 60 (1999) 63504.*
12. J. Ellis, J. Kim, and D. V. Nanopoulos, *Phys.Lett. 145B, (1984) 181.*
13. E. W. Kolb and M. Turner, *The Early Universe.* Addison-Wesley, 1990.
14. G. F. Giudice, E. W. Kolb, A. Riotto, *e-print hep-ph/0005123.*

ELSEVIER

Nuclear Physics B (Proc. Suppl.) 97 (2001) 282–285

www.elsevier.nl/locate/npe

The influence of using plastic scintillators in the determination of extensive air showers parameters

M. A. L. de Oliveira * and J. A. Chinellato † [a]

[a]Instituto de Física Gleb Wataghin
Universidade Estadual de Campinas (UNICAMP)
C. P. 6165 CEP 13083-970, Campinas, Brazil

We obtain a single particle signal for a plastic scintillator module, built in a similar way to the usual pyramid-like cosmic ray detectors . In the experimental setup, cosmic ray muons or electrons cross scintillator and also a streamer-tube chamber module (capable of tracking particles), so that one could account for different conditions of detection and efficiency. Usual electronics gave the ADC spectra, wich were used to simulate the response of such scintillator modules in typical EAS arrays. The aim is to investigate the low sampling of particles and the influence of the fluctuations introduced by the scintillator modules in the determination of EAS parameters.

1. Introduction

This work is concentrated in the study of plastic scintillators in pyramid-like detectors, as components ground-based experiments that observe cosmic ray showers. We are interested in the fluctuations in particle density measurements performed by such detectors and also in the subsequent determination of EAS features.

The experimental quantity to be analysed is the ADC spectrum. It will depend on geometrical conditions under which the particles crosses the scintillator plate — different light intensities will strongly affect the ADC readings. As most of the shower particles are relativistic their energy will not play an important role.

For a given number of particles crossing the module (e.g., in a shower), the ADC response will be a composition of one-particle characteristic spectra. This introduces fluctuations in the estimates of shower parameters. Tracking devices, like the streamer modules we used, would not have this fluctuation.

Such pyramid-like detectors are, in general, so scarcely distributed on the ground that they can detect only a very small fraction of particles from the shower front. This introduces a sampling effect that will be examined here, for three generic arrays with different number of detector stations.

*leigui@ifi.unicamp.br
†chinella@ifi.unicamp.br

2. Calibration of the single particle signals

The experimental procedures described hereafter has been done with the apparatuses of the EASCAMP experiment [1], which consists nowadays of an array of 12 scintillator detectors and 3 muon telescopes (streamer tubes modules) built by Leptons Group of the Cosmic Ray Department at UNICAMP.

The scintillator modules are made of a plastic scintillator block NE102A [2] of $100 \times 70 \times 2\,cm^3$ and a photomultiplier (Philips XP2040) [3] packed inside a wooden box in the shape of a rectangular pyramid. The scintillator lays in the bottom and the photomultiplier is in the top of the pyramid. The walls inside the wooden box were painted in white to improve the diffusion of the light towards the photomultiplier. Its signal is sent to an ADC (LeCroy 2249W), where it is converted to a number (ADC channel) proportional to the pulse charge. To obtain the single particle calibration, an experimental system has been assembled in which a scintillator module was put above a streamer tubes module, as shown in figure 1. The streamer tube module was used as a tracking device to select single charged particles that crossed the scintillator plate. In figure 1, we can see the display for a typical one-particle

PII S0920-5632(01)01284-1

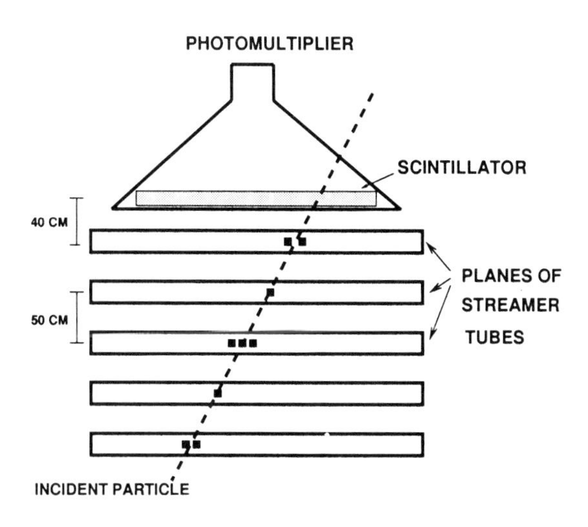

Figure 1. Experimental system of the single particle calibration.

event - they are mainly due to cosmic ray muons and a small part due to electrons, for the trigger conditions we used.

In figure 2, we show the distribution of the ADC readings, corrected to the vertical direction. Its mean value (262.3 channels in our case) has been taken as scale of conversion from ADC channel to the number of particles. The reader can notice also that the distribution of the signals fluctuates through a wide range of values around the mean, which give rise to errors in performing particle density measurements.

3. Simulation of EAS arrays and results

Suppose we have a set of N detectors that can register the particle densities $\rho_1, \ldots, \rho_N$. Our hypothesis is to use a χ^2-function, whose statistical Poisson fluctuation for each detector is taken as the square root of the experimental density:

$$\chi^2 = \sum_{i=1}^{ndet} \frac{\left(\rho^i_{obs} - \rho^i\right)^2}{\rho^i_{obs}} \tag{1}$$

here, $ndet$ is the number of detectors in the array, ρ_{obs} is the observed density and ρ is the predicted density (taken from some model) in each detector

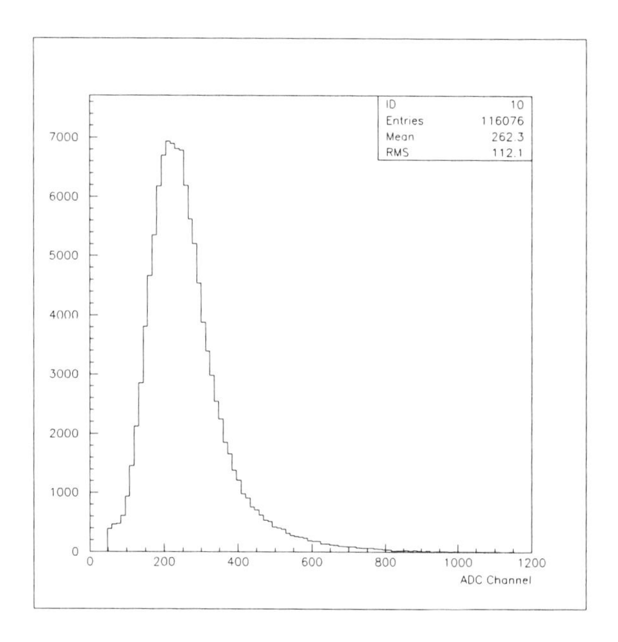

Figure 2. Distribution of single particle signals.

i. The most commonly used density distribution function is the approximation given by Greisen [4] of the Nishimura and Kamata solution [5] for the electromagnetic component:

$$\rho(r) = \frac{N_e}{r_1^2} f_{NKG} \tag{2}$$

with:

$$f_{NKG} = c(s) \left(\frac{r}{r_1}\right)^{s-2} \left(1 + \frac{r}{r_1}\right)^{s-4.5} \tag{3}$$

where r is the distance from the shower axis, r_1 is the Molière radius ($\sim 78\,m$ at sea level), N_e is the total number of electrons, s is the shower age parameter and $c(s)$ is a normalization constant. Notice that the final dependence for χ^2-function is: $\chi^2 = \chi^2(x_0, y_0, N_e, s)$, where x_0 and y_0 are the shower core coordinates.

For a given primary energy E_0 and atmospheric depth t (in units of radiation lengths), the total number of electrons N_e is given in Approximation

B [5] by:

$$N_e(E_0,t) = \frac{0.31}{\sqrt{\beta_0}} exp\left[t\left(1-\frac{3}{2}\ln s\right)\right], \quad (4)$$

while s and β_0 are given by:

$$s = \frac{3t}{t+2\beta_0} \ , \ \beta_0 = \ln\frac{E_0}{\epsilon_0} \quad (5)$$

where $\epsilon_0 = 84.2$ MeV is the critical energy. We will take t as the depth after the first interaction of the primary particle with exponential distribution with a mean free path of 80 g/cm^2.

The simulation of the shower front is naïve in these calculations. We fixed the input primary energy (E_0) and calculated the number of electrons N_e and the age s at sea level, as in equations 4 and 5. After that, we sampled the EAS core position in the array, from a uniform distribution inside a circle of one Molière radius. Then, the particle density hitting each scintillator module is taken from equation 2.

To account for a group of shower particles hitting a pyramid module, we summed up single particle signals, each one obtained from the probability density given in the histogram of figure 2. For that purpose, we used the routine HRNMD1 from Cernlib [6]. In this way, we simulate the response of the electronics and the actual number of particles is given dividing this value by the average ADC channel for single particle (262.3 channels). This procedure will be refered from now on as *signal fluctuation*. Doing so, simulating and reconstructing the shower front by fitting the same NKG function, all the deviations we have in estimating shower parameters are due to the low sampling of shower particles (small sensitive area, as in usual arrays) and the fluctuation in the ADC readings.

The determination of the EAS parameters was performed with MINUIT minimization algorithms [7] of the χ^2 function of equation 1. The initial core position has been chosen with a geometrical method [8] based on the ratios of the particle densities of three detectors to a given reference detector. The initial age is fixed in $s \sim 1.2$, which is the average value in our case and N_e is guessed through the biggest particle density for each event:

$$N_e = \frac{r_1^2 \times \rho^{i=max}}{f_{NKG}(x_0,y_0,s)} \quad (6)$$

We studied three array configurations with total area of 10000 m^2, namely, square nets of 100, 64 or 25 detectors. Here, we present our estimates of EAS core position and primary energy incertitudes (see table 1). We can examine $d90\%$, which is the distance around the simulated core position, that contains 90% of the reconstructed cores. To estimate the primary energy, we used a parametrization in β_0, given by:

$$\beta_0 = 2.9398 + 0.9 lnN_e \quad (7)$$

for N_e after the minimization. Then, we estimated the incertitude through $\Delta E/E_0$, where ΔE is the half-width of the distribution of reconstructed energies. In table 1, the quantities between parenthesis are the results of the calculations without taking into account the *signal fluctuation* - they show only the efficiency of the reconstruction algorithm, for different conditions of sampling. The quantities outside parenthesis in columns 3 and 4 include *signal fluctuation*. These values show the actual precisions for each array. The precision has been lowered, as expected, by including the fluctuation. Also, we can note the performance of the reconstruction algorithm in different situations: for the same value of primary energy, the resolution becomes better when we increase the number of scintillator modules, meaning that the sampling is better. When the primary energy grows, the reconstruction is better, as expected, because the relative fluctuations are smaller than those at lower energies.

4. Conclusions

In the first part of this work, we developed an experimental system with the aim of obtaining a calibration of the signals of a plastic scintillator detector, coupling it with a streamer tubes module. With the streamer module we selected events of single particle to get the distribution of ADC readings in the scintillator (shown in figure 2). The detectors containing plastic scintillators, used in many cosmic ray experiments, can

array($ndet$)	E_0 (eV)	$d90\%$ (m)	$\Delta E/E_0$ x 100
100	10^{15} eV	10.3 (0.004)	26% (2.7%)
100	10^{16} eV	5.65 (.0008)	12% (1.8%)
64	10^{15} eV	36.1 (0.007)	29% (3.8%)
64	10^{16} eV	9.91 (0.002)	15% (2.1%)
25	10^{15} eV	127. (0.014)	33% (5.3%)
25	10^{16} eV	17.4 (0.003)	19% (2.5%)

Table 1
Reconstruction of EAS core position and primary energy. Between parenthesis, we present the results due only to minimization method (without *signal fluctuation*).

give distributions like that of figure 2 and bring incertitudes in the measurements of EAS parameters. Such an experimental setup can be a useful tool to calibrate scintillator modules in EAS arrays. In the second part, we used the obtained distribution to simulate the fluctuation of the detectors' signals in three given arrays and reconstructed the EAS parameters in each case. In this way, we could evaluate the incertitudes for reconstructing the EAS parameters. The third column of table 1 shows that the *signal fluctuation* in scintillator modules introduce an incertitude that ranges from some to tenths of meters in the reconstructed core position. The fourth column shows a variation of 10 to 30% in the incertitude for primary energy determination. For traditional air shower arrays that operate with scintillator-based modules such those cited here, our results could be taken into account.

For a more detailed discussion of the procedures done in this work, the reader is referred to [8]. We would like to acknowledge CNPq[1] for the financial support.

REFERENCES

1. *The EASCAMP detector at Campinas*, A. R. Biral et al, 24th Int. Cosmic Ray Conference, Rome, Italy, **Vol. 1**(1995) 450-453.
2. *Scintillation Materials*, Nuclear Enterprises Technology catalog.
3. *Electron Tubes*, Philips Data Handbook, Part 9, Sept. 1982.
4. K. Greisen, *Prog. in Cosmic Ray Physics*, Amsterdam: North Holland Publ. Co., **Vol. III**(1956) 1.
5. K. Kamata and J. Nishimura, *Suppl. Prog. Theoretical Physics*, **6**(1958) 93.
6. *Cern Program Library*, at URL http://www.cern.ch
7. *MINUIT - Function Minimization and Error Analysis*, CERN Program Library Entry **D506**, Geneva (1992).
8. M. A. Leigui de Oliveira, *O Uso de Cintiladores Plásticos na Medida de Densidade de Partículas e Sua Influência da Caracterização de Chuveiros Atmosféricos Extensos*, M. Sc. Thesis, IFGW – Unicamp (1995) Brazil.

[1] Conselho Nacional de Desenvolvimento Científico e Tecnológico - Brazil.

ELSEVIER

Nuclear Physics B (Proc. Suppl.) 97 (2001) 286–289

NUCLEAR PHYSICS B
PROCEEDINGS
SUPPLEMENTS

www.elsevier.nl/locate/npe

Study of the influence of redshift on pathlengths of UHE gammas.

R. C. Rigitano
Instituto de Física 'Gleb Wataghin'
Universidade Estadual de Campinas—Unicamp
13083-970 Campinas (SP) Brasil.

The evolution of the intergalactic low-frequency radio background in ancient epochs can in principle have influence on the transparency of the universe to very-high-energy gamma rays. Calculation of path lengths of very-high-energy gammas, limited by simple pair production but taking into account the redshift, showed a small effect at extremely high energies, $\gtrsim 10^{23}$ eV. .

PACS Numbers: 95.30, 98.01A

1. Introduction and motivation

In view of the present-day still uncleared question about the nature of rather well-established observations of VHE Cosmic Ray events, above $3\,10^{19}$ eV—and in spite of the different conjectures raised for their explanation—a pure electromagnetic nature should still be taken as a possibility. Previous calculations of HE-photon mean path lengths through the cosmic low-frequency radio background were performed in the past [1–4]. In this work we have included an explicit redshift (z) dependence of the extragalactic low-frequency photon density, as calculated from the integrated contributions of the sources.

Still in this connection, it should be kept in mind that in the low-frequency region of the radio cosmic background the spectral brightness arises only from the contributions of distant sources. In other words there is no evidence for the existence of a relic fraction in that range of frequencies.

2. Integrated spectral brightness

Differently from the microwave background, which has a relic origin, the low-frequency radio background is an integrated brightness from contributions of in-principle identifiable sources. The spectral brightness at frequency ν is

$$I_\nu = \frac{1}{4\pi}\int_0^\infty\int_0^\infty \frac{L_\nu}{4\pi d^2(1+z)^{1+\alpha}}\rho(L_\nu,z)\,dV\,dL_\nu \quad , \tag{1}$$

where L_ν is source spectral luminosity, $\rho(L_\nu,z)$ is the redshift dependent luminosity function, dV stands for the differential co-moving volume of an spherical shell, equal to [5, equation (15.7)]

$$\frac{4\pi d^2 c}{H(1+z)(1+\Omega z)^{1/2}}\,dz \quad ,$$

with d being the distance, given in the Friedmann model by

$$d = \frac{cz}{H}\frac{1}{1+z}\left[1+\frac{z(1-\Omega/2)}{(1+\Omega z)^{1/2}+1+\Omega z/2}\right]$$

(quantity α is the source spectral index, H the Hubble constant, Ω the cosmological density parameter, and c is the velocity of light)[1].

We apply the luminosity functions $\rho(L_\nu,z)$ as described by Condon [6] which reproduce source counts and parameter distributions at different frequencies. The reproduction of the weighted source counts at 1400 MHz (using formula (9) of Reference [6]), and the corresponding contributions to the spectral brightness, are shown in Figures 1a and 1b, together with the contributions coming from the sources—spiral galaxies and radio galaxies—in different intervals in spectral luminosity: $20 < \log_{10}[L_{1400}(\mathrm{W/Hz})] \leq 20.5$,

[1]The present-epoch values $H = H_0$ and $\Omega = \Omega_0$ have been used througout (for the calculations which follow for redshifts different from zero this implies that distances are overestimated). We assumed in these calculations $H_0 = 100\,\mathrm{km.s^{-1}Mpc^{-1}}$ and $\Omega_0 = 1$ (flat space-time) . A mean spectral index $\alpha = 0.7$ for all sources was adopted following Condon [6].

PII S0920-5632(01)01285-3

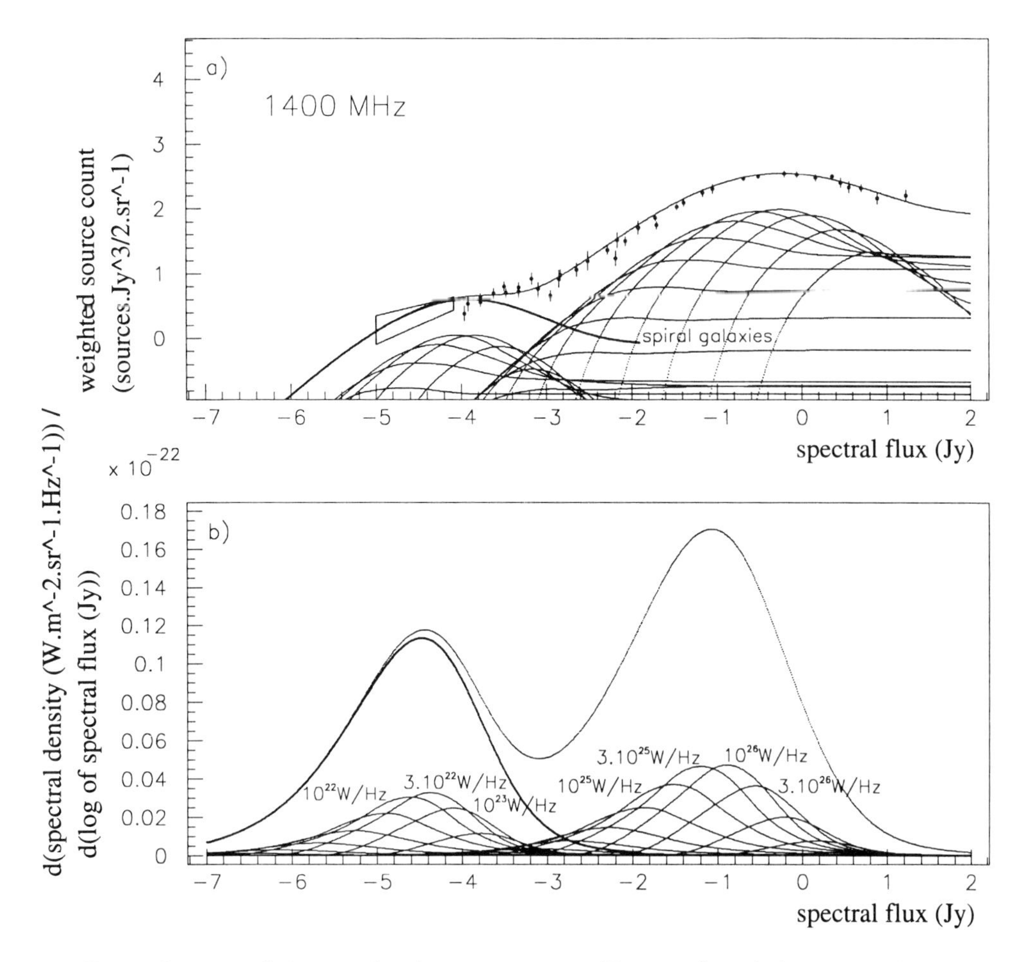

Figure 1. Reproduction of the weighted source counts (Figure a) and the spectral brightnesses per logarithm of flux (Figure b) at 1400 MHz according to the luminosity functions prescribed in Reference [6], together with the contributions due to sources in different intervals of spectral luminosity: 10^{20}–$3\,10^{20}$ W/Hz, $3\,10^{20}$–10^{21} W/Hz... and $3\,10^{23}$–10^{24} W/Hz for spiral galaxies, and 10^{23}–$3\,10^{23}$ W/Hz... to $3\,10^{27}$–10^{28} W/Hz for radio galaxies. The summed contribution (in each figure) due only to spiral galaxies has been detached. The overall integrated spectral brightness amounts to $6.5\,10^{-23}$ W.m^{-1}Hz^{-1}sr^{-1}. Ordinate: logarithm of weighted count $S_\nu^{5/2}\, n(S_\nu)$ (in Jy$^{3/2}$sr^{-1})—S_ν is the spectral flux; abscissa: logarithm of spectral flux, in janskys. In Figure b the ordinates are in W.m^{-1}Hz^{-1}sr^{-1}; abscissa: logarithm of spectral flux, in janskys.

$\ldots 27.5 < \log_{10}[L_{1400}(\mathrm{W/Hz})] \leq 28.$[2]

At the other frequencies the corresponding contributions follow profiles of spectral luminosity obtained from synchrotron emission and self-absorption processes in the sources—spiral and radio galaxies. The integration over the spectral flux at each frequency, and a sum over the intervals in source luminosity—as indicated in Expression (1), yield the radio extragalactic background, so evaluated until a low-frequency cutoff. Such a cutoff is of importance for the transparency of universe to the hypothetical extremely-high-energy photons from the distant sources.

[2] As a matter of completeness the overall integrated spectral brightness at 1400 MHz amounts to $6.5\,10^{-23}$ W.m^{-1}Hz^{-1} sr^{-1}.

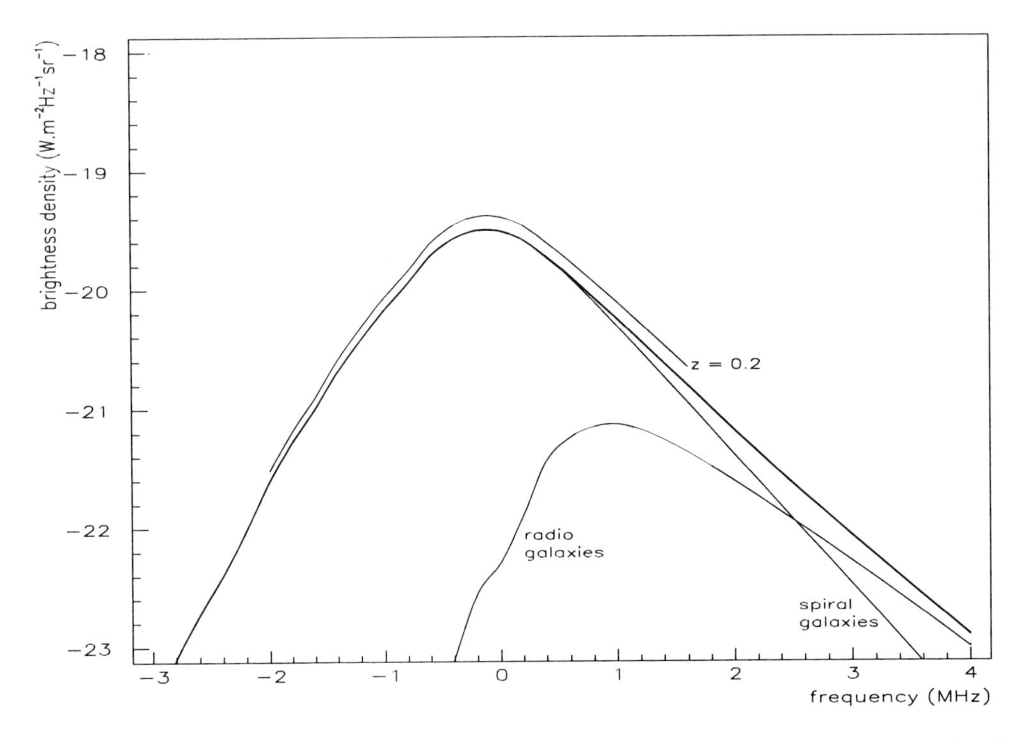

Figure 2. The contributions to the extragalactic radio spectrum coming from spiral galaxies and radio galaxies respectively, and the total spectrum, at redshift $z=0$. The spectrum changes only slightly for moderate redshifts, as shown by the shifted profile corresponding to $z = 0.2$.

3. The influence of redshift on the radio spectral density

We have considered a redshift dependent brightness $I_\nu(z_{\rm obs.})$ by introducing the difference $z - z_{\rm obs.}$ in formula (1),

$$I_\nu(z_{\rm obs.}) = \frac{1}{4\pi} \times$$

$$\int_0^\infty \int_{z_{\rm obs.}}^\infty \frac{L_\nu}{4\pi d^2(1+z-z_{\rm obs.})^{1+\alpha}} \rho(L_\nu, z)\, dV\, dL_\nu \quad ,$$

as well as in the subsequent formulas. Here, $z_{\rm obs.}$ stands for the redshift of the location the background is to be evaluated. In the ancient epoch photons evolved in a shorter interval of redshift. The result is an enhancement of the radio background in ancient epochs, as compared to the present one, which would be perceived by photons crossing large distances. It is opportune to note that a competitive effect happens between production of radio photons by the sources and the expansion of the universe, in such a way that somewhat larger densities existed in the past. In Figure 2 the calculated background is shown, for the present epoch, and a moderate redshift $z = 0.2$, and showing also the contributions arising from spiral galaxies and from the radio galaxies.

4. The mean free paths of VHE gamma rays and the redshift; conclusion

The gamma-ray mean free paths have then been obtained by means of the general algorithm mentioned previously [1,2], as a function of energy. They are shown in Figure 3. The higher is the energy of the incoming gamma ray, the more important the influence of very-low-frequency photons of the background in the opac-

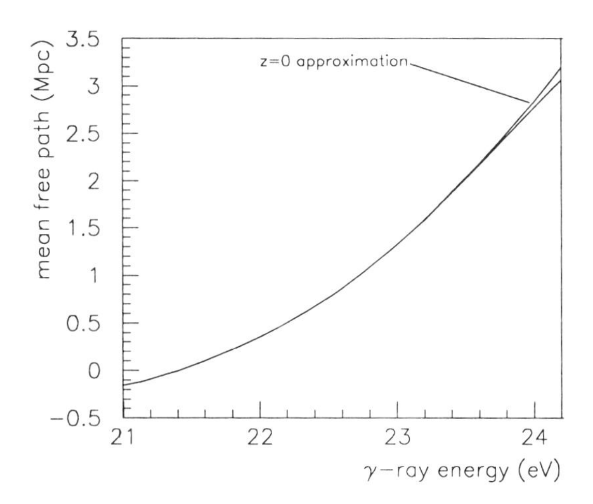

Figure 3. Mean free paths of high-energy gamma rays. Account for the redshift dependence of the radio photon density shows its effect only at extremely high energies. Ordinate: logarithm of path length in Mpc; abscissa: logarithm of energy in eV.

ity to the gamma ray[3]. Larger densities of low-frequency photons in the past enhance the opacity to gamma rays.

As result of the present calculation the relatively high background density in ancient epochs has a small—but in principle not negligible—effect on the mean free paths of photons with energy above 10^{23} eV.

If future observations will show evidence for the existence in nature of such extremely high-energy photons, the radio background evolution (not the evolution of the sources themselves) would have a chance to be identified. But on the other hand in such extreme regimes other channels also play their role, *viz.* multiple pair production and muon pair production, and their participation can have catastrofic effects on the paths of extremely high-energy gamma rays. In that case a redshift influence will then be hardly observed.

[3]The connection is due to a peak in the cross-section for pair production just above the threshold. That is the reason for the overall growth in transparency for photons more energetic than $\sim 10^{21}$ eV (they collide preferentially with the low-density radio cutoff photons).

Acknowledgements

The author was supported by FAPESP—*Fundação de Amparo à Pesquisa do Estado de São Paulo,* Brazil, grant 95/02341-0, for a two-years period in the *Laboratori Nazionali del Gran Sasso,* Italy, dedicated to the development of ideas connected to those here presented.

Special thanks are directed to prof. V. S. Berezinsky, who received and oriented the author at the *Laboratori.*

REFERENCES

1. A.I. Nikishov. Absorption of high-energy photons in the universe. *Sov. Phys. JETP*, 14:393, 1962.
2. R. J. Gould and G. P. Schreder. Pair production in photon-photon collisions. *Phys. Rev.*, 155:1404, 1967.
3. V.S. Berezinsky. Inverse compton effect, pair production, and the passage of high-energy electrons and photons through the metagalaxy. *Sov. J. Nuc. Phys.*, 11:222, 1970.
4. R.J. Protheroe and P.L. Biermann. A new estimate of the extragalactic radio background and implications for ultra-high-energy gamma-ray propagation. *Astropart. Phys.*, 6:45, 1996.
5. J.J. Condon. Radio sources and cosmology. In G.L. Verschuur and K.I. Kellermann, editors, *Galactic and Extragalactic Radio Astronomy*, chapter 15. Springer-Verlag New York Inc., 1988.
6. J.J. Condon. Cosmological evolution of radio sources. *Ap. J.*, 287:461, 1984.

ELSEVIER

Nuclear Physics B (Proc. Suppl.) 97 (2001) 290–293

www.elsevier.nl/locate/npe

Analysis of the Čerenkov time profile as a technique to distinguish the primary particle

L. V. de Souza Filho and C. Dobrigkeit [a] *

[a]Instituto de Física Gleb Wataghin, UNICAMP
CEP 13083-970, C.P. 6165, Campinas, Brasil

In this paper, we discuss the possibility of excluding the cosmic ray background in gamma ray telescope experiments using the Čerenkov photons arrival time. A method to distinguish proton and iron from gamma induced showers of energies around 50 TeV is presented. A total number of 300 simulated showers are analysed one by one and a distinction method is developed. The simulations were performed with the program CORSIKA and a detailed simulation of the telescope is applied to the surviving Čerenkov photons. Two different shapes of telescopes are tested: parabolic and Davies-Cotton, and the superiority of the parabolic mirrors concerning time information is illustrated.

1. Introduction

The time profile of the Čerenkov photons was first proposed as a technique to determine the primary particle in 1968 by D. Fegan et al.[1]. However, only recently it has been effectively studied as a viable technique.

The long time elapsed between the first proposal and the real use of the method was due to the low time resolution of electronic instruments and photomultipliers available during the 70's and 80's. Besides that, the success of the Imaging Technique [2] has contributed to postpone the implementation of the time method.

Nevertheless, the construction of the new generation of telescopes (VERITAS, HESS and MAGIC) and the advent of Flash ADC's have brought the time technique to researchers' attention again.

In the Imaging Technique, the shape and orientation of the signal in the photomultiplier camera are determined by fitting an ellipse. Cuts in the ellipse parameters allow the rejection of cosmic ray background events with an efficiency of almost 100 %. However, about 50 % of the gamma induced showers are excluded by the same cuts [3].

Despite that, this technique is only effective at small zenith angles what has limited the observations to $\theta < 30°$. On the other hand, the time method is expected to have its distinction power increasing with zenith angle.

As shown by P. Sommers and J.W. Elbert [4], large zenith angle observation would be a great improvement for gamma ray astronomy since the effective collection area and the energy threshold increase with zenith angle.

The present work shows it is possible to distinguish proton and iron from gamma induced showers for zenith angles larger than 30° and energies around 50 TeV.

2. Čerenkov light from the EM and muonic components

Previous works have studied the main features of Čerenkov photons in extensive air showers [5]. They show that electrons and muons are the particles which generate most of the Čerenkov photons detected in the telescopes.

The number of muons in hadronic showers is much larger than the number of muons in gamma induced showers of equal energy (see table 1) and most of these muons have enough energy to generate Čerenkov light through the atmosphere.

In hadronic showers, the electromagnetic component develops high in the atmosphere and creates its Čerenkov photons far away from the tele-

*The authors would like to thank CAPES and FAPESP for the financial support.

PII S0920-5632(01)01286-5

Table 1
Average number of muons in protons and gamma simulated vertical showers

Energy	Proton	Gamma
1 TeV	23.8 ± 10.3	0.2 ± 0.2
10 TeV	198.8 ± 67.1	1.7 ± 1.5
50 TeV	803.3 ± 244.5	19.8 ± 14.9

scope. On the other hand, muons traveling with velocity $v_\mu > c/n$, where c is the speed of light in vacuum and n the index of refraction of the air, create Čerenkov photons close to the observational level. This simple model predicts two separated sources of Čerenkov photons in hadronic showers which, due to the difference in distance of production and velocity of the particles, cause the detection of two distinct peaks in the time profiles.

In gamma induced showers at TeV energies the number of muons is very reduced so that the time profiles are almost completely formed by the electromagnetic component.

The effect which allows the distinction between two peaks in the hadronic showers is the interaction of particles with the air molecules that causes the differences in velocity and height of production. As it is known, the length of atmosphere crossed by the particles increases with zenith angle so that we expect a larger distinction effect for showers with large zenith angle.

For energies above 10 TeV the simple model is not completely true anymore since the number of cascades increases quickly with energy in the hadronic showers. The large number of cascades produces a great number of Čerenkov photons which hide the muonic component. However, the increasing of the effect due to the observation at large zenith angles could hopefully lead to a distinction method.

3. Monte Carlo Simulation

The results presented in the following sections are based on showers simulated by the CORSIKA [6] program with QGSJET [7] and EGS4 [8] routines.

Recently, some papers have been published

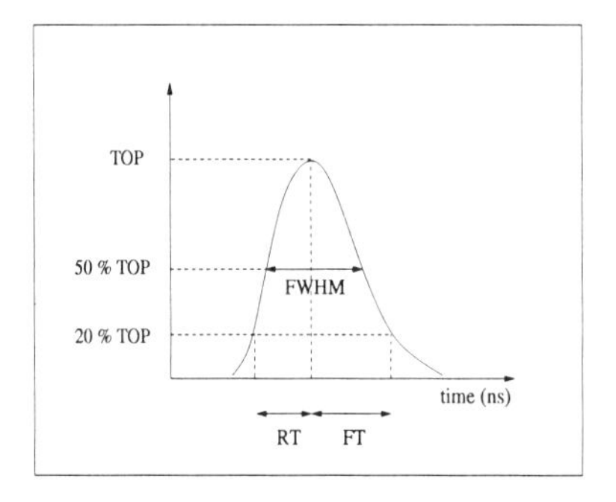

Figure 1. Definition of the parameters Rise Time (RT), Full Width at Half Maximum (FWHM) and Fall Time (FT)

confronting simulated and measured data with the aim of testing hadronic interaction models. QGSJET has shown better agreement with the experimental data [9]. In particular, the number of muons is satisfactorily simulated by QGSJET for energies below 10^{15} eV as can be seen in [10].

4. Analysis of the time profile

We begin studying a set of 50 simulated showers induced by protons and 50 simulated showers induced by gammas with energy 50 TeV and arrival direction of 0° and 30°.

The time profiles of each shower were characterized by the parameters Rise Time (RT), Full Width at Half Maximum (FWHM) and Fall Time (FT). These parameters are defined as the time intervals determined in the Čerenkov time profile as shown in figure 1.

For the first set of simulated showers we could not find any difference in the time profiles and parameters (RT, FWHM and FT) distributions for none of the two inclinations (0° and 30°).

However, for a second set of 150 showers (50 gamma + 50 proton + 50 iron induced showers) with energy 30 TeV and arrival direction of 0° and 50°, the differences in the time profiles are very clear.

A great number of individual time profiles shows an early peak which can be seen in figure 2 where we have summed all the photons in all the

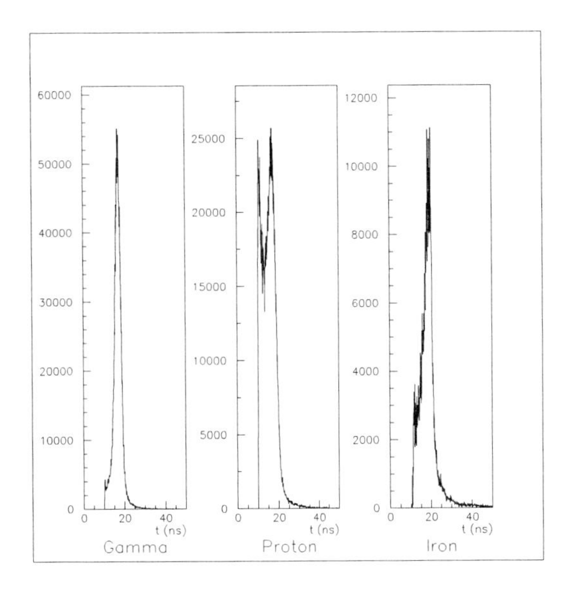

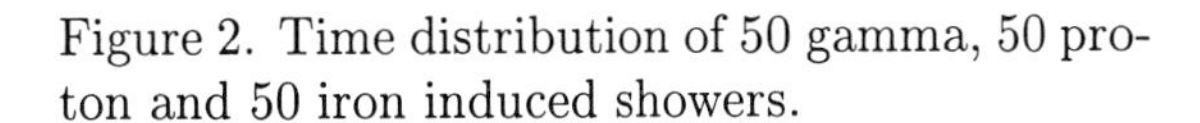

Figure 2. Time distribution of 50 gamma, 50 proton and 50 iron induced showers.

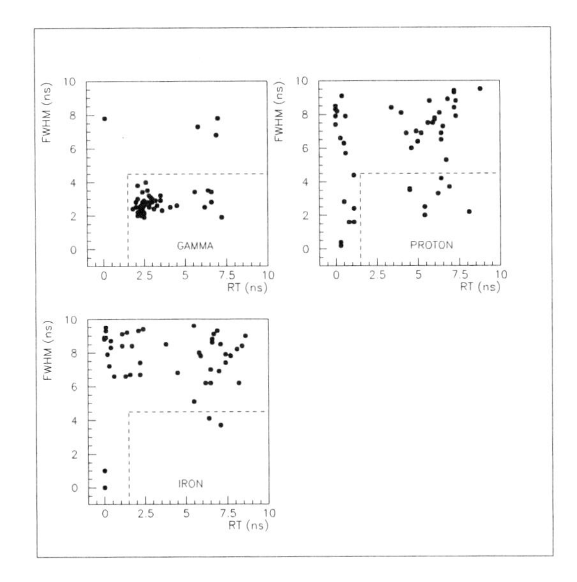

Figure 3. Illustration of the cuts proposed here. The points inside the dotted lines represent the showers identified as gamma induced showers in 50 proton, 50 iron and 50 gamma primary showers.

50 showers. The distribution for proton showers shows a very distinct early peak and although this peak is not present in the iron distribution, an excess of early Čerenkov photons is very clear in comparison to the gamma distribution.

Based on this information, we manage to elaborate cuts in the time parameters RT and FWHM which resulted in a successful distinction method. Figure 3 illustrates the cuts. All the simulated showers were analysed one by one. The gamma showers domain was defined by time profiles with $RT > 1.5$ ns and $FWHM < 4.5$ ns. The efficiency of this method can be quantified with the quality factor (q) defined in [11] and for our purpose q can be expressed as:

$$q = \frac{N_a^\gamma}{N_t^\gamma}\left(\frac{N_a^{crb}}{N_t^{crb}}\right)^{-\frac{1}{2}} \quad (1)$$

N_t^γ and N_t^{crb} are the total number of simulated showers induced by gammas and cosmic ray background events. N_a^γ is the number of showers induced by gammas and identified as gammas by the distinction method. N_a^{crb} is the number of showers induced by protons or iron identified wrongly as gamma events.

For this analysis we obtained a quality factor of 3 (hadrons:gamma). Table 2 shows detailed numerical results for the analysis.

5. A Qualitative Analysis of Different Telescopes Shapes

The results presented in the previous section do not take into account the shape of the telescopes. The photons were analysed as they arrived in the observational level.

The shape of the telescope is known to have a great influence on the time profile. Parabolic and Davies-Cotton mirrors are the most used configurations among the gamma ray experiments. In order to test the validity of the method presented here, we simulated two different shapes of mirrors and analysed the correspondent time profiles. Figure 4 shows a representative example of the time profile for three configurations of telescopes comparing to the time profile of the photons be-

Table 2
Number of showers identified as gamma or cosmic ray background from a total of 50 Gamma + 50 Proton + 50 Iron simulated showers.

Primary	Gamma	Cosmic Backgr.	Total
Gamma	45	5	50
Proton	7	43	50
Iron	2	48	50

fore they reached the observational level.

As expected, parabolic mirrors do not change the shape of the time profiles. On the other hand, present Davies-Cotton designs produce a wide spread in the time profile, which makes them not suitable for the use of the time methods based on the parameters RT, FWHM and FT here explored.

A quantitative study of the changes in the efficiency of the method is still to be done. The number of detected photons decreases very strongly if the telescope optics is simulated which implies only a small number of Čerenkov photons will hit a telescope. Therefore this kind of quantitative study is somehow prohibitive due to the computer memory necessary to store a great number of simulated showers.

6. Conclusion

As shown above, the time profile can be used as a distinction method for high energy showers at large zenith angles. Showers with energy up to 50 TeV were analysed and a simple method was demonstrated to be able to identify gamma showers from a set of gamma, proton and iron primaries.

Special cuts for the parameters Rise Time and Full Width at Half Maximum needed to be implemented. The gamma domain for showers at 50° with energy of 30 TeV was discovered to be $RT > 1.5$ ns and $FWHM < 4.5$ ns. This values are dependent on energy, core distance and inclination. Therefore they should not be considered as general cuts. However specific analysis for different measurements could be easily deduced.

Different telescope shapes also influence on the efficiency of the method and special calibrations would be necessary.

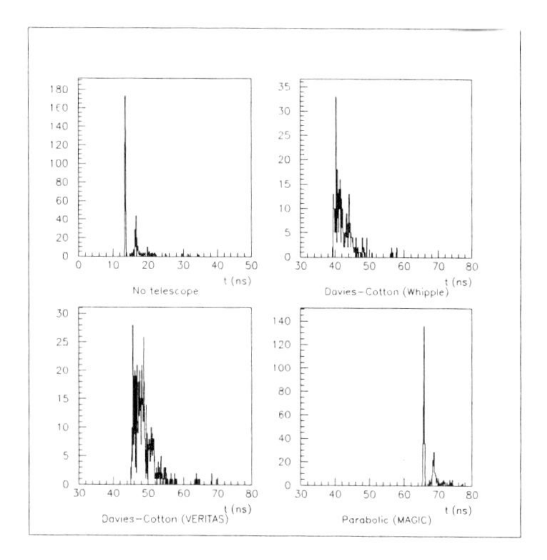

Figure 4. Example of the time profile for three different shapes of telescopes. The Whipple telescope is a Davies-Cotton with $f = 7$ m and 10 m of diameter, VERITAS is a Davies-Cotton with $f = 12$ m and 10 m of diameter and the MAGIC telescope is parabolic with aperture and focus length of 17 m.

REFERENCES

1. D. Fegan et al., Can. J. Phys., 46 433 (1968)
2. M. Hillas, Proc. XVIth ICRC, 3 445 (1985)
3. M. Hess, Astropart. Phys., 11 363 (1999)
4. P. Sommers and J.W. Elbert, J. Phys. G, 13 553 (1987)
5. M. Hillas, J. Phys. G, 8 1475 (1982)
6. D. Heck et al., Forschungszentrum Karlsruhe, FZKA 6019 (1998)
7. N.N. Kalmykov et al., Phys. At. Nucl., 56 3 (1993)
8. W.R. Nelson et al., Stanford Linear Accelerator Center, SLAC 265 (1985)
9. A. Lindner, Proceedings of the Xth Int. Symp. Very High Energy Cosmic Ray Int., (1998)
10. J. Knapp, Nuclear Physics B (Proc. Suppl.), 75A 89 (1999)
11. Ti-pei Li and Yu-qian Ma., Astrophys. J., 272 317 (1983)

ELSEVIER

Nuclear Physics B (Proc. Suppl.) 97 (2001) 294–297

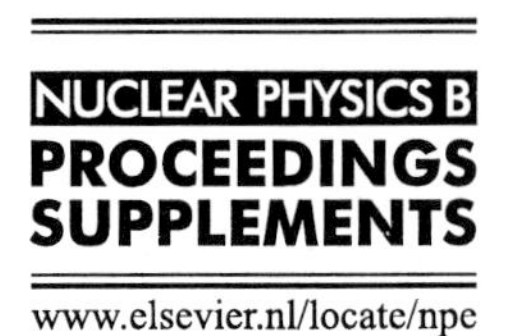

www.elsevier.nl/locate/npe

The Response of the EAS Muon Component in the GAMMA Installation

M.Z. Zazyan[a], A.P. Garyaka[a], R.M. Martirosov[a] and J. Procureur[b]

[a]Yerevan Physics Institute, Cosmic Ray Division, Alikhanian Brothers 2, 375036 Yerevan, Armenia

[b]Centre d'Etudes Nucleaires de Bordeaux-Gradignan, Université de Bordeaux 1 Rue du Solarium, 33175 Gradignan-Cedex, France

The response of the EAS muon component is studied by the detector simulation program ARES developed for the GAMMA experiment on Mt. Aragats (ANI Cosmic Ray Observatory, Armenia). Comparisons of experimental data with predictions of detector response simulations on the muon lateral distribution and the distribution of the muon size are presented.

1. Introduction

The main purpose of air shower experiments is to investigate the primary energy spectrum and mass composition of cosmic rays. The evaluation of the experimental data, however, requires a comparison of measurements with *pseudo-experimental* data, which simulate the stochastic development of shower cascade in the atmosphere and response of the detector setup. The detector simulation program ARES [1] based on the GEANT3 package [2] has been developed for the actual layout of the GAMMA array on Mt.Aragats (ANI Cosmic Ray Observatory) [3,4].

2. Experimental Setup

The GAMMA array consists of detectors in the surface part for registration of the extensive air shower (EAS) soft component, a calorimeter building and a large muon underground detector registering the muon component. The full device is described in detail elsewhere [3,4].

The muon underground detector consists of two parts: 60 plastic scintillators ($100 \times 100 \times 5$ cm^3) are placed in the underground hall below the ANI calorimeter building and another 90 detectors are installed in the tunnel connecting the underground hall with the surface. The detectors are built into aluminum or iron pyramid-shaped housings. The energy thresholds for muon detection are ≈ 5 GeV in the hall and ≈ 2.5 GeV in the tunnel.

3. EAS simulation

The development of EAS through the earth atmosphere has been performed using the CORSIKA code (version 5.20) [5] with NKG option, the high energy interaction model VENUS [6] and GHEISHA [7] for interactions at lower energies. Simulations have been carried out for the five types of primaries: P, He, O, Si and Fe. The showers were distributed in the energy range of 5×10^{14}eV - 3×10^{15}eV according to the power law with an index of -2.65. The zenith angles are $\theta < 30°$.

4. Detector simulation

The response of the GAMMA installation to the EAS muon component has been determined by the ARES program (version 1.01). The shower core positions are varying within a square of $60m \times 60m$ with the center in the middle of the calorimeter. This selection is close to the experimental shower treatment. Total number of EAS used for the present analysis is 1000. Primaries are mixed (normal mixed composition) in proportion: P:He:O:Si:Fe=40:21:14:13:12. The showers are distributed within the shower size interval of $10^5 - 10^6$.

For each simulated shower hadrons and muons are tracked through all components of the instal-

PII S0920-5632(01)01287-7

lation. The detailed simulation takes into account all relevant particle interactions, including the tracking of the secondaries produced in the setup. As output from ARES we have the energy deposits of particles hitting each individual muon detector. A rough estimation for muon number in the detector is made by dividing the sum of the energy deposits in the detector by the mean muon energy deposit 9.5 MeV [8]. This number is called "estimated" muon number. The "true" muon number is known from the simulation.

5. Muon Lateral Distribution

Muon lateral distributions for showers from each type of primaries were calculated. The results for proton and iron induced showers for the tunnel detectors case are shown in Figure 1. One

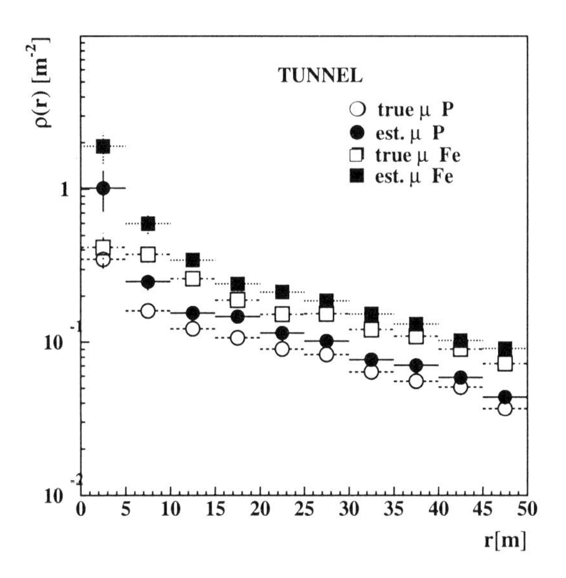

Figure 1. "Estimated" and "true" muons lateral distributions for proton and iron initiated showers (tunnel case).

can see some overestimation of muon number as far as hadrons (direct from CORSIKA and secondaries from hadron and muon interactions in absorber) as well as energetic electrons and photons produced by muons and hadrons can also contribute to the overall energy deposit in the muon detectors (see Figure 2 for primary protons).

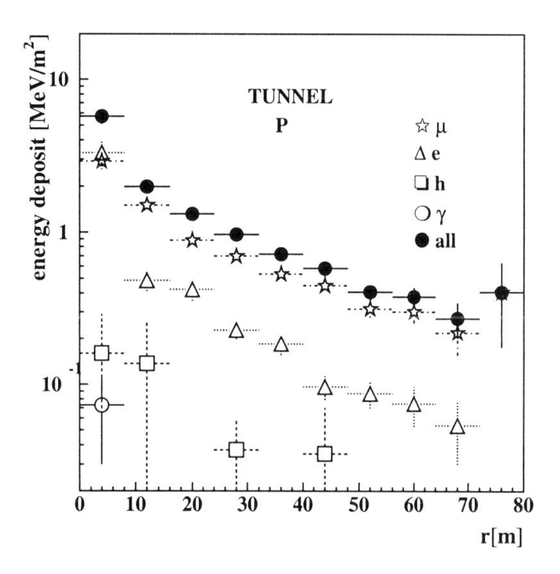

Figure 2. Simulated energy deposit per m^2 in muon detectors for all registered particles as a function of the distance to the shower core for proton induced showers (tunnel case).

To compare our simulated data with the experimental measurements, it is important to select showers according to the size. In GAMMA experiment the full N_e interval of $10^5 - 10^6$ is divided into four bins of equal logarithmic size: $10^5 - 1.78 \times 10^5$, $1.78 \times 10^5 - 3.16 \times 10^5$, $3.16 \times 10^5 - 5.62 \times 10^5$ and $5.62 \times 10^5 - 10^6$. But it is not quite correct to compare simulation results with the data in the first and the forth bins as far as in the simulation we miss some proton (for the first bin) and iron (for the forth bin) induced showers. Thus we use data of the second and third shower size intervals.

Figure 3 compares muon lateral distributions obtained from the experimental data with those from the simulated (CORSIKA + ARES) events for the shower size interval of $1.78 \times 10^5 < N_e < 3.16 \times 10^5$ for hall and tunnel detectors. Similar distributions for $3.16 \times 10^5 < N_e < 5.62 \times 10^5$ are shown in Figure 4.

The results of the pure CORSIKA simulation (without EAS propagation through the installation) for normal primary composition and the approximation of the experimental data by modified Hillas function [9] are also shown in Figures 3 and 4.

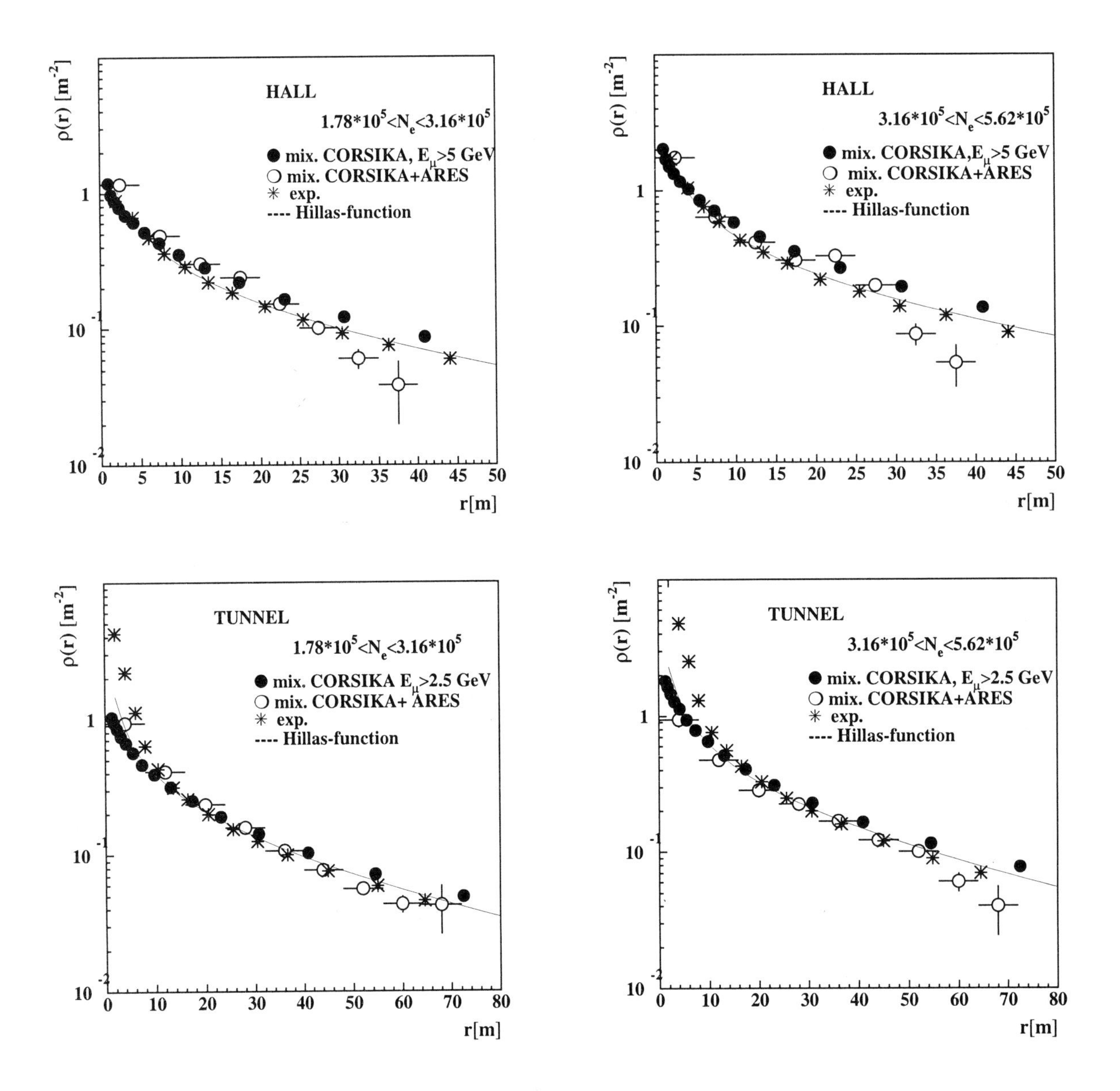

Figure 3. Muon lateral distributions for the shower size interval of $1.78 \times 10^5 < N_e < 3.16 \times 10^5$.

Figure 4. Muon lateral distributions for the shower size interval of $3.16 \times 10^5 < N_e < 5.62 \times 10^5$.

The modified Hillas function is:

$$\rho_\mu(N_e, r) = 0.952a(N_e/10^5) \times exp(-r/80)/r^{(0.75+0.06\ln(N_e/10^5))}\ \mathrm{m}^{-2} \quad (1)$$

where N_e is the shower size, r is the distance from the shower core in meters. Our calculations show [10] that for the GAMMA installation altitude, optimal values for the parameter a are 1 and 1.3, for hall and tunnel detectors, respectively. One can see a good agreement between the experimental and full simulation (CORSIKA+ARES) data. It can be noticed that in the tunnel, very close to the shower axis, the simulated densities are smaller than the experimental values because NKG option for CORSIKA is selected and so there is no electromagnetic punch-through in ARES simulation.

6. Muon size

The muon size N_μ for each event is obtained from the detected muon densities ρ_i:

$$N_\mu(r) = \sum(\rho_i/\omega_\mu(r))/k \quad (2)$$

$\omega_\mu(r)$ is a density probability obtained from the approximation (1) and k is the number of detectors ($k = 60$ for hall and $k = 90$ for tunnel). The experimental muon size is estimated by using the same procedure for the hall data. The muon size distributions for experimental and simulated showers with $1.78 \times 10^5 < N_e < 3.16 \times 10^5$ are presented in Figure 5. One can see a reasonable agreement between them.

7. Conclusion

The detector simulation program ARES for the GAMMA experiment has been developed. The muon lateral distributions and muon shower size observed by GAMMA are well reproduced by the full simulation of events with CORSIKA + ARES.

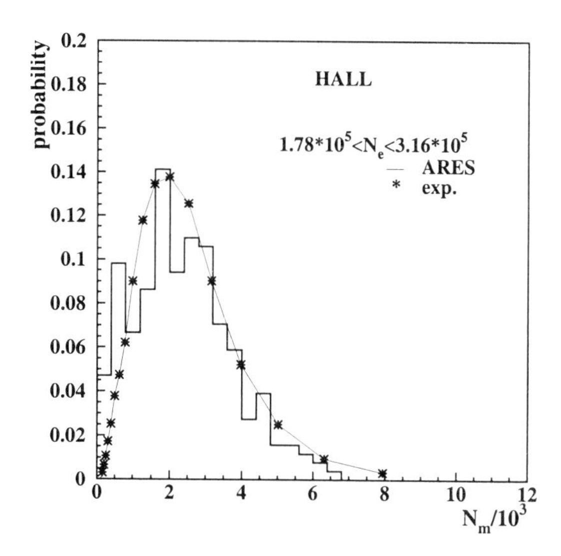

Figure 5. Muon size distributions for the shower size interval of $1.78 \times 10^5 < N_e < 3.16 \times 10^5$.

REFERENCES

1. A. Haungs, A.F. Badea, M.Z.Zazyan, Proc. of the Workshop ANI 99, FZKA 6472 (2000) 95
2. GEANT: CERN Program Library Long Writeup W5013, CERN (1993)
3. S.A. Arzumanian et al., Proc. 24th ICRC (Rome) **1** (1995) 482
4. A.A. Chilingarian et al., Proc. 16th ECRS (Alcala de Henares) (1998) 445
5. J.N. Capdevielle et al., KFK-Report 4498 (1992)
   D. Heck et al., FZK-Report 6019 (1998)
6. K. Werner, Physics Reports, 232 (1993) 87
7. H. Fesefeldt, The Simulation of Hadron Showers, RWTH Aachen Report PHITA 85/02 (1985)
8. M.Z.Zazyan, A. Haungs, Proc. of the Workshop ANI 98, FZKA 6215 (1998) 37
9. A.M. Hillas et al.,Proc. 16th ICRC (Budapest) **3** (1969) 533
10. V.S. Eganov et al., J. Phys. G : Nucl. Part. Phys. to be published

ELSEVIER

Nuclear Physics B (Proc. Suppl.) 97 (2001) 299–301

LIST OF PARTICIPANTS

Alaverdyan, Artur Y. .... P.N.Lebedev Physical Institute, Moscow, Russia
Amaral, Márcia Gonçalves do .... Universidade Federal Fluminense, Niterói, Brazil
Arteaga Velázquez, Juan Carlos .... CINVESTAV, Mexico City, Mexico
Augusto, Carlos Roberto Alves .... Universidade Federal Fluminense, Niterói, Brazil
Barroso, Sérgio Luiz Carmelo .... IFGW-UNICAMP, Campinas, Brazil
Bellandi Filho, José .... IFGW-UNICAMP, Campinas, Brazil
Borisov, Alexander S. .... P.N.Lebedev Physical Institute, Moscow, Russia
Capdevielle, Jean-Noël .... PCC, CNRS/Collège de France, Paris, France
Castellina, Antonella .... Istituto di Cosmo-Geofisica del C.N.R., Torino, Italy
Catalani, Fernando .... IFGW-UNICAMP, Campinas, Brazil
Cecchini, Stefano .... Ist. TeSRE/CNR, Bologna, Italy
Chinellato, José Augusto .... IFGW-UNICAMP, Campinas, Brazil
Chirinos Díaz, Johana Maria .... Centro Brasileiro de Pesquisas Físicas, Rio de Janeiro, Brazil
Cillis, Analia Nilda .... Universidade Nacional de La Plata, La Plata, Argentina
Ciocci, Maria Agnese .... I.N.F.N-Pisa, S.Piero a Grado (Pisa), Italy
Cronin, James W. .... University of Chicago, Chicago, USA
Dib, Claudio .... Universidade Técnica Federico Santa Maria, Valparaiso, Chile
Dobrigkeit, Carola .... IFGW-UNICAMP, Campinas, Brazil
Escobar, Carlos Ourivio .... IFGW-UNICAMP, Campinas, Brazil
Fauth, Anderson Campos .... IFGW-UNICAMP, Campinas, Brazil
Fleitas, Juliana Rita .... IFGW-UNICAMP, Campinas, Brazil
Gharakhanyan, Larissa .... P.N.Lebedev Physical Institute, Moscow, Russia
Gouffon, Philippe .... IFUSP, São Paulo, Brazil
Grieder, Peter K. F. .... Physikalisches Institut, Bern, Switzerland
Guzzo, Marcelo Moraes .... IFGW-UNICAMP, Campinas, Brazil
Hamburger, Amélia Imperio .... IFUSP, São Paulo, Brazil
Hamburger, Ernst W. .... IFUSP, São Paulo, Brazil
Haungs, Andreas .... Forschungszentrum Karlsruhe, Karlsruhe, Germany
Holanda, Pedro Cunha de .... IFGW-UNICAMP, Campinas, Brazil
Hörandel, Joerg R. .... The University of Chicago, Chicago, USA
Ivanov, Ivan A. .... P.N.Lebedev Physical Institute, Moscow, Russia
Jones, Lawrence W. .... University of Michigan, Ann Arbor, U.S.A.

Kawasumi, Norio .... Yamanashi University, Kofu, Japan
Kempa, Janusz .... University of Lodz, Lodz, Poland
Lagutin, Anatoly Alexeyevith .... Altai State University, Barnaul, Russia
Lidvansky, Aleksandr .... Institute for Nuclear Research, Moscow, Russia
Makarov, Vyacheslav Vasilyevich .... Moscow State University, Moscow, Russia
Malinowsky, Jan .... University of Lodz, Lodz, Poland
Martirosov, Romen .... Yerevan Physics Institute, Yerevan, Armenia
Matthiae, Giorgio .... University of Roma II and INFN, Roma, Italy
Menon, Márcio .... IFGW-UNICAMP, Campinas, Brazil
Menzione, Aldo .... I.N.F.N-Pisa, S.Piero a Grado (Pisa), Italy
Mikhailov, Alexei A. .... Institute of Cosmophysical Research and Aeronomy, Yakutsk, Russia
Morejon Gonzales, Danays .... Centro Brasileiro de Pesquisas Físicas, Rio de Janeiro, Brazil
Morello, Carlo .... Istituto di Cosmo-geofisica C.N.R., Torino, Italy
Mukhamedshin, Rauf A. .... Institute for Nuclear Research, Moscow, Russia
Natale, Adriano A. .... IFT/UNESP, São Paulo, Brazil
Navia Ojeda, Carlos .... Universidade Federal Fluminense, Niterói, Brazil
Nikolsky, Sergey I. .... P.N.Lebedev Physical Institute, Moscow, Russia
Nunokawa, Hiroshi .... IFGW-UNICAMP, Campinas, Brazil
Oblakov, Igor V. .... P.N.Lebedev Physical Institute, Moscow, Russia
Ochi, Nobuaki .... Okayama University, Okayama, Japan
Ohmori, Nobuharu .... Kochi University, Kochi City, Japan
Ohsawa, Akinori .... ICRR, University of Tokyo, Tanashi, Japan
Olinto, Angela .... University of Chicago, Chicago, USA
Ortiz, Jeferson Altenhofen .... IFGW-UNICAMP, Campinas, Brazil
Pemmaraju, Ammiraju .... 248 College Drive, Edison, USA
Petersen, Bert .... University of Nijmegen, Nijmegen, The Netherlands
Procureur, Jacques .... Centre d'Études Nucléaires de Bordeaux, Gradignan, France
Rigitano, Reinaldo Camargo .... Instituto de Física, UNICAMP, Campinas, Brazil
Risse, Markus .... Forschungszentrum Karlsruhe, Karlsruhe, Germany
Rosenfeld, Rogério .... IFT/UNESP, São Paulo, Brazil
Salmeron, Roberto A. .... École Polytechnique, Palaiseau, France
Schatz, Gerd .... Habichtweg 4, Bruchsal, Germany
Shellard, Ronald Cintra .... Centro Brasileiro de Pesquisas Físicas, Rio de Janeiro, Brazil
Shibuya, Edison Hiroyuki .... IFGW-UNICAMP, Campinas, Brazil
Sinitsyna, Vera G. .... P.N.Lebedev Physical Institute, Moscow, Russia

Sinitsyna, Vera Y. .... P.N.Lebedev Physical Institute, Moscow, Russia

Slavatinsky, Sergei A. .... P.N.Lebedev Physical Institute, Moscow, Russia

Souza Filho, Luiz Vitor de .... IFGW-UNICAMP, Campinas, Brazil

Stozhkov, Yuri .... P. N. Lebedev Physical Institute, Moscow, Russia

Swain, John .... Northeastern University, Boston, USA

Tamada, Masanobu .... Kinki University, Osaka, Japan

Turtelli, Armando .... IFGW-UNICAMP, Campinas, Brazil

Waxman, Eli .... Weizmann Institut of Science, Rehovot, Israel

Wilk, Grzegorz .... The Andrzej Soltan Institute for Nuclear Studies, Warsaw, Poland

Wilkens, Henric .... HEFIN -University of Nijmegen, Nijmegen, The Netherlands

Włodarczyk, Zbigniew .... Pedagogical University, Kielce, Poland

Yodh, Gaurang B. .... University of California, Irvine, USA

AUTHOR INDEX

GENERAL INFORMATION

Nuclear Physics B – Proceedings Supplements (PS) is the premier publication outlet for the proceedings of key conferences on high-energy physics and related areas. The series covers both large international conferences and more specialized topical meetings. Under the guidance of the Editorial Board, the newest discoveries and the latest developments, reported at carefully selected meetings, are published covering experimental as well as theoretical particle physics, hadronic physics, cosmology, astrophysics and gravitation, field theory and statistical systems.

Note to Conference Organizers

Organizers of upcoming meetings who are interested in exploring the possibilities of *Nuclear Physics B – Proceedings Supplements* as the publication outlet of their proceedings are invited to send full details of the planned conference to:

Drs. Machiel Kleemans
Publishing Editor
Elsevier Science B.V., P.O. Box 103, 1000 AC Amsterdam, The Netherlands
Telephone: +31 20 485 2524; Telefax: +31 20 485 2580
E-mail: M.KLEEMANS@ELSEVIER.NL

Proceedings will be produced from camera-ready copy and will be published within three months of the Publisher's receipt of the final and complete typescript. Attractive bulk-order arrangements will be offered to conference organizers for participants' copies.

Publication information

Nuclear Physics B – Proceedings Supplements (ISSN 0920-5632). For 2001, volumes 91–101 are scheduled for publication. Subscription prices are available upon request from the Publisher or from the Regional Sales Office nearest you or from this journal's website (http://www.elsevier.nl/locate/npe). A combined subscription to Nuclear Physics A volumes 679–695, Nuclear Physics B volumes 592–619, and Nuclear Physics B Proceedings Supplements volumes 91–101 is available at a reduced rate. Further information is available on this journal and other Elsevier Science products through Elsevier's website (http://www.elsevier.nl). Subscriptions are accepted on a prepaid basis only and are entered on a calender year basis. Issues are sent by standard mail (surface within Europe, air delivery outside Europe). Priority rates are available upon request. Claims for missing issues should be made within six months of the date of dispatch.

Advertising information